Communications and Control Engineering

Communications and Control Engineering is a high-level academic monograph series publishing research in control and systems theory, control engineering and communications. It has worldwide distribution to engineers, researchers, educators (several of the titles in this series find use as advanced textbooks although that is not their primary purpose), and libraries.

The series reflects the major technological and mathematical advances that have a great impact in the fields of communication and control. The range of areas to which control and systems theory is applied is broadening rapidly with particular growth being noticeable in the fields of finance and biologically inspired control. Books in this series generally pull together many related research threads in more mature areas of the subject than the highly specialised volumes of *Lecture Notes in Control and Information Sciences*. This series's mathematical and control-theoretic emphasis is complemented by *Advances in Industrial Control* which provides a much more applied, engineering-oriented outlook.

Indexed by SCOPUS and Engineering Index.

Publishing Ethics: Researchers should conduct their research from research proposal to publication in line with best practices and codes of conduct of relevant professional bodies and/or national and international regulatory bodies. For more details on individual ethics matters please see: https://www.springer.com/gp/authors-editors/journal-author/journal-author-helpdesk/publishing-ethics/14214.

Andrey Polyakov

Generalized Homogeneity in Systems and Control Volume II

Infinite-Dimensional Systems

Second Edition

Springer

Andrey Polyakov
Inria centre at the University of Lille
Lille, France

ISSN 0178-5354 ISSN 2197-7119 (electronic)
Communications and Control Engineering
ISBN 978-3-031-77495-9 ISBN 978-3-031-77496-6 (eBook)
https://doi.org/10.1007/978-3-031-77496-6

Mathematics Subject Classification: 93-02, 93C10, 93C15, 93C20, 93C25, 93C35, 93C55, 93C73

This Springer imprint is published by the registered company Springer Nature Switzerland AG
The registered company address is: Gewerbestrasse 11, 6330 Cham, Switzerland

To my family

Preface

The second volume of the monograph extends the theory of generalized homogeneity to infinite-dimensional spaces and demonstrates that many methods of finite-dimensional control homogeneous systems analysis and design are applicable to infinite-dimensional models. In particular, an input-to-state stability with respect to homogeneously involved perturbations can be analyzed considering an unperturbed homogeneous system only. Stability of infinite-dimensional homogeneous systems is characterized by a homogeneous Lyapunov function, but its convergence rate is defined by the homogeneity degree. A consistent discretization/approximation of homogeneous systems can be designed by means of homogeneity-preserving methods.

The advantages of homogeneous control discovered in Volume I for finite-dimensional systems remain valid in infinite-dimensional case. For instance, a homogeneous stabilization reduces overshoots and ensures a better regulation quality compared with the linear feedback. Finite/fixed-time homogeneous stabilizers for linear plants can be designed by solving linear equations/inequalities. A homogeneity-preserving discretization/approximation essentially improves approximation precision.

In Volume II, some bridges of the homogeneity theory with the theory of self-similar solutions of partial-differential equation are discussed. Elements of the theory of evolution equations in Banach and Hilbert spaces are surveyed. Some features of stability analysis of infinite-dimensional systems are considered. The concept of a linear dilation in a finite-dimensional Euclidean space is extended to an infinite-dimensional normed vector space. A special attention is paid to linear dilations in function spaces (e.g., Lebesgue and Sobolev spaces). Generalized homogeneous Banach and Hilbert spaces are studied. The basics of the theory of homogeneous operators in Banach and Hilbert spaces are elaborated. Homogeneous evolution equations (abstract differential equations) are studied. Homogeneity-based methods for stability and robustness analysis of abstract differential equations are developed. The so-called homogeneous Galerkin method is proposed for a consistent discretization of homogeneous infinite-dimensional system. Various homogeneous controllers are designed for linear and nonlinear plants. Homogeneous stabilization using modal decomposition is discussed. Theoretical constructions are supported with various examples. In particular, some nonlinear partial-differential equations such as Burgers, Navier–Stokes, and Saint-Venant equations are shown to be homogeneous in a generalized sense.

Lille, France
September 2024

Andrey Polyakov

Contents

Notation

- $\mathbb{N}$ (resp., $\mathbb{Z}$, $\mathbb{Q}$) is the set of natural (resp., integer, rational) numbers.
- $\mathbb{R}$ is the field of reals; $\overline{\mathbb{R}} = \mathbb{R} \cup \{-\infty\} \cup \{+\infty\}$ and $\mathbb{R}_+ = [0, +\infty)$.
- $\mathbb{C}$ is the field of complex numbers with $\mathbf{i} = \sqrt{-1}$; $\Re(\lambda)$ and $\Im(\lambda)$ denote the real and imaginary parts of complex number $\lambda \in \mathbb{C}$, respectively.
- $\mathbb{R}^n$ is the Euclidean space of real column vectors: $z = (z_1, ..., z_n)^\top \in \mathbb{R}^n$.
- $\mathbb{R}^{n\times m}$ is the space of real $n \times m$ matrices, $\mathbb{R}^n = \mathbb{R}^{n\times 1}$.
- $B^\top$ denotes the transposition of $B \in \mathbb{R}^{n\times m}$ and $A^{-\top} = (A^{-1})^\top$ for an invertible matrix $A \in \mathbb{R}^{n\times n}$.
- If $P = P^\top \in \mathbb{R}^{n\times n}$ then $P \succ 0$ (resp., $\succeq 0$) means that the matrix P is positive definite (resp., semidefinite) and $P \prec 0$ (resp., $\preceq 0$) means that the matrix P is negative definite (resp., semidefinite).
- $\lambda_{\min}(P)$ and $\lambda_{\max}(P)$ denotes minimum and maximum eigenvalues of a symmetric matrix $P = P^\top \in \mathbb{R}^n$.
- $\text{rank}(A)$ denotes the rank of the square matrix $A \in \mathbb{R}^{m\times n}$.
- $\text{tr}(A)$ denotes the trace (sum of diagonal elements) of the matrix $A \in \mathbb{R}^{n\times n}$.
- $|x| = \sqrt{x^\top x} = \sqrt{x \cdot x}$ is the Euclidean norm for $x \in \mathbb{R}^n$.
- The inequalities $y < 0$, $y \le 0$, $y > 0$, $y \ge 0$ for $y \in \mathbb{R}^n$ are understood in the component-wise sense.
- The set consisting of elements $x_1, x_2, ..., x_n$ is denoted $\{x_1, x_2, ..., x_n\}$.
- $\partial\Omega$ is the boundary of a set Ω.
- $\text{int}(\Omega)$ denotes the interior of a set Ω.
- The power set (the set of all subsets) of a set M is 2^M.
- Almost everywhere in the book, $\mathbb{B}$ is a real Banach space with a norm $\|\cdot\|$ and $\mathbb{H}$ is a real Hilbert space with an inner product $\langle\cdot,\cdot\rangle$. The complex Banach spaces are briefly considered only in Chap. 2. The Banach spaces are also denoted by $\mathbb{X}$, $\mathbb{Y}$, and $\mathbb{V}$.
- We write $f : X \mapsto Y$ if a function f maps a set X into a set Y.
- The notation $\|\cdot\|_{\mathbb{X}}$ is utilized if it is necessary to indicate that this is a norm in the space $\mathbb{X}$.
- $S = \{x \in \mathbb{B} : \|x\| = 1\}$ is the unit sphere in $\mathbb{B}$.

- $\mathcal{L}(\mathbb{X}; \mathbb{Y})$ is the space of linear bounded operators $\mathbb{X} \mapsto \mathbb{Y}$, where $\mathbb{X}$ and $\mathbb{Y}$ are Banach spaces, and

$$\|A\|_{\mathcal{L}(\mathbb{X},\mathbb{Y})} = \sup_{u\neq 0}\frac{\|Au\|_{\mathbb{Y}}}{\|u\|_{\mathbb{X}}} \text{ and } \lfloor A\rfloor_{\mathcal{L}(\mathbb{X},\mathbb{Y})} = \inf_{u\neq 0}\frac{\|Au\|_{\mathbb{Y}}}{\|u\|_{\mathbb{X}}}, A \in \mathcal{L}(\mathbb{X}; \mathbb{Y}).$$

- We also use the notation $\|A\|$ and $\lfloor A\rfloor$ for shortness if the context is clear.
- $I \in \mathcal{L}(\mathbb{B}; \mathbb{B})$ is the identity operator and $I_n \in \mathbb{R}^{n\times n}$ is the identity matrix.
- $f : \mathcal{D}(f) \subset \mathbb{B} \mapsto \mathbb{B}$ denotes a (possibly unbounded) operator/function in $\mathbb{B}$ with the domain $\mathcal{D}(f)$; we also use the notation like $x \mapsto f(x)$ for the explicit definition of a function and its argument (in this monograph, the symbol $\to$ is utilized only for limits); linear operators and set-valued mappings are usually denoted by capital letters $A : \mathcal{D}(A) \subset \mathbb{B} \mapsto \mathbb{B}$ (resp., $F : \mathbb{B} \mapsto 2^{\mathbb{B}}$).
- If $A : \mathcal{D}(A) \subset \mathbb{B} \mapsto \mathbb{B}$ is linear closed densely defined infinitesimal generator of a strongly continuous semigroup (or group) of bounded linear operators on a Banach space $\mathbb{B}$ then $\mathcal{D}_{\exp}(A^{\infty})$ denotes the set of all its entire vectors.
- $f_1(f_2)$ and $f_1 \circ f_2$ denote the composition of nonlinear operators (functions) f_1 and f_2. The composition of linear operators A and B is denoted by AB.
- $\mathrm{supp}(f)$ denotes a support $\{x \in X : f(x) \neq \mathbf{0}\}$ of the function $f : X \mapsto Y$
- $\dot{u}(t) = \lim\limits_{h\to 0} \frac{u(t+h)-u(t)}{h}$ denotes the time derivative of a function $u : \mathbb{R} \mapsto \mathbb{B}$, where the limit is understood dependently of a context. For shortness, we frequently omit the time argument t of a function $u : \mathbb{R} \mapsto \mathbb{B}$ and write, for example, $\dot{u} \in \mathbb{B}$ or $\dot{u} = f(u)$ if a context is clear.
- $\mathrm{div}(u) = \sum_{i=1}^{n} \frac{\partial u_i}{\partial z_i}$, where $u : \mathbb{R}^n \mapsto \mathbb{R}^n$ is a vector field and $\frac{\partial u}{\partial z_i}$ denotes the partial derivative of u.
- The dot product for vectors $u = (u_1, ..., u_n)^{\top}$, $v = (v_1, ..., v_n)^{\top}$ is given by $u \cdot v = \sum_{i=1}^{n} u_i v_i$, where, dependently of the nature of vectors, the symbol $u_i v_i$ may be a product of scalars or a composition of operators u_i and v_i.
- $\nabla^{\top} = \frac{\partial}{\partial z} = \left(\frac{\partial}{\partial z_1}, ..., \frac{\partial}{\partial z_n}\right)$, $\frac{\partial^k}{\partial z^k} = \frac{\partial}{\partial z}\frac{\partial^{k-1}}{\partial z^{k-1}}$ and $\Delta = \nabla \cdot \nabla = \mathrm{div}(\nabla) = \sum_{i=1}^{n} \frac{\partial^2}{\partial z_i^2}$.
- ℓ^2 is the Hilbert space of quadratically summable sequences, i.e., for $x = \{x_i\}_{i=1}^{\infty} \in \ell^2$ it holds $\|x\|_{\ell^2} = \sqrt{\sum_{i=1}^{\infty} x_i^2} < 0$. The summation and multiplication by a scalar of vectors from ℓ^2 are understood in component-wise manner. The inner product is defined as $\langle x, y\rangle = \sum_{i=1}^{\infty} x_i y_i$ for $x, y \in \ell^2$.
- $X \times Y$ denotes the Cartesian product of sets X and Y.
- $\overline{X}$ denotes the closure of the set X of a metric space.
- $C(X; Y)$ is the space of uniformly continuous functions $X \mapsto Y$ with the supremum norm, $\|f\| = \sup_{x\in X} \|f(x)\|_Y$, where $f \in C(X; Y)$ and X, Y are normed vector spaces (or their subsets).
- $C_c^{\infty}(\Omega; \mathbb{R}^m)$ is the space of infinitely differentiable (smooth) functions having compact support in Ω, where $\Omega \subset \mathbb{R}^m$ is a set with non-empty interior.
- $C_0^{\infty}(\Omega; \mathbb{R}^m)$ is the space of infinitely differentiable (smooth) functions vanishing on the boundary of Ω.
- Let $\mathbb{S}(\mathbb{R}^n; \mathbb{R}^m)$ be the Schwartz space, i.e., $x \in \mathbb{S}(\mathbb{R}^n; \mathbb{R}^m) \Rightarrow x \in C_0^{\infty}(\mathbb{R}^n; \mathbb{R}^m)$: $\sup_{z\in\mathbb{R}^n} |z|^k |\nabla^p x(z)| < \infty$ for any $k \geq$ and any $p \geq 0$.

- $L^p(\Omega;\mathbb{R}^m)$, $1 \le p < \infty$ is the Lebesgue space of functions $x : \Omega \mapsto \mathbb{R}^m$

$$\int_{\Omega} |x(z)|^p dz < +\infty$$

with the norm

$$\|x\|_{L^p} := \left(\int_{\Omega} |x(z)|^p dz\right)^{1/p}, 0 < p < \infty.$$

- $L^\infty(\Omega;\mathbb{R}^m)$, $1 \le p < \infty$ is the Lebesgue space of uniformly essentially bounded functions $x : \Omega \mapsto \mathbb{R}^m$ with the norm

$$\|x\|_{L^\infty} := \text{ess sup}|x(z)|, p = \infty.$$

- The Lebesgue space $L^2(\Omega;\mathbb{R}^m)$ is also a Hilbert space with the inner product

$$\langle u, v\rangle_{L^2} = \int_{\Omega} u(x) \cdot v(x) dx.$$

- $W^{k,p}(\Omega;\mathbb{R}^m)$ is the Sobolev space of weakly differentiable functions from $L^p(\Omega;\mathbb{R}^m)$, which weak derivatives up to the order k belong to L^p space as well (see Appendix D for more details).
- The Sobolev space $H^k(\Omega;\mathbb{R}^m) = W^{k,2}(\Omega;\mathbb{R}^m)$ is a Hilbert space of functions $\Omega \mapsto \mathbb{R}^m$ with the inner product

$$\langle u, v\rangle_{H^k} := \sum_{i=0}^{k} \langle \nabla^i u, \nabla^i v\rangle_{L^2}$$

and the norm $\|\cdot\|_{H^k} = \sqrt{\langle\cdot,\cdot\rangle_{H^p}}$.
- $H_0^k(\Omega;\mathbb{R}^m)$ is the completion of $C_c^\infty(\Omega;\mathbb{R}^m)$ with respect to $\|\cdot\|_{H^k}$.
- If $\Omega_1 \subset \mathbb{B}$ and $\Omega_2 \subset \mathbb{B}$ then, by definition, the identity $\Omega_1 = \Omega_2$ means $\Omega_1 \subset \Omega_2$ and $\Omega_2 \subset \Omega_1$.
- The geometric sum of two sets is denoted "$\dot{+}$" and given by

$$\mathbf{M}_1 \dot{+} \mathbf{M}_2 = \bigcup_{x_1 \in \mathbf{M}_1, x_2 \in \mathbf{M}_2} x_1 + x_2, \tag{1}$$

where $\mathrm{M}_1 \subset \mathbb{B}$, $\mathrm{M}_2 \subset \mathbb{B}$.
- The product of a scalar $\alpha \in \mathbb{R}$ and a set $\mathbf{M} \subset \mathbb{B}$ is defined as follows:

$$\alpha\mathbf{M} = \mathbf{M}\alpha = \bigcup_{x \in \mathbf{M}} \alpha x. \tag{2}$$

- The product of a set $\mathbf{N} \subset \mathbb{R}$ and a set $\mathbf{M} \subset \mathbb{B}$ is defined as follows:

$$\mathbf{N} \cdot \mathbf{M} = \bigcup_{\alpha \in \mathbf{N}} \alpha \mathbf{M}. \tag{3}$$

- An application of an operator $f : \mathbb{B} \mapsto \mathbb{X}$ to a set $\mathbf{M} \subset \mathbb{B}$ is given by

$$f(\mathbf{M}) := \bigcup_{x \in \mathbf{M}} f(x). \tag{4}$$

- $B(r) := \{x \in \mathbb{B} : \|x\| < r\}$ is the open ball in $\mathbb{B}$ of the radius $r \in \mathbb{R}_+$ with the center at the origin. Under introduced notation, $B(y, r) = y \dot{+} B(\varepsilon)$ is an open ball of the radius $\varepsilon > 0$ centered at $y \in \mathbb{B}$.
- $\text{span}\{e_1, e_2, ...\}$ denotes a linear hull, i.e.,

$$\text{span}\{e_1, e_2, ...\} = \{y : y = \alpha_1 e_1 + \alpha_2 e_2 + ..., \alpha_1, \alpha_2, ... \in \mathbb{R}\}.$$

- The convex hull of $A \subset \mathbb{B}$, denoted $co(A)$, is the smallest convex set containing A. The closed convex hull of A, denoted $\overline{co}(A)$, is the smallest closed convex set containing A.
- The sign function is defined by

$$\text{sign}_\sigma(\rho) := \begin{cases} 1 & \text{if } \rho > 0, \\ -1 & \text{if } \rho < 0, \\ \sigma & \text{if } \rho = 0, \end{cases} \tag{5}$$

  where $\sigma \in \mathbb{R} : -1 \leq \sigma \leq 1$. By default, $\text{sign}(\rho) = \text{sign}_0(\rho)$.
- $x^{\lfloor \alpha \rceil} := |x|^\alpha \text{sign}[x]$ is a power operation, which preserves the sign of the number $x \in \mathbb{R}$. For example, $(-2)^{\lfloor 2 \rceil} = -4$ and $2^{\lfloor 2 \rceil} = 4$.
- A continuous function $\sigma : \mathbb{R}_+ \mapsto \mathbb{R}_+$ belongs to the class $\mathcal{K}$ if $\sigma(0) = 0$ and σ is strictly increasing ($t_1 < t_2 \Rightarrow \sigma(t_1) < \sigma(t_2)$).
- A function $\sigma \in \mathcal{K}$ belongs to the class $\mathcal{K}^\infty$ if $\sigma(t) \to +\infty$ as $t \to +\infty$.
- A continuous function $\xi : \mathbb{R}_+ \times \mathbb{R}_+ \mapsto \mathbb{R}_+$ belongs to the class $\mathcal{KL}$ if $\xi(\cdot, t) \in \mathcal{K}$ for any fixed $t \geq 0$ and $\xi(s, \cdot) : \mathbb{R}_+ \mapsto \mathbb{R}_+$ is strictly decreasing to zero for any fixed $s \geq 0$.
- $\mathbf{d}$ denotes a linear dilation in a normed vector space.
- $\mathcal{H}_\mathbf{d}(\mathbb{B})$ is a set of $\mathbf{d}$-homogeneous functionals $\mathbb{B} \mapsto \mathbb{R}$ and $\deg_\mathbf{d}(h) \in \mathbb{R}$ is a homogeneity degree of $h \in \mathcal{H}_\mathbf{d}(\mathbb{B})$.
- $\mathcal{F}_\mathbf{d}(\mathbb{B})$ is a set of $\mathbf{d}$-homogeneous operators $\mathbb{B} \mapsto \mathbb{B}$ and $\deg_\mathbf{d}(f) \in \mathbb{R}$ is a homogeneity degree of $f \in \mathcal{F}_\mathbf{d}(\mathbb{B})$.

Chapter 1
Introduction to Abstract Homogeneity

This chapter presents an introduction to generalized homogeneous systems in abstract (e.g., Banach and Hilbert) spaces. A relation of the homogeneity theory with the theory of self-similar solutions of Partial Differential Equations (PDEs) is considered. Examples of generalized dilations in Lebesgue spaces are presented. Basic ideas of infinite-dimensional homogeneous control system design are discussed.

1.1 Dilation Symmetries of Physical Models

1.1.1 Infinite-Dimensional Dynamical Models

In Volume I of this monograph, various homogeneous control systems with state vectors belonging to the finite-dimensional Euclidean space $\mathbb{R}^n$ have been considered. However, many models of mathematical physics have a distributed state representing, for example, a temperature distribution in a building or a velocity field in a water channel. The corresponding dynamical models have state vectors belonging to function spaces being infinite-dimensional vector spaces, e.g., $C(\mathbb{R}^n; \mathbb{R}^m)$ or $L^2(\mathbb{R}^n; \mathbb{R}^m)$. For more details about infinite-dimensional spaces such as Banach, Hilbert, and various other function spaces, we refer the reader to monographs [1–4] and/or to Appendix A. The so-called *evolution equations* (also known as *abstract differential equations*) describe dynamics of such systems [5–8].

Let us consider an abstract differential equation:

$$\dot{x}(t) = g(x(t)), \quad t > t_0, \quad x(t) \in \mathbb{B}, \tag{1.1}$$

$$x(t_0) = x_0, \tag{1.2}$$

A. Polyakov, *Generalized Homogeneity in Systems and Control Volume II*,
Communications and Control Engineering,
https://doi.org/10.1007/978-3-031-77496-6_1

where the (possibly time-dependent) operator $g : \mathcal{D}(g) \subset \mathbb{B} \mapsto \mathbb{B}$ maps a domain $\mathcal{D}(g)$ of a Banach space $\mathbb{B}$ into $\mathbb{B}$, and the time derivative $\dot{x}(t) = \lim_{h\to 0} \frac{x(t+h)-x(t)}{h}$ is defined in some sense (e.g., in a weak topology of $\mathbb{B}$). The evolution equation (1.1) (resp., an *evolution inclusion* like $\dot{x}(t) \in g(x(t))$, $x(t) \in \mathbb{B}$) is a generalization of an *ordinary differential equation (ODE)* (resp., a *differential inclusion (DI)*) studied in Volume I of this monograph. Taking $\mathbb{B} = \mathbb{R}^n$ in (1.1) we derive an ODE (resp., DI). The evolution equation (1.1) may define a *partial differential equation (PDE)* if $\mathbb{B}$ is a function space.

Let $x(t)$ denote a distribution of a temperature on the real line $\mathbb{R}$ at the time instant $t \geq 0$. This means that, at each instant of time, the state $x(t)$ is a function $x(t, \cdot) : \mathbb{R} \mapsto \mathbb{R}$ such that $x(t, z)$ defines a temperature at time $t \in \mathbb{R}$ at the point $z \in \mathbb{R}$ of the real line. Therefore, $x(t) \in \mathbb{B}$ for all $t \geq 0$, where, for example, $\mathbb{B} = L^2(\mathbb{R}; \mathbb{R})$ or $\mathbb{B} = C(\mathbb{R}; \mathbb{R})$. An evolution of the temperature distribution x in time is modeled by the well-known heat equation

$$\frac{\partial x}{\partial t} = \rho \frac{\partial^2 x}{\partial z^2}, \quad t > 0, \quad x(0, z) = x_0(z), \quad z \in \mathbb{R}, \tag{1.3}$$

where $\rho > 0$ is a thermal diffusivity constant. The considered PDE can be represented as the linear evolution equation

$$\dot{x}(t) = Ax(t), \quad t > 0, \quad x(0) = x_0 \in \mathbb{B}, \tag{1.4}$$

where $A : \mathcal{D}(A) \subset \mathbb{B} \mapsto \mathbb{B}$ is the linear differential operator

$$A = \rho \frac{\partial^2}{\partial z^2} \tag{1.5}$$

with a domain $\mathcal{D}(A)$ dependent on $\mathbb{B}$. The differential operator A is defined only on some subspace of $\mathbb{B}$. More precisely, we have $\mathcal{D}(A) = C^2(\mathbb{R}; \mathbb{R}) \subset C(\mathbb{R}; \mathbb{R})$ in the case of strong (conventional) derivatives or $\mathcal{D}(A) = H^2(\mathbb{R}; \mathbb{R}) \subset L^2(\mathbb{R}; \mathbb{R})$ in the case of weak derivatives (see Appendix D).

The existence and uniqueness of solutions to the abstract differential equation (1.1) in a Banach/Hilbert space is a nontrivial problem in the general case (see, e.g., [5–7, 9–12]). Some difficulties appear even if we try to define formally a solution to the above simple PDE. Indeed, let us also consider the integral equation

$$x(t) = x_0 + \int_0^t Ax(\tau)d\tau, \tag{1.6}$$

which can be considered as an alternative representation of the abstract differential equation (1.4). The integral in (1.6) is understood in the sense of Bochner (see Appendix C). A function x satisfying the integral equation (1.6) is called a *mild solution* of the linear abstract differential equation (1.4) (see, e.g., [7, page 146]).

On the one hand, $Ax(\tau)$ should exist and belong to $\mathbb{B}$ for $\tau > 0$ in order to fulfill (1.6) and (1.4). Therefore, to be a solution of the evolution equation, $x(\tau)$ should belong to $\mathcal{D}(A)$ for $\tau > t_0$. On the other hand, even if $x(\tau) \in \mathcal{D}(A)$ the term $Ax(\tau)$ as well as $\int_0^t Ax(\tau)d\tau$ belongs to $\mathbb{B}$ (but not to $\mathcal{D}(A)$ in the general case). Therefore, the terms on the left-hand side and on the right-hand side of the above differential and integral equations may not belong to the same function space. So, it is not obvious that a function $x \in C([t_0, T); \mathbb{B})$ satisfying the above differential or integral equation may exist.

In Chap. 2, the problem of the existence and uniqueness of solutions of the evolution equation (1.1) in a Banach (or a Hilbert) space is studied using the theory of strongly continuous semigroups of linear bounded operators. This approach is well developed in the literature (see, e.g., [6, 7, 9, 13]) for abstract differential equations of the form

$$\dot{x} = Ax + f(t, x), \quad t > t_0, \quad x(t_0) = x_0 \in \mathbb{B}, \tag{1.7}$$

where $A : \mathcal{D}(A) \subset \mathbb{B} \mapsto \mathbb{B}$ is the so-called closed linear (possibly unbounded) operator on $\mathbb{B}$, and $f : \mathcal{D}(f) \subset (\mathbb{R} \times \mathbb{B}) \mapsto \mathbb{B}$ is a regular (in a certain sense) function of the second argument.

1.1.2 Self-Similar Solutions

Geometric properties of the operator g on right-hand side of (1.1) should be inherited by solutions of the abstract differential equation. In particular, similar to the finite-dimensional case, we may expect that a homogeneity (dilation symmetry) of an operator g should be inherited somehow by solutions of (1.1). In fact, this property has been already discovered in various infinite-dimensional models of mathematical physics.

For instance, dilation symmetries of mathematical models of fluid dynamics are studied since 1940s, when the so-called self-similar solutions have been discovered. Some results about self-similarity in fluids have been originally reported by Karl Bechert (see [14–16]). The mentioned results had inspired mathematical theory of blast waves developed by Geoffrey Taylor, John von Neumann, Leonid Sedov, Kirill Stanyukovich, and others in the 1940s (see, e.g., [17–20]).

The self-similarity of solutions can be discovered even for the heat system (1.3), which has the unique solution given by (see, e.g., [7])

$$x_{x_0}(t, z) = \frac{1}{\sqrt{4\pi\rho t}} \int_{\mathbb{R}} e^{\frac{-(z-y)^2}{4\rho t}} x_0(y)dy, \quad \forall t > 0, \tag{1.8}$$

where the index x_0 highlights the dependence of the solution x_{x_0} on the initial state x_0 being an element of a properly defined function space $\mathbb{B}$. Notice that $x_{x_0}(t) \to x_0$ as $t \to 0$ in the norm topology of the space.

Making the change of the integration variable, we derive

$$x_{x_0}(\gamma^2 t, \gamma z) = \frac{1}{\sqrt{4\pi\gamma^2 t}} \int_{\mathbb{R}} e^{\frac{-(\gamma z - y)^2}{4\gamma^2 t}} x_0(y) dy$$
$$= \frac{1}{\sqrt{4\pi t}} \int_{\mathbb{R}} e^{\frac{-(z-\tilde{y})^2}{4t}} x_0(\gamma\tilde{y}) d\tilde{y} = x_{\tilde{x}_0}(t, z), \quad \forall t > 0, \quad \forall z \in \mathbb{R}, \quad (1.9)$$

where $\gamma > 0$ is a scaling factor and $\tilde{x}_0$ is the "scaled" initial state given by

$$\tilde{x}_0(z) = x_0(\gamma z), \quad z \in \mathbb{R}.$$

Notice that we scale only the z-argument of the initial condition, but this implies the scaling of both z-argument and the time evolution of the solution. These kinds of symmetries of solutions can also be discovered for various nonlinear models of fluid dynamics such as Burgers equation or Navier–Stokes equations (see, e.g., [21] and references therein).

The identity (1.9) is similar to the property of finite-dimensional homogeneous dynamical systems discovered in Volume I of the monograph:

> a solution of a system with a "scaled" initial condition can be obtained by a "scaling" of a solution for the original (non-scaled) initial condition.

Multiplication by a positive scalar is the standard (isotropic) dilation, so, on the one hand, the identity (1.9) is a dilation symmetry of solutions of the heat system. On the other hand, as mentioned above, the heat system can be interpreted as a linear evolution equation. For $\mathbb{B} = \mathbb{R}^n$, the evolution equation (1.1) corresponds to an ODE. The $\mathbf{d}$-homogeneous ODE always has (see Volume I of this monograph) the following dilation symmetry of solutions:

$$x_{\mathbf{d}(s)x_0}(t) = \mathbf{d}(s)x_{x_0}(e^{\mu s}t), \quad \forall t \geq 0, \quad \forall s \in \mathbb{R}, \quad (1.10)$$

where $x_{\tilde{x}}$ denotes a solution of the system (1.1) with the initial condition $x(0) = \tilde{x}$, $\mathbf{d}$ is a linear dilation, and $\mu \in \mathbb{R}$ is the homogeneity degree of the system. Below we show that, in fact, *the identity* (1.9) *is equivalent to* (1.10) *under a proper selection of a linear dilation* $\mathbf{d}$ *in a Banach space* $\mathbb{B}$ (being the Lebesgue space L^p in the case of the considered heat system). Therefore, various systems having self-similar solutions are expected to be homogeneous in a generalized sense.

1.2 Infinite-Dimensional Homogeneity

1.2.1 Linear Dilations in Abstract Spaces

Inspired by [22], a dilation in a Banach space $\mathbb{B}$ can be introduced as

a one-parameter family $\mathbf{d}(s), s \in \mathbb{R}$ of homeomorphisms on $\mathbb{B}$ satisfying the *group property*:

- $\mathbf{d}(0)x = x, \quad \mathbf{d}(s+t)x = \mathbf{d}(s) \circ \mathbf{d}(t)x \quad \forall t, s \in \mathbb{R}, \quad \forall x \in \mathbb{B},$

and the *limit property*:

- $\limsup\limits_{s \to -\infty} \|\mathbf{d}(s)x\| = 0, \quad \liminf\limits_{s \to +\infty} \|\mathbf{d}(s)x\| = +\infty, \quad \forall x \in \mathbb{B} \backslash \{\mathbf{0}\}.$

The limit property characterizes groups being dilations. This property is a version of the so-called *modified Terasaka's condition* [23] for normed vector spaces. The introduced definition simply repeats the definition of a dilation in a finite-dimensional space by replacing $\mathbb{B}$ with $\mathbb{R}^n$ (see Volume I or [22]).

In this monograph, we consider only *linear strongly continuous dilations* assuming that $\mathbf{d}(s)$ is a linear bounded invertible operator in $\mathbb{B}$, and the mapping $s \mapsto \mathbf{d}(s)x$ is continuous for any $x \in \mathbb{B}$. By analogy with the finite-dimensional case, the *standard (Euler) dilation* in $\mathbb{B}$ is the linear dilation $\mathbf{d}(s) = e^s I$ introduced as a scalar multiplication of a vector $x \in \mathbb{B}$ by a positive scalar:

$$x \mapsto e^s x, \quad s \in \mathbb{R}.$$

By definition, any other dilation in $\mathbb{B}$ is *generalized*. For example, scaling the argument of a function $x \in L^p(\mathbb{R}; \mathbb{R})$ with $p \in [1, +\infty)$

$$(\mathbf{d}(s)x)(z) = x(e^{-s}z), \qquad s \in \mathbb{R}, \quad z \in \mathbb{R}, \tag{1.11}$$

is a linear dilation in $L^p(\mathbb{R}; \mathbb{R})$ in the sense of the above definition. Indeed, making the change of the integration variable in the Lebesgue integral, we derive

$$\|\mathbf{d}(s)x\|_{L^p(\mathbb{R};\mathbb{R})} = \left(\int_{\mathbb{R}} x^p(e^{-s}z)dz\right)^{\frac{1}{p}} = \left(e^s \int_{\mathbb{R}} x^p(\tilde{z})d\tilde{z}\right)^{\frac{1}{p}} = e^{\frac{s}{p}} \|x\|_{L^p(\mathbb{R};\mathbb{R})}.$$

So, $\mathbf{d}(s)$ is a linear bounded invertible operator for any $s \in \mathbb{R}$ and the limit property is fulfilled in $L^p(\mathbb{R}; \mathbb{R})$. The family $\mathbf{d}$ of homeomorphisms on $L^p(\mathbb{R}; \mathbb{R})$, obviously, satisfies the group property as well. Notice that, being a dilation in $L^p(\mathbb{R}; \mathbb{R})$, the group (1.11) is not a dilation in $C(\mathbb{R}; \mathbb{R})$, since the scaling of the argument of a function $x \in C(\mathbb{R}; \mathbb{R})$ does not change the C-norm of the function, so the limit property is not fulfilled in $C(\mathbb{R}; \mathbb{R})$.

Linear dilations in Banach and Hilbert spaces are studied in Chap. 4, where special attention is paid to linear dilations in the Lebesgue spaces $L^p(\mathbb{R}^n;\mathbb{R}^k)$ and the Sobolev spaces $W^{m,p}(\mathbb{R}^n;\mathbb{R}^k)$.

1.2.2 Homogeneous Evolution Equation

Let us consider again the evolution equation (1.1). By definition (see Volume I for $\mathbb{B}=\mathbb{R}^n$), the operator $g:\mathcal{D}(g)\subset\mathbb{B}\mapsto\mathbb{B}$ is $\mathbf{d}$-homogeneous of degree $\mu\in\mathbb{R}$ with respect to a linear dilation $\mathbf{d}$ in $\mathbb{B}$ if

$$g(\mathbf{d}(s)x)=e^{\mu s}\mathbf{d}(s)g(x),\quad \forall x\in\mathcal{D}(g),\quad \forall s\in\mathbb{R}. \tag{1.12}$$

Under this condition, all solutions of the system (1.1) satisfy the identity (1.10) in $\mathbb{B}=\mathbb{R}^n$. In Chap. 6, we show that this conclusion remains valid for evolution equations in abstract spaces [24].

For example, in the first section of this chapter, we have discovered that the identity (1.9) represents a dilation symmetry of solutions of the heat system (1.3). Considering the differential operator $A=\rho\frac{\partial^2}{\partial z^2}$ as an operator in $L^2(\mathbb{R};\mathbb{R})$, we conclude that the identity (1.9) coincides with the identity (1.10) for

$$\mu=-2$$

provided that a linear dilation $\mathbf{d}$ in $L^2(\mathbb{R};\mathbb{R})$ is given by (1.11). Therefore, we expect that the differential operator

$$A: H^2(\mathbb{R};\mathbb{R})\subset L^2(\mathbb{R};\mathbb{R})\mapsto L^2(\mathbb{R};\mathbb{R})$$

given by (1.5) is $\mathbf{d}$-homogeneous of degree -2. Indeed, for any $x\in H^2(\mathbb{R};\mathbb{R})$ we have $\mathbf{d}(s)x\in H^2(\mathbb{R};\mathbb{R})$ and

$$(A\circ\mathbf{d}(s)x)(z)=\rho\frac{\partial^2}{\partial z^2}x(e^{-s}z)=\rho e^{-2s}\left.\frac{\partial^2 x(\tilde{z})}{\partial\tilde{z}^2}\right|_{\tilde{z}=e^{-s}z}=e^{-2s}(\mathbf{d}(s)\circ Ax)(z),$$

i.e., the identity (1.12) holds for $\mu=-2$. This example shows that linear differential operators in L^p may have nonzero homogeneity degree. In the finite-dimensional case $\mathbb{B}=\mathbb{R}^n$, linear homogeneous operators with nonzero degrees are characterized by nilpotent matrices. They form an important class of homogeneous control plants (see Volume I). Homogeneous operators in Banach and Hilbert spaces are studied in Chap. 5, where, in particular, a class of linear generalized homogeneous differential operators in Sobolev spaces is introduced.

The so-called homogeneous Lie groups underlie various methods of algebraic and geometric analysis of differential operators on manifolds developed since 1970s (see, e.g., [25–27]). We refer the reader to the fundamental monograph [28] for more details about homogeneous Lie groups and their applications in operator theory, harmonic analysis, abstract algebra, etc.

The aim of this monograph is to develop homogeneity-based methods for analysis and design of control systems.

1.3 Abstract Homogeneous Control Systems

Finite-dimensional homogeneous control systems have some important advantages in comparison with the linear control systems:

- *faster convergence* (finite-time and fixed-time stability);
- *better robustness* (input-to-state stability of homogeneous systems and rejection of a larger class of perturbations by homogeneous feedback);
- *smaller overshoots* (better regulation under state and time constraints).

In Volume I of the monograph, these properties of homogeneous systems have been proven theoretically, supported by various control experiments and numerical simulations. In Volume II, we show that all mentioned important properties remain valid for the infinite-dimensional setting. Table 1.1 presents a summary of the properties being equivalent for homogeneous systems in $\mathbb{R}^n$ and $\mathbb{B}$.

Table 1.1 Comparison of properties of homogeneous systems in finite- and infinite-dimensional spaces

	$\dot{x} = g(x), \quad t > 0,$ $g(\mathbf{d}(s)x) = e^{\mu s}\mathbf{d}(s)g(x)$	
	$x(t) \in \mathbb{R}^n$	$x(t) \in \mathbb{B}$
Symmetry of solutions	✓	✓
Local property⇔ Global property	✓	✓
Homogeneous Lyapunov function	✓	✓
Stability ⇔ Robustness	✓	✓
Finite/Fixed-time stability dependently of the degree $\mu \neq 0$	✓	✓
Consistent discretization (preserves convergence rates)	✓	✓
Algebraic design of homogeneous stabilizers (for linear plants modeled by PDEs)	✓	✓

1.3.1 Convergence Rates

Finite-time and fixed-time stability is a well-known property of homogeneous ODEs. In Chap. 6, it is shown that the uniform asymptotic stability of the homogeneous system in $\mathbb{B}$ implies its finite-time stability for negative homogeneity degree, nearly fixed-time stability for the positive homogeneity degree, and exponential stability for the zero homogeneity degree. This conclusion corresponds to the results discovered in Volume I for systems in $\mathbb{R}^n$. *However, an evolution equation* (1.7) *in a Banach space may be fixed-time stable even if A is linear time-invariant operator and $f = \mathbf{0}$.* Indeed, in the theory of evolution semigroups, the fixed-time stability of the system (1.7) with $f = \mathbf{0}$ corresponds to the case when the operator A is the generator of the so-called nilpotent semigroup [13]. For example, the transport equation

$$\frac{\partial x}{\partial t} = -\frac{\partial x}{\partial z}, \quad t > 0, \quad x(0, z) = x_0(z), \quad z \in (0, 1), \tag{1.13}$$

with the homogeneous Dirichlet boundary condition

$$x(t, 0) = 0, \tag{1.14}$$

has the unique solution

$$x_{x_0}(t, z) = \begin{cases} x_0(z - t) & \text{for } z - t > 0, \\ 0 & \text{for } z - t \leq 0, \end{cases} \quad t \geq 0, \quad z \in [0, 1], \tag{1.15}$$

which becomes the zero function of z after the time instant $t = 1$ independently of x_0. Such a property of evolution systems is known as finite-time extinction [29–31]. Taking into account $\|x_{x_0}(t)\|_{\mathbb{B}} \leq \|x_0\|_{\mathbb{B}}$ for both $\mathbb{B} = C([0, 1]; \mathbb{R})$ and $\mathbb{B} = L^p((0, 1); \mathbb{R})$, we conclude that the zero solution of the considered transport equation is Lyapunov stable and, consequently, fixed-time stable in C and in L^p, simultaneously. Notice that the considered system is standard homogeneous of the zero degree. Therefore, in contrast to the finite-dimensional case, *the nonzero homogeneity degree is not a necessary condition on a homogeneous system to be finite-time or nearly fixed-time stable.* In the finite-dimensional space, such a property was discovered for homogeneous systems, whose right-hand side depends on the initial state (see Chap. 6 of Volume I). Such a system can be considered as a system with a time-varying delay. Time-delay systems are infinite-dimensional [32]. Therefore, the possibility of fixed-time stability of homogeneous systems with zero degree is a specific feature of the infinite-dimensional case.

The Lyapunov stability of an abstract differential equation in L^p may not be equivalent to Lyapunov stability in C. For example, the inviscid Burgers equation

$$\frac{\partial x}{\partial t} + \frac{\partial (x^2/2)}{\partial z} = 0$$

with $x(0) = x_0 \in L^2$ is an example of the so-called energy conservation law (i.e., the L^2-norm of the state $x(t)$ is time-invariant). This means that the zero solution of the system is Lyapunov stable in L^2, namely $\|x_{x_0}(t)\|_{L^2} = \|x_0\|_{L^2}$ for any $x_0 \in L^2$. However, the solution $x_{x_0}(t)$ may have the so-called shocks (discontinuities) which appear in a finite time even if $x_0 \in C$, so the stability analysis in C is simply impossible. In this monograph, most examples of stability analysis are developed for evolution systems in Lebesgue and Sobolev spaces. Elements of stability theory of evolution equations/inclusions are presented in Chap. 3.

Notice also that the self-similarity of solutions allows essential simplification of the analysis of nonlinear dynamics. Since 1940s, this property is known to be useful for approximation of solutions and an analysis of nonlinear PDEs (see, e.g., [18, 19]). In Chap. 7, we show the homogeneity allows the so-called consistent discretization of the infinite-dimensional system to be realized. Namely, a homogeneous evolution system can be approximated (using a Galerkin-like method) by a homogeneous ODE of the same degree with respect to a (possibly non-diagonalizable) linear dilation in a Euclidean space such that stability (as well as a convergence rate) of the original system is preserved by the approximated dynamics. The examples presented in Chap. 7 illustrate the importance of the linear dilations with non-diagonalizable generators for numerical methods.

1.3.2 Robustness

One of most important properties of homogeneous control systems is their robustness (e.g., *Input-to-State Stability*) with respect to a rather large class of perturbations. Being originally introduced for the finite-dimensional system [33], the Input-to-State Stability (ISS) analysis of infinite-dimensional system forms an important research direction nowadays [34–38].

Let $\mathbb{B}$ and $\mathbb{V}$ be Banach spaces. Let us consider

$$\dot{x}(t) = Ax(t) + f(x(t), q(t)), \quad t > t_0, \quad x(t_0) = x_0, \tag{1.16}$$

where, as before, $x(t) \in \mathbb{B}$ is a system state, $A : \mathcal{D}(A) \subset \mathbb{B} \mapsto \mathbb{B}$ is a linear (possibly) unbounded operator satisfying certain assumptions, $q \in L^\infty(\mathbb{R}; \mathbb{V})$ is an exogenous input (perturbation), $f : \mathbb{B} \times \mathbb{V} \mapsto \mathbb{B}$ is a nonlinear operator being locally Lipschitz continuous in the first argument and Hölder continuous on the second argument, and $\mathbb{V}$ is a normed vector space. The system (1.16) is said to be ISS if there exists $\gamma \in \mathcal{K}$ and $\beta \in \mathcal{KL}$ such that

$$\|x_{x_0,q}(t)\| \leq \beta(\|x_0\|, t - t_0) + \gamma(\|q\|_{L^\infty((t_0,t);\mathbb{V})}), \quad \forall t \geq t_0, \forall x_0 \in \mathbb{B}, \forall q \in L^\infty(\mathbb{R}; \mathbb{V}), \tag{1.17}$$

where $x_{x_0,q}$ is any solution of (1.16). For $q = \mathbf{0}$, the inequality (1.17) implies that the unperturbed system (1.16) is globally uniformly asymptotically stable. It is well known (see Volume I or [39–41]) that the converse result holds for finite-dimensional homogeneous system: *any unperturbed uniformly asymptotically stable homogeneous system is ISS with respect to homogeneously involved perturbations*. In Chap. 6 we extended this result to homogeneous evolution equations [42]. In this chapter, we also show that the homogeneous domination approach (see Volume I or [43, 44]) is also valid for the infinite-dimensional setting. The corresponding analysis is essentially based on homogeneous converse Lyapunov function theorem, which is proven in this monograph for both homogeneous and locally homogeneous systems.

1.3.3 Homogeneous Control

In Volume I, an efficient algorithm for homogeneous stabilization of a controllable linear plant

$$\dot{x} = Ax + Bu, \quad t > 0, \tag{1.18}$$

is designed for the finite-dimensional case:

$$x(t) \in \mathbb{R}^n, \quad u(t) \in \mathbb{R}^n, \quad A \in \mathbb{R}^{n\times n}, \quad B \in \mathbb{R}^{n\times n}.$$

In Chap. 8, a similar scheme [45] is developed for the infinite-dimensional setting:

$$x(t) \in \mathbb{B}, \quad u(t) \in \mathbb{U}, \quad A \in \mathcal{D}(A) \subset \mathbb{B} \mapsto \mathbb{B}, \quad B \in \mathcal{L}(\mathbb{U}; \mathbb{B}),$$

where $\mathbb{B}$ and $\mathbb{U}$ are Banach spaces and A is a linear closed densely defined operator. The resulting feedback law has the form

$$u = K_0 x + \|x\|_{\mathbf{d}}^{\mu} \tilde{\mathbf{d}}(\ln \|x\|_{\mathbf{d}}) K \mathbf{d}(-\ln \|x\|_{\mathbf{d}})x, \quad K_0, K \in \mathcal{L}(\mathbb{B}; U), \tag{1.19}$$

where $\mathbf{d}$ is a dilation in $\mathbb{B}$, $\|\cdot\|_{\mathbf{d}}$ is a canonical homogeneous norm in $\mathbb{B}$ (introduced in Chap. 4), and $\tilde{\mathbf{d}}$ is a group of linear bounded operators in $\mathbb{U}$. The structure of the controller perfectly fits its finite-dimensional analog (see Chaps. 9 and 13 of Volume I). As before, the linear component $K_0 x$ of the feedback is aimed at a generalized homogenization of the operator

$$A_0 = A + BK_0$$

with respect to the dilation $\mathbf{d}$, while the linear operator K is selected such that the linear operator $A + B(K + K_0)$ is a generator of a strongly continuous semigroup of constructions (in a certain norm in $\mathbb{B}$). Such a design repeats the finite-dimensional constructions (see Volume I). For a class of differential operators A in L^2, the procedure of control parameter tuning is shown to be algebraic and similar to the finite-dimensional case.

Concluding Remarks

Homogeneity (dilation symmetry) can be introduced for infinite-dimensional systems in abstract (Hilbert and Banach) spaces. A linear dilation in a Banach space can be defined by a group of linear bounded operators satisfying a certain limit property. Homogeneity of solutions of infinite-dimensional PDEs can be considered as a particular case of the so-called self-similarity of solutions, which is well-studied in systems theory and widely utilized in mathematical physics. The homogeneity also underlies various methods of algebraic and geometric analysis of differential operators and the so-called homogeneous Lie groups. The aim of Volume II of the monograph is to demonstrate that most methods of analysis and design of homogeneous control systems in abstract spaces can be developed similar to the finite-dimensional case studied in Volume I. The infinite-dimensional homogeneous systems considered in Volume II are modeled by differential equations in Banach and Hilbert spaces. An introduction to generalized homogeneity of abstract dynamical systems can be found in [46].

References

1. Rudin W (1991) Functional analysis. McGraw-Hill, Inc.
2. Conway JB (1994) A course in functional analysis. Springer-Verlag
3. Driver BK (2003) Analysis tools with applications. Springer
4. Bogachev VI (2007) Measure theory. Springer
5. Dalecky L, Krein MG (1974) Stability of solutions of differential equations in Banach spaces. American Mathematical Society, Providence, Rhode Island
6. Pazy A (1983) Semigroups of linear operators and applications to partial differential equations. Springer
7. Engel K-J, Nagel R (2000) One-parameter semigroups for linear evolution equations. Springer Verlag, Berlin, Heidelberg, New York
8. Tucsnak M, Weiss G (2009) Observation and control for operator semigroups. Birkhauser
9. Yosida K (1980) Functional analysis. Springer Verlag, Berlin, Heidelberg, New York
10. Krein SG, Khazan MI (1985) Differential equations in a Banach space. J Sov Math 30(3):2154–2239
11. Tolstonogov A (2000) Differential inclusions in a Banach space. Springer
12. Curtain RF, Zwart H (2012) An introduction to infinite-dimensional linear systems theory. Springer Science & Business
13. Hille E, Phillips R (1957) Functional analysis and semi-groups. AMS
14. Bechert K (1940) Zur theorie ehener storunyen in reibungsfreien gasen. Ann d Physik 37(2):89–122
15. Bechert K (1940) Zur theorie ehener storunyen in reibungsfreien gasen ii. Ann d Physik 38(1):1–25
16. Bechert K (1941) Uber die differentialgleichungen der wellenausbreitung in gasen. Ann d Physik 39(5):357–372
17. Taylor G (1944) The propagation and decay of blast waves. Technical report, British Civilian Defence Research Committee
18. von Neumann J (1943) Proposal and analysis of a new numerical method for the treatment of hydrodynamic shock problems. Technical report, National Defense Research Committee

19. Sedov L (1945) Certain unsteady motions of a compressible liquid. Doklady Academii Nauk SSSR, XLVI I(2):94–96
20. Stanyukovich K (1945) Self-similar solutions of equations of hydromechanics. Doklady Academii Nauk SSSR 48(5):331–333
21. Stanyukovich KP (1960) Unsteady motion of continuous media. Pergamon Press, Oxford
22. Kawski M (1991) Families of dilations and asymptotic stability. In: Analysis of Controlled Dynamical Systems, pp 285–294
23. Husch LS (1970) Topological characterization of the dilation and the translation in frechet spaces. Math Ann 190:1–5
24. Polyakov A, Efimov D, Fridman E, Perruquetti W (2016) On homogeneous distributed parameters equations. IEEE Trans Autom Control 61(11):3657–3662
25. Kohn J (1973) Pseudo-differential operators and hypoellipticity. In: Proceedings of symposia in pure mathematics, University of California, Berkeley, California, vol 23. American Mathematical Society, Providence, Rhode Island, pp 61–69
26. Folland G (1975) Subelliptic estimates and function spaces on nilpotent Lie groups. Arkiv for Matematik 13(1–2):161–207
27. Rothschild L, Stein E (1976) Hypoelliptic differential operators and nilpotent groups. Acta Math 137(3–4):247–320
28. Fischer V, Ruzhansky M (2016) Quantization on nilpotent Lie groups. Springer
29. Majda A (1975) Disappearing solutions for the dissipative wave equation. Indiana Univ Math J 24(12):1119–1133
30. Perrollaz V, Rosier L (2014) Finite-time stabilization of 2x2 hyperbolic systems on tree-shaped networks. SIAM J Control Optim 52(1):143–163
31. Galaktionov VA, Vazquez JL (1995) Necessary and sufficient conditions for complete blow-up and extinction for one-dimensional quasilinear heat equations. Arch Rat Mech Anal 129(3):225–244
32. Fridman E (2014) Introduction to time-delay systems: analysis and control. Springer
33. Sontag ED (1989) Smooth stabilization implies coprime factorization. IEEE Trans Autom Control 34:435–443
34. Jayawardhana B, Logemann H, Ryan EP (2008) Infinite-dimensional feedback systems: the circle criterion and input-to-state stability. Commun Inf Syst 8:403–404
35. Dashkovskiy S, Mironchenko A (2011) Local ISS of reaction-diffusion systems. In: Proceedings of the 18th IFAC world congress
36. Karafyllis I, Krstic M (2018) Input-to-state stability for PDEs. Springer
37. Mironchenko A, Prieur C (2020) Input-to-state stability of infinite dimensional systems: recent results and open questions. SIAM Rev 62(3):529–614
38. Mironchenko A (2023) Input-to-state stability: theory and applications. Springer
39. Ryan EP (1995) Universal stabilization of a class of nonlinear systems with homogeneous vector fields. Syst & Control Lett 26:177–184
40. Hong Y (2001) H_∞ control, stabilization, and input-output stability of nonlinear systems with homogeneous properties. Automatica 37(7):819–829
41. Andrieu V, Praly L, Astolfi A (2008) Homogeneous approximation, recursive observer design, and output feedback. SIAM J Control Optim 47(4):1814–1850
42. Polyakov A (2021) Input-to-State Stability of homogeneous infinite dimensional systems with locally Lipschitz nonlinearities. Automatica 129:109615
43. Zubov VI (1958) On systems of ordinary differential equations with generalized homogeneous right-hand sides. Izvestia vuzov Mathematica 1:80–88. http://mi.mathnet.ru/ivm2874
44. Qian C (2005) A homogeneous domination approach for global output feedback stabilization of a class nonlinear systems. In: American control conference, pp 4708–4715
45. Polyakov A, Coron J-M, Rosier L (2018) On homogeneous finite-time control for linear evolution equation in hilbert space. IEEE Trans Autom Control 63(9):3143–3150
46. Cheban D (2023) Global asymptotic stability of generalized homogeneous dynamical systems. Buletinul academiei de stiinte a republicii moldova matematica 102(2):52–82

Chapter 2
Abstract Differential Equations

The evolution equations (also known as abstract differential equations) propose a unified approach to the modeling and analysis of dynamical systems governed by various differential laws. Solutions of abstract differential equations are introduced based on the theory of continuous semigroups of bounded linear operators. This chapter surveys some elements of this theory. Many popular infinite-dimensional models of control systems and mathematical physics can be represented as abstract differential equations. Some of them are considered in the chapter as illustrative examples.

2.1 Groups and Semigroups of Linear Bounded Operators

Let $\mathbb{B}$ be a Banach space (over the field $\mathbb{R}$ or $\mathbb{C}$) with the norm $\|\cdot\|$ and $\mathcal{L}(\mathbb{B};\mathbb{B})$ be a space of linear bounded operators $\mathbb{B} \mapsto \mathbb{B}$ with the operator norm

$$\|A\| = \sup_{x\neq 0} \frac{\|Ax\|}{\|x\|}, \quad A \in \mathcal{L}(\mathbb{B};\mathbb{B}), \quad x \in \mathbb{B}.$$

Definition 2.1 A one-parameter family $\Phi(t), t \in \mathbb{R}_+$ (resp., $t \in \mathbb{R}$) of linear bounded operators from $\mathbb{B}$ into $\mathbb{B}$ is called a semigroup (resp., group) of linear bounded operators on $\mathbb{B}$ if

- $\Phi(0) = I$, where I is the identity operator on $\mathbb{B}$;
- $\Phi(t+s) = \Phi(t)\Phi(s) = \Phi(s)\Phi(t)$ for every $s \geq 0$ and every $t \geq 0$ (resp., for every $s \in \mathbb{R}$ and every $t \in \mathbb{R}$).

A. Polyakov, *Generalized Homogeneity in Systems and Control Volume II*,
Communications and Control Engineering,
https://doi.org/10.1007/978-3-031-77496-6_2

If Φ is a group of linear bounded operators on $\mathbb{B}$ then, obviously, $\Phi(t)$ is invertible and

$$[\Phi(t)]^{-1} = \Phi(-t), \quad \forall t \in \mathbb{R}.$$

Let us introduce the notation

$$\lfloor \Phi(s) \rfloor = \inf_{x \neq 0} \frac{\|\Phi(s)x\|}{\|x\|}.$$

Corollary 2.1 *If Φ is a group of linear bounded operators on $\mathbb{B}$ then*

$$\|\Phi(s)\| \cdot \lfloor \Phi(-s) \rfloor = 1, \quad \forall s \in \mathbb{R}. \tag{2.1}$$

Proof Since $\|x\| = \|\Phi(s)\Phi(-s)x\| \leq \|\Phi(s)\| \cdot \|\Phi(-s)x\|$ and

$$\|x\| = \|\Phi(-s)\Phi(s)x\| \geq \lfloor \Phi(-s) \rfloor \cdot \|\Phi(s)x\|$$

then

$$1 \leq \|\Phi(s)\| \cdot \inf_{x \neq 0} \frac{\|\Phi(-s)x\|}{\|x\|} = \|\Phi(s)\| \cdot \lfloor \Phi(-s) \rfloor$$

and

$$1 \geq \lfloor \Phi(-s) \rfloor \sup_{x \neq 0} \frac{\|\Phi(s)x\|}{\|x\|} = \lfloor \Phi(-s) \rfloor \cdot \|\Phi(s)\|,$$

respectively. The obtained inequalities imply the identity (2.1). □

2.1.1 *Uniformly Continuous Groups and Semigroups*

Definition 2.2 A semigroup (group) of bounded linear operators Φ is *uniformly continuous* if

$$\|\Phi(s) - I\| \to 0 \quad \text{as} \quad s \to 0^+$$

(resp., as $s \to 0$).

The above condition and the semigroup property imply continuity of the mapping

$$\Phi : [0, +\infty) \mapsto \mathcal{L}(\mathbb{B}; \mathbb{B})$$

in the operator norm. The linear operator A defined as follows:

$$\mathcal{D}(A) = \left\{ x \in \mathbb{B} : \lim_{s \to 0+} \frac{\Phi(s)x - x}{s} \text{ exists} \right\}$$

$$Ax := \lim_{s\to 0^+} \frac{\Phi(s)x - x}{s} \quad \text{for} \quad x \in \mathcal{D}(A)$$

is called the infinitesimal generator of the strongly continuous semigroup Φ and $\mathcal{D}(A)$ is the domain of A. For groups, we take $s \to 0$ in the above definitions. The following lemma (inspired by [1, page 3]) provides a useful characterization of the uniformly continuous semigroups.

Lemma 2.1 *A linear operator $A : \mathcal{D}(A) \subset \mathbb{B} \mapsto \mathbb{B}$ is an infinitesimal generator of a uniformly continuous semigroup (group) Φ of linear bounded operators on $\mathbb{B}$ if and only if A is a bounded linear operator, $\mathcal{D}(A) = \mathbb{B}$.*

Proof The proof is given for the semigroup Φ. The group Φ can be studied similarly. *Necessity.* If $A \in \mathcal{L}(\mathbb{B}; \mathbb{B})$ then the series

$$e^{At} := \sum_{n=0}^{+\infty} \frac{(tA)^n}{n!} \tag{2.2}$$

is convergent in $\mathcal{L}(\mathbb{B}; \mathbb{B})$ and for any $t \geq 0$ it defines a linear bounded operator $\Phi(t) := e^{tA}$ for any $t \geq 0$. Obviously, $\Phi(0) = I$. Moreover, it can be shown that $\Phi(t+s) = \Phi(s)\Phi(t)$ and

$$\|\Phi(t) - I\| \leq t\, e^{t\|A\|}\, \|A\|, \quad \left\| \frac{\Phi(t) - I}{t} - A \right\| \leq \|A\|\, \|\Phi(t) - I\|.$$

This means that Φ is the uniformly continuous semigroup and A is its infinitesimal generator.

Sufficiency. Let Φ be a uniformly continuous semigroup of linear bounded operators on $\mathbb{B}$ and let $\rho > 0$ be small enough such that

$$\left\| I - \frac{1}{\rho} \int_0^\rho \Phi(s)ds \right\| < 1.$$

Hence, the operators $\frac{1}{\rho}\int_0^\rho \Phi(s)ds$ and $\int_0^\rho \Phi(s)ds$ are invertible, so

$$(\Phi(h) - I) \int_0^\rho \Phi(s)ds = \int_0^\rho \Phi(s+h)ds - \int_0^\rho \Phi(s)ds = \int_h^{\rho+h} \Phi(s)ds - \int_0^\rho \Phi(s)ds$$

$$= \int_h^{\rho+h} \Phi(s)ds - \int_0^\rho \Phi(s)ds + \int_0^h \Phi(s)ds - \int_0^h \Phi(s)ds = \int_\rho^{\rho+h} \Phi(s)ds - \int_0^h \Phi(s)ds$$

$$= \int_0^h \Phi(\rho + s)ds - \int_0^h \Phi(s)ds = (\Phi(\rho) - I) \int_0^h \Phi(s)ds.$$

Hence, we conclude that

$$h^{-1}(\Phi(h) - I) = h^{-1} \int_0^h \Phi(s)ds \, (\Phi(\rho) - I) \left(\int_0^\rho \Phi(s)ds \right)^{-1}.$$

Since

$$h^{-1} \int_0^h \Phi(s)ds = I + h^{-1} \int_0^h (\Phi(s) - I)ds$$

and

$$\left\| h^{-1} \int_0^h (\Phi(s) - I)ds \right\| \leq h^{-1} \int_0^h \|\Phi(s) - I\| \, ds \to 0 \text{ as } h \to 0^+$$

then

$$h^{-1}(\Phi(h) - I) \to A := (\Phi(\rho) - I) \left(\int_0^\rho \Phi(s)ds \right)^{-1} \quad \text{as } h \to 0^+.$$

Obviously, the obtained operator A is linear and bounded by construction. □

The following corollary summarizes the most important properties of uniformly continuous semigroups of linear bounded operators [1, p. 3].

Corollary 2.2 *Let Φ be a uniformly continuous semigroup of linear bounded operators on $\mathbb{B}$. Then there exists a linear bounded operator $A \in \mathcal{L}(\mathbb{B}; \mathbb{B})$ such that*

- $\Phi(t) = e^{At} = \sum_{i=0}^{+\infty} \frac{t^i A^i}{i!}$;
- *A is the infinitesimal generator of Φ;*
- $\|\Phi(t)\| \leq e^{t\|A\|}$ *for* $t \geq 0$;
- *the map $t \mapsto \Phi(t)$ is differentiable for all $t \geq 0$:*

$$\dot{\Phi}(t) = A\Phi(t) = \Phi(t)A.$$

Proof Let A be an infinitesimal generator of Φ. Let us consider the semigroup $S(t) = e^{tA}$ given by (2.2). Its infinitesimal generator is also A, i.e., the semigroups Φ and S have the same infinitesimal generator A. This means that $\Phi(t) = S(t)$ for $t \geq 0$. All other properties are straightforward. □

The following properties of uniformly continuous groups are straightforward consequences of Corollary 2.2.

Corollary 2.3 *Let Φ be a uniformly continuous group of linear bounded operators on $\mathbb{B}$. Then*

- *the operator $\Phi(t)$ is invertible and $(\Phi(t))^{-1} = \Phi(-t)$ for any $t \in \mathbb{R}$;*
- *there exists a linear bounded operator $A \in \mathcal{L}(\mathbb{B};\mathbb{B})$ such that*
  - $\Phi(t) = e^{At} = \sum_{i=0}^{+\infty} \frac{t^i A^i}{i!}$ *for all $t \in \mathbb{R}$;*
  - *A is the infinitesimal generator of Φ;*
  - *$(-A)$ is the infinitesimal generator of the group $\Psi(t) = \Phi(-t)$, $t \in \mathbb{R}$;*
  - $\|\Phi(t)\| \le e^{|t|\|A\|}$ *for $t \in \mathbb{R}$;*
  - *the mapping $t \mapsto \Phi(t)$ is differentiable for all $t \in \mathbb{R}$:*

$$\dot{\Phi}(t) = A\Phi(t) = \Phi(t)A.$$

If linear bounded operators A and B commute $AB = BA$ then the semigroup generated by $A + B$ can be found using semigroups generated by A and B, respectively [2].

Corollary 2.4 *Let linear bounded operators $A, B \in \mathcal{L}(\mathbb{B};\mathbb{B})$ be an infinitesimal generators of the uniformly continuous semigroups (groups) Φ and Ψ. If A and B commute then the semigroup Θ generated by the linear bounded operator $A + B$ is given by*

$$e^{t(A+B)} = e^{tA}e^{tB} = e^{tB}e^{tA}, \quad \forall t \in \mathbb{R}.$$

The proof follows the conventional power series arguments

$$\sum_{i=0}^{\infty} \frac{(tA + tB)^i}{i!} = \sum_{i=0}^{\infty} \frac{\sum_{j=0}^{i} \frac{i!(tA)^j (tB)^{i-j}}{j!(i-j!)}}{i!} = \sum_{i=0}^{\infty} \frac{(tA)^i}{i!} \sum_{k=0}^{\infty} \frac{(tB)^k}{k!} = e^{tA}e^{tB}$$

provided that A and B commute.

Example 2.1 Let $\mathbb{B} = \mathbb{R}^2$ and

$$\Phi(s) = e^s \begin{pmatrix} \cos(s) & \sin(s) \\ -\sin(s) & \cos(s) \end{pmatrix}.$$

Obviously, $\Phi(0) = I$ and $\|\Phi(s) - I\| \to 0$ as $t \to 0^+$. Let us find a generator of the semigroup Φ:

$$\frac{1}{s}(\Phi(s) - I) = \frac{1}{s}\left(e^s \begin{pmatrix} \cos(s) & \sin(s) \\ -\sin(s) & \cos(s) \end{pmatrix} - I\right) =$$

$$\frac{e^s - 1}{s} \begin{pmatrix} \cos(s) & \sin(s) \\ -\sin(s) & \cos(s) \end{pmatrix} + \frac{1}{s} \begin{pmatrix} \cos(s)-1 & \sin(s) \\ -\sin(s) & \cos(s)-1 \end{pmatrix} \to A = \begin{pmatrix} 1 & 1 \\ -1 & 1 \end{pmatrix} \quad \text{as} \quad s \to 0^+.$$

Since $\dot{\Phi} = A\Phi = \Phi A$ then Φ is the so-called fundamental system of the linear ODE

$$\dot{x} = Ax$$

and $x(t) = \Phi(t)x_0$ is a solution of the system for any $x_0 \in \mathbb{R}^2$.

2.1.2 Strongly Continuous Groups and Semigroups

Definition 2.3 *(Strongly continuous (semi)group)* A semigroup (group) Φ of linear bounded operators on $\mathbb{B}$ is *strongly continuous* if

$$\Phi(t)x \to x \quad \text{as} \quad t \to 0^+ \quad \text{for every} \quad x \in \mathbb{B}$$

(resp., as $t \to 0$).

A strongly continuous semigroup (group) is shortly denoted by C_0-semigroup (resp., C_0-group). *Any strongly continuous semigroup of linear bounded operators on $\mathbb{B} = \mathbb{R}^n$ is uniformly continuous.* The strongly continuous semigroup (group) of linear bounded operators in an infinite-dimensional space is an important notion for the theory of abstract differential equations in the view of the following theorem (inspired by [[1], p. 4]).

Theorem 2.1 *If Φ is a strongly continuous semigroup of linear bounded operators on $\mathbb{B}$ and $A : \mathcal{D}(A) \subset \mathbb{B} \mapsto \mathbb{B}$ is its infinitesimal generator then*

(1) there exist $M \geq 1$ and $\omega \in \mathbb{R}$ such that $\|\Phi(t)\| \leq Me^{\omega t}$ for $t \geq 0$;
(2) the mapping $t \mapsto \Phi(t)x$ is continuous for any $x \in \mathbb{B}$;
(3) $h^{-1}\int_t^{t+h} \Phi(s)x\,ds \to \Phi(t)x$ as $h \to 0^+$ for any $x \in \mathbb{B}$;
(4) $\int_0^t \Phi(s)x\,ds \in \mathcal{D}(A)$ and $A\left(\int_0^t \Phi(s)x\,ds\right) = \Phi(t)x - x$ for any $x \in \mathbb{B}$;
(5) $\Phi(t)x \in \mathcal{D}(A)$ and $\frac{d}{dt}\Phi(t)x = A\Phi(t)x = \Phi(t)Ax$, $\forall x \in \mathcal{D}(A)$, $\forall t \geq 0$;
(6) $\Phi(t)x - \Phi(s)x = \int_s^t A\Phi(\sigma)x\,d\sigma = \int_s^t \Phi(\sigma)Ax\,d\sigma$ for any $x \in \mathcal{D}(A)$ and any $t, s \in \mathbb{R}^+$.

Proof (1) Notice that there exists a number $\rho > 0$ such that $\|\Phi(t)\|$ is uniformly bounded on $t \in [0, \rho]$. Indeed, otherwise there exists a sequence $\{t_n\}$ such that $t_n \geq 0$, $t_n \to 0$ as $n \to +\infty$ and the family $\{\Phi(t_n)\}$ of linear bounded operator is unbounded, i.e., $\lim_{n\to+\infty} \|\Phi(t_n)\| = +\infty$. Hence, by Banach–Steinhaus theorem (see Appendix B), we have $\{x \in \mathbb{B} : \lim_{n\to+\infty} \|\Phi(t_n)x\| = +\infty\} \neq \emptyset$, but this contradicts the property of strong continuity. Thus, there exists $M \geq 0$ such that $\|\Phi(t)\| \leq M$ for $t \in [0, \rho]$. Since $\Phi(0) = I$ then $M \geq 1$. Let $\omega = \rho^{-1} \ln M \geq 0$ and $t = n\rho + \delta$, $n \in \{0, 1, 2, ...\}$. Then

$$\|\Phi(t)\| = \|\Phi^n(\rho)\Phi(\delta)\| \le M^{n+1} \le MM^{t/\rho} = Me^{\omega t}.$$

(2) Since $\|\Phi(t+h)x - \Phi(t)x\| \le \|\Phi(t)\|\|\Phi(h)x - x\| \le Me^{\omega t}\|\Phi(h)x - x\|$ for any $h > 0$ then strong continuity immediately implies continuity of the mapping $t \mapsto \Phi(t)x$.
(3) Since $\int_t^{t+h} \Phi(s)x\, ds = \int_0^h \Phi(s)\Phi(t)x\, ds = h\Phi(t)x + \int_0^h (\Phi(s) - I)\Phi(t)x\, ds$ then the third property immediately follows from the continuity of the mapping $s \mapsto \Phi(s)y$ for any $y \in \mathbb{B}$.
(4) If $x \in \mathbb{B}$ and $h > 0$ then

$$\frac{\Phi(h)-I}{h}\int_0^t \Phi(s)x\, ds = \frac{1}{h}\int_h^{t+h} \Phi(s)x\, ds - \frac{1}{h}\int_0^t \Phi(s)x\, ds + \frac{1}{h}\int_0^h \Phi(s)x\, ds - \frac{1}{h}\int_0^h \Phi(s)x\, ds$$

$$= \frac{1}{h}\int_t^{h+t} \Phi(s)x\, ds - \frac{1}{h}\int_0^h \Phi(s)x\, ds \to \Phi(t)x - x \quad \text{as} \quad h \to 0^+.$$

(5) Let $x \in \mathcal{D}(A)$ and $h > 0$. Then

$$\frac{\Phi(h)-I}{h}\Phi(t)x = \Phi(t)\frac{\Phi(h)-I}{h}x \to \Phi(t)Ax \quad \text{as} \quad h \to 0^+.$$

Thus $\Phi(t)x \in \mathcal{D}(A)$ and $A\Phi(t)x = \Phi(t)Ax = \frac{d^+}{dt}\Phi(t)$, where $\frac{d^+}{dt}\Phi(t)x := \lim_{h\to 0^+} \frac{\Phi(t+h)x - \Phi(t)x}{h}$ is the right-hand derivative of $\Phi(t)x$. Let us show that the left-hand derivate also exists and coincides with $\Phi(t)Ax$. Let $0 < h < t$. Indeed,

$$\left\|\frac{\Phi(t)x - \Phi(t-h)x}{h} - \Phi(t)Ax\right\| = \left\|\Phi(t-h)\left(\frac{\Phi(h)x - x}{h} - Ax\right) + \Phi(t-h)Ax - \Phi(t)Ax\right\|$$

$$\le \|\Phi(t-h)\|\left\|\frac{\Phi(h)x - x}{h} - Ax\right\| + \|\Phi(t-h)Ax - \Phi(t)Ax\| \to 0 \quad \text{as} \quad h \to 0^+.$$

(6) The sixth property is obtained by an integration of the fifth identity. □

The following corollary characterizes a class of linear evolution equations, which admit solutions given by a strongly continuous semigroup of linear bounded operators on $\mathbb{B}$.

Let us recall a couple of notions from set theory and operator theory (see Appendix B for more details):

- *a set Q is dense in $\mathbb{B}$ if any neighborhood of $x \in \mathbb{B}$ contains points from Q;*
- *a linear operator $A : \mathcal{D}(A) \subset \mathbb{B} \mapsto \mathbb{B}$ is said to be closed if the convergence $x_n \to x$ and $Ax_n \to y \in \mathbb{B}$ as $n \to \infty$ implies $x \in \mathcal{D}(A)$ and $Ax = y$.*

Generators of strongly continuous groups and semigroups are closed densely defined linear operators [1, p. 5].

Corollary 2.5 *Let $A : \mathcal{D}(A) \subset \mathbb{B} \mapsto \mathbb{B}$ be an infinitesimal generator of a strongly continuous semigroup of linear bounded operators on $\mathbb{B}$. Then the domain $\mathcal{D}(A)$ is dense in $\mathbb{B}$ and A is a closed linear operator.*

Proof Let us denote $x_t = \frac{1}{t}\int_0^t \Phi(s)x\,ds$. Since for any $x \in \mathbb{B}$ we have $x_t \in \mathcal{D}(A)$ and $x_t \to x$ as $t \to 0^+$ then $\mathcal{D}(A)$ is dense in $\mathbb{B}$. Let us prove that the linear operator A is closed. Let a sequence $\{x_n\}$ and an element $x \in \mathbb{B}$ be such that $x_n \in \mathcal{D}(A)$, $x_n \to x$ and $Ax_n \to y$ as $n \to +\infty$. From Theorem 2.1 we have

$$\Phi(t)x_n - x_n = \int_0^t \Phi(s)Ax_n\,ds.$$

The term $\Phi(s)Ax_n$ converges to $\Phi(s)y$ uniformly on $s \in [0, t]$. Hence,

$$\frac{\Phi(t)x - x}{t} = \frac{1}{t}\int_0^t \Phi(s)y\,ds,$$

and letting $t \to 0^+$ we derive $Ax = y$ and $x \in \mathcal{D}(A)$. The proof is complete. □

Example 2.2 Let us consider the family Φ of linear operators in $\mathbb{B} = L^2((0, 1); \mathbb{R})$ defined as follows:

$$(\Phi(t)x)(y) = 2\sum_{n=1}^{\infty} e^{-n^2\pi^2 t}\sin(\pi n y)\int_0^1 \sin(\pi n z)x(z)dz, \quad t \geq 0.$$

Since $\{\sqrt{2}\sin(\pi n y)\}_{i=1}^{\infty}$ is an orthonormal basis in $L^2((0, 1); \mathbb{R})$, then, using Parseval's identity (see Appendix A), we derive

$$\begin{aligned}\|\Phi(t)x\|^2 &= \sum_{n=1}^{\infty} e^{-2n^2\pi^2 t}\left(\int_0^1 \sqrt{2}\sin(\pi n z)x(z)dz\right)^2 \\ &\leq e^{-2\pi^2 t}\sum_{i=1}^{\infty}\left(\int_0^1 \sqrt{2}\sin(\pi n z)x(z)dz\right)^2 = e^{-2\pi^2 t}\|x\|^2.\end{aligned}$$

Therefore, Φ is a family of linear bounded operators on $L^2((0, 1); \mathbb{R})$. Obviously, $\Phi(0)x$ simply gives representation of x in the form of Fourier series, so $\Phi(0)x = x$ for $x \in L^2((0, 1); \mathbb{R})$ and $\Phi(t)x \to x$ as $t \to 0^+$. Moreover, $\Phi(t + s) = \Phi(t)\Phi(s)$. Therefore, Φ is a semigroup of linear bounded operators on $\mathbb{B}$. If $x \in H^2((0, 1); \mathbb{R}) \cap H_0^1((0, 1); \mathbb{R})$ then using integration by parts we derive

$$(t^{-1}(\Phi(t) - I)x)(y) = 2\sum_{n=1}^{\infty}\frac{e^{-n^2\pi^2 t} - 1}{-n^2\pi^2 t}\sin(\pi n y)\int_0^1 (-n^2\pi^2)\sin(\pi n z)x(z)dz$$

$$2\sum_{n=1}^{\infty}\frac{e^{-n^2\pi^2 t}-1}{-n^2\pi^2 t}\sin(\pi ny)\int_0^1 n\pi\cos(\pi nz)\frac{\partial x}{\partial z}dz$$

$$=2\sum_{n=1}^{\infty}\frac{e^{-n^2\pi^2 t}-1}{-n^2\pi^2 t}\sin(\pi ny)\int_0^1 \sin(\pi nz)\frac{\partial^2 x}{\partial z^2}dz.$$

Hence, for $t \to 0^+$ we conclude that the generator of the strongly continuous semigroup Φ is given by the differential operator

$$Au := \frac{\partial^2 x}{\partial z^2}$$

with

$$x \in \mathcal{D}(A) := H^2((0,1);\mathbb{R}) \cap H_0^1((0,1);\mathbb{R}).$$

The following properties of C_0-groups immediately follow from the corresponding properties of the semigroups.

Corollary 2.6 *If Φ is a strongly continuous group of linear bounded operators on $\mathbb{B}$ and $A : \mathcal{D}(A) \subset \mathbb{B} \mapsto \mathbb{B}$ is its infinitesimal generator then*

(1) there exist $M \geq 1$ and $\omega \in \mathbb{R}$ such that $\|\Phi(t)\| \leq Me^{\omega|t|}$ for $t \in \mathbb{R}$;
(2) the mapping $t \mapsto \Phi(t)x$ is continuous for any $x \in \mathbb{B}$;
(3) $h^{-1}\int_t^{t+h}\Phi(s)x ds \to \Phi(t)x$ as $h \to 0$ for any $x \in \mathbb{B}$;
(4) $\int_{t_2}^{t_1}\Phi(s)x\, ds \in \mathcal{D}(A)$ and $A\left(\int_{t_2}^{t_1}\Phi(s)x\, ds\right) = \Phi(t_1)x - \Phi(t_2)x$ for any $x \in \mathbb{B}$ and any $t_1, t_2 \in \mathbb{R}$;
(5) $\Phi(t)x \in \mathcal{D}(A)$ and $\frac{d}{dt}\Phi(t)x = A\Phi(t)x = \Phi(t)Ax$, $\forall x \in \mathcal{D}(A)$, $\forall t \in \mathbb{R}$;
(6) $\Phi(t)x - \Phi(s)x = \int_s^t A\Phi(\sigma)x\, d\sigma = \int_s^t \Phi(\sigma)Ax\, d\sigma$ for any $x \in \mathcal{D}(A)$ and any $t, s \in \mathbb{R}$.

The following result [1, p. 6] highlights the uniqueness of infinitesimal generators of semigroups (groups).

Theorem 2.2 *Let Φ_1 and Φ_2 be strongly continuous semigroups (groups) of bounded linear operators on $\mathbb{B}$ with infinitesimal generators A_1 and A_2, respectively. If $A_1 = A_2$ then $\Phi_1(t) = \Phi_2(t)$ for all $t \geq 0$ (resp., for all $t \in \mathbb{R}$).*

Proof Let $x \in \mathcal{D}(A_1) = \mathcal{D}(A_2)$. For $t \geq s$ the function $s \mapsto \Phi_1(t-s)\Phi_2(s)x$ is differentiable, so that

$$\frac{d}{ds}\Phi_1(t-s)\Phi_2(s)x = -A_1\Phi_1(t-s)\Phi_2(s)x + \Phi_1(t-s)A_2\Phi_2(s)x =$$

$$-\Phi_1(t-s)A_1\Phi_2(s)x + \Phi_1(t-s)A_2\Phi_2(s)x = 0$$

since $A_1 = A_2$. Therefore the considered function is a constant and, in particular, its values at $s = 0$ and $s = t$ are the same, i.e., $\Phi_1(t)x = \Phi_2(t)x$ for any $x \in \mathcal{D}(A_1) = \mathcal{D}(A_2)$. Since the linear operators $\Phi_1(t)$ and $\Phi_2(t)$ are bounded and the set $\mathcal{D}(A_1) = \mathcal{D}(A_2)$ is dense in $\mathbb{B}$ then $\Phi_1(t)x = \Phi_2(t)x$ for any $x \in \mathbb{B}$. The proof is complete.□

Example 2.3 The family of operators (see [2, p. 69])

$$(\Phi(t)x)(z) = \frac{1}{\sqrt{4\pi t}} \int_0^{+\infty} e^{-\frac{(z-y)^2}{4t}} x(y)dy, \quad x \in L^2(\mathbb{R}; \mathbb{R}), \quad z, y \in \mathbb{R} \tag{2.3}$$

is a strongly continuous semigroup of linear bounded operators on $L^2(\mathbb{R}; \mathbb{R})$ and $\|\Phi(t)x\|_{L^2} \leq \|x\|_{L^2}$ for any $x \in L^2$ and any $t \geq 0$. The operator

$$\frac{\partial^2}{\partial z^2} : H^2(\mathbb{R}; \mathbb{R}) \subset L^2(\mathbb{R}; \mathbb{R}) \mapsto L^2(\mathbb{R}; \mathbb{R})$$

is the infinitesimal generator of Φ.

Entire Vectors

By Corollaries 2.2 and 2.3, for uniformly continuous semigroups and groups, the exponential formula

$$\Phi(t) = e^{At} = \sum_{i=0}^{+\infty} \frac{t^i A^i}{i!} \tag{2.4}$$

takes the place, where an infinitesimal generator $A \in \mathcal{L}(\mathbb{B}; \mathbb{B})$ is a bounded linear operator. For practical issues, it is important to know if a similar formula can be established for strongly continuous (semi)groups with unbounded infinitesimal generators.

Corollary 2.7 *Let $A : \mathcal{D}(A) \subset \mathbb{B} \mapsto \mathbb{B}$ be an infinitesimal generator of a strongly continuous semigroup (or group) Φ of linear bounded operators on $\mathbb{B}$. Then the set*

$$\mathcal{D}(A^\infty) := \bigcap_{i=1}^{\infty} \mathcal{D}(A^i)$$

is dense in $\mathbb{B}$.

Proof The proof is inspired by [1]. Let $x \in \mathbb{B}$ and $\phi \in C_c^\infty((0, +\infty); \mathbb{C})$ (resp., $\phi \in C_c^\infty((0, +\infty); \mathbb{R})$ if $\mathbb{B}$ is a real Banach space). Since ϕ is compactly supported on $(0, +\infty)$ then there exist $s_{\min} > 0$ and $s_{\max} > s_{\min}$ such that

$$\int_0^{+\infty} \phi(\tau)\Phi(\tau)x \, d\tau = \int_{s_{s\min}}^{s_{\max}} \phi(\tau)\Phi(\tau)x \, d\tau.$$

Since $\phi \in C_c^\infty$ is uniformly bounded, then, by Theorem 2.1, we have

$$\int_{s_{\min}}^{s_{\max}} |\phi(\tau)| \cdot \|\Phi(\tau)x\| d\tau \le M\|\phi\|_C \int_{s_{\min}}^{s_{\max}} e^{\omega\tau} d\tau < +\infty$$

for some $M \ge 1, \omega \in \mathbb{R}$. This means (see Appendix C) that the Bochner integral

$$y = y(x, \phi) := \int_0^{+\infty} \phi(\tau)\Phi(\tau)x \, d\tau$$

is convergent and $y \in \mathbb{B}$. If $0 < h < s_{\min}$ then

$$\frac{\Phi(h) - I}{h} y = \int_0^{+\infty} \phi(\tau) \frac{\Phi(\tau + h) - \Phi(\tau)}{h} x \, d\tau$$

and

$$\int_0^{+\infty} \frac{\phi(\tau - h) - \phi(\tau)}{h} \Phi(\tau) x d\tau \to - \int_0^{+\infty} \dot{\phi}(\tau)\Phi(\tau)x \, d\tau \quad \text{as} \quad h \to 0^+.$$

Therefore, we conclude

$$Ay = - \int_0^{+\infty} \dot{\phi}(\tau)\Phi(\tau)x.$$

Since $\dot{\phi} \in C_c^\infty$ then $Ay \in \mathbb{B}$ (i.e., $y \in \mathcal{D}(A)$) and repeating the above considerations we derive

$$A^i y = (-1)^i \int_0^{+\infty} \phi^{(i)}(\tau)\Phi(\tau)x \, d\tau \quad \forall i \in \mathbb{N}$$

and, consequently, $y \in \mathcal{D}(A^\infty)$.

Let $\mathcal{D} = \{y(x, \phi) : x \in \mathbb{B}, \phi \in C_c^\infty\}$. By construction $\mathcal{D}$ is a linear subspace of $\mathbb{B}$. Let us show that $\mathcal{D}$ is dense in $\mathbb{B}$. Suppose the contrary, i.e., $\mathcal{D}$ is not dense in $\mathbb{B}$. In this case, we have $\operatorname{int}(\mathbb{B} \backslash \mathcal{D}) \ne \emptyset$ and $\operatorname{int}(\mathbb{B} \backslash \mathcal{D}) \cap \mathcal{D} = \emptyset$. By Hahn–Banach Separation theorem (see Appendix A), there exists a linear continuous functional $f : \mathbb{B} \mapsto \mathbb{R}$ such that

$$f(u) < \inf_{y \in \mathcal{D}} f(y), \quad \forall u \in \operatorname{int}(\mathbb{B} \backslash \mathcal{D}). \tag{2.5}$$

Since f is a linear continuous functional then

$$\int_0^\infty \phi(s) f(\Phi(s)x) ds = f\left(\underbrace{\int_0^\infty \phi(s)\Phi(s)x \, ds}_{=y} \right)$$

for every $x \in \mathbb{B}$ and every $\phi \in C_c^\infty$. Therefore, we have

$$f(u) < \inf_{y \in \mathcal{D}} f(y) = \inf_{x \in \mathbb{B}, \phi \in C_c^\infty} \int_0^\infty \phi(s) f(\Phi(s)x) ds, \quad \forall u \in \text{int}(\mathbb{B} \backslash \mathcal{D}).$$

Hence, taking into account that $f(\Phi(s)\,\mathbf{0}) = 0$ for all $s \in \mathbb{R}$, we conclude

$$f(u) < 0, \quad \forall u \in \text{int}(\mathbb{B} \backslash \mathcal{D}).$$

Since Φ is a strongly continuous semigroup of linear bounded operators on $\mathbb{B}$, then the function $s \mapsto f(\Phi(s)u)$ is continuous and $f(\Phi(s)u) \to f(u)$ as $s \to 0$. Hence, for any $u \in \text{int}(\mathbb{B} \backslash \mathcal{D})$, there exists $\phi \in C_c^\infty$ such that

$$f\left(\underbrace{\int_0^\infty \phi(s)\Phi(s)x\, ds}_{=y}\right) = \int_0^\infty \phi(s) f(\Phi(s)x) ds \leq 2f(u) < f(u) < 0.$$

This contradicts the inequality (2.5). Therefore, the set $\mathcal{D}$ is dense in $\mathbb{B}$. Taking into account $\mathcal{D} \subset \mathcal{D}(A^\infty)$ we complete the proof. □

Definition 2.4 Let $A : \mathcal{D}(A) \subset \mathbb{B} \mapsto \mathbb{B}$ be an infinitesimal generator of a strongly continuous semigroup (or a group) Φ of linear bounded operators on $\mathbb{B}$. A vector $y \in \mathcal{D}(A^\infty)$ is said to be an **entire vector** of Φ if the identity

$$\Phi(t)y = \sum_{i=0}^{+\infty} \frac{t^i A^i y}{i!} \tag{2.6}$$

holds for any $t \geq 0$ (resp., any $t \in \mathbb{R}$).

In the view of all claims given above, the properties of strongly continuous semigroups and groups are very similar. However, the following characterization of the space of entire vectors (see also [3], [2, page 81]) highlights the essential difference between groups and semigroups of linear bounded operators in a Banach space.

Corollary 2.8 *If $A : \mathcal{D}(A) \subset \mathbb{B} \mapsto \mathbb{B}$ is an infinitesimal generator of a strongly continuous* ***group*** *Φ of linear bounded operators on $\mathbb{B}$ then* ***the set*** *$\mathcal{D}_{\exp}(A^\infty)$* ***of all entire vectors of Φ is dense in*** *$\mathbb{B}$.*

Proof For $x \in \mathbb{B}$ let us define

$$x_n = \sqrt{\frac{1}{\pi}} \int_{-\infty}^{+\infty} e^{-\tau^2} \Phi\left(\frac{\tau}{\sqrt{n}}\right) x\, d\tau. \tag{2.7}$$

First, let us show that $x_n \in \mathcal{D}(A^\infty)$ for any $n \in \mathbb{B}$. Indeed, by Theorem 2.15, $\exists M \geq 1$ and $\exists \omega \geq 0$ such that $\|\Phi(t)\| \leq Me^{\omega|t|}$ for all $t \in \mathbb{R}$ and

$$\sqrt{\frac{1}{\pi}} \int_{-\infty}^{+\infty} \left\| e^{-\tau^2} \Phi\left(\frac{\tau}{\sqrt{n}}\right) x \right\| d\tau \leq M\|x\| \sqrt{\frac{1}{\pi}} \int_{-\infty}^{+\infty} e^{-\tau^2 + \frac{\omega|\tau|}{\sqrt{n}}} d\tau < +\infty.$$

Then the Bochner integral on the right-hand side of (2.7) is convergent in $\mathbb{B}$ for any $n \in \mathbb{N}$, so $x_n \in \mathbb{B}$. Moreover,

$$\begin{aligned} \frac{\Phi(h) - I}{h} x_n &= \sqrt{\frac{1}{\pi}} \int_{-\infty}^{+\infty} e^{-\tau^2} \frac{\Phi\left(\frac{\tau}{\sqrt{n}} + h\right) - \Phi\left(\frac{\tau}{\sqrt{n}}\right)}{h} x \, d\tau \\ &= \sqrt{\frac{1}{\pi}} \int_{-\infty}^{+\infty} \frac{e^{-(\tau - \sqrt{n}h)^2} - e^{-\tau^2}}{h} \Phi\left(\frac{\tau}{\sqrt{n}}\right) x \, d\tau, \end{aligned}$$

so, tending h to 0, we derive

$$Ax_n = -\sqrt{\frac{n}{\pi}} \int_{-\infty}^{+\infty} \frac{de^{-\tau^2}}{d\tau} \Phi\left(\frac{\tau}{\sqrt{n}}\right) x \, d\tau.$$

Repeating the above considerations we conclude that

$$A^i x_n = (-1)^i \sqrt{\frac{n^i}{\pi}} \int_{-\infty}^{+\infty} \frac{d^i e^{-\tau^2}}{d\tau^i} \Phi\left(\frac{\tau}{\sqrt{n}}\right) x \, d\tau = \sqrt{\frac{n^i}{\pi}} \int_{-\infty}^{+\infty} h_i(\tau) e^{-\tau^2} \Phi\left(\frac{\tau}{\sqrt{n}}\right) x \, d\tau,$$

where $i = 0, 1, 2, \ldots$ and

$$h_i(\tau) = (-1)^i e^{\tau^2} \frac{d^i e^{-\tau^2}}{d\tau^i}$$

is the ith Hermite polynomial (see Appendix E). Since

$$\int_{-\infty}^{+\infty} h_i^2(\tau) e^{-\tau^2} \, d\tau = 2^i i! \sqrt{\pi}$$

and

$$C_n = \int_{-\infty}^{+\infty} e^{-\tau^2 + \frac{2\omega|\tau|}{\sqrt{n}}} d\tau \leq \int_{-\infty}^{+\infty} e^{-\tau^2 + 2\omega|\tau|} d\tau < +\infty$$

then

$$\|A^i x_n\| \le M\|x\|\sqrt{\frac{n^i}{\pi}}\int_{-\infty}^{+\infty} \overbrace{|h_i(\tau)|e^{\frac{\omega|\tau|}{\sqrt{n}}}}^{\le\frac{h_i^2(\tau)+e^{\frac{2\omega|\tau|}{\sqrt{n}}}}{2}} e^{-\tau^2}\,d\tau \le$$

$$M\|x\|\frac{n^{i/2}}{2\sqrt{\pi}}\left(\int_{-\infty}^{+\infty} h_i^2(\tau)e^{-\tau^2}\,d\tau + \int_{-\infty}^{+\infty} e^{-\tau^2+\frac{2\omega|\tau|}{\sqrt{n}}}\,d\tau\right)$$

$$= M\|x\|n^{i/2}\frac{2^i i!\sqrt{\pi}+C_n}{2\sqrt{\pi}} < +\infty.$$

This means that $x_n \in \mathcal{D}(A^i)$ for any $i \in \mathbb{N}$ and any $n \in \mathbb{N}$, i.e., $x_n \in \mathcal{D}(A^\infty)$ for any $n \in \mathbb{N}$. Let us show that for any $t \in \left(-\frac{1}{2\sqrt{n}}, \frac{1}{2\sqrt{n}}\right)$ we have

$$\sqrt{\frac{1}{\pi}}\int_{-\infty}^{+\infty}\sum_{i=k+1}^{\infty}\frac{n^{i/2}t^i h_i(\tau)}{i!}e^{-\tau^2}\Phi\left(\frac{\tau}{\sqrt{n}}\right)x\,d\tau \to 0 \quad \text{as} \quad k\to+\infty.$$

Indeed, since

$$\left|\sqrt{\frac{1}{\pi}}\int_{-\infty}^{+\infty}\sum_{i=k+1}^{\infty}\frac{n^{i/2}t^i h_i(\tau)}{i!}e^{-\tau^2}\Phi\left(\frac{\tau}{\sqrt{n}}\right)x\,d\tau\right| \le$$

$$M\|x\|\sqrt{\frac{1}{\pi}}\sum_{i=k+1}^{\infty}n^{i/2}|t|^i\int_{-\infty}^{+\infty}\frac{|h_i(\tau)|e^{\frac{\omega|\tau|}{\sqrt{n}}}}{i!}e^{-\tau^2}\,d\tau$$

$$\le M\|x\|\sum_{i=k+1}^{+\infty}\frac{n^{i/2}|t|^i(2^i i!\sqrt{\pi}+C_n)}{2\sqrt{\pi}i!} \to 0 \quad \text{as} \quad k\to\infty$$

for any $t \in \left(-\frac{1}{2\sqrt{n}}, \frac{1}{2\sqrt{n}}\right)$. Taking into account (see Appendix E)

$$\sum_{i=0}^{\infty}\frac{n^{i/2}t^i h_i(\tau)}{i!} = e^{2\tau\sqrt{n}t-nt^2}, \quad \forall\tau, t\in\mathbb{R}$$

we derive

$$\sum_{i=0}^{k}\frac{t^i A^i x_n}{i!} - \sqrt{\frac{1}{\pi}}\int_{-\infty}^{+\infty} e^{2\sqrt{n}\tau t-nt^2}e^{-\tau^2}\Phi\left(\frac{\tau}{\sqrt{n}}\right)x\,d\tau =$$

$$\sqrt{\frac{1}{\pi}}\int_{-\infty}^{+\infty}\sum_{i=k+1}^{\infty}\frac{n^{i/2}t^i h_i(\tau)}{i!}e^{-\tau^2}\Phi\left(\frac{\tau}{\sqrt{n}}\right)x\,d\tau \to 0 \quad \text{as} \quad k\to+\infty$$

for any $t \in \left(-\frac{1}{2\sqrt{n}}, \frac{1}{2\sqrt{n}}\right)$. Hence, we conclude that

$$\sum_{i=0}^{+\infty} \frac{t^i A^i x_n}{i!} = \sqrt{\frac{1}{\pi}} \int_{-\infty}^{+\infty} e^{-(\tau-\sqrt{n}t)^2} \Phi\left(\frac{\tau}{\sqrt{n}}\right) x d\tau =$$

$$\sqrt{\frac{1}{\pi}} \int_{-\infty}^{+\infty} e^{-s^2} \Phi\left(\frac{s}{\sqrt{n}} + t\right) x d\tau = \Phi(t) x_n$$

for all $t \in \left(-\frac{1}{2\sqrt{n}}, \frac{1}{2\sqrt{n}}\right)$. Let us introduce the set

$$\mathcal{D} = \left\{ x_n \in \mathbb{B} : x_n = \sqrt{\frac{1}{\pi}} \int_{-\infty}^{+\infty} e^{-\tau^2} \Phi\left(\frac{\tau}{\sqrt{n}}\right) x \, d\tau,\ x \in \mathbb{B},\ n \in \mathbb{N} \right\} \subset \mathcal{D}_{\exp}(A^{\infty}).$$

If $t \in \left(-\frac{1}{2\sqrt{n}}, \frac{1}{2\sqrt{n}}\right)$ then for $\tilde{x} = \Phi(t)x$ we have

$$\tilde{x}_n = \Phi(t) x_n = \sqrt{\frac{1}{\pi}} \int_{-\infty}^{+\infty} e^{-\tau^2} \Phi\left(\frac{\tau}{\sqrt{n}}\right) \tilde{x} \, d\tau \in \mathcal{D}$$

and

$$\Phi(2t) x_n = \Phi(t) \tilde{x}_n = \sum_{i=0}^{\infty} \frac{t^i A^i \Phi(t) x_n}{i!} = \sum_{i=0}^{\infty} \sum_{j=0}^{\infty} \frac{t^{i+j} A^{i+j} x_n}{i! j!} =$$

$$\sum_{k=0}^{\infty} \frac{\left(\sum_{m=0}^{k} \frac{k!}{m!(k-m)!}\right) t^k A^k x_n}{k!} = \sum_{k=0}^{\infty} \frac{\left(\sum_{m=0}^{k} \frac{k! t^m t^{k-m}}{m!(k-m)!}\right) A^k x_n}{k!} = \sum_{k=0}^{\infty} \frac{(t+t)^k A^k x_n}{k!}.$$

Repeating the same consideration we derive

$$\Phi(2^r t) x_n = \sum_{k=0}^{\infty} \frac{(2^r t)^k A^k x_n}{k!}, \quad \forall t \in \left(-\frac{1}{2\sqrt{n}}, \frac{1}{2\sqrt{n}}\right), \forall r \in \mathbb{N}.$$

This means that the identity $\Phi(t) x_n = \sum_{k=0}^{\infty} \frac{t^k A^k x_n}{k!}$ holds for all $t \in \mathbb{R}$.

Let us show that for any $x \in \mathbb{B}$ we have

$$x_n \to x \quad \text{as} \quad n \to +\infty.$$

Indeed, using the identity $\sqrt{\frac{1}{\pi}} \int_{-\infty}^{+\infty} e^{-\tau^2} d\tau = 1$ we derive

$$x_n - x = \sqrt{\frac{1}{2\pi}}\int_{-\infty}^{+\infty} e^{-\frac{\tau^2}{2}}\Phi\left(\tfrac{\tau}{\sqrt{n}}\right)x\,d\tau - \left(\sqrt{\frac{1}{2\pi}}\int_{-\infty}^{+\infty} e^{-\frac{\tau^2}{2}}d\tau\right)x =$$

$$\sqrt{\frac{1}{2\pi}}\int_{-\infty}^{+\infty} e^{-\frac{\tau^2}{2}}\left(\Phi\left(\tfrac{\tau}{\sqrt{n}}\right)x - x\right)ds = \sqrt{\frac{1}{2\pi}}\int_{-n^{1/4}}^{n^{1/4}} e^{-\frac{\tau^2}{2}}\left(\Phi\left(\tfrac{\tau}{\sqrt{n}}\right)x - x\right)ds$$

$$+\underbrace{\sqrt{\tfrac{1}{2\pi}}\int_{n^{1/4}}^{+\infty} e^{-\frac{\tau^2}{2}}\left(\Phi\left(\tfrac{\tau}{\sqrt{n}}\right)x - x\right)d\tau}_{\to 0 \text{ as } n\to+\infty} + \underbrace{\sqrt{\tfrac{1}{2\pi}}\int_{-\infty}^{-n^{1/4}} e^{-\frac{\tau^2}{2}}\left(\Phi\left(\tfrac{\tau}{\sqrt{n}}\right)x - x\right)d\tau}_{\to 0 \text{ as } n\to+\infty}.$$

Taking into account

$$\left\|\sqrt{\frac{1}{2\pi}}\int_{-n^{1/4}}^{n^{1/4}} e^{-\frac{\tau^2}{2}}\left(\Phi\left(\tfrac{\tau}{\sqrt{n}}\right)x - x\right)d\tau\right\| \le \underbrace{\sup_{s\in\left[-\frac{1}{n^{1/4}},\frac{1}{n^{1/4}}\right]} \|\Phi(s)x - x\|}_{\to 0 \text{ as } n\to\infty},$$

we conclude $x_n \to x$ as $n \to \infty$ for any $x \in \mathbb{B}$. This means that the set $\mathcal{D} \subset \mathcal{D}_{\exp}(A^\infty)$ is dense in $\mathbb{B}$. The proof is complete. □

> In general, Corollary 2.8 is not valid for strongly continuous semigroups.

Example 2.4 The translation semigroup (see, e.g., [4], page 537)

$$(\Phi(t)x)(z) = \begin{cases} x(z-t) & \text{if } z \ge t, \\ 0 & \text{if } z < t \end{cases}$$

on $\mathbb{B} = L^2((0,1);\mathbb{R})$ is the example of the so-called nilpotent semigroup:

$$\Phi(t)x = \mathbf{0} \quad \forall t \ge 1, \quad \forall x \in \mathbb{B}.$$

The only entire vector of such a semigroup is $x = \mathbf{0}$. Indeed, if $y \in \mathcal{D}_{\exp}(A^\infty)$ is the entire vector of Φ then, due to nilpotence, we have

$$\Phi(t)y = \sum_{i=0}^{\infty}\frac{t^i A^i y}{i!} = \mathbf{0}, \quad \forall t \ge 1,$$

where $A = -\frac{\partial}{\partial x}$ with $\mathcal{D}(A) = \{x \in H^1((0,1);\mathbb{R}) : x(0) = 0\}$ is the infinitesimal generator of Φ. This is possible only for $y = \mathbf{0}$, i.e., $\mathcal{D}_{\exp}(A^\infty) = \{\mathbf{0}\}$.

The following example (inspired by [2, p. 66]) demonstrates an essential dependence of properties of the operator $\frac{\partial}{\partial z}$ on the domain. Namely, if, in the above example, the space $L^2((0,1);\mathbb{R})$ is replaced with $L^2(\mathbb{R};\mathbb{R})$ then the same differential operator becomes the generator of the strongly continuous *group* of translations.

Example 2.5 The differential operator $A = -\frac{\partial}{\partial z}$ with $\mathcal{D}(A) = H^1(\mathbb{R};\mathbb{R})$ is the infinitesimal generator a strongly continuous *group* Φ of linear bounded operators on $L^2(\mathbb{R};\mathbb{R})$ defined as follows:

$$(\Phi(t)x)(z) = x(z-t), \quad x \in L^2(\mathbb{R};\mathbb{R}), \quad z, t \in \mathbb{R}. \tag{2.8}$$

By Corollary 2.8, the space of all entire vectors of Φ is dense in $L^2(\mathbb{R};\mathbb{R})$.

This example also highlights essential dependence of properties of (semi)groups dependently of the order of the differential operators being their infinitesimal generators. Indeed, the first-order differential operator $\frac{\partial}{\partial z} : H^1(\mathbb{R};\mathbb{R}) \subset L^2(\mathbb{R};\mathbb{R}) \mapsto L^2(\mathbb{R};\mathbb{R})$ is the generator of the strongly continuous translation *group* (2.8), while the operator $\frac{\partial^2}{\partial z^2} : H^2(\mathbb{R};\mathbb{R}) \subset L^2(\mathbb{R};\mathbb{R}) \mapsto L^2(\mathbb{R};\mathbb{R})$ is the generator of the strongly continuous heat *semigroup* (2.3).

2.1.3 Characterization of Generators

For simplicity of proofs, in this subsection, $\mathbb{B}$ denotes a complex Banach space (see Appendix A), but the results given below remain true for real Banach spaces as well. Let

$$R(\lambda, A) = (\lambda I - A)^{-1}, \quad \lambda \in \mathbb{C}$$

denote the *resolvent* of the operator $A : \mathcal{D}(A) \subset \mathbb{B} \mapsto \mathbb{B}$ and

$$\rho(A) = \{\lambda \in \mathbb{C} : \lambda I - A \text{ is invertible}\}$$

be the resolvent set of A. The following important result (see, e.g., [2, p. 73]) is well-known since the late 1940s.

Theorem 2.3 (Hille–Yosida theorem) *A linear operator $A : \mathcal{D}(A) \subset \mathbb{B} \mapsto \mathbb{B}$ is an infinitesimal generator of a strongly continuous semigroup Φ of contractions (i.e., $\|\Phi(t)\| \leq 1$ for $t \geq 0$) if and only if*

i) the operator A is closed and $\mathcal{D}(A)$ is dense in $\mathbb{B}$;
ii) the resolvent set $\rho(A)$ contains $(0, +\infty)$ and

$$\|R(\lambda, A)\| \leq \frac{1}{\lambda} \text{ for any real } \lambda > 0 \text{ and any } n = 1, 2, \ldots$$

Proof *Necessity*. The property (i) is proven in Corollary 2.5, so we need to show only (ii). Let us show

$$R(\lambda, A) = R(\lambda)x := \int_0^{+\infty} e^{-\lambda s}\Phi(s)x\, ds. \tag{2.9}$$

Since $\|\Phi(t)\| \leq 1$ then $R(\lambda)$ is well-defined for every real $\lambda > 0$ and

$$\|R(\lambda)x\| \leq \int_0^{+\infty} e^{-\lambda s}\|\Phi(s)x\|\, ds \leq \frac{1}{\lambda}\|x\|,$$

i.e., $\|R(\lambda)\| \leq \frac{1}{\lambda}$. Moreover, for $h > 0$ one has

$$\tfrac{\Phi(h)-I}{h} R(\lambda) = \frac{1}{h}\int_0^{+\infty} e^{-\lambda s}(\Phi(s+h) - \Phi(s))x\, ds = \tfrac{e^{\lambda h}-1}{h} R(\lambda) - \tfrac{e^{\lambda h}}{h}\int_0^h e^{-\lambda s}\Phi(s)x\, ds$$

and taking the limit on both sides as $h \to 0^+$ we derive $AR(\lambda)x = \lambda R(\lambda)x - x$ for every $x \in \mathbb{B}$ and every $\lambda > 0$, or, equivalently, $(\lambda I - A)R(\lambda) = I$. Using Theorem 2.1 and the closedness of the operator A, we conclude

$$R(\lambda)Ax = \int_0^{+\infty} e^{-\lambda s}\Phi(s)Ax\, ds = A\left(\int_0^{+\infty} e^{-\lambda s}\Phi(s)x\, ds\right) = AR(\lambda)x$$

for any $x \in \mathcal{D}(A)$. Thus $R(\lambda)(\lambda I - A)x = x$ for $x \in \mathcal{D}(A)$ and $R(\lambda) = R(\lambda, A)$ for any $\lambda > 0$. This means that the resolvent set $\rho(A)$ contains $(0, +\infty)$.

Sufficiency. First, let us notice that

$$\lim_{\lambda\to+\infty} \lambda R(\lambda, A)x = x \quad \text{for} \quad x \in \mathbb{B}. \tag{2.10}$$

Indeed, for $x \in \mathcal{D}(A)$ we have

$$\|\lambda R(\lambda, A)x - x\| = \|R(\lambda, A)Ax\| \leq \frac{1}{\lambda}\|Ax\| \to 0 \text{ as } \lambda \to +\infty.$$

But $\mathcal{D}(A)$ is dense in $\mathbb{B}$ and $\|\lambda R(\lambda, A)\| \leq 1$, so $\lambda R(\lambda, A)x \to x$ as $\lambda \to +\infty$ for every $x \in \mathbb{B}$. Now let $A_\lambda : \mathbb{B} \mapsto \mathbb{B}$ be defined as follows:

$$A_\lambda := \lambda AR(\lambda, A) = \lambda R(\lambda, A)A = \lambda^2 R(\lambda, A) - \lambda I.$$

Using (2.10) for any $x \in \mathcal{D}(A)$ we derive

$$\lim_{\lambda\to+\infty} A_\lambda x = \lim_{\lambda\to+\infty} \lambda AR(\lambda, A)x = \lim_{\lambda\to+\infty} \lambda R(\lambda, A)Ax = Ax.$$

Since $\|A_\lambda\| \leq \lambda^2\|R(\lambda, A)\| + \lambda \leq 2\lambda$ then A_λ is the bounded linear operator, so it generates a uniformly continuous semigroup e^{tA_λ} for any $\lambda > 0$. It is nothing to show that for any $\mu > 0$ and any $\lambda > 0$ the operators A_μ, A_λ, e^{tA_μ}, and e^{tA_λ} commute with each other. Consequently, from Corollary 2.2 we conclude

$$\|e^{tA_\lambda}x - e^{tA_\mu}x\| = \left\| \int_0^1 \frac{d}{ds} e^{tsA_\lambda} e^{t(1-s)A_\mu} x\, ds \right\| =$$

$$\left\| \int_0^1 e^{tsA_\lambda} e^{t(1-s)A_\mu} t(A_\lambda - A_\mu)x\, ds \right\| \le t\|A_\lambda x - A_\mu x\|$$

where the property

$$\|e^{tA_\lambda}\| = e^{-\lambda t}\|e^{t\lambda^2 R(\lambda,A)}\| \le e^{-\lambda t} e^{t\lambda^2 \|R(\lambda,A)\|} \le 1, \quad \forall \lambda > 0, \quad \forall t \ge 0$$

is utilized on the last step. Hence, for any $x \in \mathcal{D}(A)$ we have

$$\|e^{tA_\lambda}x - e^{tA_\mu}x\| \le t\|A_\lambda x - Ax\| + t\|Ax - A_\mu x\|$$

and $e^{tA_\lambda}x$ converges to a vector in $\mathbb{B}$ as $\lambda \to +\infty$ uniformly on bounded intervals. Finally, since $D(A)$ is dense in $\mathbb{B}$ and $\|e^{tA_\lambda}\| \le 1$ then $e^{tA_\lambda}x$ with $x \in \mathbb{B}$ also converges to a vector in $\mathbb{B}$ as $\lambda \to +\infty$ and again uniformly on bounded intervals.

Let us consider now the family of operators $T(t) : \mathbb{B} \mapsto \mathbb{B}$, $t \ge 0$ defined as follows:

$$\Phi(t)x := \lim_{\lambda \to +\infty} e^{tA_\lambda}x, \quad x \in \mathbb{B}.$$

Obviously, $\|\Phi(t)\| \le 1$, $\Phi(0) = I$ and Φ satisfies the semigroup property. The function $t \mapsto \Phi(t)x$ is continuous as a uniform limit of the continuous functions $t \mapsto e^{tA_\lambda}$. Thus Φ is a strongly continuous semigroup.

Let $B : \mathcal{D}(B) \subset \mathbb{B} \mapsto \mathbb{B}$ be an infinitesimal generator of Φ. For any $x \in \mathcal{D}(A)$ we have

$$\frac{\Phi(t)x - x}{t} = \frac{1}{t} \lim_{\lambda \to +\infty} e^{tA_\lambda}x - x = \frac{1}{t} \lim_{\lambda \to +\infty} \int_0^t e^{sA_\lambda} A_\lambda x = \frac{1}{t} \int_0^t \Phi(s)Ax\, ds.$$

As $t \to 0^+$ we derive $Ax = Bx$ for any $x \in \mathcal{D}(A)$, i.e., $\mathcal{D}(A) \subset \mathcal{D}(B)$. Since B is the infinitesimal generator of Φ then $1 \in \rho(B)$. Due to assumption ii) we have $1 \in \rho(A)$. Finally, since $\mathcal{D}(B) = (I - B)^{-1}X$ and $(I - B)\mathcal{D}(A) = (I - A)\mathcal{D}(A) = X$ then $\mathcal{D}(A) = \mathcal{D}(B)$, i.e., $A = B$. The proof is complete. □

Hille–Yosida theorem is one of the fundamental results in the theory of semigroups. It inspires (see, e.g., [1], p. 20], [2, p. 77] and references therein) a complete spectral characterization of an infinitesimal generator of a strongly continuous semigroup of linear bounded operators on $\mathbb{B}$.

Corollary 2.9 *A linear operator $A : \mathcal{D}(A) \subset \mathbb{B} \mapsto \mathbb{B}$ is an infinitesimal generator of a strongly continuous semigroup Φ satisfying $\|\Phi(t)\| \leq Me^{\omega t}$ for $t \geq 0$ if and only if*

(i) A is closed operator and $\mathcal{D}(A)$ is dense in $\mathbb{B}$;
(ii) the resolvent set $\rho(A)$ contains $(\omega, +\infty)$ and

$$\|R(\lambda, A)^n\| \leq \frac{M}{(\lambda-\omega)^n} \text{ for any real } \lambda > \omega \text{ and any } n = 1, 2, ...$$

Proof Let us consider the semigroup $\tilde{\Phi} = e^{-\omega t}\Phi(t)$ then $\|\tilde{\Phi}(t)\| \leq M$ for all $t \geq 0$ and A is an infinitesimal generator of Φ if and only if $A - \omega I$ is an infinitesimal generator of $\tilde{\Phi}$ (due to Theorem 2.2). Let us introduce the new norm in $\mathbb{B}$:

$$\|x\|_{new} = \sup_{t\geq 0} \|\tilde{\Phi}(t)x\|, \quad x \in \mathbb{B}.$$

Obviously, $\|x\| \leq \|x\|_{new} \leq M\|x\|$, i.e., $\|\cdot\|_{new}$ is an equivalent norm in $\mathbb{B}$. Furthermore, for any $s \geq 0$ we derive

$$\|\tilde{\Phi}(s)x\|_{new} = \sup_{t\geq 0} \|\tilde{\Phi}(t+s)x\| = \sup_{\sigma\geq s} \|\tilde{\Phi}(\sigma)x\| \leq \sup_{\sigma\geq 0} \|\tilde{\Phi}(\sigma)x\| = \|x\|_{new}.$$

By Hille–Yosida theorem, the operator $A - \omega I$ generates a semigroup of contractions $\|\tilde{\Phi}(t)x\|_{new} \leq \|x\|_{new}$ if and only if A is closed, $\mathcal{D}(A)$ dense in $\mathbb{B}$ and $\|R(\lambda, A - \omega I)x\|_{new} \leq \|x\|_{new}/\lambda$ for any $\lambda > 0$ or, equivalently, $\|R(\lambda, A)x\|_{new} \leq \|x\|_{new}/(\lambda - \omega)$ for any $\lambda > \omega$. Finally, taking into account

$$\|R^n(\lambda, A)x\| \leq \|R^n(\lambda, A)x\|_{new} \leq \frac{\|x\|_{new}}{(\lambda-\omega)^n} \leq \frac{M\|x\|}{(\lambda-\omega)^n}$$

we complete the proof. □

As a straightforward consequence of the obtained results, we derive the following analog of the exponential formula for C_0-semigroups:

$$\Phi(t)x := \lim_{\lambda\to+\infty} e^{tA_\lambda}x, \quad x \in \mathbb{B}, \tag{2.11}$$

where $A_\lambda = \lambda AR(\lambda, A) = \lambda R(\lambda, A)A = \lambda^2 R(\lambda, A) - \lambda \in \mathcal{L}(\mathbb{B}, \mathbb{B})$.

2.1.4 Perturbation of Generators

For an analysis of infinite-dimensional models, it is important to know if the bounded perturbations do not destroy generative properties of operators. Inspired by [2, p. 158] and [1, p. 76], the following theorem characterizing semigroups generated by perturbed generators can be proven.

Theorem 2.4 *Let* $A : \mathcal{D}(A) \subset \mathbb{B} \mapsto \mathbb{B}$ *be an infinitesimal generator of a strongly continuous semigroup* Φ *of linear bounded operators on* $\mathbb{B}$ *and* $\Delta A : \mathbb{B} \mapsto \mathbb{B}$ *be a bounded linear operator on* $\mathbb{B}$*. Then the operator* $A + \Delta A$ *with* $\mathcal{D}(A + \Delta A) := \mathcal{D}(A)$ *is an infinitesimal generator of a strongly continuous semigroup* $\tilde{\Phi}$ *of linear bounded operators on* $\mathbb{B}$*. Moreover, if* $\|\Phi(t)\| \le Me^{t\omega}$ *then*

$$\|\tilde{\Phi}(t)\| \le Me^{t(\omega+M\|\Delta A\|)} \quad for \ t \ge 0 \tag{2.12}$$

and

$$\|R(\lambda, A + \Delta A)\| \le \frac{M}{\lambda-(\omega+M\|\Delta A\|)} \quad for \quad \lambda > \omega + M\|\Delta A\|.$$

Proof In the proof of Corollary 2.9 it has been shown that there exists *an equivalent norm* in $\mathbb{B}$ such that $\|\Phi(t)\| \le e^{t\omega}$ and $\|R(\lambda, A)\| \le \frac{1}{\lambda-\omega}$ for any $\lambda > \omega$, where the operator norm $\|\cdot\|$ is induced by the mentioned equivalent norm in $\mathbb{B}$. In this case, the operator $\Delta AR(\lambda, A)$ is bounded and satisfies $\|\Delta AR(\lambda, A)\| < 1$ for any $\lambda > \omega + \|\Delta A\|$. Thus, the operator $I - \Delta AR(\lambda, A)$ is invertible for any $\lambda > \omega + \|\Delta A\|$ and

$$(I - \Delta AR(\lambda, A))^{-1} = \sum_{k=0}^{\infty}[\Delta AR(\lambda, A)]^k.$$

Let us denote

$$R := R(\lambda, A)(I - \Delta AR(\lambda, A))^{-1}.$$

Since

$$(\lambda I - A - \Delta A)R = (I - \Delta AR(\lambda, A))^{-1} - \Delta AR(\lambda, A)(I - \Delta AR(\lambda, A))^{-1} = I$$

and

$$R(\lambda I - A - \Delta A)x = R(\lambda, A)\sum_{k=0}^{\infty} R(\lambda, A)[\Delta AR(\lambda, A)]^k(\lambda I - A - \Delta A)x =$$

$$R(\lambda, A)(\lambda I - A - \Delta A)x + \sum_{k=1}^{\infty} R(\lambda, A)[\Delta AR(\lambda, A)]^k(\lambda I - A - \Delta A)x =$$

$$x - R(\lambda, A)\Delta Ax + \sum_{k=1}^{\infty}[R(\lambda, A)\Delta A]^k R(\lambda, A)(\lambda I - A - \Delta A)x =$$

$$x - R(\lambda, A)\Delta Ax + \sum_{k=1}^{\infty}[R(\lambda, A)\Delta A]^k x - \sum_{k=1}^{\infty}[R(\lambda, A)\Delta A]^{k+1}x = x$$

for every $x \in \mathcal{D}(A)$ then $R(\lambda, A + \Delta A) = R$ exists for $\lambda > \omega + \|\Delta A\|$ and

$$\|(\lambda I - A - \Delta A)^{-1}\| = \left\| \sum_{k=0}^{\infty} R(\lambda, A)[\Delta A R(\lambda, A)]^k \right\| \leq$$

$$\frac{1}{(\lambda-\omega)(1-\|\Delta A R(\lambda, A)\|)} \leq \frac{1}{\lambda-\omega-\|\Delta A\|}.$$

Using Corollary 2.9, we conclude that $A + \Delta A$ is the infinitesimal generator of a strongly continuous semigroup $\tilde{\Phi}$ satisfying $\|\tilde{\Phi}(t)\| \leq e^{t(\omega+\|\Delta A\|)}$. For the original norm in $\mathbb{B}$ we have the estimate (2.12) (see the proof Corollary 2.9 for more details). □

Example 2.6 Considering the operator $A = \frac{\partial^2}{\partial z^2}$ with $\mathcal{D}(A) = H^2((0, 1); \mathbb{R}) \cap H_0^1((0, 1); \mathbb{R}) \subset \mathbb{B} := L^2((0, 1); \mathbb{R})$ studied in Example 2.2, we conclude that the operator $A + \Delta A$ is also a generator of a strongly continuous semigroup provided that $\Delta A : \mathbb{B} \mapsto \mathbb{B}$ is a bounded operator, e.g., $(\Delta A x)(z) = z x(z)$ for $x \in \mathbb{B}, z \in (0, 1)$.

The following result about perturbation of generators is a corollary of [4, Theorem 12.2.2].

Corollary 2.10 *Let $A : \mathcal{D}(A) \subset \mathbb{B} \mapsto \mathbb{B}$ be an infinitesimal generator of a strongly continuous semigroup (or group) Φ of linear bounded operators on $\mathbb{B}$. Then for any $\mu \in \mathbb{R}$ the linear closed densely defined operator $A_\mu = A + \mu I : \mathcal{D}(A_\mu) = \mathcal{D}(A) \subset \mathbb{B} \mapsto \mathbb{B}$ is an infinitesimal generator of the strongly continuous semigroup (resp., group) Φ_μ of linear bounded operators on $\mathbb{B}$ given by*

$$\Phi_\mu(t) = e^{\mu t}\Phi(t), \quad \forall t \geq 0, \tag{2.13}$$

(resp., $t \in \mathbb{R}$). Moreover, $\mathcal{D}(A) = \mathcal{D}(A_\mu)$, $\mathcal{D}(A_\mu^\infty) = \mathcal{D}(A^\infty)$, and $\mathcal{D}_{\exp}(A_\mu^\infty) = \mathcal{D}_{\exp}(A^\infty)$ for any $\mu \in \mathbb{R}$.

Proof Since Φ is a semigroup (resp., group) then, obviously, Φ_μ is a semigroup (resp., group) as well. For any $\mu \in \mathbb{R}$ we have $\mathcal{D}(A) = \mathcal{D}(A_\mu)$ and for any $x \in \mathcal{D}(A_\mu)$ we derive

$$\lim_{s\to 0^+} \frac{e^{\mu s}\Phi(s)x - x}{s} = \lim_{s\to 0^+} \frac{(e^{\mu s} - 1)\Phi(s)x}{s} + \lim_{s\to 0^+} \frac{e^{\mu s}(\Phi(s)x - x)}{s}$$

$$= (\mu I + A)x = \tilde{A}x,$$

i.e., A_μ is an infinitesimal generator of a strongly continuous semigroup (resp., group) Φ_μ. The equivalence of the domains of $\mathcal{D}(A^\infty)$ and $\mathcal{D}(A_\mu^\infty)$ is obvious. But the equivalence of the sets of all entire vectors $\mathcal{D}_{\exp}(A^\infty)$ and $\mathcal{D}_{\exp}(A_\mu^\infty)$ follows from conventional analysis for the power series of commuting operators A and μI:

$$\sum_{i=0}^{\infty} \frac{(\mu I + A)^i y}{i!} = \sum_{i=0}^{\infty} \frac{\sum_{j=0}^{i} \frac{i!}{j!(i-j)!} \mu^j A^{i-j} y}{i!} = \sum_{j=0}^{\infty} \sum_{k=0}^{\infty} \frac{\mu^j A^k y}{j!k!} = e^{\mu t} \sum_{k=0}^{\infty} \frac{A^k y}{k!}.$$

The proof is complete. □

2.1.5 *Fractional Power Operators*

In this subsection, we deal with a special class of semigroups introduced by the following assumption.

Assumption 2.1 Let a closed linear densely defined operator $A : \mathcal{D}(A) \subset \mathbb{B} \mapsto \mathbb{B}$ be an infinitesimal generator of a strongly continuous semigroup Φ of linear bounded operators on a real Banach space $\mathbb{B}$. Let us assume that $0 \in \rho(A)$, for any $x \in \mathbb{B}$ the mapping $t \mapsto \Phi(t)x$ is differentiable for all $t > 0$ and there exist $\delta > 0$, $M \geq 1$ such that

$$\|\Phi(t)\| \leq M e^{-\delta t}, \quad \forall t \geq 0. \tag{2.14}$$

Let $\Gamma : \mathbb{R}_+ \mapsto \mathbb{R}$ be the so-called Γ-function (see Appendix E) defined as follows:

$$\Gamma(\alpha) = \int_0^{+\infty} t^{\alpha-1} e^{-t} dt, \quad \alpha > 0.$$

For $\alpha > 0$ let us consider the operator

$$A^{-\alpha} := \frac{1}{\Gamma(\alpha)} \int_0^{+\infty} t^{\alpha-1} \Phi(t) dt, \quad \alpha > 0 \tag{2.15}$$

and assign $A^0 = I_n$. Inspired by ideas proposed in [1, Sect. 2.6], we prove the following lemma.

Lemma 2.2 *Under Assumption 2.1, the family of operators $A^{-\alpha}, \alpha \geq 0$ is a strongly continuous semigroup of linear bounded operators on $\mathbb{B}$.*

Proof (1) Let us show that $A^{-\alpha}$ is bounded. Indeed, since

$$\left\| \int_0^{\sigma} t^{\alpha-1} \Phi(t) dt \right\| \leq M \int_0^{\sigma} t^{\alpha-1} e^{-\delta t} dt \to \frac{M \Gamma(\alpha)}{\delta^{\alpha}} \quad \text{as} \quad \sigma \to +\infty$$

then $A^{-\alpha} \leq \frac{M}{\delta^{\alpha}} < +\infty$ for any $\alpha \in (0, +\infty)$. Moreover, the above inequality means that

$$\exists C > 0 \quad : \quad \|A^{-\alpha}\| \leq C, \quad \forall \alpha \in (0, 1).$$

(2) Let us prove the semigroup property:

$$
\begin{aligned}
A^{-\alpha}A^{-\beta} &= \frac{1}{\Gamma(\alpha)\Gamma(\beta)} \int_0^{+\infty} \int_0^{+\infty} t^{\alpha-1} s^{\beta-1} \Phi(t)\Phi(s)dtds \\
&= \frac{1}{\Gamma(\alpha)\Gamma(\beta)} \int_0^{+\infty} t^{\alpha-1} \int_t^{+\infty} (u-t)^{\beta-1} \Phi(u)dudt \\
&= \frac{1}{\Gamma(\alpha)\Gamma(\beta)} \int_0^{+\infty} \left(\int_0^u t^{\alpha-1}(u-t)^{\beta-1}dt \right) \Phi(u)du \\
&= \frac{\int_0^1 v^{\alpha-1}(1-v)^{\beta-1}dv}{\Gamma(\alpha)\Gamma(\beta)} \int_0^{+\infty} u^{\alpha+\beta-1}\Phi(u)du \\
&= \frac{1}{\Gamma(\alpha+\beta)} \int_0^{+\infty} u^{\alpha+\beta-1}\Phi(u)du = A^{-(\alpha+\beta)},
\end{aligned}
$$

where the properties of Γ function are utilized on the last step (see Appendix E).

(3) Let us show that for any $x \in \mathcal{D}(A)$ we have

$$
A^{-\alpha}x \to x \quad \text{as} \quad \alpha \to 0^+.
$$

Since $0 \in \rho(A)$ then $x = A^{-1}y$ for some $y \in \mathbb{B}$. Hence, we derive

$$
A^{-\alpha}x - x = A^{-(1+\alpha)}y - A^{-1}y = \int_0^{+\infty} \left(\frac{t^\alpha}{\Gamma(1+\alpha)} - 1 \right) T(t)y\,dt,
$$

where the identity (2.9) with $\lambda = 0$ is utilized on the last step. Using (2.21) we conclude

$$
\|A^{-\alpha}x - x\| \le C_1 \int_0^k \left| \frac{t^\alpha}{\Gamma(1+\alpha)} - 1 \right| e^{-\delta t}dt + C_2 \int_k^{+\infty} te^{-\delta t}dt
$$

for any $\alpha \in (0, 1)$ and any $k > 0$. For any $\epsilon > 0$ we can take $k > 0$ such that the second integral is less than $\epsilon/0$. Next we can select α small enough to guarantee that the first integral is less than $\epsilon/2$. Thus $A^{-\alpha}x \to x$ as $\alpha \to 0^+$ for any $x \in \mathcal{D}(A)$. Since $\mathcal{D}(A)$ is dense in $\mathbb{B}$ and $\|A^{-\alpha}\| \le C$ for $\forall \alpha \in (0, 1)$ then the same limit takes the place for any $x \in \mathbb{B}$. The proof is complete. □

The following theorem (partially inspired by [1, page 83]) introduces the so-called *fractional power operators*.

Theorem 2.5 *Under Assumption 2.1, the operator $A^{-\alpha}$ is invertible for any $\alpha \geq 0$ and*

- *the linear operator*
$$A^{\alpha} = (A^{-\alpha})^{-1} \tag{2.16}$$
is closed and $\mathcal{D}(A^{\alpha})$ = range of $A^{-\alpha}$;
- *$\mathcal{D}(A^{\alpha}) \subset \mathcal{D}(A^{\beta})$ for $\alpha \geq \beta > 0$;*
- *$\mathcal{D}(A^{\alpha})$ is dense in $\mathbb{B}$ for any $\alpha \geq 0$;*
- *for any $\alpha, \beta \in \mathbb{R}$ it holds*
$$A^{\alpha+\beta}x = A^{\alpha}A^{\beta}x = A^{\beta}A^{\alpha}x, \quad \forall x \in \mathcal{D}(A^{\gamma}), \tag{2.17}$$
where $\gamma = \max\{\alpha, \beta, \alpha+\beta\}$;
- *if the inequality (2.21) is fulfilled then for any $\alpha > 0$ the estimate*
$$\|A^{\alpha}\Phi(t)\| \leq \tfrac{M_{\alpha}}{t^{\alpha}}, \quad \forall t > 0 \tag{2.18}$$
holds for some $M_{\alpha} \geq 1$ dependent on α.

Proof (1) Let us show that $A^{-\alpha}$ is invertible (one-to-one). Notice that A^{-1} is one-to-one. Indeed, otherwise, there exist $x_1, x_2 \in \mathbb{B} : x_1 \neq x_2$ such that

$$\int_0^{+\infty} \Phi(t)(x_1 - x_2)dt = \mathbf{0}, \quad \forall s \geq 0.$$

On the one hand, we have

$$\int_0^{+\infty} \Phi(t)(x_1 - x_2)dt = \mathbf{0} \quad \Rightarrow \quad \int_0^{+\infty} A\Phi(t)(x_1 - x_2)dt = \mathbf{0}.$$

On the other hand, since Φ is differentiable then $\frac{d}{dt}\Phi(t)x = A\Phi(t)x$ for all $x \in \mathbb{B}$ and taking into account $\|\Phi(s)\| \leq Me^{-\sigma s}$ we derive the contradiction

$$\mathbf{0} = \int_0^{+\infty} A\Phi(t)(x_1 - x_2)dt = \int_0^{+\infty} \dot{\Phi}(t)(x_1 - x_2)dt = x_2 - x_1 \neq \mathbf{0}.$$

Since A^{-1} is, indeed, one-to-one then A^{-i} is one-to-one for any $i \in \mathbb{N}$. If $A^{-\alpha}x = \mathbf{0}$ with $\alpha \geq 0$ then $A^{-n}x = A^{-n+\alpha}A^{-\alpha}x = \mathbf{0}$. Since A^{-n} is one-to-one then $x = \mathbf{0}$. So, $A^{-\alpha}$ is one-to-one. Therefore, $A^{-\alpha}$ is invertible, the domain of A^{α} is the range of $A^{-\alpha}$ and A^{α} is closed for any $\alpha \geq 0$.

(2) Since for any $\alpha \geq \beta \geq 0$ we have $A^{-\alpha} = A^{-\beta}A^{-(\alpha-\beta)}$, then the range of $A^{-\alpha}$ belongs to the range of $A^{-\beta}$ or, equivalently, $\mathcal{D}(A^{\alpha}) \subset \mathcal{D}(A^{\beta})$, $\forall \alpha \geq \beta \geq 0$.

(3) Since, by Corollary 2.7, the domain $\mathcal{D}(A^{n})$ is dense in $\mathbb{B}$ and $\mathcal{D}(A^{n}) \subset \mathcal{D}(A^{\alpha})$ for any $\alpha \in [0, n)$ then $\mathcal{D}(A^{\alpha})$ is dense in $\mathbb{B}$ for any $\alpha \geq 0$.

(4) If $\alpha > 0$, $\beta > 0$ and $x \in \mathcal{D}(A^\alpha A^\beta)$ then $x \in \mathcal{D}(A^\alpha)$ and $A^\beta x \in \mathcal{D}(A^\alpha)$. For $y = A^\alpha A^\beta x$ we have $A^\beta x = A^{-\alpha} y$ and $x = A^{-\alpha} A^{-\beta} x = A^{-(\alpha+\beta)} x$. Hence, $x \in \mathcal{D}(A^{\alpha+\beta})$ and $A^{\alpha+\beta} x = A^\alpha A^\beta x$. Similarly if $x \in \mathcal{D}(A^{\alpha+\beta})$ then $x \in \mathcal{D}(A^\alpha A^\beta)$ and $A^{\alpha+\beta} x = A^\alpha A^\beta x$.

(5) Using (2.21) for a natural $n > \alpha$ we derive

$$\|A^\alpha \Phi(t)\| = \|A^{-(n-\alpha)} A^n \Phi(t)\| \leq \frac{1}{\Gamma(n-\alpha)} \int_0^{+\infty} s^{n-\alpha-1} \|A^n \Phi(t+s)\| ds \leq$$

$$\frac{1}{\Gamma(n-\alpha)} \int_0^{+\infty} s^{n-\alpha-1} \left\|A\Phi\left(\tfrac{t+s}{n}\right)\right\|^n ds \leq \frac{M_A^n n^n}{\Gamma(n-\alpha)} \int_0^{+\infty} s^{n-\alpha-1} \frac{1}{(t+s)^n} ds =$$

$$\frac{M_A^n n^n}{\Gamma(n-\alpha) t^\alpha} \int_0^{+\infty} \frac{u^{n-\alpha-1}}{(1+u)^n} du \leq \frac{M_\alpha e^{-\delta t}}{t^\alpha}, \quad \forall t > 0.$$

The proof is complete. □

The operators A^α defined in the above theorem are known as a *fractional power operator* for the operator A. Below we utilize these operators for the well-posedness analysis of infinite-dimensional nonlinear systems (see Example 2.14).

Example 2.7 It is well known (see, e.g., [2, page 69]) that the Laplace operator

$$\Delta = \sum_{i=1}^n \frac{\partial^2}{\partial z_i^2} : H^2(\mathbb{R}^n; \mathbb{R}) \subset L^2(\mathbb{R}^n; \mathbb{R}) \mapsto L^2(\mathbb{R}^n; \mathbb{R})$$

is the generator of the strongly continuous heat semigroup

$$(\Phi(t)x)(z) = \frac{1}{(4\pi t)^{n/2}} \int_{\mathbb{R}^n} e^{-\frac{|z-y|^2}{4t}} x(y) dy, \quad x \in L^2(\mathbb{R}^n; \mathbb{R}), \quad z \in \mathbb{R}^n$$

of linear bounded operators in L^2 satisfying $\|\Phi(t)x\|_{L^2} \leq \|x\|_{L^2}$. Moreover,

$$\Phi(t)\mathbb{S}(\mathbb{R}^n; \mathbb{R}) \subset \mathbb{S}(\mathbb{R}^n; \mathbb{R}), \quad \forall t \geq 0,$$

where $\mathbb{S}(\mathbb{R}^n; \mathbb{R})$ is the *Schwartz space* (see Appendix D). This semigroup is differentiable for $t > 0$. In the view of Corollary 2.10, for any $\delta > 0$ the operator $\Delta - \delta I$ generates the semigroup $\tilde{\Phi}(t) = e^{-\delta t}\Phi(t)$. So, for any $\delta > 0$ Assumption 2.1 is fulfilled and the operator $A = \Delta - \delta I$ induces a group of fractional power operators A^α with $\alpha \in \mathbb{R}$. In fact, for $\delta = 0$ the fractional power operator $\Delta^{-\alpha}$, $\alpha \geq 0$,

$$\Delta^{-\alpha} x = \frac{1}{\Gamma(\alpha)} \int_0^{+\infty} t^{\alpha-1} \Phi(t) x dt, \quad x \in L^2(\mathbb{R}^n; \mathbb{R})$$

is also well defined. Changing the order of integration and the integration variable $\tau = \frac{|z-y|}{4t}$, we derive the so-called *Riesz potential* (see, e.g., [5]):

$$(\Delta^{-\alpha}x)(z) = c_\alpha \int_{\mathbb{R}^n} \frac{x(y)}{|z-y|^{n-2\alpha}} dy \tag{2.19}$$

where $c_\alpha = \frac{\Gamma(n/2-\alpha)}{2^\alpha \pi^{n/2}\Gamma(\alpha)}$ and $\alpha \in (0, n/2)$. Its inverse defines the so-called *fractional Laplacian* $\Delta^\alpha = \left(\Delta^{-\alpha}\right)^{-1}$, $\alpha \in (0, 1)$ given by (see, e.g., [6])

$$\Delta^\alpha x = \frac{1}{\Gamma(-\alpha)} \int_0^{+\infty} t^{-1-\alpha}(\Phi(t)x - x)dt, \quad x \in \mathcal{D}(\Delta^\alpha), \tag{2.20}$$

where $\mathcal{D}(\Delta^\alpha) = H^{2\alpha}(\mathbb{R}^n; \mathbb{R})$ is the fractional Sobolev space (see Appendix D).

The following lemma introduces the estimates by fractional power operators.

Lemma 2.3 *Let* $B : \mathcal{D}(B) \subset \mathbb{B} \mapsto \mathbb{B}$ *be a closed linear operator satisfying* $\mathcal{D}(A^\alpha) \subset \mathcal{D}(B)$ *for some* $\alpha \in (0, 1)$ *then* $BA^{-\alpha}$ *is bounded and there exists* $C \geq 1$ *such that*

$$\|Bx\| \leq C\|A^\alpha x\|, \quad \forall x \in \mathcal{D}(A^\alpha).$$

Proof Since $\mathcal{D}(A^{-\alpha}) = \mathbb{B}$ and a range of $A^{-\alpha} = \mathcal{D}(A^\alpha)$ then $\mathcal{D}(BA^{-\alpha}) = \mathbb{B}$. In this case, the closed operator $BA^{-\alpha} : \mathbb{B} \mapsto \mathbb{B}$ is bounded due to Closed Graph theorem (see Appendix B). We complete the proof using the identity $A^{-\alpha}A^\alpha = x$, which holds for all $x \in \mathcal{D}(A^\alpha)$. The proof is complete. □

Example 2.8 The operator

$$B_i = \frac{\partial}{\partial z_i} : \mathcal{D}(B_i) \subset L^2(\mathbb{R}^n; \mathbb{R}) \mapsto L^2(\mathbb{R}^n; \mathbb{R})$$

is linear and closed with the domain $\mathcal{D}(B_i)$ being a closure of the Schwartz space $\mathbb{S}(\mathbb{R}^n; \mathbb{R})$ with respect to the norm $\|x\|_{B_i} = \|x\|_{L^2(\mathbb{R}^n;\mathbb{R})} + \|B_i x\|_{L^2(\mathbb{R}^n;\mathbb{R})}$. Since $\mathcal{D}(B_i) \subset \mathcal{D}(\Delta^{1/2}) = H^1(\mathbb{R}^n; \mathbb{R})$ then, by Lemma 2.3, $\frac{\partial}{\partial z_i}\Delta^{-1/2}$ is a bounded operator on $L^2(\mathbb{R}^n; \mathbb{R})$.

The results presented above are fulfilled as well for the so-called analytic semigroups, which, in the context of Assumption 2.1, correspond to the case $\delta = 0$ and satisfy the additional restriction

$$\|A\Phi(t)\| \leq \frac{M_A}{t}, \quad \forall t > 0 \tag{2.21}$$

for some $M_A \geq 1$ (please see, e.g., [1, Section 2.4] for more details about analytic semigroups).

2.2 Evolution Equations

This section investigates the solution existence and uniqueness problem for an abstract differential equation (*evolution equation*) in a Banach or a Hilbert space. The theory of semigroups of linear bounded operators (briefly surveyed above) is utilized for this purpose. Both linear and nonlinear (quasi-linear) equations are studied.

2.2.1 Linear Systems with Time-Invariant Operators

Classical Solutions of Inhomogeneous Linear Evolution Equations

Let us consider the following evolution equation:

$$\dot{x} = Ax + f(t), \quad t > 0, \tag{2.22}$$

where $A : \mathcal{D}(A) \subset \mathbb{B} \mapsto \mathbb{B}$ is an infinitesimal generator of a strongly continuous semigroup Φ of linear bounded operators on $\mathbb{B}$ and $f : [0, T] \mapsto \mathbb{B}$. The notion of the classical solution of ODE (see Volume I of this monograph) should be properly adapted to the abstract differential equation since the term Ax on the right-hand side of (2.22) is defined only for $x \in \mathcal{D}(A)$, where the domain $\mathcal{D}(A)$ is a dense set in $\mathbb{B}$ (see Theorem 2.1).

Definition 2.5 A function $x : [0, T) \mapsto \mathbb{B}$ is said to be a *classical solution* of (2.22) if x is continuous on $[0, T)$, continuously differentiable on $(0, T)$, $x(t) \in \mathcal{D}(A)$ for $t \in (0, T)$, and the equation (2.22) is fulfilled on $(0, T)$.

Inspired by [1], page 102], [2, page 145], we present the following characterization of operators allowing the existence of classical solutions.

Theorem 2.6 *Let $A : \mathcal{D}(A) \subset \mathbb{B} \mapsto \mathbb{B}$ be a densely defined linear operator on $\mathbb{B}$ with $\rho(A) \neq \emptyset$. The homogeneous ($f = \mathbf{0}$) initial value problem* (2.22) *has a unique classical solution for every $x(0) \in \mathcal{D}(A)$ if and only if A is the infinitesimal generator of a strongly continuous semigroup Φ of linear bounded operators on $\mathbb{B}$.*

Proof *Sufficiency.* If A is the infinitesimal generator of a strongly continuous semigroup Φ of bounded linear operators on $\mathbb{B}$ then from Theorem 2.1 we conclude that $x(t) = \Phi(t)x_0$ with $x_0 \in \mathcal{D}(A)$ is, indeed, a unique classical solution of (2.22) with $f = \mathbf{0}$.

Necessity. Let the homogeneous ($f = \mathbf{0}$) initial value problem (2.22) $x(0) = x_0$ have a unique classical solution $x_{x_0}(t)$ for every $x_0 \in \mathcal{D}(A)$. Let us define the graph norm as follows: $\|x\|_A = \|x\| + \|Ax\|$ for any $x \in \mathcal{D}(A)$. Since $\rho(A) \neq \emptyset$ and the operator A is closed then $\mathbb{Y} = \mathcal{D}(A)$ with the norm $\|\cdot\|_A$ is a Banach space.

Let $C([0, t_0]; \mathbb{Y})$ be the Banach space of continuous mappings $[0, t_0] \mapsto \mathbb{Y}$ with the supremum norm, i.e., $\|y\|_C = \sup_{[0,t_0]} \|y\|_A$ for $y \in C([0, t_0]; \mathbb{Y})$. Let us consider the mapping $S : \mathbb{Y} \mapsto C([0, t_0]; Y)$ defined as follows $S(x_0) = x_{x_0}(t)$ for $t \in [0, t_0]$. Linearity of (2.22) with $f = \mathbf{0}$ implies linearity of S. Since A is closed then S is closed as well. From Closed Graph theorem (see Appendix B) we conclude that S is bounded. Hence, for any $t \in [0, t_0]$ the linear operator $\Phi(t) : \mathbb{Y} \mapsto \mathbb{Y}$ given by $\Phi(t)x_0 = x_{x_0}(t)$ is bounded as well. Obviously, by construction, $\Phi(t)$ has the semigroup property $\Phi(t+s) = \Phi(s)\Phi(t)$ provided that $s, t, t+s \in [0, t_0]$. Moreover, $\Phi(t)$ is uniformly bounded, so it can be extended by $\Phi(t) := \Phi(t - t_0)\Phi(t_0)^n x$ for $t \in [nt_0, (n+1)t_0]$ to a semigroup of bounded linear operators on $\mathbb{Y}$ satisfying $\|\Phi(t)x\|_A \le Me^{\omega t}\|x\|_A$.

Let us show that $A\Phi(t)x_0 = \Phi(t)Ax_0$ for $x_0 \in D(A^2)$. Indeed, let us consider the function v given by

$$t \overset{v}{\mapsto} x_0 + \int_0^t x_{Ax_0}(s)ds,$$

where x_{Ax_0} is a unique classical solution of the system with the initial condition $x(t) = Ax_0$. Since v is continuously differentiable then using the Newton–Leibniz formula for Bochner integrals (see Appendix C), we derive

$$\dot{v}(t) = x_{Ax_0}(t) = Ax_0 + \int_0^t \dot{x}_{Ax_0}(s)ds = A\left(x_0 + \int_0^t x_{Ax_0}(s)ds\right) = Av(t).$$

Since $v(0) = x_0$ then the uniqueness of solutions yields $v(t) = x_{x_0}(t)$ and $Ax_{x_0}(t) = \dot{v}(t) = x_{Ax_0}(t)$, i.e., $A\Phi(t)x_0 = \Phi(t)Ax_0$.

Since $\mathcal{D}(A)$ is dense in $\mathbb{B}$ and $\rho(A) \neq \emptyset$ then $\mathcal{D}(A^2)$ is also dense in $\mathbb{B}$. Let $\lambda_0 \in \rho(A)$, $\lambda_0 \neq \mathbf{0}$, $y \in \mathcal{D}(A)$ and $x = (\lambda_0 I - A)y$. The operators A and $\Phi(t)$ commute and

$$\|\Phi(t)x\| = \|(\lambda_0 I - A)\Phi(t)y\| \le C_1\|\Phi(t)y\|_A \le C_1 Me^{\omega t}\|y\|_A.$$

Since $\lambda_0 \in \rho(A)$ then $(\lambda_0 I - A)$ is invertible and $R(\lambda_0, A) = (\lambda_0 I - A)^{-1}$ is a bounded linear operator. Hence we derive

$$AR(\lambda_0, A) = (A - \lambda_0 I)R(\lambda_0, A) + \lambda_0 R(\lambda_0, A) = \lambda_0 R(\lambda_0, A) - \lambda_0 I,$$

i.e., $AR(\lambda_0, A)$ is a bounded operator and

$$\|y\|_A = \|y\| + \|Ay\| = \|R(\lambda_0, A)x\| + \|AR(\lambda_0, A)x\| \le C_2\|x\|.$$

Therefore, $\|\Phi(t)x\| \le C_1 C_2 e^{\omega t}\|x\|$ provided that $x = (\lambda_0 I - A)y$ with $y \in \mathcal{D}(A^2)$. The obtained inequality can be extended to whole $\mathbb{B}$ taking into account continuity of the linear operator $\Phi(t)$. This implies that $\Phi(t)$ is a strongly continuous semigroup on $\mathbb{B}$ as well.

Let $\tilde{A}$ be a generator of Φ. Let us show that $A = \tilde{A}$. On the one hand, by assumption $\dot{\Phi}(t)x = A\Phi(t)x$ for any $x \in \mathcal{D}(A)$. Since for $t = 0$ we have $\tilde{A}x = \dot{\Phi}(0)x = A\Phi(0)x = Ax$ then $\mathcal{D}(A) \subset \mathcal{D}(\tilde{A})$. Since

$$A\Phi(t)y = \frac{d}{dt}\Phi(t)y = \Phi(t)\tilde{A}y$$

for any $y \in \mathcal{D}(A)$ then for $\lambda \geq \omega$ we have

$$A\int_0^{+\infty} e^{-\lambda s}\Phi(s)yds = \int_0^{+\infty} e^{-\lambda s}\Phi(s)\tilde{A}yds,$$

or, equivalently, $AR(\lambda, \tilde{A})y = R(\lambda, \tilde{A})\tilde{A}y$. On the other hand, $R(\lambda, \tilde{A})\tilde{A}y = \tilde{A}R(\lambda, \tilde{A})y$, therefore $AR(\lambda, \tilde{A})y = \tilde{A}R(\lambda, \tilde{A})y$ for any $y \in \mathcal{D}(A^2)$. Since $\mathcal{D}(A^2)$ is dense in $\mathbb{B}$ then $AR(\lambda, \tilde{A})x = \tilde{A}R(\lambda, \tilde{A})x$ for any $x \in \mathbb{B}$. This means that $\mathcal{D}(\tilde{A})$ is the range of $R(\lambda, \tilde{A})$ and $\mathcal{D}(\tilde{A}) \subset \mathcal{D}(A)$. Consequently, $\mathcal{D}(A) = \mathcal{D}(\tilde{A})$ and $A = \tilde{A}$. The proof is complete. □

Let us establish one more property of a strongly continuous semigroup of linear bounded operators, which can be proven using the above theorem.

Theorem 2.7 *Let a closed linear operator $A : \mathcal{D}(A) \subset \mathbb{B} \mapsto \mathbb{B}$ be a generator of a strongly continuous semigroup Φ of linear bounded operators on $\mathbb{B}$ and a closed linear operator $B : \mathcal{D}(B) \subset \mathbb{B} \mapsto \mathbb{B}$ be such that $\mathcal{D}(A) \subset \mathcal{D}(B)$ and $ABx = BAx$ for all $x \in \mathcal{D}$, where $\mathcal{D} \subset \mathcal{D}(A^2)$ is a dense set in $\mathbb{B}$ such that $\Phi(t)\mathcal{D} \subset \mathcal{D}$ for all $t \geq 0$. Then*

$$B\Phi(t)x = \Phi(t)Bx, \quad \forall t \geq 0, \quad \forall x \in \mathcal{D}(B).$$

Proof Notice that for any $x \in \mathcal{D}(A^2)$ we have $\Phi(t)x \in \mathcal{D}(A^2)$ for any $t \geq 0$. Indeed, if $x \in \mathcal{D}(A^2)$ then $Ax \in \mathcal{D}(A)$, but according to Theorem 2.1 we have $A\Phi(t)x = \Phi(t)Ax \in \mathcal{D}(A)$ provided that $x \in \mathcal{D}(A^2)$. Hence, for any $x \in \mathcal{D}$ we derive

$$\frac{d}{dt}B\Phi(t)x = \lim_{h\to 0}\frac{B\Phi(t+h)x - B\Phi(t)x}{h} = \lim_{h\to 0}\tfrac{1}{h}B\int_t^{t+h} A\Phi(s)xds =$$

$$\lim_{h\to 0}\tfrac{1}{h}\int_t^{t+h} BA\Phi(s)xds = BA\Phi(t)x = AB\Phi(t)x, \quad \forall t \geq 0,$$

where the operator B can be taken under $\int_t^{t+h}$ since $A\Phi(\cdot)x \in C([0, +\infty); \mathbb{B})$ and $A\Phi(s)x \in \mathcal{D}(A) \subset \mathcal{D}(B)$ for all $s \geq 0$ (see Appendix C). Denoting $y(t) = B\Phi(t)x$ we conclude $\dot{y}(t) = Ay(t)$, $\forall t \geq 0$. Since by Theorem 2.6 $y(t) = \Phi(t)y(0)$ is the unique solution of the evolution equation $\dot{y}(t) = Ay(t)$ then

$$y(t) = B\Phi(t)x = \Phi(t)y(0) = \Phi(t)Bx, \quad \forall t \geq 0, \quad \forall x \in \mathcal{D}.$$

Since $\Phi(t)$ is a bounded linear operator, B is a closed linear operator and the domain $\mathcal{D}$ is dense in $\mathbb{B}$ then the obtained identity holds for $x \in \mathcal{D}(B)$ as well. The proof is complete. □

Example 2.9 The operators $A = \Delta : H^2(\mathbb{R}^n; \mathbb{R}) \subset L^2(\mathbb{R}^n; \mathbb{R}) \mapsto L^2(\mathbb{R}^n; \mathbb{R})$ and $B_i = \frac{\partial}{\partial z_i} : \mathcal{D}(B_i) \subset L^2(\mathbb{R}^n; \mathbb{R}) \mapsto L^2(\mathbb{R}^n; \mathbb{R})$ commute

$$AB_i x = B_i A x, \quad \forall x \in \mathcal{D} = \mathbb{S}(\mathbb{R}^n; \mathbb{R}),$$

where the domain $\mathcal{D}(B_i)$ is a closure of $\mathcal{D}$ with respect to the norm $\|x\|_{B_i} = \|x\|_{L_2(\mathbb{R}^n;\mathbb{R})} + \|B_i x\|_{L^2(\mathbb{R}^n;\mathbb{R})}$. In the view of Theorem 2.7 we have

$$B_i \Phi(t) x = \Phi(t) B_i x, \quad \forall t \geq 0, \quad \forall x \in \mathcal{D}(B_i), \tag{2.23}$$

where Φ is the heat semigroup (2.3). Notice also that B_i are also generators of the translation groups

$$(\Psi_i(t)x)(z) = x(z_1, \ldots, z_i + t, \ldots, z_n), \quad t \in \mathbb{R},$$

where $z = (z_1, \ldots, z_n)^\top \in \mathbb{R}^n$. We, obviously, have

$$\Psi_i(s)\Psi_j(t) = \Psi_j(t)\Psi_i(s), \quad s, t \in \mathbb{R}$$

and, in fact, simple calculations show

$$\Phi(t)\Psi_i(s) = \Psi_i(s)\Phi(t), \quad s \in \mathbb{R}, \quad t \geq 0. \tag{2.24}$$

Moreover, the formulas (2.15), (2.19), and (2.23) imply that

$$B_i(\Delta - \delta I)^{-\alpha} x = (\Delta - \delta I)^{-\alpha} B_i x, \quad x \in \mathcal{D}(B_i), \quad \forall \alpha \geq 0, \quad \forall \delta \geq 0, \tag{2.25}$$

where $(\Delta - \delta I)^{-\alpha}$ is the factional operator.

If x is a classical solution of (2.22) then the function $g : [0, t] \mapsto \mathbb{B}$ defined as $g(s) = \Phi(t - s)x(s)$ satisfies the identity

$$\frac{dg}{ds} = -A\Phi(t - s)x(s) + \Phi(t - s)\dot{x}(s) = \Phi(t - s)f(s).$$

Thus, for $f \in L^1((0, T); \mathbb{B})$, the mapping $s \mapsto \Phi(t - s)f(s)$ is integrable and

$$x(t) = \Phi(t)x(0) + \int_0^t \Phi(t - s)f(s)ds. \tag{2.26}$$

In other words, any classical solution of (2.22) (if it exists) must satisfy the above identity. Obviously, if $f \equiv 0$ then $x(t) = \Phi(t)x_0$ is a classical solution to (2.22) for any $x_0 \in \mathcal{D}(A)$ due to Theorem 2.1.

In general, the continuity of f is not sufficient for the existence of a classical solution of the infinite-dimensional evolution equation (2.22).

Indeed, if $f(t) = \Phi(t)z$ where $z \in \mathbb{B}$ then f is continuous for all $t \geq 0$ and according to the formula (2.26), we derive $x(t)=\Phi(t)x_0+t\Phi(t)z$ that may be non-differentiable if $z \notin \mathcal{D}(A)$.

A characterization of functions f allowing the existence of classical solutions is given by the following theorem inspired by [1, page 107].

Theorem 2.8 *Let A be an infinitesimal generator of a strongly continuous semigroup Φ of bounded linear operators on $\mathbb{B}$. Let $f \in L^1((0, T); \mathbb{B})$ be continuous and the function $v : (0, T) \mapsto \mathbb{B}$ be defined as*

$$v(t) = \int_0^t \Phi(t-s)f(s)ds, \quad 0 < t < T.$$

Then the following claims are equivalent:

(1) the evolution equation (2.22) *has a classical solution for any $x(0) \in \mathcal{D}(A)$;*
(2) the function v is continuously differentiable;
(3) $v(t) \in \mathcal{D}(A)$ for $0 < t < T$ and the function $Av : (0, T) \mapsto \mathbb{B}$ is continuous.

Proof **1)⇒2)** If x is a classical solution to (2.22) then this solution is given by (2.26), i.e., $v(t) = x(t) - \Phi(t)x_0$. Since the second term in the latter identity is continuously differentiable (see Theorem 2.1) and x is the classical solution then the claim 2) holds.
2)⇒3) It is easy to verify for $h > 0$ the following identity:

$$\frac{\Phi(h) - I}{h} v(t) = \frac{v(t+h) - v(t)}{h} - \frac{1}{h} \int_t^{t+h} \Phi(t+h-s)f(s)ds.$$

If v is continuously differentiable then tending $h \to 0^+$ we derive $v(t) \in \mathcal{D}(A)$ for $0 < t < T$ and $Av(t) = \dot{v}(t) - f(t)$.
3)⇒1) If $v(t) \in \mathcal{D}(A)$ then due to the above identity v is differentiable from the right and $D^+v(t) = Av(t) + f(t)$. Since Av and f are continuous then v is continuously differentiable and $\dot{v}(t) = Av(t) + f(t)$. Since $v(0) = \mathbf{0}$ then $x(t) = \Phi(t)x_0 + v(t)$ is the classical solution to (2.22) for any $x_0 \in \mathcal{D}(A)$. The proof is complete. □

Strong Solutions of Inhomogeneous Linear Evolution Equations

Similar to the finite-dimensional case, the notion of a strong solution is introduced by the next definition.

Definition 2.6 A function $x : [0, T] \mapsto \mathbb{B}$ is said to be a strong solution to (2.22) if x is continuous on $[0, T]$, differentiable almost everywhere on $(0, T)$, $x(t) \overset{a.e.}{\in} \mathcal{D}(A)$, $\dot{x} \in L^1((0, T); \mathbb{B})$, and x satisfies (2.22) almost everywhere on $(0, T)$.

The class of functions f allowing the existence of strong solutions is characterized by the following theorem inspired by [1, page 109].

Theorem 2.9 *Let A be an infinitesimal generator of a strongly continuous semigroup Φ of bounded linear operators on $\mathbb{B}$. Let $f \in L^1((0, T); \mathbb{B})$ and the function $v : (0, T) \mapsto \mathbb{B}$ be defined as*

$$v(t) = \int_0^t \Phi(t - s) f(s) ds, \quad 0 < t < T.$$

Then the following claims are equivalent:

(1) the evolution equation (2.22) has a strong solution for any $x(0) \in \mathcal{D}(A)$;
(2) the function v is differentiable almost everywhere and $\dot{v} \in L^1((0, T); \mathbb{B})$;
(3) $v(t) \in \mathcal{D}(A)$ almost everywhere on $0 < t < T$ and $Av \in L^1((0, T); \mathbb{B})$.

The proof of this theorem repeats the proof of Theorem 2.8.

A sufficient condition of existence of classical/strong solutions of the system (2.22) can be obtained as a straightforward corollary of Theorems 2.8 and 2.9.

Corollary 2.11 *Let A be an infinitesimal generator of a strongly continuous semigroup Φ of bounded linear operators on $\mathbb{B}$. If f is differentiable (almost everywhere) on $(0, T)$ and $\dot{f} \in C([0, T); \mathbb{B})$ (resp., $\dot{f} \in L^1((0, T); \mathbb{B})$) then the evolution equation (2.22) has a unique classical (resp., strong) solution on $[0, T)$.*

Proof Since f is differentiable (almost everywhere) on $(0, T)$ then f is continuous on $(0, T)$ and

$$v(t) = \int_0^t \Phi(t - s) f(s) ds = \int_0^t \Phi(s) f(t - s) ds \quad \text{for} \quad t \in (0, T)$$

and

$$\dot{v}(t) = \Phi(t) f(0) + \int_0^t \Phi(s) \dot{f}(t - s) ds = \Phi(t) f(0) + \int_0^t \Phi(t - s) \dot{f}(s) ds$$

(almost) everywhere on $(0, T)$. So the existence of the unique classical (strong) solution follows from Theorem 2.8 (resp., Theorem 2.9). The proof is complete. □

Mild Solutions

The strong solution of the evolution equation (1.1) assumes that the derivative $\dot{x}(t)$ exists almost everywhere on $(0, T)$. The strong solutions of ODEs (see Volume I of this monograph) are defined by absolutely continuous functions satisfying the differential equation almost everywhere. In the finite-dimensional case, any absolutely continuous function is differentiable almost everywhere. Unfortunately, this property may not hold for infinite-dimensional spaces in the general case. For instance, there exist Banach spaces in which we can discover nowhere differentiable globally Lipschitz functions (see, e.g., Theorem 5.21 from [7]). This motivates various notions of generalized solutions of abstract differential equation.

Definition 2.7 A continuous function $x : [0, T) \mapsto \mathbb{B}$ is said to be a *mild solution* of (2.22) if it satisfies (2.26) for all $t \in [0, T)$.

A mild solution of the finite-dimensional system (2.22) ($\mathbb{B} = \mathbb{R}^n$) is strong if $f \in L^1$. For the infinite-dimensional case, the function defined by (2.26) may be nowhere differentiable. Fortunately, any mild solution can be approximated (with arbitrary high precision) by means of classical solutions (see, e.g., [1, page 108]).

Theorem 2.10 *Let A be an infinitesimal generator of a strongly continuous semigroup Φ of linear bounded operators on $\mathbb{B}$ and $f \in L^1_{loc}((0, T); \mathbb{B})$. If x is a mild solution of (2.22) on $[0, T)$ then x is the uniform limit of classical solutions on $[0, T']$, $\forall T' \in (0, T)$.*

Proof Let $M \geq 1$ and $\omega \geq 0$ be such that $\|\Phi(t)\| \leq Me^{\omega t}$. Let a sequence $x_n \in \mathcal{D}(A)$ be such that $x_n \to x(0) \in \mathbb{B}$ as $n \to +\infty$ and let a sequence $f_n \in C^1([0, T']; \mathbb{B})$ be such that $f^n \to f \in L^1((0, T'); \mathbb{B})$ as $n \to +\infty$. From Corollary 2.11 we conclude that the Cauchy problem

$$\dot{x}^n = Ax^n + f^n, \qquad x^n(0) = x_n$$

has a unique classical solution x^n satisfying

$$x^n(t) = \Phi(t)x_n + \int_0^t \Phi(t-s)f^n(s)ds.$$

For the mild solution x we derive

$$\|x(t) - x^n(t)\| \leq Me^{\omega t}\|x(0) - x_n\| + \int_0^t Me^{\omega(t-s)}\|f(s) - f^n(s)\|\,ds$$

$$\leq Me^{\omega T}\left(\|x(0) - x_n\| + \int_0^T \|f(s) - f^n(s)\|\,ds\right) \to 0 \text{ as } n \to \infty$$

uniformly on t. □

Example 2.10 Let us consider the model of a heat system with Dirichlet boundary conditions

$$\frac{\partial x}{\partial t} = \frac{\partial^2 x}{\partial z^2}, \quad t > 0, \quad z \in (0, 1),$$

$$x(t, 0) = 0, \quad x(t, 1) = \gamma(t), \quad x(0, z) = x_0(z),$$

where $\gamma : [0, +\infty) \mapsto \mathbb{R}$ is a twice continuously differentiable function. It is easy to verify that

$$w(t, z) = x(t, z) - \gamma(t)z$$

satisfies

$$\frac{\partial w}{\partial t} = \frac{\partial^2 x}{\partial z^2} - \dot{\gamma}(t)z, \quad w(t, 0) = 0, \quad w(t, 1) = 0.$$

Denoting $A = \frac{\partial^2}{\partial x^2}$ and

$$f(t) \in \mathbb{B} = L^2((0, 1); \mathbb{R}) \text{ with } (f(t))(z) = -\dot{\gamma}(t)z, \quad z \in (0, 1),$$

we conclude (using Example 2.2 and Corollary 2.11) that the transformed system has a classical solution for any initial condition

$$w(0, z) = w_0(z),$$

where

$$w_0 \in \mathcal{D}(A) = H^2((0, 1); \mathbb{R}) \cap H_0^1((0, 1); \mathbb{R}).$$

Thus the original system also has a classical solution provided that

$$x_0 - \gamma(0)z \in \mathcal{D}(A).$$

The latter inclusion defines the so-called *comparability condition*:

$$x_0(0) = 0 \text{ and } x_0(1) = \gamma(0).$$

Notice that if $\dot{\gamma}$ is not continuously differentiable but an absolutely continuous function then the existence of strong solutions can be proven similarly. Finally, if $\dot{\gamma}z \in L^1((0, T); \mathbb{B})$ then the original system has just a unique mild solution for any initial condition $x_0 \in \mathbb{B}$.

2.2.2 Linear Systems with Time-Varying Operators

Let us consider now the time-varying evolution equation

$$\dot{x} = (A + \Delta A(t))x + f(t), \quad t > s \geq 0, \tag{2.27}$$

where $A : \mathcal{D}(A) \subset \mathbb{B} \mapsto \mathbb{B}$ is a linear (possibly) unbounded operator, ΔA is an operator-valued function such that $\Delta A(t) \in \mathcal{L}(\mathbb{B}, \mathbb{B})$, and $f : \mathbb{R} \mapsto \mathbb{B}$ is as before.

Definitions of classical and strong solutions of the evolution equation (2.27) are similar to the linear system (2.22) with a time-invariant operator. Mild solutions are introduced as follows.

Under some restrictions on ΔA and f, any solution of the considered system with the initial condition

$$x(s) = x_s \in \mathbb{B}$$

can be presented as

$$x(t) = U(t,s)x_s + \int_s^t U(t,\sigma) f(\sigma)\, d\sigma \tag{2.28}$$

where U is the so-called *evolution family* (or the fundamental system) generated by $A + \Delta A(t)$, that is a two-parameter family of bounded linear operators on $\mathbb{B}$ satisfying the following conditions:

- $U(s,s) = I, \quad U(t,r)U(r,s) = U(t,s)$ for $0 \leq s \leq r \leq t \leq T$;
- the mapping $(t,s) \to U(t,s)$ is strongly continuous for $0 \leq s \leq t \leq T$.

The evolution family is a semigroup-based representation of solutions of the system $\dot{x} = (A + \Delta A(t))x$ in the case of a time-varying generator. If $\Delta A(t) = \Delta A$ is time-invariant and Φ is a strongly continuous semigroup generated by $A + \Delta A$ (see Theorem 2.4) then

$$U(t,s) := \Phi(t-s), \qquad 0 \leq s \leq t \leq T.$$

Recall that, in the time-invariant case, the semigroup Φ always defines a classical solution $\Phi(t)x_0$ provided that $x_0 \in \mathcal{D}(A + \Delta A)$. This is not true in the time-varying case even if $\Delta A : [0,T) \mapsto \mathcal{L}(\mathbb{B}, \mathbb{B})$ is a uniformly continuous family of linear bounded operators on $\mathbb{B}$. The corresponding example can be found in [2, page 490]. In the general case, the evolution family U given by the following theorem defines just a mild solution (2.28) of the system (2.27) even if $f \equiv 0$ (see, e.g., [1], page 135]).

Theorem 2.11 *Let A be an infinitesimal generator of a strongly continuous semigroup of linear bounded operators on $\mathbb{B}$ and the mapping*

$$\Delta A : [0, T) \mapsto \mathcal{L}(\mathbb{B}; \mathbb{B})$$

be uniformly continuous (i.e., continuous in the operator norm). Then there exists a unique evolution family U such that

e1) $\|U(t, s)\| \leq Ce^{\gamma(t-s)}$ *for some* $C \geq 1$ *and some* $\gamma \in \mathbb{R}$*;*

e2) $\frac{\partial^+}{\partial t} U(t, s)v\Big|_{t=s} = (A + \Delta A(t))v$ *for* $v \in \mathcal{D}(A)$ *and* $0 \leq s \leq t < T$*;*

e3) $\frac{\partial}{\partial s} U(t, s)v = -U(t, s)(A + \Delta A(s))v$ *for* $v \in \mathcal{D}(A)$ *and* $0 \leq s \leq t < T$*.*

Proof **I.** Since the semigroup Φ generated by A is strongly continuous then for a certain equivalent norm in $\mathbb{B}$ (see the proof of Corollary 2.9), we have

$$\|\Phi(s)\| \leq e^{\omega s}, \quad s \geq 0 \quad \text{and} \quad \|R(\lambda, A)\| \leq \frac{1}{\lambda - \omega}, \quad \forall \lambda > \omega,$$

where $\omega \in \mathbb{R}$ is some constant dependent of Φ. Since ΔA is continuous in the operator norm then there exists $K > 0$ such that

$$\sup_{t \in [0,T]} \|\Delta A(t)\| \leq K.$$

Due to Theorem 2.4 the operator $A + \Delta A(t) : \mathcal{D}(A + \Delta A(t)) \subset \mathbb{B} \mapsto \mathbb{B}$ with $\mathcal{D}(A + \Delta A(t)) := \mathcal{D}(A)$ generates a strongly continuous semigroup $\tilde{\Phi}_t$ of linear bounded operators in $\mathbb{B}$ for any $t \in [0, T]$ and $\|\tilde{\Phi}_t(s)\| \leq e^{s(\omega + \|\Delta A(t)\|)}$ for $s \geq 0$,

$$\|R(\lambda, A + \Delta A(t))\| \leq \frac{1}{\lambda - \omega - \|\Delta A(t)\|} \quad \text{for} \quad \lambda > \omega + \|\Delta A(t)\|.$$

Hence, we conclude

$$\left\| \prod_{j-1}^{k} R(\lambda, A + \Delta A(t_j)) \right\| \leq (\lambda - \omega - K)^{-k}$$

and (due to Corollary 2.9)

$$\left\| \prod_{j=1}^{k} \tilde{\Phi}_{t_j}(s_j) \right\| \leq \exp\left((\omega + K) \sum_{j=1}^{k} s_j \right) \tag{2.29}$$

for any $k = 1, 2, 3, ...,$ any $s_j \geq 0$, any $t_j \in [0, T]$, $j = 1, 2, ..., k$ and any $\lambda > \omega + K$. Taking into account $\mathcal{D}(A + \Delta A(t)) = \mathcal{D}(A)$ from Theorem 2.1, we conclude that $\tilde{\Phi}_t(s)\mathcal{D}(A) \subset \mathcal{D}(A)$ for any $s \geq 0$.

II. Let $t_j^n = jT/n$, $j = 0, 1, 2, ..., n$ and $A_n(t)$ be defined as follows:

$$\Delta A_n(t) = \Delta A(t_k^n) \quad \text{for} \quad t_j^n \leq t \leq t_{j+1}^n,$$

with $\Delta A_n(T) = \Delta A(T)$. Obviously,

$$\|\Delta A(t) - \Delta A_n(t)\|_{\mathcal{L}(\mathbb{B};\mathbb{B})} \to 0 \text{ as } n \to +\infty$$

uniformly on $t \in [0, T]$. Let us define a two-parameter family of bounded operators

$$U_n(t,s) = \begin{cases} \tilde{\Phi}_{t_j^n}(t-s) & \text{if } t_j^n \leq s \leq t \leq t_{j+1}^n, \\ & k > l, \\ \tilde{\Phi}_{t_k^n}(t - t_k^n)\left[\prod_{j=m+1}^{k-1} \tilde{\Phi}_{t_j^n}\left(\frac{T}{n}\right)\right]\tilde{\Phi}_{t_m^n}(t_{m+1}^n - s) \text{ if } & t_k^n \leq t \leq t_{k+1}^n, \\ & t_l^n \leq s \leq t_{l+1}^n. \end{cases}$$

It is easy to verify that

$$U_n(s,s) = I, \quad U_n(t,s) = U_n(t,r)U_n(r,s), \quad 0 \leq s \leq r \leq t \leq T.$$

and using (2.29) we derive $\|U_n(t,s)\| \leq e^{(\omega+K)(t-s)}$, i.e.,

$$(s,t) \mapsto U_n(s,t) \text{ is strongly continuous on } 0 \leq s \leq t \leq T.$$

Since $\tilde{\Phi}_t(s)\mathcal{D}(A) \subset \mathcal{D}(A)$ then we have $U_n(t,s)\mathcal{D}(A) \subset \mathcal{D}(A)$ and for any $v \in \mathcal{D}(A)$, we derive

$$\begin{cases} \frac{\partial}{\partial t}U_n(t,s)v = (A + \Delta A_n(t))U_n(t,s)v & \text{for } t \neq t_j^n, \quad j = 0, 1, ..., n, \\ \frac{\partial}{\partial s}U_n(t,s)v = -U_n(t,s)(A + \Delta A_n(s))v & \text{for } s \neq t_j^n, \quad j = 0, 1, ..., n. \end{cases}$$

In addition, the mapping $r \mapsto U_n(t,r)U_m(r,s)v$ is differentiable on $[s,t]$ (except a finite set of points $\{t_j^n\}$) and

$$U_n(t,s)v - U_m(t,s)v = -\int_s^t \frac{\partial}{\partial r}U_n(t,r)U_m(r,s)v\,dr$$

$$= \int_s^t U_n(t,r)(\Delta A_n(r) - \Delta A_m(r))U_m(r,s)v\,dr.$$

Hence, for any $v \in \mathcal{D}(A)$ we have

$$\|U_n(t,s)v - U_m(t,s)v\| \le e^{(\omega+K)(t-s)}\|v\| \int_s^t \|\Delta A_n(r) - \Delta A_m(r)\| dr.$$

Since $\mathcal{D}(A)$ is dense in $\mathbb{B}$ then the above inequality is also fulfilled for $v \in \mathbb{B}$, so $U_n(t,s)$ converges in $\mathcal{L}(\mathbb{B};\mathbb{B})$ uniformly on $0 \le s \le t \le T$ as $n \to +\infty$. Let us denote

$$U(t,s) = \lim_{n\to+\infty} U_n(t,s).$$

The uniform convergence guarantees that U is an evolution family as well, i.e., $U(s,s) = I, U(t,r)U(r,s) = U(t,s)$ for $0 \le s \le t < T$ and the mapping $(t,s) \mapsto U(t,s)$ is strongly continuous for $0 \le s \le t < T$.

III. Since for any $\tau \ge 0$ and for any $v \in \mathcal{D}(A)$ we have

$$U_n(t,s)v - \tilde{\Phi}_\tau(t-s)v = -\int_s^t \frac{\partial}{\partial r} U_n(t,r)\,\tilde{\Phi}_\tau(r-s)v\,dr =$$

$$= \int_s^t U_n(t,r)(\Delta A_n(r) - \Delta A(\tau))\tilde{\Phi}_\tau(r-s)v\,dr,$$

taking the limit on both sides for $n \to +\infty$ we derive

$$\|U(t,s)v - \tilde{\Phi}_\tau(t-s)v\| \le e^{(\omega+K)(t-s)}\|v\| \int_s^t \|\Delta A(r) - \Delta A(\tau)\|\,dr.$$

Again since $\mathcal{D}(A)$ is dense in $\mathbb{B}$ then the above inequality also holds for $v \in \mathbb{B}$, so

$$\lim_{t-s\to 0^+} \frac{\|U(t,s)u - \tilde{\Phi}_\tau(t-s)u\|}{t-s} = 0 \quad \text{for} \quad \tau \in \{s,t\}$$

uniformly on $u \in \mathbb{B}$. This immediately implies

$$\left.\frac{\partial^+}{\partial t} U(t,s)v\right|_{t=s} = (A + \Delta A(t))v \text{ for } v \in Y \text{ and } 0 \le s \le t \le T$$

and

$$\left.\frac{\partial^-}{\partial s} U(t,s)v\right|_{s=t} = (A + \Delta A(s))v \text{ for } v \in Y \text{ and } 0 \le s \le t \le T.$$

Taking into account the strong continuity of U we derive

$$\frac{\partial +}{\partial s} U(t,s)v = \lim_{h\to 0^+} U(t,s+h)\frac{v - U(s+h,s)v}{h} = -U(t,s)(A + \Delta A(s))v$$

and

$$\frac{\partial-}{\partial s}U(t,s)v = \lim_{h\to 0^+} U(t,s)\frac{v-U(s,s-h)v}{h} = -U(t,s)(A+\Delta A(s))v,$$

i.e., the mapping $r \mapsto U(t,r)v$ is continuously differentiable on $[0,t], t < T$.

IV. To prove the uniqueness of U let us suppose that there exists another evolution family V satisfying e1)–e3). In this case, we have

$$V(t,s)v - U_n(t,s)v = -\int_s^t \frac{\partial}{\partial r}V(t,r)\,\tilde{\Phi}_\tau(r-s)v\,dr =$$

$$= \int_s^t V(t,r)(\Delta A(r) - \Delta A_m(r))U_m(r,s)v\,dr$$

and repeating the above considerations we conclude $V(t,s) = \lim_{n\to+\infty} U_n(t,s)$, i.e., $V(t,s) = U(t,s)$. The proof is complete. □

The regularity condition for the time-varying linear equation (2.27) (i.e., the existence of classical/strong solutions) can be obtained only under additional restrictions on $\Delta A(t)$. The following result is partially inspired by ideas discussed in [1, page 143].

Theorem 2.12 *Let $A : \mathcal{D}(A) \subset \mathbb{B} \mapsto \mathbb{B}$ be an infinitesimal generator of a strongly continuous semigroup Φ of linear bounded operators on $\mathbb{B}$.*

Let the mapping $\Delta A : [0,T] \mapsto \mathcal{L}(\mathbb{B};\mathbb{B})$ be continuous in the operator norm and

$$\Delta A(t)\mathcal{D}(A) \subset \mathcal{D}(A), \quad t \in [0,T].$$

Then there exists a unique evolution family $U(t,s)$ such that

E1) $\|U(t,s)\| \le Ce^{\gamma(t-s)}$ *for some* $C \ge 1$ *and some* $\gamma \in \mathbb{R}$;
E2) $\frac{\partial}{\partial t}U(t,s)v = (A+\Delta A(t))U(t,s)v$ *for* $v \in \mathbb{Y}$ *and* $0 \le s \le t \le T$;
E3) $\frac{\partial}{\partial s}U(t,s)v = -U(t,s)(A+\Delta A(s))v$ *for* $v \in \mathbb{Y}$ *and* $0 \le s \le t \le T$;
E4) $U(t,s)\mathbb{Y} \subset \mathbb{Y}$ *for* $0 \le s \le t \le T$;
E5) *for* $v \in \mathbb{Y}$ *the mapping* $(t,s) \mapsto U(t,s)v$ *is continuous in* $\mathbb{Y}$ *for* $0 \le s \le t \le T$,

where $\mathbb{Y} = \mathcal{D}(A)$ *is a Banach space with the graph norm* $\|v\|_{\mathbb{Y}} = \|v\| + \|Av\|$.

Proof **I.** Let us consider the family of linear operators $\Delta A_0(t)$ defined as follows:

$$\Delta A_0(t) := R^{-1}(\lambda_0, A)\Delta A(t)R(\lambda_0, A), \quad t \in [0,T]$$

with $\lambda_0 > \omega$, where ω is defined in Corollary 2.9.

Let us show that the operators $\Delta A_0(t)$ are bounded. First of all, notice that $R(\lambda_0, A) : \mathbb{B} \mapsto \mathcal{D}(A)$ is defined on whole $\mathbb{B}$ provided that $\lambda_0 > \omega$. Since $\Delta A(t)$

$\mathcal{D}(A) \subset \mathcal{D}(A)$ then $\Delta A_0(t) : \mathbb{B} \mapsto \mathbb{B}$ is also defined on whole $\mathbb{B}$. According to Closed Graph theorem (see Appendix B), if $\Delta A_0(t)$ is a closed operator then it is bounded. Consider a sequence $x_n \in \mathbb{B}$ which converges to some element $x \in \mathbb{B}$. Since the linear operators $R(\lambda_0, A)$ and $\Delta A(t)$ are bounded (continuous) then the sequence $z_n := \Delta A(t)R(\lambda_0, A)x_n \in \mathcal{D}(A)$ converges to $z := \Delta A(t)R(\lambda_0, A)x \in \mathcal{D}(A)$. Finally, since the operator $R^{-1}(\lambda_0, A) = \lambda_0 I - A$ is closed then the sequence $y_n := R^{-1}(\lambda_0, A)z_n \in \mathbb{B}$ converges to $y := R^{-1}(\lambda_0, A)z \in \mathbb{B}$. Therefore, the graph of the operator $\Delta A_0(t) : \mathbb{B} \mapsto \mathbb{B}$ is closed, and this operator is bounded.

Let us show that the mapping $t \mapsto A_0(t)$ is strongly continuous. Let $x \in \mathbb{B}$, $t_0 \in [0, T]$, and $t_i \in [0, T]$ be an arbitrary sequence such that $t_i \to t_0$ as $i \to +\infty$. Let us denote $y_i := \Delta A(t_i)R(\lambda_0, A)x$. Since $\Delta A(t)$ is a uniformly continuous family of linear bounded operators and $R(\lambda_0, A)$ is a bounded operator then

$$y_i \to y_0 := \Delta A(t_0)R(\lambda_0, A)x \quad \text{as} \quad i_k \to +\infty.$$

On the other hand, $\Delta A(t)R(\lambda_0, A)\mathbb{B} = \mathcal{D}(A)$, i.e., $y_0, y_i \in \mathcal{D}(A)$. Since the operator A is closed then

$$Ay_i \to Ay_0 \in \mathbb{B} \quad \text{as} \quad i \to +\infty$$

and

$$\Delta A_0(t_{i_k})x = (\lambda_0 I - A)y_{i_k} \to \lambda_0 y_0 - Ay_0 = \Delta A_0(t_0)x \in \mathbb{B} \quad \text{as} \quad i \to +\infty,$$

i.e., the mapping $t \mapsto \Delta A_0(t)$ is, indeed, strongly continuous. In this case, from Banach–Steinhaus theorem (see Appendix B) we conclude

$$\sup_{t\in[0,T]} \|\Delta A_0(t)\| = K_0 < +\infty.$$

Let us denote also $K = \sup_{t\in[0,T]} \|\Delta A(t)\|$.

II. Since $\Delta A(t)$ satisfies all conditions of Theorem 2.11, then there exists a unique evolution family U satisfying the properties e1)–e3) proven in Theorem 2.11. Obviously, if $H = Me^{|\gamma|T}$ then $\|U(t, s)\| \leq H$ for $0 \leq s \leq t < T$. Let us show that there exists a unique evolution family W satisfying

$$W(t, s)x - U(t, s)x + \int_s^t W(t, r)(\Delta A_0(r) - \Delta A(r))U(r, s)x\, dr, \tag{2.30}$$

where $x \in \mathbb{B}$ and $0 \leq s \leq t \leq T$. Let us denote $W_0(t, s) = U(t, s)$ and define

$$W_m(t, s)x = \int_s^t W_{m-1}(t, r)(\Delta A_0(r) - \Delta A(r))U(r, s)x\, ds.$$

Since $\|\Delta A_0(r) - \Delta A(r)\| \leq \|\Delta A_0(r)\| + \|\Delta A(r)\| \leq K_0 + K$ for all $r \in [0, T]$ then $\|W_m(t, s)\| \leq \frac{H^{m+1}(K_0+K)^m(t-s)^m}{m!}$. We conclude that the series

$$W(t,s) = \sum_{i=0}^{\infty} W_i(t,s)$$

converges in $\mathcal{L}(\mathbb{B};\mathbb{B})$. Thus $W(t,s)$ is defined and strongly continuous. Obviously, it satisfies the integral operator equation (2.30). The uniqueness of W can be easily proven in the standard way (by contradiction).

III. Let us show that $U(t,s) = R(\lambda_0, A)W(t,s)R^{-1}(\lambda_0, A)$. This would immediately imply the properties E4) and E5) since $R(\lambda_0, A)\mathbb{B} = \mathcal{D}(A)$. From the property e3) of Theorem 2.11, we derive

$$\frac{\partial}{\partial s} U(t,r)R(\lambda_0, A)x = -U(t,r)(A + \Delta A(r))R(\lambda_0, A)x$$

$$= -U(t,r)R(\lambda_0, A)(A + \Delta A_0(r))x$$

for any $x \in \mathbb{B}$ and $0 \le r \le t < T$.

Let U_n be the operators constructed in Theorem 2.11. Then for $v \in \mathcal{D}(A)$ we have

$$\frac{\partial}{\partial s} U(t,r)R(\lambda_0, A)U_n(r,s)v = -U(t,r)R(\lambda_0, A)(\Delta A_0(r) - \Delta A_n(r))U_n(r,s)v.$$

Integrating the latter identity from $r = s$ to $r = t$, we derive

$$R(\lambda_0, A)U_n(t,s)v - U(t,s)R(\lambda_0, A)v =$$

$$-\int_s^t U(t,r)R(\lambda_0, A)(\Delta A_0(r) - \Delta A_n(r))U_n(r,s)v\,dr.$$

Since the estimate 2.29 implies $\|U_n(r,s)\| \le e^{(\omega+K)(t-s)}$ then

$$\left\| \int_s^t U(t,r)R(\lambda_0, A)(\Delta A(r) - \Delta A_n(r))U_n(r,s)v\,dr \right\| \le$$

$$\tfrac{He^{(\omega+K)(t-s)}}{\lambda_0-\omega} \|v\| \int_s^t \|\Delta A(r) - \Delta A_n(r)\|\,dr.$$

Therefore for $n \to \infty$, taking into account that $\mathcal{D}(A)$ is dense in $\mathbb{B}$, we obtain

$$R(\lambda_0, A)U(t,s)v - U(t,s)R(\lambda_0, A)v =$$

$$-\int_s^t U(t,r)R(\lambda_0, A)(\Delta A_0(r) - \Delta A(r))U(r,s)v\,dr$$

for any $v \in \mathbb{B}$. On the other hand, multiplying (2.30) from the left by $R(\lambda_0, A)$ we derive

$$R(\lambda_0, A)W(t,s)x =$$

$$R(\lambda_0, A)U(t,s)x + \int_s^t R(\lambda_0, A)W(t,r)(\Delta A_0(r) - \Delta A(r))U(r,s)x\,dr.$$

Due to the uniqueness of W we conclude

$$U(t,s)R(\lambda_0, A) = R(\lambda_0, A)W(t,s).$$

Thus $U(t,s) = R(\lambda_0, A)W(t,s)R^{-1}(\lambda_0, A)$ and the properties E4) and E5) hold. Hence, using the properties e2) and e3) from Theorem 2.11 we derive

$$\frac{\partial^+}{\partial t}U(t,s)v = \lim_{h\to 0^+} \tfrac{U(t+h,t)-I}{h} U(t,s)v = (A + \Delta A(t))U(t,s)v$$

and

$$\frac{\partial^-}{\partial t}U(t,s)v = \lim_{h\to 0^+} \tfrac{U(t,t-h)-I}{h} U(t-h,s)v = (A + \Delta A(t))U(t,s)v$$

for any $v \in \mathcal{D}(A)$, i.e., E2) holds. The property E3) has been established in Theorem 2.11. The proof is complete. □

Obviously, the properties E2), E4), and E5) guarantee the existence of the classical solution of (2.27) in the form

$$x(t) = U(t,s)x_s$$

with $x(t_s) = x_s \in \mathcal{D}(A)$.

Corollary 2.12 *Let all conditions of Theorem 2.12 hold and* $f : [0,T) \mapsto \mathbb{B}$ *be differentiable (almost everywhere) on* $[0,T)$ *and* $\frac{d}{dt} f \in C([0,T); \mathbb{B})$ *(resp,.* $\frac{d}{dt} f \in L^1((0,T); \mathbb{B})$*) then the system* (2.27) *has a unique classical (resp., strong) solution on* $[0,T)$ *given by* (2.28).

The proof of this corollary literally repeats the proof of Corollary 2.11.

Example 2.11 Let us consider the following hyperbolic partial differential equation:

$$\frac{\partial^2 u}{\partial t^2} = \frac{\partial^2 u}{\partial z^2} + \lambda(t)u, \quad u(t,0) = 0, \quad u(t,1) = 0, \quad t > 0$$

$$u(0,z) = u_0(z), \quad \left.\frac{\partial u}{\partial t}\right|_{t=0} = u_0'(z), \quad z \in [0,1],$$

where $\lambda : \mathbb{R} \mapsto \mathbb{R}$ is a continuous function of time. The system can be rewritten in the form (2.27) considering the Banach space $\mathbb{B} = H^1((0,1);\mathbb{R}) \times L^2((0,1);\mathbb{R})$ and the operators

$$A = \begin{pmatrix} \mathbf{0} & I \\ \frac{\partial^2}{\partial z^2} & \mathbf{0} \end{pmatrix} : \mathcal{D}(A) \subset \mathbb{B} \mapsto \mathbb{B} \quad \text{and} \quad \Delta A(t) = \begin{pmatrix} \mathbf{0} & \mathbf{0} \\ \lambda(t) I & \mathbf{0} \end{pmatrix} : \mathbb{B} \mapsto \mathbb{B}$$

where

$$D(A) = \{(u, v) \in \mathbb{B} : u \in H^2((0,1);\mathbb{R}) \cap H_0^1((0,1);\mathbb{R});\ v \in H_0^1((0,1);\mathbb{R})\},$$

$\mathbf{0}$ is the zero operator in $L^2((0,1);\mathbb{R})$, and I is the identity operator in $L^2((0,1);\mathbb{R})$. The operator A generates a strongly continuous semigroup of linear bounded operators on $\mathbb{B}$ (see, e.g., page 219, [1]). Since the mapping $t \mapsto \Delta A(t)$ is continuous in $\mathcal{L}(\mathbb{B};\mathbb{B})$ and $\Delta A(t)\mathcal{D}(A) \subset \mathcal{D}(A)$ then according to Theorem 2.12 the considered system has a unique classical solution in $\mathbb{B}$.

In all above cases, we have considered additive perturbations of generators. The following theorem deals with a scalar multiplicative perturbation.

Theorem 2.13 *Let $A : \mathcal{D}(A) \subset \mathbb{B} \mapsto \mathbb{B}$ be an infinitesimal generator of a strongly continuous semigroup Φ of linear bounded operators on $\mathbb{B}$, and a function $\gamma : [0, T] \mapsto (0, +\infty)$ be continuous. Then the evolution family U generated by the operator $\gamma(t)A$ is given by*

$$U(t, s) = \Phi\left(\int_s^t \gamma(\sigma)d\sigma\right), \qquad 0 \le s \le t \le T$$

and it satisfies the properties E1)–E5) of Theorem 2.12.

Proof The proof uses the properties of strongly continuous semigroups of linear bounded operators on $\mathbb{B}$ given by Theorem 2.1.

E1) Since Φ is a strongly continuous semigroup then there exists $C \ge 1$ and $\omega > 0$ such that

$$\|\Phi(\tau)\| \le Ce^{\omega\tau}, \quad \forall \tau \ge 0.$$

Hence, using continuity and positivity of γ we derive

$$\|U(t, s)\| \le Ce^{\omega \int_s^t \gamma(\sigma)d\sigma} \le Ce^{(t-s)\omega \max\limits_{t\in[0,T]} \gamma(t)}$$

E2) and E3). Since for any $v \in \mathcal{D}(A)$ we have

$$\frac{d}{d\tau}\Phi(\tau) = A\Phi(\tau) = \Phi(\tau)A, \quad \tau \ge 0$$

then

$$\frac{\partial}{\partial t}U(t,s)v = \gamma(t)AU(t,s)v, \qquad 0 \le s \le t \le T$$

and

$$\frac{\partial}{\partial s}U(t,s)v = -\gamma(s)U(t,s)Av, \qquad 0 \le s \le t \le T.$$

The property $E4$) immediately follows from $\Phi(\tau)\mathcal{D}(A) \subset \mathcal{D}(A)$ for any $\tau \ge 0$. Finally, taking into account

$$U(t,s)v - U(t_0,s_0)v = \Phi\left(\int_s^t \gamma(\sigma)d\sigma\right)v - \Phi\left(\int_{s_0}^{t_0} \gamma(\sigma)d\sigma\right)v =$$

$$\Phi\left(\int_{s_0}^{t_0} \gamma(\sigma)d\sigma\right)\left(\Phi\left(\int_s^t \gamma(\sigma)d\sigma - \int_{s_0}^{t_0} \gamma(\sigma)d\sigma\right) - I\right)v$$

and $A\Phi(\tau) = \Phi(\tau)A$ we derive

$$\|U(t,s)v - U(t_0,s_0)v\|_Y = \|U(t,s)v - U(t_0,s_0)v\| + \|A(U(t,s)v - U(t_0,s_0)v)\| =$$

$$\|U(t,s)v - U(t_0,s_0)v\| + \|(U(t,s) - U(t_0,s_0))Av\|.$$

Using strong continuity of Φ, we conclude $\|U(t,s)v - U(t_0,s_0)v\|_Y \to \mathbf{0}$ as $(t,s) \to (t_0,s_0)$ with $t \ge s$ and $t_0 \ge s_0$. The proof is complete. □

2.2.3 Nonlinear Evolution Equations

Let us consider now the nonlinear evolution equation

$$\dot{x} = Ax + f(t,x), \quad t > t_0, \tag{2.31}$$

where $t_0 \in \mathbb{R}$ is an initial instant of time, $A : \mathcal{D}(A) \subset \mathbb{B} \mapsto \mathbb{B}$ is a generator of a strongly continuous semigroup Φ of linear bounded operators on $\mathbb{B}$, and $f : [t_0, +\infty) \times \mathbb{B} \mapsto \mathbb{B}$ is a nonlinear mapping.

Definitions of *classical and strong solutions* of nonlinear evolution equation (2.31) are literally the same as for linear evolution equations (see Definitions 2.5 and 2.6). Inspired by [1, p. 184], a mild solution of nonlinear evolution equation is introduced as follows.

Definition 2.8 A continuous function $x : [t_0, t_0 + T) \mapsto \mathbb{B}$ is said to be a *mild solution* to (2.31) if it satisfies the following integral equation:

$$x(t) = \Phi(t - t_0)x(0) + \int_{t_0}^{t} \Phi(t - s) f(s, x(s))\, ds \tag{2.32}$$

for all $t \in [t_0, t_0 + T)$, where $0 < T \leq +\infty$.

To guarantee the existence of solutions of (2.31), the nonlinear function f must satisfy some conditions, which are more restrictive than conditions imposed to the right-hand sides of ODEs. Indeed, since the Lebesgue measure does not exist in Banach spaces, the Filippov's theory of differential equations with discontinuous right-hand sides cannot be straightforwardly transferred to the evolution equations. Moreover, the classical solutions of (2.31) may not exist even for a continuous function f independent of x (see Sect. 2.2.1).

Mild Solutions of Nonlinear Evolution Equations

The existence of a mild solution of the evolution equation (2.31) can still be proven for Lipschitz continuous nonlinearities. The following theorem is inspired by ideas presented in [1, page 184].

Theorem 2.14 *Let $A : \mathcal{D}(A) \subset \mathbb{B} \mapsto \mathbb{B}$ be an infinitesimal generator of a strongly continuous semigroup Φ of linear bounded operators on $\mathbb{B}$. If $f : [t_0, t_0 + T] \times \mathbb{B} \mapsto \mathbb{B}$ satisfies the following Lipschitz condition on x uniformly on $t \in [t_0, t_0 + T]$ for some $T \in (0, +\infty)$:*

$$\exists L = L(T) > 0 : \|f(t, x) - f(t, y)\| \leq L\|x - y\|,\ x, y \in \mathbb{B},\ t \in [t_0, t_0 + T]$$

and $f(\cdot, x(\cdot)) \in L^1((t_0, t_0 + T); \mathbb{B})$ for any $x \in C([t_0, t_0 + T]; \mathbb{B})$, then the system (2.31) *has a unique mild solution*

$$x_{t_0,x_0} : [t_0, t_0 + T] \mapsto \mathbb{B}$$

for any initial condition $x(t_0) = x_0 \in \mathbb{B}$.

Proof Let us consider the metric space

$$\mathbf{X} = \left\{ x \in C([t_0, t_0 + T]; \mathbb{B}) : x(t_0) = x_0 \text{ and } \sup_{t \in [t_0, t_0 + T]} \|x(t)\| \leq +\infty \right\},$$

with $T > 0$ and the metric $\rho(x_1, x_2) = \sup_{t \in [t_0, t_0 + T]} \|x_1(t) - x_2(t)\|$. Let us consider the operator S defined as follows:

$$(Sx)(t) = \Phi(t - t_0)x_0 + \int_{t_0}^{t} \Phi(t - \tau) f(\tau, x(\tau)) d\tau.$$

First, let us notice $S : \mathbf{X} \mapsto \mathbf{X}$. Indeed,

$$\|(Sx)(t)\| \leq M\|x_0\| + ML \int_{t_0}^{t} \|x(t)\| d\tau,$$

where $M = \max_{t \in [t_0, t_0+T]} \|\Phi(t)\| < +\infty$ (see Theorem 2.1). Hence,

$$\sup_{t \in [t_0, t_0+T]} \|(Sx)(t)\| \leq M\|x_0\| + MLT \sup_{t \in [t_0, t_0+T]} \|x(t)\| < +\infty.$$

To complete the proof let us show that the operator S has the unique fixed point $S(x^*) = x^*$ with $x^* \in \mathbf{X}$. For any $x_1, x_2 \in \mathbf{X}$ we have

$$\|S(x_1)(t) - S(x_2)(t)\| \leq M \int_{t_0}^{t} \|f(\tau, x_1(\tau)) - f(\tau, x_2(\tau))\| d\tau$$

$$\leq ML(t - t_0) \sup_{t \in [t_0, t_0+T]} \|x_1(\tau) - x_2(\tau)\| = ML(t - t_0)\rho(x_1, x_2).$$

Hence, by induction, we derive

$$\|S^n(x_1)(t) - S^n(x_2)(t)\| \leq \frac{(ML(t - t_0))^n}{n!} \rho(x_1, x_2)$$

and $\rho(S^n x_1, S^n x_2) \leq \frac{(MLT)^n}{n!} \rho(x_1, x_2)$. Hence, for sufficiently large $n \in \mathbb{N}$, we have $q := \frac{(MLT)^n}{n!} < 1$ and

$$\|S^{nk}(x_1)(t) - S^{nk}(x_2)(t)\| \leq q^k \rho(x_1, x_2), \quad k \in \mathbb{N}.$$

Using the contraction principle (see Banach Fixed-Point theorem in Appendix B) we conclude the existence of a unique fixed point x^* of the operator S in $\mathbf{X}$, which, obviously, corresponds to a mild solution of the nonlinear system (2.31). The proof is complete. □

The evolution equation like (2.31) is a rather universal model of an infinite-dimensional system. Examples of PDEs described by (2.31) have been considered above. In the following example, we show that the same model can be utilized for differential equations with time delays.

Example 2.12 Let us consider the functional differential equation

$$\dot{y}(t) = g(t, y_h(t)), \quad t > 0$$

that is equivalent to the integral equation

$$y(t) = y(0) + \int_0^t g(s, y_h(s))\,ds, \tag{2.33}$$

where $y(t) \in \mathbb{R}^n$ and $y_h(t) \in L^2((-h, 0); \mathbb{R}^n)$ is defined as $(y_h(t))(\theta) = y(t+\theta)$, $\theta \in [-h, 0]$, and $g : (0, +\infty) \times L^2((-h, 0); \mathbb{R}^n) \mapsto \mathbb{R}^n$ is a nonlinear operator that satisfies in $L^2([-h, 0]; \mathbb{R}^n)$ the Lipschitz condition uniformly on t. The initial condition of the latter system is given by

$$y(t) = y^0(t) \text{ for } t \in [-h, 0] \quad (\text{or, equivalently, } y_h(0) = y^0 \in L^2((-h, 0); \mathbb{R}^n)).$$

Let us also consider the nonlinear evolution system (2.31) in the Banach space

$$\mathbb{B} = \mathbb{R}^n \times L^2((-h, 0); \mathbb{R}^n)$$

with the operator $A = \begin{pmatrix} \mathbf{0} & \mathbf{0} \\ \mathbf{0} & \frac{\partial}{\partial z} \end{pmatrix}$ having the domain

$$\mathcal{D}(A) = \left\{x = (x_1, x_2) \in \mathbb{B} : x_2 \in H^1((-h, 0); \mathbb{R}^n) \text{ and } x_1 = x_2(0)\right\},$$

and $f(t, x) = \begin{pmatrix} g(t, x_2) & \mathbf{0} \end{pmatrix}$. The operator A generates the strongly continuous semigroup of linear bounded operators in $\mathbb{B}$ given by

$$(\Phi(t)x)(z) = \begin{cases} (x_1, x_2(t+z)) & \text{if } -h \le t+z \le 0, \\ (x_1, x_1) & \text{if } \quad t+z > 0, \end{cases}$$

$$t \ge 0, \quad z \in [-h, 0], \quad x = (x_1, x_2) \in \mathbb{B}.$$

From Theorem 2.14 we conclude that the evolution system has a mild solution (2.32) for any $x(0) \in \mathbb{B}$, in particular, for $x(0) = (y^0(0), y^0)$. From (2.32) we derive

$$(x_2(t))(z) = (\Phi(t)x_0)(z) + \int_0^t g(s, x_2(s))\,ds, \quad z \in (-h, 0)$$

and for $z \to 0$ the latter coincides with (2.33). Thus the original functional differential equation has a solution. Taking into account a continuity of the function $t \mapsto \int_0^t g(s, x_2(s))\,ds$ we conclude that the solution is continuous for $t > 0$.

Mild solutions of (2.31) with a locally Lipschitz continuous mapping f may exist locally and may be prolonged up to a "boundary" of a regularity domain.

Corollary 2.13 *Let $A : \mathcal{D}(A) \subset \mathbb{B} \mapsto \mathbb{B}$ be an infinitesimal generator of a strongly continuous semigroup Φ of linear bounded operators on $\mathbb{B}$. If $f : [t_0, +\infty) \times \mathbb{B} \mapsto \mathbb{B}$ satisfies the Lipschitz condition in any set*

$$K(r) = \{x \in \mathbb{B} : 0 < r \leq \|x\| \leq 1/r\}, \quad r > 0$$

uniformly on compact intervals of time, i.e., for any $r > 0$ and any $T > 0$ there exists $\exists L(T, r) > 0$ such that

$$\|f(t, x_1) - f(t, x_2)\| \leq L(T, r)\|x_1 - x_2\| \tag{2.34}$$

for all $t \in [t_0, t_0 + T]$ and for all $x_1, x_2 \in K(r)$, and $f(\cdot, x(\cdot)) \in L^1((t_0, t_0 + T); \mathbb{B})$ for any $x \in C([t_0, t_0 + T]; \mathbb{B})$, then for any initial condition $x(t_0) = x_0 \in \mathbb{B}\backslash\{\mathbf{0}\}$ there exists $t_{\max} > t_0$ such that the system (2.31) *has a unique mild solution $x_{t_0,x_0} : [t_0, t_{\max}) \mapsto \mathbb{B}$, where*

$$t_{\max} = +\infty \qquad or \quad t_{\max} < +\infty : \lim_{t \to t_{\max}} \|x_{t_0,x_0}(t)\| \in \{+\infty, 0\}.$$

If, additionally, $f(t, \mathbf{0}) = \mathbf{0}$ then the system has (probably nonunique) zero solution for $x_0 = \mathbf{0}$.

Proof If $f(t, \mathbf{0}) = \mathbf{0}$ then the system has a zero solution $x(t) = \mathbf{0}, t \geq t_0$ for the initial condition $x(t_0) = \mathbf{0}$.

Let $a \in C^\infty([0, +\infty); [0, +\infty))$ be the so-called "cut-off" function such that $a(s) = 0$ for $s \in [0, \delta] \cap [1/\delta, +\infty)$ and $a(s) = 1$ for $t \in [2\delta, 1/(2\delta), \delta \in (0, 1/2)$. In this case, $f^\delta(t, x) = a_\delta(\|x\|)f(t, x)$ satisfy Theorem 2.14. For any initial condition $x(0) = x_0 \in \mathbb{B}$ and any $T > 0$ there exists a unique mild solution $x^\delta(\cdot, x_0) : [t_0, T] \mapsto \mathbb{B}$ of the system (2.31) with the nonlinearity f^δ. Given $x_0 \in \mathbb{B} : 2\delta < \|x_0\| < 1/(2\delta)$, let us select $t^\delta > 0$ such that $2\delta < \|x^\delta(t, x_0)\| < 1/(2\delta)$ for all $t \in (t_0, t^\delta)$. If these inequalities hold for all $t > t_0$ then $t^\delta = +\infty$. Obviously, the restriction of x^δ to the interval $[t_0, t_\delta]$ is the mild solution of the system (2.31) with the original nonlinearity f. Tending $\delta \to 0$ we complete the proof. □

Example 2.13 Let us consider the following evolution equation:

$$\dot{x} = Ax + f(x), \quad t > 0, \quad x(0) = x_0,$$

where A generates a strongly continuous semigroup of linear bounded operators on $\mathbb{B}$ and $f : \mathbb{B} \mapsto \mathbb{B}$ is given by $f(\mathbf{0}) = \mathbf{0}$,

$$f(x) := \alpha \frac{x}{\sqrt{\|x\|}}, \qquad \alpha \in \mathbb{R}, \quad x \neq \mathbf{0}.$$

It can be easily shown that f is locally Lipschitz continuous on $\mathbb{B}\backslash\{\mathbf{0}\}$. Therefore, the considered evolution equation has mild solutions on $\mathbb{B}$.

Let us consider the system (2.31) with unbounded nonlinearity f satisfying the special restriction.

Definition 2.9 A closed densely defined nonlinear operator $f : \mathbb{R} \times \mathcal{D}(f) \mapsto \mathbb{B}$ with $\mathcal{D}(f) \subset \mathbb{B}$ is said to be M-regular if there exists a linear closed operator $M : \mathcal{D}(M) \subset \mathbb{B} \mapsto \mathbb{B}$ with a bounded inverse $M^{-1} : \mathbb{B} \mapsto \mathcal{D}(f)$ such that the nonlinear mapping $x \mapsto f(t, M^{-1}x)$ is locally Lipschitz continuous in $x \in \mathbb{B}\backslash\{\mathbf{0}\}$ uniformly in t. More precisely, for any $r > 0$ and any $T > 0$ there exists $L = L(r, T) > 0$ such that

$$\|f(t, M^{-1}x_1) - f(t, M^{-1}x_2)\| \leq L\|x_1 - x_2\| \tag{2.35}$$

for all $x_i \in K(r)$ and all $t \in [-T, T]$ where $i = 1, 2$.

The M-regular operator may be unbounded. To guarantee the existence of mild solutions of the system (2.31) with M-regular f, the operators A and f have to be consistent in some sense.

Theorem 2.15 *Let $A : \mathcal{D}(A) \subset \mathbb{B} \mapsto \mathbb{B}$ be an infinitesimal generator of a strongly continuous semigroup Φ of linear bounded operators on $\mathbb{B}$ and an operator $f : \mathbb{R} \times \mathcal{D}(f) \mapsto \mathbb{B}$ with $\mathcal{D}(f) \subset \mathbb{B}$ be M-regular and $f(\mathbf{0}) = \mathbf{0}$. If*

$$M^{-1}\Phi(t)Mx_0 = \Phi(t)x_0 \quad \forall x_0 \in \mathcal{D}(M) \quad \text{and} \quad \forall t \geq 0,$$

$$\|M\Phi(\cdot)\| \in L^1_{loc}(\mathbb{R}_+; \mathbb{R}_+) \quad \text{and} \quad \int_0^t \|M\Phi(t)\| ds \to 0 \quad \text{as} \quad t \to 0^+, \tag{2.36}$$

then the evolution equation (2.31) with the initial condition $x(t_0) = x_0 \in \mathcal{D}(M)$, $t_0 \in \mathbb{R}$ has a mild solution $x \in C([t_0, t_1]; \mathbb{B})$ such that $x(t) \in \mathcal{D}(M)$ for $t \in [t_0, t_1]$, where $t_1 > t_0$ is some number. If $x_0 \neq \mathbf{0}$ then this solution is unique as long as $0 < \|x_{t_0,x_0}(t)\| < +\infty$.

Proof Let $t_1' > t_0$, $K = \max_{t \in [t_0, t_1']} \|f(t, M^{-1}x_0)\|$, $x_0 \in \mathcal{D}(M)$, and V_{x_0} be a neighborhood of x_0, where the Lipschitz condition (2.35) holds for all $t \in [t_0, t_1]$. Given $\delta > 0$ let $t_1 \in (t_0, t_1']$ be selected such that

$$\|(\Phi(t - t_0) - I)Mx_0\| < \frac{\delta}{2}, \quad \forall t \in [t_0, t_1)$$

and

$$(L\delta + K) \int_0^{t_1 - t_0} \|M\Phi(\tau)\| \, d\tau < \frac{\delta}{2}.$$

Due to strong continuity of Φ and due to the condition (2.36), such a selection of $t_1 > t_0$ is always possible.

Let $\mathbb{X}$ be the Banach space $C([t_0, t_1]; \mathbb{B})$ with the usual supremum norm $\|y\|_{\mathbb{X}} = \max_{t\in[t_0,t_1]} \|y(t)\|$ for $y \in \mathbb{X}$.

Let us define the mapping S on $\mathbb{X}$ as follows:

$$(Sy)(t) = \Phi(t - t_0)Mx_0 + \int_{t_0}^{t} M\Phi(t - \tau)f\left(\tau, M^{-1}y(\tau)\right) d\tau.$$

Due to (2.35) and continuity of y we have $f(\cdot, M^{-1}y(\cdot)) \in C([t_0, t_1]; \mathbb{B})$. Since (2.36) holds then the function $\tau \mapsto M\Phi(t - \tau)f(\tau, M^{-1}y(\tau))$ is Bochner integrable and $S : \mathbb{X} \mapsto \mathbb{X}$. Obviously, $(Sy)(t_0) = Mx_0$ for any $y \in \mathbb{X}$.

Let B_δ be defined as

$$B_\delta = \{y \in \mathbb{X} : y(t_0) = Mx_0, \|y(t) - Mx_0\| \leq \delta, \forall t \in [t_0, t_1]\}.$$

Obviously, B_δ is a non-empty, closed bounded subset of $\mathbb{X}$ and for any $y \in B_\delta$ we have

$$\|(Sy)(t) - Mx_0\| \leq \|(\Phi(t - t_0) - I)Mx_0\| +$$

$$\int_{t_0}^{t} \|M\Phi(t - \tau)(f(\tau, M^{-1}y(\tau)) - f(\tau, x_0))\| \, d\tau + \int_{t_0}^{t} \|M\Phi(t - \tau)f(\tau, x_0)\| \, d\tau \leq$$

$$\delta/2 + (L\delta + K) \int_{0}^{t_1 - t_0} \|M\Phi(\tau)\| \, d\tau \leq \delta,$$

i.e., $S : B_\delta \mapsto B_\delta$.

Furthermore, since M^{-1} is a bounded operator then for a sufficiently small $\delta > 0$ we have $My(t) \in V_{x_0}$ for all $t \in [t_0, t_1]$. In this case, for $y_1, y_2 \in B_\delta$ we derive

$$\|(Fy_1)(t) - (Fy_2)(t)\| \leq \int_{t_0}^{t} \|M\Phi(t - \tau)\| \cdot \|f(\tau, M^{-1}y_1(\tau)) - f(\tau, M^{-1}y_2(\tau))\| d\tau$$

$$\leq L \int_{t_0}^{t_1} \|M\Phi(t - \tau)\| \cdot \|y_1(\tau) - y_2(\tau)\| d\tau \leq L \int_{t_0}^{t_1} \|M\Phi(t - \tau)\| d\tau \cdot \|y_1 - y_2\|_{\mathbb{X}} \leq$$

$$\tfrac{L\delta}{2L\delta + 2K} \|y_1 - y_2\|_{\mathbb{X}} \leq \tfrac{1}{2} \|y_1 - y_2\|_{\mathbb{X}},$$

i.e., $\|Fy_1 - Fy_2\|_{\mathbb{X}} \leq \frac{1}{2}\|y_1 - y_2\|_{\mathbb{X}}$.

Hence, using Banach Fixed-Point theorem (see Appendix B) we conclude that there exists a unique fixed point $Fy = y \in B_\delta$, i.e.,

$$y(t) = \Phi(t)Mx_0 + \int_{t_0}^{t} M\Phi(t - \tau)f(\tau, M^{-1}y(\tau)) \, d\tau.$$

Since the bounded operator $M^{-1} : \mathbb{B} \mapsto \mathbb{B}$ commutes with Bochner integral $M^{-1} \int = \int M^{-1}$ (see Appendix C) then

$$M^{-1}y(t) = M^{-1}\Phi(t)Mx_0 + M^{-1}\int_{t_0}^{t} M\Phi(t-\tau)f(\tau, M^{-1}y(\tau))\,d\tau =$$

$$\Phi(t)x_0 + \int_{t_0}^{t} \Phi(t-\tau)f(\tau, M^{-1}y(\tau))\,d\tau,$$

i.e., $x(t) := M^{-1}y(t) \in \mathcal{D}(M)$ is a mild solution of the evolution equation (2.31). This solution is unique as long as $0 < \|x_{t_0,x_0}\| < +0$. Since a mild solution x_{t_0,x_0} reaches $\mathbf{0}$ in a finite time t^*, then, due to $f(t, \mathbf{0}) = \mathbf{0}$, this solution can be prolonged to zero as follows: $x_{t_0,x_0}(t) = \mathbf{0}$ for $t \geq t^*$. The proof is complete. □

Notice that, under conditions of the above theorem, the prolongation of solution to zero may be nonunique in the general case. The uniqueness of the prolonged solutions can be guaranteed if the origin of the system is Lyapunov stable (see Chap. 3) or if the nonlinear mapping $f \circ M^{-1}$ is Lipschitz continuous in a neighborhood of zero.

The above theorem (with a slightly more strong conditions on f) can be proven for the so-called analytic semigroup Φ (see, e.g., [1, page 196]) such that the existence of strong (or even classical) solutions can be guaranteed. The operator M is frequently represented by a fractional power operator of A.

Example 2.14 Let us consider the viscid Burgers equation of the form

$$\dot{x} = \nu\Delta x - (x \cdot \nabla)x, \quad t \geq 0, \quad x(t) \in L^2(\mathbb{R}^n; \mathbb{R}^n),$$

where $\nu > 0$ is the viscosity coefficient, $\Delta : H^2(\mathbb{R}^n; \mathbb{R}^n) \subset L^2(\mathbb{R}^n; \mathbb{R}^n) \mapsto L^2(\mathbb{R}^n; \mathbb{R}^n)$ is the Laplace operator, and the nonlinear operator $f : \mathcal{D}(f) \subset L^2(\mathbb{R}^n; \mathbb{R}^n) \mapsto L^2(\mathbb{R}^n; \mathbb{R}^n)$ is given by

$$f(x) = -(x \cdot \nabla)x.$$

Let us show that f is a nonlinear M-regular operator with

$$M := \Delta^\gamma : H^{2\gamma}(\mathbb{R}^n; \mathbb{R}^n) \subset L^2(\mathbb{R}^n; \mathbb{R}^n) \mapsto L^2(\mathbb{R}^n; \mathbb{R}^n)$$

being the fractional Laplacian (see Example 2.7), where

$$\gamma = 1/2 + n/8$$

and $H^{2\gamma}(\mathbb{R}^n; \mathbb{R}^n)$ is a fractional Sobolev space (see Appendix D). Indeed, since for the norms in $\mathbb{R}^n$ we have $|\cdot|_2 \leq |\cdot|_1 \leq \sqrt{n}|\cdot|_2$ then for $\nabla u \in L^4$ and $v \in L^4$ using the Hölder inequality we derive

$$
\begin{aligned}
\|(v\cdot\nabla)u\|_{L^2} &= \sqrt{\sum_{i=1}^{n}\int_{\mathbb{R}^n}\left(\sum_{j=1}^{n} v_j\frac{\partial u_i}{\partial z_j}\right)^2} \le \sqrt{\sum_{i=1}^{n}\int_{\mathbb{R}^n}\left(\sum_{j=1}^{n}\left|v_j\frac{\partial u_i}{\partial z_j}\right|\right)^2}\\
&\le \sqrt{n\sum_{i=1}^{n}\int_{\mathbb{R}^n}\left(\sum_{j=1}^{n}\left|v_j\frac{\partial u_i}{\partial z_j}\right|^2\right)} \le \sqrt{n\sum_{i=1}^{n}\int_{\mathbb{R}^n}\sum_{j=1}^{n} v_j^2\left|\frac{\partial u_i}{\partial z_j}\right|^2}\\
&\le \sqrt{n\sum_{i=1}^{n}\int_{\mathbb{R}^n}\sqrt{\sum_{j=1}^{n} v_j^4}\sqrt{\sum_{j=1}^{n}\left|\frac{\partial u_i}{\partial z_j}\right|^4}} \le \sqrt{n\int_{\mathbb{R}^n}\sqrt{\sum_{j=1}^{n} v_j^4}\sum_{i=1}^{n}\sqrt{\sum_{j=1}^{n}\left|\frac{\partial u_i}{\partial z_j}\right|^4}}\\
&\le \sqrt{n\int_{\mathbb{R}^n}\sqrt{\sum_{j=1}^{n} v_j^4\sum_{i=1}^{n}\sum_{j=1}^{n}\left|\frac{\partial u_i}{\partial z_j}\right|^2}} \le \sqrt{n}\|u\|_{L^4}\left(\int_{\mathbb{R}^n}\left(\sum_{i=1}^{n}\sum_{j=1}^{n}\left|\frac{\partial u_i}{\partial z_j}\right|^2\right)^2\right)^{\frac{1}{4}}\\
&\le n\|v\|_{L^4}\|\nabla u\|_{L_4}.
\end{aligned}
$$

By the Hardy–Littlewood–Sobolev fractional integration theorem (see Appendix D or [5, Page 119]) we have

$$
\|\Delta^{-n/8}y\|_{L^4}\le C\|y\|_{L^2},\qquad \forall y\in L^2.
$$

Since $\Delta^{-\gamma}=\Delta^{-n/8}\Delta^{-1/2}$ and the operators $\frac{\partial}{\partial z_i}$, $\Delta^{-n/8}$ commute (see Example 2.9) then

$$
\|\nabla\Delta^{-\gamma}u\|_{L^4}\le C\|\nabla\Delta^{-1/2}u\|_{L^2}\le \tilde{C}\cdot\|u\|_{L^2},
$$

where the boundedness of the operator $\frac{\partial}{\partial z_i}\Delta^{-1/2}$ is utilized on the last step (see Example 2.8). The M-regularity of f follows from the above inequality. Indeed, for any $u_1, u_2\in L^2(\mathbb{R}^n;\mathbb{R}^n)$, we have

$$
\begin{aligned}
\|f(\Delta^{-\gamma}u_1)-f(\Delta^{-\gamma}u_2)\|_{L^2} \le& n\|\Delta^{-\gamma}u_1\|_{L^4}\|\nabla\Delta^{-\gamma}(u_1-u_2)\|_{L^4}\\
&+n\|\Delta^{-\gamma}(u_1-u_2)\|_{L^4}\|\nabla\Delta^{-\gamma}u_2\|_{L^4}\\
\le& nC\tilde{C}\|\Delta^{-1/2}u_1\|_{L^2}\|u_1-u_2\|_{L^2}\\
&+nC\tilde{C}\|\Delta^{-1/2}(u_1-u_2)\|_{L^2}\|u_2\|_{L^2}\\
\le& \tilde{L}(\|u_1\|_{L_2}+\|u_2\|_{L_2})\|u_1-u_2\|_{L^2},
\end{aligned}
$$

where $\tilde{L}=nC\tilde{C}\|\Delta^{-1/2}\|$. Notice that the condition (2.36) is fulfilled for $n\in\{1,2,3\}$ since $\|\Delta^{\gamma}\Phi(t)\|_{L^2}\le\frac{C_\Phi}{t^\gamma}$ for $\gamma\in(0,1)$ (see the formula (2.18)). Therefore, by Theorem 2.15, for any initial condition $x(t_0)=x_0\in H^{2\gamma}(\mathbb{R}^n;\mathbb{R}^n)$ the considered PDE with $n \geq 3$ has a unique mild solution $x : [t_0, t_0 + T) \mapsto H^{2\gamma}(\mathbb{R}^n; \mathbb{R}^n)$ defined, at least, locally in time.

Continuous Dependence of Solutions on Initial Conditions

In addition to the existence and uniqueness of solutions, a continuous dependence of solutions on initial conditions is required for any well-posed (in Hadamard's sense) problem.

Corollary 2.14 *If all conditions of Theorem 2.14 hold then all mild solutions of the system* (2.31) *depend continuously on initial conditions:*

for any $T > 0$ and any $r > 0$ there exists $\tilde{L} = \tilde{L}(T, r) > 0$ such that the inequality

$$\|x_{t_0,x_0^1}(t) - x_{t_0,x_0^2}(t)\| \leq \tilde{L}\|x_0^1 - x_0^2\|, \quad \forall t \in [t_0, t_0 + T] \tag{2.37}$$

holds for any $x_0^j \in B(r)$, where $j = 1, 2$.

Proof Let us consider two mild solutions x_{t_0,x_0^1} and x_{t_0,x_0^2} of the system (2.51) with $x_0^1, x_0^2 \in B(r)$. Let $M = \sup_{s\in[t_0,t_0+T]} \|\Phi(s)\| > 0$ and $\delta = \frac{T}{N}$, where $N > 2MTL$ is a natural number and $L = L(T, r) > 0$ is a Lipschitz constant for f (see Theorem 2.14). Let us denote

$$t_i = t_0 + i\delta, \quad x_i^1 = x_{t_0,x_0^2}(t_i), \quad x_i^2 = x_{t_0,x_0^2}(t_i), \quad i = 0, 1, ..., N - 1.$$

In this case we have

$$x_{t_0,x_0^j}(t) = \Phi(t - t_i)x_i^j + \int_{t_i}^{t} \Phi(t - s)f(s, x_{t_0,x_0^j}(s))ds, \quad j = 1, 2, \quad i = 0, 1, ..., N$$

and

$$\|x_{t_0,x_0^1}(t) - x_{t_0,x_0^2}(t)\| = \left\| \Phi(t - t_i)(x_i^1 - x_i^2) + \int_{t_i}^{t} \Phi(t - s)\Delta f(s)ds \right\|,$$

where $\Delta f(s) = f(s, x_{t_0,x_0^1}(s)) - f(s, x_{t_0,x_0^2}(s))$. Since

$$\|\Delta f(s)\| \leq L\|x_{t_0,x_0^1}(s) - x_{t_0,x_0^2}(s)\|$$

then

$$\sup_{t\in[t_i,t_{i+1}]} \|x_{t_0,x_0^1}(t) - x_{t_0,x_0^2}(t)\| \leq M\|x_1^i - x_2^i\| + \delta ML \sup_{s\in[t_i,t_{i+1}]} \|x_{t_0,x_1}(s) - x_{t_0,x_2}(s)\|.$$

Hence, using $\delta < \frac{1}{2ML}$ we derive

$$\sup_{t\in[t_i,t_{i+1}]} \|x_{t_0,x_1}(t) - x_{t_0,x_2}(t)\| \le \frac{M}{1-\delta ML}\|x_1^i - x_2^i\|.$$

Taking into account $\|x_1^{i+1} - x_2^{i+1}\| \le \sup_{t\in[t_i,t_{i+1}]}\|x_{t_0,x_1}(t) - x_{t_0,x_2}(t)\|$ we derive

$$\sup_{t\in[t_0,t_i]} \|x_{t_0,x_0^1}(t) - x_{t_0,x_0^2}(t)\| \le \left(\frac{M}{1-\delta ML}\right)^i \|x_0^1 - x_0^2\|, i = 1, 2, ..., N.$$

For $i = N$ we have $t_N = t_0 + T$ and taking $\tilde{L} = (\frac{M}{1-\delta ML})^N$ we complete the proof. □

The property of continuous dependence of solutions on parameters is useful for stability and robustness analysis of evolution control systems.

Strong Solutions of Nonlinear Evolution Equations

Let us study the case when mild solutions become strong. The definition of a strong solution to the system (2.31) is literally the same as for the linear evolution equation (see Definition 2.6). The following result holds also for any reflexive Banach space (see, e.g., [1, Theorem 1.6, Chap. 6]).

Theorem 2.16 *Let A be a generator of a strongly continuous group of linear bounded operators on a Hilbert space $\mathbb{H}$. If $f : [t_0, t_0 + T] \times \mathbb{H} \mapsto \mathbb{H}$ is locally Lipschitz continuous in both variables then any mild solution of the system* (2.31) *with $x(0) = x_0 \in \mathcal{D}(A)$ is a locally Lipschitz continuous strong solution. Moreover, any mild solution of the system* (2.31) *with $x(t_0) = x_0 \in \mathbb{H}$ is a uniform limit of strong solutions with $x(t_0) = x_i \in \mathcal{D}(A)$ such that $x_i \to x_0$ as $i \to +\infty$.*

Proof First, let us show that any mild solution $x : [t_0, t_0 + T] \mapsto \mathbb{H}$ with $x(0) = x_0 \in \mathcal{D}(A)$ is a locally Lipschitz continuous function of time. Indeed, let $\|\Phi(t)\| \le M$ and $\|f(t, x(t))\| \le N$ for $t_0 \le t \le t_0 + T$. For $0 < h < t - t_0$ we have

$$x(t+h) - x(t) = \Phi(t+h-t_0)x_0 - \Phi(t-t_0)x_0 + \int_{t_0}^{t_0+h} \Phi(t+h-s)f(s, x(s))ds +$$

$$\int_{t_0}^{t} \Phi(t-s)(f(s+h), x(s+h) - f(s, x(s))ds.$$

Hence, using $\Phi(t+h-t_0)x_0 - \Phi(t-t_0)x_0 = \int_{t-t_0}^{t+h-t_0} \Phi(t+h-t_0-\tau)Ax_0 d\tau$ we derive

$$\|x(t+h)-x(t)\| \leq Mh\|Ax_0\| + hMN + MC\int_{t_0}^{t} \|x(s+h)-x(s)\|ds$$

and, using Grönwall–Bellman inequality (see Appendix E), we conclude locally Lipschitz continuity of x on time variable. In this case, the function $t \mapsto f(t, x(t))$ is also locally Lipschitz continuous. In the case of the Hilbert space, this implies (see Radon–Nikodym property in Appendix C) that this function is differentiable almost everywhere and belongs to $L^1((0, T); \mathbb{H})$. Applying Corollary 2.11 we conclude that x is a strong solution.

Since $\mathcal{D}(A)$ is dense in $\mathbb{H}$ then for any $x_0 \in \mathbb{H}$ there exists $x_i \in \mathcal{D}(A)$ such that $x_i \to x_0$ as $i \to +\infty$. Finally, the continuous dependence of solutions on initial conditions implies that any mild solution is a uniform limit of strong solutions. The proof is complete. □

The above result remains valid for the system with a nonlinearity being locally Lipschitz continuous on $[t_0, +\infty) \times (\mathbb{H}\backslash\{\mathbf{0}\})$.

Corollary 2.15 *Let all conditions of Corollary 2.13 be fulfilled for a Hilbert space* $\mathbb{H}$ *and* f *be locally Lipschitz continuous on time. Then any mild solution of the system* (2.31) *with* $x(0) = x_0 \in \mathcal{D}(A)\backslash\{\mathbf{0}\}$ *is a locally Lipschitz continuous strong solution as long as* $0 < \|x(t)\| < +\infty$. *Moreover, any nonzero mild solution of the system* (2.31) *with* $x(t_0) = x_0 \in \mathbb{H}\backslash\{\mathbf{0}\}$ *is a uniform limit of strong solutions with* $x(t_0) = x_i \in \mathcal{D}(A)\backslash\{\mathbf{0}\}$ *such that* $x_i \to x_0$ *as* $i \to +\infty$.

Proof If $a_\delta \in C^\infty$ is such that $a_r(1) = 1$ for $r \in [\delta, 1/\delta]$ and $a_\delta(r) = 0$ for $r \in [0, 0.5\delta] \cap [1/(2\delta), +\infty)$ with $\delta \in (0, 1)$ then the operator

$$f_\delta(t, x) = a_\delta(\|x\|_{\mathbf{d}}) f(t, x)$$

satisfies all conditions of Theorem 2.16. Any solution of (2.31) with $\delta < \|x_0\| < 1/\delta$ coincides with the solution of

$$\dot{x} = Ax + f_\delta(t, x), \quad t > t_0, \quad x(t_0) = x_0$$

as long as $\delta < \|x(t)\| < 1/\delta$. Tending δ to 0 we complete the proof. □

Classical Solutions of Nonlinear Evolution Equations

The nonlinear system (2.31) may also have classical solutions under some additional conditions to the nonlinear mapping f. The definition of a classical solution to the equation (2.31) repeats Definition 2.5. The following theorem (partially inspired by [1, page 187]) gives a sufficient condition for existence of classical solutions of nonlinear evolution equations.

Theorem 2.17 *Let $A : \mathcal{D}(A) \subset \mathbb{B} \mapsto \mathbb{B}$ be an infinitesimal generator of a strongly continuous semigroup Φ of linear bounded operators on $\mathbb{B}$ and a mapping $f : [t_0, t_0 + T] \times \mathbb{B} \mapsto \mathbb{B}$ be continuously differentiable. Then the system* (2.31) *has a unique classical solution $x_{t_0,x_0} : [t_0, t_0 + T] \mapsto \mathbb{B}$ for any initial condition $x(t_0) = x_0 \in \mathcal{D}(A)$. Moreover, any mild solution of the system* (2.31) *with $x(t_0) = x_0 \in \mathbb{B}$ is a uniform limit of classical solutions with $x(t_0) = x_i \in \mathcal{D}(A)$, such that $x_i \to x_0$ as $i \to +\infty$.*

Proof First, let us note that if f is continuously Frechét differentiable then it satisfies all conditions of Corollary 2.13. Therefore, the Cauchy problem (2.31) has a mild solution $x_{t_0,x_0} : [t_0, T] \mapsto \mathbb{B}$ for any $x_0 \in \mathbb{B}$. Let us show that for $x_0 \in \mathcal{D}(A)$ this mild solution is the classical one.

Let $\frac{\partial}{\partial s} f(s, x) \in \mathbb{B}$ and $\frac{\partial}{\partial x} f(s, x) \in \mathcal{L}(\mathbb{B}; \mathbb{B})$ denote partial Frechét derivatives of f at the point $(t, x) \in [t_0, t_0 + T] \times \mathbb{B}$.

Under our assumption the function $g : [t_0, t_0 + T] \mapsto \mathbb{B}$ defined as follows:

$$g(t) = \Phi(t - t_0) f(t_0, x_0) + A\Phi(t - t_0)x_0 + \int_{t_0}^{t} \Phi(t - s) \frac{\partial f(s, x_{t_0,x_0}(s))}{\partial s} ds,\ t \in [t_0, t_0 + T]$$

is continuous, i.e., $g \in C([t_0, t_0 + T]; \mathbb{B})$, but the function $h : [t_0, t_0 + T] \times \mathbb{B} \mapsto \mathbb{B}$ defined as

$$p(s, w) = \frac{\partial}{\partial x} f(s, x_{t_0,x_0}(s))w$$

is continuous on s and Lipschitz continuous on w, since the mapping $[t_0, t_0 + T] \mapsto \mathcal{L}(\mathbb{B}; \mathbb{B})$ defined as $s \mapsto \frac{\partial}{\partial x} f(s, x_{t_0,x_0}(s))$ is continuous in the operator norm. Similar to the proof of Theorem 2.14, one can be shown that the integral equation

$$w(t) = g(t) + \int_{t_0}^{t} \Phi(t - s)p(s, w(s))ds$$

has a unique solution $w \in C([t_0, t_0 + T]; \mathbb{B})$.

On the other hand, simple calculations show

$$\begin{aligned} x_{t_0,x_0}(t + h) - x_{t_0,x_0}(t) &= \Phi(t + h - t_0)x_0 - \Phi(t - t_0)x_0 \\ &+ \textstyle\int_{t_0}^{t_0+h} \Phi(t + h - s) f(s, x_{t_0,x_0}(s))ds \\ &+ \textstyle\int_{t_0}^{t} \Phi(t - s)(f(s + h, x_{t_0,x_0}(s + h)) - f(s, x_{t_0,x_0}(s)))ds. \end{aligned}$$

Hence, for

$$w_h(t) = \frac{x_{t_0,x_0}(t + h) - x_{t_0,x_0}(t)}{h} - w(t)$$

using continuous differentiability of f we derive

$$w_h(t) - \int_{t_0}^{t} \Phi(t-s)p(s, w_h(s))ds =$$

$$\tfrac{\Phi(t+h-t_0)x_0 - \Phi(t-t_0)x_0}{h} - A\Phi(t-t_0)x_0$$

$$+\frac{1}{h}\int_{t_0}^{t_0+h} \Phi(t+h-s)f(s, x_{t_0,x_0}(s))ds - \Phi(t-t_0)f(t_0, x_0)$$

$$+\int_{t_0}^{t} \Phi(t-s)\left(\tfrac{\partial}{\partial s}f(s, x_{t_0,x_0}(s+h)) - \tfrac{\partial}{\partial s}f(s, x_{t_0,x_0}(s))\right)ds + O(h),$$

where all on the right-hand side converge to $\mathbf{0}$ as $h \to 0$ uniformly on t. This yields

$$\|w_h(t)\| \le \varepsilon(h) + M\int_{t_0}^{t} \|w_h(s)\|ds, \text{ where } \varepsilon(h) \to 0 \text{ as } h \to 0$$

and $\|w_h(t)\| \to 0$ as $h \to 0$ uniformly on t, or, equivalently, $w(t) = \frac{d}{dt}x_{t_0,x_0}(t)$. Since $x \in C([t_0, t_0+T]; \mathbb{B})$ then x_{t_0,x_0} is continuously differentiable. Finally, using Corollary 2.11 we derive that x_{t_0,x_0} is the unique classical solution of the system (2.31).

Since $\mathcal{D}(A)$ is dense in $\mathbb{B}$ then for any $x_0 \in \mathbb{B}$ there exists $x_i \in \mathcal{D}(A)$ such that $x_i \to x_0$ as $i \to +\infty$. Finally, continuous dependence of solutions on initial conditions implies that any mild solution is a uniform limit of classical solutions. The proof is complete. □

Example 2.15 Let us consider the following evolution equation:

$$\dot{x} = Ax + f(x), \quad t > 0, \quad x(0) = x_0, \quad \alpha \in \mathbb{R},$$

where A generates a strongly continuous semigroup of linear bounded operators on a Hilbert space $\mathbb{H}$ with a norm induced by an inner product $\|x\| = \sqrt{\langle x, x\rangle}$ and $f : \mathbb{H} \mapsto \mathbb{H}$ is defined as follows:

$$f(x) = \frac{\rho\left(\frac{\|x\|^4}{\delta^4}\right)}{\|x\|^\alpha}x,$$

where $\delta > 0$, $\alpha \in \mathbb{R}$ and $\rho : [0, +\infty) \mapsto [0, 1]$ is the "cut-off" function given by

$$\rho(r) = \begin{cases} 0 & \text{if} \quad r = 0, \\ e^{-\frac{1}{r}}(e - e^{-\frac{1}{1-r}}) & \text{if } \ 0 < r < 1, \\ 1 & \text{if} \quad r > 1. \end{cases}$$

The function ρ is smooth and $0 = \rho'(0) = \rho'(r)$ for $r \geq 1$. Since the norm is continuously Frechét differentiable on $\mathbb{H}\backslash\{\mathbf{0}\}$

$$(D\|x\|)(h) = \frac{\langle h, x\rangle}{\|x\|}, \quad x \in \mathbb{H}\backslash\{\mathbf{0}\}, \quad h \in \mathbb{H},$$

we derive the Frechét derivative $Df(x) \in \mathcal{L}(\mathbb{H}; \mathbb{H})$ of f as follows:

$$(Df(x))(h) = \frac{4\langle h,x\rangle\rho'\left(\frac{\|x\|^4}{\delta^4}\right)\|x\|^{2-\alpha}}{\delta^4}x - \frac{\langle h,x\rangle\rho\left(\frac{\|x\|^4}{\delta^4}\right)}{2\|x\|^{2+\alpha}}x + \frac{\rho\left(\frac{\|x\|^4}{\delta^4}\right)}{\|x\|^{\alpha}}h.$$

Since, ρ is a "cut-off" function then the functional $g : \mathbb{H} \mapsto [0, +\infty)$ given by $g(x) := \|x\|^{\gamma}\rho\left(\frac{\|x\|^4}{\delta^4}\right)$ with $\gamma \in \mathbb{R}$ is continuous on $\mathbb{H}$, $g(\mathbf{0}) = \mathbf{0}$, so Df is continuous with respect to x in the operator norm. Therefore, the considered evolution equation has classical solutions on $\mathcal{D}(A)$ and mild solutions on $\mathbb{H}$.

On Nonlinear Evolution Equations with Perturbations

Similar to the linear evolution system, the case of a time-varying generator can be treated as a perturbation of a time-invariant operator. The existence of solutions for a nonlinear system in this case can be proven similarly.

Corollary 2.16 *Let A and ΔA satisfy Theorem 2.12 or Theorem 2.13. If f satisfies Theorem 2.14 (resp., Theorem 2.17) then the system*

$$\dot{x} = (A + \Delta A(t))x + f(t, x), \quad x(t_0) = x_0 \tag{2.38}$$

has a unique mild (resp., classical) solution given by

$$x(t) = U(t, t_0)x_0 + \int_{t_0}^{t} U(t, s)f(s, x(s))ds,$$

where U is an evolution family generated by $A + \Delta A$.

The proof of this corollary repeats the proof of Theorem 2.14 (resp., Theorem 2.17) replacing $\Phi(t - s)$ with $U(t, s)$.

Corollary 2.17 *Let A be a generator of a strongly continuous semigroup of linear bounded operators on $\mathbb{B}$ and $\gamma : [0, T] \mapsto (0, +\infty)$ be a continuous positive function. Then the following claims hold:*

- *any mild solution of the evolution equation*

$$\dot{x} = \gamma(t)(Ax + f(t, x)), \quad x(t_0) = x_0 \tag{2.39}$$

is given by

$$x(t) = \Phi\left(\int_{t_0}^{t} \gamma(\sigma)d\sigma\right)x_0 + \int_{t_0}^{t} \Phi\left(\int_{s}^{t} \gamma(\sigma)d\sigma\right)\gamma(s)f(s, x(s))ds;$$

- *if f satisfies Theorem 2.14 (resp., Theorem 2.17) then the system* (2.39) *has a unique mild (resp., classical) solution;*
- *if $\mathbb{B} = \mathbb{H}$ and f satisfies Theorem 2.16 then the system* (2.39) *has a unique strong solution;*
- *if Φ and f satisfy Theorem 2.15 then the system has a mild solution x for any initial condition $x_0 \in \mathcal{D}(M)$ and $x(t) \in \mathcal{D}(M)$ for $t > 0$ (see Theorem 2.15 for the definition of the operator M).*

The proof of the corollary immediately follows from Theorem 2.13, which shows that an evolution family $U(t, s)$ (see Sect. 2.2.2) generated by $\gamma(t)A$ is given by

$$U(t, s) := \Phi\left(\int_{s}^{t} \gamma(\sigma)d\sigma\right).$$

Motivated by the above results, we introduce the following definition of a mild solution of the evolution equation:

$$\dot{x} = h(x)(Ax + f(x)), \quad t > 0, \quad x(0) = x_0, \tag{2.40}$$

where A, f, x, x_0 are, as before, the functional $h : \mathbb{B} \mapsto \mathbb{R}_+$ is continuous on $\mathbb{B}\backslash\{\mathbf{0}\}$ and $h(x) > 0$ for $x \neq \mathbf{0}$.

Definition 2.10 A continuous function $x : [0, \bar{\tau}) \mapsto \mathbb{B}$ is said to be a mild solution of (2.40) if it satisfies the following integral equation:

$$x(\tau) = \Phi\left(\int_{0}^{\tau} h(x(\sigma))d\sigma\right)x_0 + \int_{0}^{\tau} \Phi\left(\int_{s}^{\tau} h(x(\sigma))d\sigma\right)h(x(s))f(x(s))ds \tag{2.41}$$

for all $t \in [0, \bar{t})$.

Strong and classical solutions to (2.40) are defined in the usual way. It is easy to show that the introduced mild solutions satisfy the so-called semigroup property (see **Concluding Remarks** of this chapter).

Lemma 2.4 *Let A be a generator of a strongly continuous semigroup of linear bounded operators on $\mathbb{B}$, $f : \mathbb{B} \mapsto \mathbb{B}$, a functional $h : \mathbb{B} \mapsto \mathbb{R}$ be continuous on $\mathbb{B}\backslash\{\mathbf{0}\}$ and*

$$0 < \inf_{\|x\|=r} h(x), \qquad \sup_{\|x\|=r} h(x) < +\infty, \quad \forall r > 0. \tag{2.42}$$

Then the following two claims are equivalent:

(1) any mild solution y of the evolution equation

$$\dot{y} = Ay + f(y), \quad y(0) = x_0 \neq \mathbf{0} \tag{2.43}$$

exists on a time interval $[0, t_{\max})$, where $t_{\max} = +\infty$ or $t_{\max} \leq +\infty$ such that

$$\lim_{t \to t_{\max}} \|y(t)\| = +\infty \quad or \quad \lim_{t \to t_{\max}} \|y(t)\| = 0; \tag{2.44}$$

(2) any mild solution x of (2.40) is given by (2.41) and exists on $[0, \tau_{\max})$, where $\tau_{\max} = \int_0^{t_{\max}} h^{-1}(y(\sigma))d\sigma$ and $\tau_{\max} = +\infty$ or $\tau_{\max} < +\infty$ such that

$$\lim_{\tau \to \tau_{\max}} \|x(\tau)\| = 0 \; or \; \lim_{t \to \tau_{\max}} \|x(\tau)\| = +\infty.$$

Moreover, for any mild solution y of the system (2.43) there exists a mild solution x of the system (2.40) given by (2.41) such that the identity

$$y(t) = x\left(\int_0^t \frac{d\sigma}{h(y(\sigma))}\right) \tag{2.45}$$

holds, and, inversely, for any mild solution x of the system (2.40) there exists a mild solution y of the system (2.43) such that the identity

$$x(\tau) = y\left(\int_0^\tau h(x(\sigma))d\sigma\right) \tag{2.46}$$

holds.

Proof By assumption, the system

$$\dot{y} = Ay + f(y), \quad y(t_0) = x_0 \neq \mathbf{0}$$

has a mild solution $y(t) = \Phi(t)x_0 + \int_0^t \Phi(t-s)f(y(s))ds$ with $t \in [0, t_{\max})$. Let $t_{\max} > 0$ be such that $0 < \|y(t)\| < +\infty$ on $[0, t_{\max})$. In this case, $0 < h(y(t)) < +\infty$ for all $t \in [0, t_{\max})$ and $t \mapsto h(y(t))$ is continuous on $[0, t_{\max})$.

Let us denote $\tau := \int_0^t h^{-1}(y(\sigma))d\sigma$ and show that $\tau \mapsto x(\tau) \in \mathbb{B}$ with $\tau \in [t_0, \tau_{\max})$, $\tau_{\max} = \int_0^{t_{\max}} h^{-1}(y(\sigma))d\sigma$ given by $x(\tau) := y(t)$ is a mild solution of the Cauchy problem (2.40). First of all, notice that $t = \int_0^\tau h(x(\sigma))d\sigma$. Hence, using the representation of y and the notation $U(t,s) = \Phi\left(\int_s^t h(x(\sigma))d\sigma\right)$ we derive

$$x(\tau) = y\left(\int_0^\tau h(x(\sigma))d\sigma\right) = U(\tau, t_0)x_0 +$$

$$\int_0^{\int_0^\tau h(x(\sigma))d\sigma} \Phi\left(\int_0^\tau h(x(\sigma))d\sigma - s\right) f(y(s))ds.$$

Making the change of the integration variable $s = \int_0^{s_\tau} \frac{d\sigma}{h(x(\sigma))}$ we derive

$$x(\tau) = U(\tau, 0)x_0 + \int_0^\tau U(\tau, s_\tau) f\left(y\left(\int_0^{s_\tau} h(x(\sigma))d\sigma\right)\right)h(x(s_\tau))ds_\tau =$$

$$U(\tau, 0)x_0 + \int_0^\tau U(\tau, s_\tau)h(x(s_\tau)) f(x(s_\tau))ds_\tau,$$

i.e., we have shown that there exists x satisfying (2.41).

Conversely, if we assume that $x(\tau)$, $\tau \geq 0$ is a solution of (2.40) given by (2.41) then we assign $y(t) := x(\tau)$, $t := \int_0^\tau h(x(\sigma))d\sigma$ and, repeating the above considerations, we show that the constructed y is a mild solution of (2.43).

1) $\Rightarrow$ 2) Now we are ready to prove that the first claim implies the second one. Suppose the contrary, namely there exists a mild solution $x : [0, \tau_{\max}) \mapsto \mathbb{B}\backslash\{\mathbf{0}\}$ such that $\tau_{\max} < +\infty$

$$0 < \underline{c} := \liminf_{\tau \to \tau_{\max}} \|x(\tau)\| \leq \bar{c} := \limsup_{\tau \to \tau_{\max}} \|x(\tau)\| < +\infty$$

and x cannot be prolonged by continuity to $\tau = \tau_{\max}$, i.e., $\forall x^* \in \mathbb{B}$ we have $\|x(\tau) - x^*\| \nrightarrow 0$ as $\tau \to \tau_{\max}$.

In this case, the continuous function $h(x(\tau)) : [0, \tau_{\max}) \mapsto \mathbb{R}_+$ is separated from zero and uniformly bounded. Consequently, the function

$$\tau \mapsto \int_0^\tau h(x(\sigma))d\sigma$$

is continuous at $\tau = \tau_{\max}$.

Let y be a solution of (2.43) such that the formula (2.59) holds for all $\tau \in [0, \tau_{\max})$. Hence, $0 < \underline{c} \leq \|y(t)\| \leq \bar{c} < +\infty$ for all $t \in [0, t^*)$, $t^* = \int_0^{\tau_{\max}} h(x(\sigma))d\sigma < +\infty$. By assumption 1), any solution of (2.43) is defined as long as $0 < \|y(t)\| < +\infty$, i.e., the considered solution y can be prolonged by continuity to $t = t^*$ and

$$x(\tau) = y\left(\int_0^\tau h(x(\sigma))d\sigma\right) \to y(t^*) \in \mathbb{B}\backslash\{\mathbf{0}\} \quad \text{as} \quad \tau \to \tau_{\max}.$$

We obtain the contradiction.

The case 2) ⇒ 1) can be treated similarly. The proof is complete. □

Example 2.16 Let us consider the evolution system

$$\dot{x} = (1 + \|x\|)Ax, \quad t > 0, \quad x(0) = x_0 \in \mathbb{B}$$

where $A : \mathcal{D}(A) \subset \mathbb{B} \mapsto \mathbb{B}$ is a generator of a strongly continuous semigroup Φ of linear bounded operators on $\mathbb{B}$. According to Lemma 2.4 the considered system has a solution given by

$$x(t) = \Phi\left(\int_0^t (1 + \|x(s)\|)ds\right)x_0.$$

If $x_0 \in \mathcal{D}(A)$ then from Theorem 2.1 we derive $x(t) \in \mathcal{D}(A)$, x is differentiable, and

$$\dot{x}(t) = \frac{d}{dt}\Phi\left(\int_0^t (1 + \|x(s)\|)ds\right)x_0 = (1 + \|x(t)\|)\left.\frac{d}{d\tau}\Phi(\tau)\right|_{\tau=\int_0^t (1+\|x(s)\|}x_0 =$$

$$(1 + \|x(t)\|)A\Phi\left(\int_0^t (1 + \|x(s)\|)ds\right)x_0 = (1 + \|x\|)Ax,$$

i.e., x is, in fact, a classical solution of the considered system.

From the repeating considerations of the latter example, we prove the following corollary.

Corollary 2.18 *Let A be a generator of a strongly continuous semigroup of linear bounded operators on $\mathbb{B}$ and a functional $h : \mathbb{B} \mapsto [0, +\infty)$ be continuous on $\mathbb{B}\backslash\{\mathbf{0}\}$ such that*

$$0 < \inf_{\|u\|=r} h(u) \quad \text{and} \quad \sup_{\|u\|=r} h(u) < +\infty, \quad \forall r > 0.$$

Then the nonlinear system

$$\dot{x} = h(x)Ax, \quad t > 0, \quad x(0) = x_0 \in \mathcal{D}(A)$$

has a classical solution for any $x_0 \in \mathcal{D}(A)$.

Obviously that solutions of systems studied in the above corollaries are unique if $h \equiv \text{const}$.

If an exogenous disturbance does not perturb the generator A, the problem of the existence of solutions of the perturbed system can be studied by means of Theorem 2.15.

2.2.4 Nonlinear Implicit Evolution Equations

Let $\tilde{\mathbb{X}} = \mathbb{B} \times \mathbb{X}$, where $\mathbb{B}$ is Banach space and $\mathbb{X}$ is a vector space. Let us consider the nonlinear implicit evolution equation

$$\begin{aligned} \frac{du}{dt} &= Au + G(u, p), \\ \mathbf{0} &= Q(u, p), \end{aligned} \qquad t > 0, \tag{2.47}$$

where a densely defined closed linear operator

$$A : \mathcal{D}(A) \subset \mathbb{B} \mapsto \mathbb{B}$$

generates a strongly continuous semigroup Φ of linear bounded operators on $\mathbb{B}$, and

$$G : \mathcal{D}(G) \subset \tilde{\mathbb{X}} \mapsto \mathbb{B} \quad \text{and} \quad Q : \mathcal{D}(Q) \subset \tilde{\mathbb{X}} \mapsto \mathbb{X}$$

are a (possibly) nonlinear operators.

Definition 2.11 A pair (u, p) with $u : [0, T) \mapsto \mathbb{B}$ and $p : [0, T) \mapsto \mathbb{X}$ is said to be

- a *mild* solution of the implicit evolution equation (2.47) if $u \in C([0, T); \mathbb{B})$, $G(u, p) \in L^1((0, T); \mathbb{B})$, and

$$\begin{aligned} u(t) &= \Phi(t)u(0) + \int_0^t (\Phi(t-\tau) \circ G)(u(\tau), p(\tau))\, d\tau, \\ \mathbf{0} &\overset{a.e.}{=} Q(u(t), p(t)), \end{aligned} \qquad t \in (0, T); \tag{2.48}$$

- a *strong* solution of the implicit evolution equation (2.47) if $u \in C([0, T); \mathbb{B})$, u is differentiable almost everywhere on $(0, T)$, $G(u, p) \in L^1((0, T); \mathbb{B})$, $\frac{du}{dt} \in L^1((0, T); \mathbb{B})$ and the equation (2.47) is satisfied almost everywhere on $(0, T)$;
- a *classical* solution of the evolution equation (2.47) if $u \in C([0, T); \mathbb{B})$, $\frac{du}{dt} \in C((0, T); \mathbb{B})$, $u(t) \in \mathcal{D}(A)$, and $(u(t), p(t)) \in \mathcal{D}(G) \cap \mathcal{D}(Q)$ for $t \in (0, T)$ and (2.47) is satisfied on $(0, T)$.

To guarantee the existence of solutions of the implicit evolution equation, a certain regularity of Q and G is required.

Theorem 2.18 *Let a densely defined closed linear operator $A : \mathcal{D}(A) \subset \mathbb{B} \mapsto \mathbb{B}$ generate a strongly continuous semigroup Φ of linear bounded operators on $\mathbb{B}$.*

Let us assume that there exist operators $M_1 : \mathcal{D}(M_1) \subset \mathbb{B} \mapsto \mathbb{X}$ and $M_2 : \mathcal{D}(M_2) \subset \mathbb{B} \mapsto \mathbb{B}$ such that

(1) for any $u \in \mathcal{D}(M_1)$ we have $(u, M_1(u)) \in \mathcal{D}(G)$;
(2) $\exists \tilde{H} \subset \mathcal{D}(M_1) \cap \mathcal{D}(M_2)$ such that

$$(u, M_1(u)) \in \mathcal{D}(Q) \quad \text{and} \quad \mathbf{0} = Q(u, M_1(u)), \qquad \forall u \in \tilde{H};$$

(3) the operator M_2 has a bounded right inverse $M_2^{-1} \in \mathcal{L}(\mathbb{B}; \mathbb{B})$ satisfying

$$M_2^{-1}\Phi(t)M_2 u = \Phi(t)u, \qquad \forall u \in \mathcal{D}(M_2) \subset \mathcal{D}(M_1), \qquad \forall t \geq 0,$$

$$\|M\Phi(\cdot)\| \in L^1_{loc}(\mathbb{R}_+; \mathbb{R}_+) \ \text{and} \ \int_0^t \|M_2\Phi(t)\|\, d\tau \to 0 \text{ as } t \to 0; \tag{2.49}$$

(4) the operator $G(M_2^{-1}(\cdot), M_1 \circ M_2^{-1}(\cdot)) : \mathbb{B} \mapsto \mathbb{B}$ is locally Lipschitz continuous on $\mathbb{B}$.

If there exists an operator $M_3 \in \mathcal{L}(\mathbb{B}; \mathbb{B})$ such that

- *M_3 is a projector of $\mathcal{D}(M_2)$ on $\tilde{H}$, i.e.,*

$$M_3 u \in \tilde{H}, \ \forall u \in \mathcal{D}(M_2) \quad \text{and} \quad M_3 u_0 = u_0, \ \forall u_0 \in \tilde{H};$$

- *$M_3\Phi(t) = \Phi(t)M_3, \quad \forall t \geq 0$;*
- *$M_3 G(u, M_1(u)) = G(u, M_1(u))$ for all $\forall u \in \mathcal{D}(M_2)$,*

then for any $u_0 \in \tilde{H}$ there exists a mild solution (u, p) of the implicit evolution equation (2.47) defined on $(0, T)$ such that $u(0) = u_0$, $p(0) = M_1(u_0)$ and $u(t) \in \tilde{H}$ for all $t \in [0, T)$ and for some $0 < T \leq +\infty$.

Moreover, if the operator M_1 is unique then the mild solution is unique as well.

Proof Using Theorem 2.15, we conclude that the evolution equation

$$\dot{u} = Au + G(u, M_1 u), \quad u(0) = u_0 \in \mathcal{D}(M_2)$$

has a unique mild solution $u(t) \in \mathcal{D}(M_2)$ given by

$$u(t) = \Phi(t)u_0 + \int_0^t \Phi(t - \tau)G\left(u(\tau), M_1 u(\tau)\right) d\tau$$

for some $T \leq +\infty$. Since M_3 projects $\mathcal{D}(M_2)$ into $\tilde{H}$ then $M_3 u(t) \in \tilde{H}$ and $M_3 u_0 = u_0$ for any $u_0 \in \tilde{H}$. Since M_3 is a linear bounded operator which commutes with Φ then for any $u_0 \in \tilde{H}$, we derive

$$M_3 u(t) = M_3 \Phi(t) u_0 + M_3 \int_0^t \Phi(t-\tau) G\left(u(\tau), M_1(u(\tau))\right) d\tau =$$

$$\Phi(t) M_3 u_0 + \int_0^t \Phi(t-\tau) M_3 G\left(u(\tau), M_1(u(\tau))\right) d\tau =$$

$$\Phi(t) u_0 + \int_0^t \Phi(t-\tau) G\left(u(\tau), M_1(u(\tau))\right) d\tau = u(t).$$

Therefore, $\tilde{u}(t) = M_3 u(t) \in \tilde{H} \subset \mathcal{D}(M_1)$, $p(t) = M_1(u(t))$ is a mild solution of the implicit evolution equation (2.47) with the initial condition $u(0) = u_0 \in \tilde{H}$ and $p(0) = M_1(u_0)$. □

Example 2.17 *(Navier–Stokes equations)* The system of Navier–Stokes equations

$$\begin{aligned} \frac{\partial u}{\partial t} &= \nu \Delta u - (u \cdot \nabla) u - \nabla p, \\ \mathbf{0} &= \operatorname{div} u \end{aligned} \tag{2.50}$$

where u denotes the velocity of a fluid in $\mathbb{R}^3$, p denotes the scalar pressure, and $\nu > 0$ denotes viscosity of the fluid, is the classical model of the flow of an incompressible viscous fluid. The existence of local-in-time solutions of the Navier–Stokes equations is studied, for example, in [8]. It can also be proven using Theorem 2.18. Indeed, the equation (2.50) can be represented in the form (2.47), where

$$\mathbb{B} = L^2(\mathbb{R}^3; \mathbb{R}^3), \quad \mathbb{X} = L^1_{loc}(\mathbb{R}^3; \mathbb{R}),$$

$$A := \Delta : H^2(\mathbb{R}^3; \mathbb{R}^3) \subset \mathbb{B} \mapsto \mathbb{B}$$

is the Laplace operator in a space of functions $\mathbb{R}^3 \mapsto \mathbb{R}^3$, the operator $G(u, p) : \mathcal{D}(G) \subset \mathbb{B} \times \mathbb{X} \mapsto \mathbb{B}$ is given by

$$G(u, p) = -(u \cdot \nabla) u - \nabla p,$$

and the operator $Q : \mathcal{D}(Q) \subset \mathbb{B} \times \mathbb{X} \mapsto \mathbb{X}$ is given by

$$Q(u, p) = \operatorname{div} u$$

in the domain $\mathcal{D}(Q) = \mathcal{D}(G) = H^1(\mathbb{R}^3; \mathbb{R}^3) \times \mathbb{X}$.

1) Let $\tilde{H}$ be the set of divergent-free vector fields defined as follows:

$$\tilde{H} = \left\{ u \in H^{\frac{7}{4}}(\mathbb{R}^3; \mathbb{R}^3) : \operatorname{div} u = \mathbf{0} \right\}.$$

Obviously, $\tilde{H}$ is a linear subspace of $H^{\frac{7}{4}}(\mathbb{R}^3; \mathbb{R}^3)$. By Plancherel lemma (see Appendix D or, for example, [8]) the linear operator

$$f \mapsto \frac{\partial^2}{\partial z_i \partial z_j} \Delta^{-1} f, \quad i, j \in \{1, ..., 3\}$$

is bounded as an operator from $H^k(\mathbb{R}^3; \mathbb{R}^3)$ to $H^k(\mathbb{R}^3; \mathbb{R}^3)$ for any $k \geq 0$, where Δ^{-1} is the inverse Laplacian given by

$$(\Delta^{-1} f)(z) = -\int_{\mathbb{R}^3} \frac{f(y)}{4\pi |z - y|} dy, \quad z \in \mathbb{R}^3.$$

Let us define the operator $M_1 : H^{\frac{7}{4}}(\mathbb{R}^3; \mathbb{R}^3) \subset \mathbb{B} \mapsto \mathbb{B}$ as follows:

$$M_1(u) = \operatorname{div}(\Delta^{-1}(u \cdot \nabla)u).$$

In this case, we have $G(u, M_1(u)) = -\nabla M_1(u) - (u \cdot \nabla)u$.

2) Let the operator $M_2 : H^{\frac{7}{4}}(\mathbb{R}^3; \mathbb{R}^3) \subset \mathbb{B} \mapsto \mathbb{B}$ be defined as follows: $M_2 = \Delta^{7/8}$. It satisfies the condition 3) of the above theorem (see Example 2.14 for more details).

3) Since $\operatorname{div}\Delta^{-1}$ is a bounded operator on L^2 then using Example 2.14, we conclude that the nonlinear operator $\tilde{G} : \mathbb{B} \mapsto \mathbb{B}$ given by

$$\tilde{G}(u) = G(M_2^{-1} u, M_1 \circ M_2^{-1}(u)), \quad u \in \mathbb{B}$$

is locally Lipschitz continuous.

4) Let $M_3 \in \mathcal{L}(\mathbb{B}; \mathbb{B})$ be defined as follows:

$$M_3(u) = u - \nabla \operatorname{div}(\Delta^{-1} u).$$

The operator M_3 is a projector of $H^{\frac{7}{4}}(\mathbb{R}^3; \mathbb{R}^3)$ onto $\tilde{H}$. Indeed, for any $u \in H^{\frac{7}{4}}(\mathbb{R}^3; \mathbb{R}^3)$ we have

$$\operatorname{div}(M_3(u)) = \operatorname{div} u - \Delta \operatorname{div}(\Delta^{-1} u) = \operatorname{div} u - \operatorname{div}(\Delta \Delta^{-1} u) = \mathbf{0},$$

i.e., $M_3 u \in \tilde{H}$. For any $u = (u_1, u_2, u_3) \in \tilde{H}$ we have $\operatorname{div} u = \sum_{i=1}^{3} \frac{\partial u_i}{\partial z_i} = \mathbf{0}$ and using integration by parts we derive

$$
\begin{aligned}
\operatorname{div}(\Delta^{-1}u) &= -\int_{\mathbb{R}^3}\sum_{i=1}^{3} u_i(y)\frac{\partial}{\partial z_i}\frac{1}{4\pi|z-y|}dy \\
&= \int_{\mathbb{R}^3}\sum_{i=1}^{3} u_i(y)\frac{\partial}{\partial y_i}\frac{1}{4\pi|z-y|}dy \\
&= -\int_{\mathbb{R}^3}\frac{\sum_{i=1}^{3}\frac{\partial u_i(y)}{\partial y_i}}{4\pi|z-y|}dy \\
&= \Delta^{-1}\operatorname{div}(u_0) = \mathbf{0}\,.
\end{aligned}
$$

Therefore, we have

$$M_3(u) = u \quad \text{if} \quad u_0 \in \tilde{H}.$$

Since the operator Δ^{-1} satisfies all conditions of Theorem 2.7 we have $\Delta^{-1}\Phi(t)v = \Phi(t)\Delta^{-1}v$ for all $v \in \mathbb{B}$. Moreover, since $\Phi(t)v \in \mathcal{D}(A) = H^2(\mathbb{R}^3;\mathbb{R}^3)$ for all $v \in H^2(\mathbb{R}^3;\mathbb{R}^3)$ then $\Delta^{-1}\Phi(t)u = \Phi(t)\Delta^{-1}u \in H^2(\mathbb{R}^3;\mathbb{R}^3)$ for any $u \in \mathbb{B}$. Similarly, since the operator Δ commutes with $\nabla\operatorname{div}$ then, by Theorem 2.7, we have $\nabla\operatorname{div}(\Phi(t)v) = \Phi(t)\nabla\operatorname{div}(v)$ for any $v \in H^2(\mathbb{R}^3;\mathbb{R}^3)$. Hence, we conclude that

$$M_3\Phi(t) = \Phi(t)M_3, \quad \forall t \geq 0.$$

Finally, for any $u \in H^{\frac{7}{4}}(\mathbb{R}^3;\mathbb{R}^3)$ we have

$$\operatorname{div}(\Delta^{-1}G(u, M_1(u))) = M_1(u) - \operatorname{div}(\Delta^{-1}(u\cdot\nabla)u) = \mathbf{0}$$

and

$$M_3(G(u, M_1(u))) = G(u, M_1(u)).$$

Therefore, all conditions of Theorem 2.18 are fulfilled and the system of Navier–Stokes equations has a mild solution $u(t) \in \tilde{H}$, $p(t) = M_1(u(t))$ defined on some interval of time for any initial condition $u(0) = u_0 \in \tilde{H}$ with $p(0) = M_1(u_0)$. For more details about the local-in-time and global-in-time existence of solutions of Navier–Stokes equations we, refer the reader to [8, 9].

2.3 Evolution Inclusions

2.3.1 Selector-Based Mild, Strong, and Classical Solutions

An analog of the ordinary differential inclusion studied in Volume I can be introduced in the Banach space $\mathbb{B}$ (see, e.g., [10, 11]). Let us consider the following evolution inclusion:

$$\dot{x} - Ax \in G(t, x), \quad t > t_0, \tag{2.51}$$

where, as before, the system state x is an element of a Banach space $\mathbb{B}$, $t_0 \in \mathbb{R}$ is an initial instant of time, A is a generator of a strongly continuous semigroup Φ of linear bounded operators on $\mathbb{B}$, and $G : \mathbb{R} \times \mathbb{B} \mapsto 2^{\mathbb{B}}$ is a set-valued mapping.

We can define a mild (strong, classical) solution of (2.51) by means of a *selector* (a single-valued mapping) of the set-valued mapping F.

Definition 2.12 A continuous function $x : [t_0, t_0 + T) \mapsto \mathbb{B}$ is said to be a *mild solution* to (2.51) if there exists a selector $f \in L^1_{loc}((t_0, t_0 + T); \mathbb{B})$ such that

$$f(s) \in G(s, x(s)) \quad \text{for almost all} \quad s \in (t_0, t_0 + T)$$

and

$$x(t) = \Phi(t - t_0)x_0 + \int_{t_0}^{t} \Phi(t - s) f(s) ds, \quad \forall t \in [t_0, t_0 + T),$$

where $0 < T \leq +\infty$. If this mild solution satisfies (2.51) for (almost) all $t \in (0, T)$ then x is called a classical (resp., strong) solution of (2.51).

In Volume II we deal mainly with evolution inclusions induced by some evolution equations. The next trivial corollary becomes useful in this context.

Corollary 2.19 *If there exists* $f : [t_0, t_0 + T) \times \mathbb{B} \mapsto \mathbb{B}$ *such that*

$$f(t, x) \in G(t, x), \quad \forall t \in [t_0, t_0 + T), \quad \forall x \in \mathbb{B}$$

and f *satisfies all conditions of Theorem 2.15 then the evolution inclusion* (2.51) *has a mild solution for any* $x_0 \in \mathbb{B}$.

The proof of this corollary is obvious. Indeed, if F admits the existence of a selector $f(t, x)$ satisfying all conditions of Theorem 2.15 then the evolution equation

$$\dot{x} = Ax + f(t, x), \quad t > t_0, \quad x(t_0) = x_0$$

has a unique mild solution x_{t,x_0}, which, obviously, satisfies Definition 2.12.

Similar to the finite-dimensional case, a differential inclusion can be utilized for modeling of system with perturbations and uncertainties. Indeed, together with the system

$$\dot{x} = Ax + f(t, x, d(t)), \quad \|d(t)\| < D$$

we may always consider

$$\dot{x} - Ax \in G(t, x), \quad G(t, x) := \bigcup_{d \,:\, \|d\| \leq D} f(t, x, d)$$

under some reasonable assumptions on f.

2.3.2 On Filippov-Like Regularization

Lebesgue measure does not exist in Banach spaces and the conventional Filippov's regularization (see Volume I of this monograph) cannot be applied for an abstract evolution system (2.31) with a discontinuous f. However, in some cases Filippov's method still can be utilized in $\mathbb{B}$ (see, e.g., [12]). In this monograph we deal only with the case when f may have a discontinuity only at the origin. For more details about Filippov-like regularizations of sliding-mode control models in $\mathbb{B}$, we refer the reader, for example, to [13].

Definition 2.13 Let $f : [t_0, +\infty) \times \mathbb{B}\backslash\{\mathbf{0}\} \mapsto \mathbb{B}$ be a single-valued mapping such that

$$\sup_{0<x<r, t\in[t_0,t_0+T]} \|f(t,x)\| < +\infty$$

for any $r \in (0, +\infty)$ and any $T \in (0, +\infty)$. A continuous function $x : [t_0, t_0 + T) \mapsto \mathbb{B}$ is said to be a *mild Filippov solution* of (2.31) if x is a mild solution of the evolution inclusion (2.51) with

$$G(t,x) := \bigcap_{\varepsilon>0} \text{co}\, f(t, x \dot{+} B(\varepsilon)\backslash\{\mathbf{0}\}). \tag{2.52}$$

If f satisfies conditions of Corollary 2.13, the mild Filippov solution coincides with mild solution on $\mathbb{B}\backslash\{\mathbf{0}\}$. If $\mathbf{0} \in G(t, \mathbf{0})$ then, using Definition 2.12, any mild Filippov solution $x(t)$ converging to $\mathbf{0}$ as $t \to t^*$ can be prolonged to the origin as follows $x(t) = \mathbf{0}$, $t \geq t^*$. This prolongation may be unique provided that the origin is Lyapunov stable (see Chap. 3).

Example 2.18 Consider the evolution equation

$$\dot{x} = Ax + f(x), \quad t > 0, \quad x(0) = x_0, \quad f(x) = \alpha\frac{x}{\|x\|}, \quad \alpha \in \mathbb{R},$$

where A generates a strongly continuous semigroup of linear bounded operators on a Banach space $\mathbb{B}$. Since f is Lipschitz continuous on $\mathbb{B}\backslash\{\mathbf{0}\}$ then, by Corollary 2.13, the considered equation has mild solutions on $\mathbb{B}\backslash\{\mathbf{0}\}$. Since $f(\mathbf{0})$ is not defined and f is discontinuous at $\mathbf{0}$, we cannot use the same theorem for analysis of existence of solutions of the system with $x_0 \in \mathbf{0}$. Applying Filippov's method we derive

$$G(x) = \bigcap_{\varepsilon>0} \overline{\text{co}} f(t, x \dot{+} B(\varepsilon)\backslash\{\mathbf{0}\}) = \alpha \cdot \begin{cases} \frac{x}{\|x\|}, & \text{if } x \neq \mathbf{0}, \\ B(1), & \text{if } x = \mathbf{0}. \end{cases}$$

Since $\mathbf{0} \in G(\mathbf{0})$ then the obtained evolution inclusion

$$\dot{x} - Ax \in G(x)$$

has mild Filippov solutions on whole $\mathbb{B}$ (including $\mathbf{0}$). Indeed, if $x(t') = \mathbf{0}$ then, according to Definition 2.13, $x(t) \equiv \mathbf{0}, t > t'$ is a possible prolongation of a solution converging to zero in a finite time.

A continuous dependence of solutions of evolution inclusions on parameters can be established as follows.

Theorem 2.19 *Let the evolution inclusion* (2.51) *have mild solutions defined on* $[t_0, t_0 + T]$ *for any initial condition* $x(t_0) = x_0 \in B(r_0)$*, where* $r_0 > 0$ *is a positive number possibly depended on* $T \in (0, +\infty)$*, and all mild solutions with initial conditions in the ball* $B(r_0)$ *are bounded by some* $r^{\max} > r_0$ *uniformly on* $t \in [t_0, t_0 + T]$.

Let there exist $K > 0$ *such that*

$$\sup_{f_1 \in F(t,x)} \sup_{f_2 \in F(t,x)} \|f_1 - f_2\| \leq K, \quad \forall x \in B(r^{\max}), \quad \forall t \in [t_0, t_0 + T].$$

Then for any $\varepsilon > 0$ *there exists* $\delta > 0$ *such that the inequalities*

$$x_0^1, x_0^2 \in B(r_0) \quad : \|x_0^1 - x_0^2\| \leq \delta, \quad \text{and} \quad |t - t_0| \leq \delta$$

imply

$$\sup_{t \in [t_0, t_0+T]} \|x_{t_0, x_0^1}(t) - x_{t_0, x_0^2}(t)\| \leq \varepsilon$$

for any pair of solutions x_{t_0,x_0^1} *and* x_{t_0,x_0^2} *(even if solutions are not unique).*

Proof Let x_{t_0,x_0^1} and x_{t_0,x_0^2} be two mild solutions of the system (2.51) with $x_0^1, x_0^2 \in B(r_0)$. Then

$$x_{t_0,x_0^j}(t) = \Phi(t - t_0)x_0^j + \int_{t_0}^{t} \Phi(t - s) f_j(s) ds,$$

where Φ is a strongly continuous semigroup generated by A and

$$f_j(s) \in F(s, x_{t_0,x_0^j}(s)), \qquad j = 1, 2$$

are some selectors.

Let $M = \sup_{s \in [t_0, t_0+T]} \|\Phi(s)\| > 0$ and $\delta = \varepsilon/(M + 2MK)$, $\varepsilon > 0$. In this case we have

$$x_{t_0,x_0^j}(t) = \Phi(t - t_i)x_0^j + \int_{t_0}^{t} \Phi(t - s) f_j(s) ds, \quad j = 1, 2$$

and

$$\|x_{t_0,x_0^1}(t) - x_{t_0,x_0^2}(t)\| = \left\| \Phi(t - t_0)(x_i^1 - x_i^2) + \int_{t_0}^{t} \Phi(t - s)(f_1(s) - f_2(s)) ds \right\|.$$

Since

$$\|f_1(s) - f_2(s)\| \leq \sup_{f_1 \in F(s, x_{t_0, x_0^1}(s))} \sup_{f_2 \in F(s, x_{t_0, x_0^2}(s))} \|f_1 - f_2\| \leq 2K$$

then

$$\sup_{t \in [t_i, t_{i+1}]} \|x_{t_0, x_0^1}(t) - x_{t_0, x_0^2}(t)\| \leq M\|x_1^0 - x_2^0\| + 2MK(t - t_0) \leq \varepsilon$$

provided that $\|x_1^0 - x_2^0\| \leq \delta$ and $t - t_0 < \delta$. □

Example 2.19 Let

$$G(x) = \alpha \cdot \begin{cases} \frac{x}{\|x\|}, & \text{if } x \neq \mathbf{0}, \\ B(1), & \text{if } x = \mathbf{0}, \end{cases}$$

where $x \in \mathbb{B}$. Obviously, G is not Lipschitz continuous on $\mathbb{B}$. However, it is satisfies the above theorem, so a certain continuous dependence of solutions of an initial data is preserved.

2.3.3 Generalized Solution

Let us consider an evolution inclusion

$$\dot{x}(t) \in F(t, x(t)), \quad t > t_0, \quad x(0) = x_0 \tag{2.53}$$

where $F : \mathcal{D}(F) \mapsto 2^{\mathbb{B}}$ is a nonlinear (possibly) set-valued mapping with $\mathcal{D}(F) \subset \mathbb{R} \times \mathbb{B}$. Classical and strong solutions of (2.53) are defined in the usual way (see Definitions 2.6 and 2.5). A generalized solution of (2.53) can be introduced as a limit of Euler-like approximates in $\mathbb{B}$.

Definition 2.14 A function $x_\varepsilon : [t_0, t_N] \mapsto \mathbb{B}$ given by

$$x_\varepsilon(t) := \frac{t - t_{i-1}}{t_i - t_{i-1}} x_i + \frac{t_i - t}{t_i - t_{i-1}} x_{i-1}, \quad \text{for} \quad t \in [t_i, t_{i-1}),$$

with

$$t_i : 0 < t_i - t_{i-1} \leq \varepsilon, \quad i = 1, 2, ..., N$$

and $(t_i, x_i) \in \mathcal{D}(F)$ satisfying

$$\sum_{t_0 < t_i \leq t_0 + T'} \inf_{f_i \in F(t_i, x_i)} \|x_i - x_{i-1} - (t_i - t_{i-1}) f_i\| \leq \varepsilon$$

is called an *approximate solution* of (2.53) on $[t_0, t_0 + T']$ if $t_N \geq T'$.

A continuous function $x : [t_0, t_0 + T) \mapsto \mathbb{B}$ is said to be a *generalized solution* of (2.53) if for any $T' \in (0, T)$ its restriction $x : [t_0, t_0 + T'] \mapsto \mathbb{B}$ is a uniform limit of approximate solutions x_ε on $[t_0, t_0 + T']$, i.e.,

$$\lim_{\varepsilon \to 0} \sup_{t \in [t_0, t_0 + T']} \|x_\varepsilon(t) - x(t)\| = 0.$$

The presented definition [10] introduces an analog of mild solutions for systems with set-valued right-hand sides.

Example 2.20 Let us consider the linear operator $A : \mathcal{D}(A) \subset L^2((0, 1); \mathbb{R}) \mapsto L^2((0, 1); \mathbb{R})$ given by

$$Ax = \frac{\partial^2 x}{\partial z^2}, \quad z \in (0, 1), \quad x(0) = x(1) = 0$$

for

$$x \in \mathcal{D}(A) := H^2((0, 1); \mathbb{R}) \cap H_0^1((0, 1); \mathbb{R}).$$

In Example 2.2 it is shown that A generates a strongly continuous semigroup Φ of linear bounded operators on $\mathbb{B} = L^2((0, 1); \mathbb{R})$.

Let us consider the initial value problem

$$\dot{x} \in F(t, x) := A(x - q(t)) \dot{+} [-1, 1]f, \quad t > 0, \quad x(0) = x_0 \in \mathcal{D}(A),$$

where $x(t) \in \mathbb{B}$, $q \in L^\infty((0, +\infty); \mathbb{B})$ is given by

$$(q(t))(z) = \sin(t)\,\mathrm{sign}(z - 0.5),$$

and f is the time-invariant vector from $D(A)$ defined as follows:

$$f(z) = e^{-z^2}, \quad z \in (0, 1).$$

Obviously, $q(t) \in \mathbb{B}$ and $q(t) \notin H^2((0, 1); \mathbb{R})$ for all $t \neq \pi n$, where n is an arbitrary integer number. Formally, the considered system does not admit the representation (2.51), so it does not have a mild solution in the sense of Definition 2.12.

Let us consider the system

$$\dot{y} \in \{Ay - \dot{q}(t)\} \dot{+} [-1, 1]f, \quad t > 0, \quad y(0) = x(0) - q(0) = x_0 \in \mathcal{D}(A)$$

with

$$(\dot{q}(t))(z) = \cos(t)\,\mathrm{sign}(z - 0.5).$$

Since $\dot{q} \in C^1(\mathbb{R}; \mathbb{B})$ then this system has classical solutions $y(t) \in \mathcal{D}(A)$.

On the other hand, the domain of F is defined as follows:

$$\mathcal{D}(F) = \{(t, x) \in \mathbb{R} \times \mathbb{B} : x = q(t) + y,\ y \in \mathcal{D}(A)\}$$

and it is easy to see that

$$x(t) = q(t) + y(t)$$

is a generalized solution to the original system, which satisfies Definition 2.14 with approximate solutions generated by $x_i = x(t_i) \in \mathcal{D}(F)$.

Finally, we present a simple result about generalized solutions of perturbed evolution inclusions.

Theorem 2.20 *Let $F : \mathcal{D}(F) \subset \mathbb{B} \mapsto \mathbb{B}$ be such that for any $x_0 \in \mathbb{B}\backslash\{\mathbf{0}\}$ and a functional $h : \mathbb{B} \mapsto \mathbb{R}$ be continuous on $\mathbb{B}\backslash\{\mathbf{0}\}$ and*

$$0 < \inf_{\|x\|=r} h(x), \qquad \sup_{\|x\|=r} h(x) < +\infty, \qquad \forall r > 0. \tag{2.54}$$

Then the following two claims are equivalent:

(1) there exists a generalized solution $y : [0, t_{\max}) \mapsto \mathbb{B}$ of the evolution inclusion

$$\dot{y} \in F(y), \quad y(0) = x_0 \neq \mathbf{0}, \tag{2.55}$$

where $t_{\max} = +\infty$ or $t_{\max} \leq +\infty$ such that

$$\lim_{t\to t_{\max}} \|y(t)\| = +\infty \text{ or } \lim_{t\to t_{\max}} \|y(t)\| = 0; \tag{2.56}$$

(2) there exists a generalized solution $x : [0, \tau_{\max}) \mapsto \mathbb{B}$ of the evolution inclusion

$$\dot{x} \in h(x)F(x), \quad x(0) = x_0 \neq \mathbf{0}, \tag{2.57}$$

where $\tau_{\max} = +\infty$ or $\tau_{\max} < +\infty$ such that

$$\lim_{\tau\to \tau_{\max}} \|x(\tau)\| = 0 \text{ or } \lim_{t\to \tau_{\max}} \|x(\tau)\| = +\infty.$$

Moreover, for any generalized solution $y : [0, t_{\max}) \mapsto \mathbb{B}$ of the system (2.43) *there exists a generalized solution $x : [0, \tau_{\max}) \mapsto \mathbb{B}$ of the system* (2.40) *given by* (2.41) *such that the identities*

$$y(t) = x\left(\int_0^t \frac{d\sigma}{h(y(\sigma))}\right), t \in [0, t_{\max}), \quad t_{\max} = \int_0^{\tau_{\max}} h(x(\sigma))d\sigma, \tag{2.58}$$

hold, and, inversely, for any generalized solution $x : [0, \tau_{\max}) \mapsto \mathbb{B}$ *of the system* (2.40) *there exists a generalized solution* $y : [0, t_{\max}) \mapsto \mathbb{B}$ *of the system* (2.43) *such that the identities*

$$x(\tau) = y\left(\int_0^{\tau} h(x(\sigma))d\sigma\right), \tau \in [0, \tau_{\max}), \quad \tau_{\max} = \int_0^{t_{\max}} \frac{d\sigma}{h(y(\sigma))} \tag{2.59}$$

hold.

Proof Let us denote

$$\tau := \int_0^t h^{-1}(y(\sigma))d\sigma.$$

Let us show that $x(\tau) \in \mathbb{B}\backslash\{\mathbf{0}\}$ with $\tau \in [0, \tau_{\max})$ given by

$$x(\tau) := y(t)$$

is a solution of the Cauchy problem (2.57).

Let $T'_y < t_{\max}$. By Definition 2.14, y is a uniform limit of approximate solutions y_ε on $[0, T'_y]$ with

$$\sum_{t_0 < t_i < T'_y} \inf_{f_i \in F(y_i)} \|y_i - y_{i-1} - (t_i - t_{i-1}) f_i\| \leq \varepsilon, \quad i = 1, 2, ..., N.$$

Let us denote $x_i = y_i$ and

$$\tau_0 = 0, \quad \tau_1 = h^{-1}(x_1)t_1, \quad \tau_i = \sum_{j=1}^{i} h^{-1}(x_i)(t_j - t_{j-1}), \quad i = 2, ..., N.$$

Then $h(x_i) = h(y_i)$ and

$$\sum_{t_0 < t_i < T'_y} \inf_{f_i \in F(y_i)} \|y_i - y_{i-1} - (t_i - t_{i-1}) f_i\| =$$

$$\sum_{t_0 < t_i < T'_y} \inf_{f_i \in F(y_i)} \left\|y_i - y_{i-1} - h^{-1}(x_i)(t_i - t_{i-1})h(x_i) f_i\right\| =$$

$$\sum_{t_0 < t_i < T'_y} \inf_{\tilde{f}_i \in h(x_i)F(x_i)} \left\|x_i - x_{i-1} - (\tau_i - \tau_{i-1})\tilde{f}_i\right\| \leq \varepsilon.$$

Since $0 < \|y(t)\| < +\infty$ for all $t \in [0, T'_y]$ and $h \circ y \in C([0, T'_y]; (0, +\infty))$ then $M = \max_{[0,T']} h^{-1}(y(t)) < +\infty$ and $\tau_i - \tau_{i-1} < M\varepsilon$.

Therefore, the evolution inclusion (2.57) has the approximate solution $x_{\bar{\varepsilon}}$ generated by y_ε with $\bar{\varepsilon} = \varepsilon \max\{1, M\}$. Notice that

$$\sum_{t_0 < t_i \le T'_y} \tau_i - \tau_{i-1} = \sum_{t_0 < t_i \le T'_y} h^{-1}(y_i)(t_i - t_{i-1}) \to T'_x := \int_0^{T'_y} \frac{d\sigma}{h(y(\sigma))} \quad \text{as} \quad \varepsilon \to 0.$$

Since $y_\varepsilon \to y$ as $\varepsilon \to 0$ uniformly on $[0, T'_y]$ then $x_{\bar{\varepsilon}} \to x$ as $\varepsilon \to 0$ uniformly on $[0, T'_x]$. All construction can be made in the inverse way to show that any solution of x system corresponds to a solution of y system. Taking into account $T'_x \to \tau_{\max} \Leftrightarrow T'_y \to t_{\max}$ we complete the proof. □

An advanced analysis of evolution inclusions in the general case goes out of the scope of this monograph. We refer the reader to [2, 10] for more details about solution definition, existence, and uniqueness problems for abstract evolution inclusions.

Concluding Remarks

PDEs and time-delay models can be treated as evolution equations (abstract differential equations) in Banach (or Hilbert) spaces. In this chapter, various evolution systems are considered. The existence and uniqueness of solutions of linear and nonlinear systems are studied based on theory of strongly continuous groups and semigroups. We complete this chapter with the following simple remark.

Remark 2.1 (*Semigroup property of solutions*) Let $x_{t_0,x_0}(t)$ with $t \ge t_0$ denote a solution of the system (2.53) (or the system (2.51)) with the initial condition $x(t_0) = x_0$. If the solution is forward unique then the identity

$$x_{t_0,x_0}(t) = x_{t_1,x_1}(t) \tag{2.60}$$

holds for

$$x_1 = x_{t_0,x_0}(t_1), \qquad t \ge t_1 \ge t_0 \tag{2.61}$$

as long as solution exists. This identity is known as *semigroup property of solutions*. The same semigroup property holds for systems with nonunique solutions in the set-theoretic sense, i.e., assuming that $x_{t_0,x_0}(t)$ denotes the set of all solutions initiated in the set $x_0 \subset \mathbb{B}$.

Below we use the group theory in order to introduce dilations in abstract (Banach and Hilbert) space. We refer the reader to [1, 2] for more details about semigroup approach to the analysis of abstract evolution equations. More advanced results about evolution inclusions in systems and control theory can be also found, for example, in [13, 14].

References

1. Pazy A (1983) Semigroups of linear operators and applications to partial differential equations. Springer
2. Engel K-J, Nagel R (2000) One-parameter semigroups for linear evolution equations. Springer, Berlin, Heidelberg, New York
3. Radyno Y (1985) Vectors of exponential type in the operator calculus, and differential equations. Differ Uravn. (in Russian) 21(9):1559–1569
4. Hille E, Phillips R (1957) Functional analysis and semi-groups. AMS
5. Stein E (1970) Singular integrals and differentiability properties of functions. Princeton University Press
6. Kwasnicki M (2017) Ten equivalent definitions of the fractional laplace operator. Fract Calc Appl Anal 20:7–51
7. Benyamini Y, Lindenstauss J (2000) Geometric nonlinear functional analysis: Vol.1. AMS, Providence, Rhode Island
8. Ozanski W, Pooley B (2018) Leray's fundamental work on the Navier–Stokes equations: a modern review of "Sur le mouvement d'un liquide visqueux emplissant l'espace", pages 11–203. London Mathematical Society Lecture Note Series: Partial Differential Equations in Fluid Mechanics, Series Number 452. Cambridge University Press
9. Leray J (1938) Sur le mouvement d'un liquide visqueux que limitent des parois. Acta Math 63:193–248
10. Krein SG, Khazan MI (1985) Differential equations in a banach space. J Sov Math 30(3):2154–2239
11. Tolstonogov A (2000) Differential inclusions in a banach space. Springer
12. Orlov Y (2008) Discontinuous systems: Lyapunov analysis and robust synthesis under uncertainty conditions. Springer
13. Orlov Y (2020) Nonsmooth lyapunov analysis in finite and infinite dimensions. Springer
14. Levaggi L (2013) Existence of sliding motions for nonlinear evolution equations in banach space. Discrete and continuous dynamical systems, special issue: 477–487

Chapter 3
Lyapunov Methods for Abstract Differential Equations

This chapter introduces the elements of stability theory for evolution equations in abstract spaces. Special attention is paid to an analysis of exponential, finite-, and fixed-time stability being typical for homogeneous systems. Contrary to the finite-dimensional case, the finite/fixed-time stability in the infinite-dimensional setting can be discovered even for linear systems. The Lyapunov function method is adapted for the analysis of stability and convergence rate of an abstract differential equation. Lyapunov theorems are obtained for various systems.

3.1 Stability of Infinite-Dimensional Systems

3.1.1 Lyapunov Stability

Let us consider again the nonlinear system

$$\dot{x} \in F(t, x), \quad t > t_0, \quad x(t_0) = x_0, \tag{3.1}$$

where $t_0 \in \mathbb{R}$ is an initial instant of time, $x(t) \in \mathbb{B}$ is the system state, $x_0 \in \mathbb{B}$ is the initial state, $\mathbb{B}$ is a real Banach space, and

$$F : \mathbb{R} \times \mathcal{D}(F) \mapsto 2^{\mathbb{B}}$$

is a (possibly unbounded and possibly set-valued) mapping such that $\mathbf{0} \in F(t, \mathbf{0})$. The set $\mathcal{D}(F) \subset \mathbb{B}$ specifies the domain of F, which is assumed to be dense in $\mathbb{B}$.

A. Polyakov, *Generalized Homogeneity in Systems and Control Volume II*, Communications and Control Engineering,
https://doi.org/10.1007/978-3-031-77496-6_3

Below we deal with the evolution system (3.1) satisfying one of the following assumptions.

Assumption 3.1 We assume that for any $x_0 \in \mathbb{B}$ the system (3.1) has a solution

$$x_{t_0,x_0} \in C([0, t_{\max}); \mathbb{B}),$$

where

$$t^{\max} = +\infty \text{ or } t^{\max} < +\infty : \lim_{t \to t^{\max}} \|x_{t,x_0}(t)\| = +\infty.$$

The time instant $t^{\max}$ may depend on x_0 and on a concrete solution x_{t_0,x_0} if solutions are not unique.

The above assumption does not specify a type of solution of the system (3.1) asking only its continuity in $\mathbb{B}$. We refer the reader to Chap. 2 for more details about solutions of evolution equations/inclusions. *By default, Assumption* 3.1 *is supposed to be fulfilled in all considerations of this chapter.*

The above assumption is sufficient for the introduction of all stability definitions and, even, for the proof of some Lyapunov function theorems. However, the Lyapunov function method becomes practically useful if a stability or an instability of the system can be established using the right-hand side of differential equation/inclusion only. In the case of the abstract system, the corresponding analysis can be provided under the following assumption.

Assumption 3.2 Let Assumption 3.1 be fulfilled, and let any solution

$$x_{t_0,x_0} \in C([t_0, T]; \mathbb{B})$$

with $x_0 \in \mathbb{B}\backslash\{\mathbf{0}\}$ and $x_{t_0,x_0}(t) \neq \mathbf{0}$, $t \in [t_0, T]$ be a uniform limit of locally Lipschitz continuous strong solutions x_{t_0,x_i} of (3.1) with initial conditions $x(0) = x_i \in \mathcal{D}(F)$ such that $x_i \to x_0$ as $i \to \infty$.

This assumption is sufficient to expand the conventional Lyapunov theorems to abstract differential equations. The existence of strong solutions for the evolution system (3.1) is studied in Chap. 2, where, in particular, it is shown that the local Lipschitz continuity of strong solutions can be proven in case of reflexive Banach (e.g., Hilbert) spaces. Indeed, if

$$F(t, x) = Ax + f(t, x),$$

where A is a generator of a strongly continuous semigroup of linear bounded operators on a Hilbert space $\mathbb{H}$ and f is a locally Lipschitz continuous function, then, in view of Theorems 2.10, 2.16 and Corollary 2.14, the system (3.1) satisfies Assumption 3.2.

Definition 3.1 *(Lyapunov stability)* The system (3.1) is said to be locally (globally) *Lyapunov stable* if there exists $\varepsilon_{t_0} \in \mathcal{K}_\infty$ such that

$$\|x_{t_0,x_0}(t)\| \leq \varepsilon_{t_0}(\|x_0\|), \qquad t > t_0$$

for any solution x_{t_0,x_0} of (3.1) and any $x_0 \in U$, where $U \subset \mathbb{B}$ is a neighborhood of the origin (resp., $U = \mathbb{B}$).

If the function ε_{t_0} does not depend on t_0 then the origin is *uniformly Lyapunov stable*. Obviously, if $F(t, x)$ is independent of t (*autonomous* case) and the zero solution of the evolution equation/inclusion is Lyapunov stable, then it is uniformly Lyapunov stable. The system is *unstable* if it does not satisfy Definition 3.1.

Proposition 3.1 *If the system* (3.1) *is Lyapunov stable then the zero solution* $x_{t_0,\mathbf{0}}(t) \equiv \mathbf{0}$ *of the system* (3.1) *is unique.*

Proof Suppose that there exists a solution $x_{t_0,\mathbf{0}}$ and a time instant $t' > t_0$ such that $x_{t_0,\mathbf{0}}(t') \neq \mathbf{0}$. Denote $\varepsilon := \|x_{t_0,\mathbf{0}}(t')\|/2 > 0$. From the definition of Lyapunov stability, we have that for any $\varepsilon > 0$ there exits $\delta > 0$ such that the inequality $\|x_0\| < \delta$ always implies $\|x_{t_0,x_0}(t)\| < \varepsilon$ for all $t > t_0$. Taking into account that, in our case, $\|x_0\| = \|\mathbf{0}\| = 0 < \delta$, we obtain the contradiction. □

Definition 3.2 A positive definite functional $V : \mathbb{R} \times \mathbb{B} \mapsto [0, +\infty)$ is said to be

- **locally proper** (proper at $\mathbf{0}$) if there exists $\underline{V}, \overline{V} \in \mathcal{K}_\infty$ such that

  $$\underline{V}(\|x\|) \leq V(t, x) \leq \overline{V}(\|x\|) \quad \text{for} \quad x \in \Omega\backslash\{\mathbf{0}\}, \quad \forall t \in \mathbb{R},$$

  where $\Omega \subset \mathbb{B}$ is a neighborhood of $\mathbf{0}$ (i.e., Ω contains a ball of a positive radius centered at zero);
- **globally proper** if the above inequality holds for $\Omega = \mathbb{B}$.

The locally and globally proper positive definite functions are conventional Lyapunov function candidates for stability analysis of dynamical systems (see, e.g., [1–4] and Volume I of this monograph). To analyze a decay of a non-smooth function V along trajectories of a system, the generalized derivatives can be utilized.

Lyapunov stability theorem [1] and converse Lyapunov stability theorem [5] can be proven in a similar way even for infinite-dimensional systems [6, p. 205]. Recall that the mild solution is introduced as a continuous function satisfying an integral equation (see Chap. 2), which corresponds to (3.1) in a certain sense. A continuous (and even Lipschitz continuous) function $\mathbb{R} \mapsto \mathbb{B}$ may be nowhere differentiable. The theorems given below deal with *generalized (contingent) derivatives* $D_{\mathbb{K}}$ studied in Chap. 4 of Volume I.

Theorem 3.1 *Let Assumption 3.1 hold. The system* (3.1) *is locally (globally) uniformly Lyapunov stable if and only if there exists a locally (globally) proper positive definite functional* $V : \mathbb{R} \times \mathbb{B} \mapsto \mathbb{R}_+$ *such that the inequality*

$$D_{\mathbb{K}} V(t, x_{t_0,x_0}(t)) \leq 0, \qquad t > t_0, \tag{3.2}$$

holds for any $t_0 \in \mathbb{R}$ *and any solution* x_{t_0,x_0} *of* (3.1) *as long as*

$$x_{t_0,x_0}(t) \in \Omega \backslash \{\mathbf{0}\},$$

where $\Omega \subset \mathbb{B}$ *is a neighborhood of the origin (resp.,* $\Omega = \mathbb{B}$*).*

Proof *Sufficiency*. Given $x_0 \in \Omega$, let us consider an arbitrary solution $x_{t_0,x_0}(t)$ of (3.1). The inequality (3.2) and Lemma 4.1 of Volume I imply that the function $t \mapsto V(x_{t_0,x_0}(t))$ is decreasing for $t > t_0$ as long as $x_{t_0,x_0}(t) \in \Omega \backslash \{\mathbf{0}\}$. The properness of V yields

$$\underline{V}(\|x_{t_0,x_0}(t)\|) \leq V(t, x_{t_0,x_0}(t)) \leq V(t_0, x_0) \leq \overline{V}(\|x_0\|). \tag{3.3}$$

Since $\underline{V} \in \mathcal{K}^\infty$ then there exists $\underline{V}^{-1} \in \mathcal{K}^\infty$. Hence, for $\varepsilon := \underline{V}^{-1}(\overline{V}) \in \mathcal{K}^\infty$ we have

$$\|x_{t_0,x_0}(t)\| \leq \varepsilon(|x_0|)$$

as long as $x_{t_0,x_0}(t) \in \Omega \backslash \{\mathbf{0}\}$ (resp., $x_{t_0,x_0} \neq \mathbf{0}$). Therefore, the function ε (independent of t_0) satisfies Definition 3.1 for any $x_0 \in B(R)$ (resp., $x_0 \in \mathbb{B}$), where $R > 0$ is such that $B(\varepsilon(R)) \subset \Omega$. If $\exists t' \geq t_0 : x_{t_0,x_0}(t') = \mathbf{0}$ then repeating the proof of Proposition 3.1 we derive $x_{t_0,x_0}(t) = \mathbf{0}, \forall t \geq t'$.

Necessity. Let S_{t_0,x_0} denote a set of all solutions of the system (3.1) with the initial condition $x(t_0) = x_0$. Let the origin of the system (3.1) be uniformly locally (globally) Lyapunov stable. In this case, the functional $V : \mathbb{R} \times \mathbb{B} \mapsto \mathbb{R}_+$ given by

$$V(t_0, x_0) := \sup_{t_0 \in \mathbb{R}, x_{t_0,x_0} \in S_{t_0,x_0}} \sup_{t \geq t_0} \|x_{t_0,x_0}(t)\|$$

is locally (globally) proper. Indeed, by construction, V is nonnegative and $\underline{V}(\|x_0\|) := \|x_0\| \leq V(t_0, x_0)$ for all $x_0 \in \mathbb{B}$. The definition of local (global) uniform Lyapunov stability implies that $V(t_0, x_0) \leq \varepsilon(\|x_0\|)$ for all $x_0 \in \Omega$ (resp., $\Omega = \mathbb{B}$), where $\varepsilon \in \mathcal{K}_\infty$. Hence, we conclude that V is locally (globally) proper. Finally, due to uniform Lyapunov stability of the origin, $V(x_{t_0,x_0}(t))$ is decreasing for any $x_0 \in \Omega$ and, by Lemma 4.1 of Volume I, $D_{\mathbb{K}} V(x_{t_0,x_0}(t)) \leq 0$. The proof is complete. □

Formally, the above theorem needs a knowledge of solutions of the evolution system in order to analyze the Lyapunov stability. In some cases, the condition (3.2) can be checked without an exact knowledge of solutions even if the function V is not differentiable (see Chap. 4 of Volume I).

Under additional restrictions to the system (3.1), the stability analysis can be done using the operator F only.

Corollary 3.1 *Let Assumption 3.2 hold. If there exists a locally (globally) proper locally Lipschitz continuous on $\Omega\backslash\{\mathbf{0}\}$ positive definite functional $V : \mathbb{B} \mapsto \mathbb{R}_+$ such that*

$$\sup_{y\in F(t,x)} D^+V(x; y) \leq 0, \quad \forall x \in \mathcal{D}(F) \cap \Omega\backslash\{\mathbf{0}\}, \ \forall t \in \mathbb{R}, \tag{3.4}$$

then the origin of the system (3.1) is locally (globally) uniformly Lyapunov stable, where $\Omega \subset \mathbb{B}$ is a neighborhood of the origin (resp., $\Omega = \mathbb{B}$) and

$$D^+V(x; y) = \limsup_{h\to 0^+} \frac{V(x + hy) - V(x))}{h} \tag{3.5}$$

is the right-hand upper Gateaux derivative of V at the point $x \in \mathcal{D}(F)$ in the direction y.

Proof For any locally Lipschitz continuous strong solution x_{t_0,x_i} with $x(0) = x_i \in \mathcal{D}(F)$, we have

$$x_{t_0,x_i}(t) \overset{a.e.}{\in} \mathcal{D}(F), \dot{x}_{t_0,x_i}(t) \overset{a.e.}{\in} F(t, x_{t_0,x_i}(t)).$$

Since V is locally Lipschitz continuous and x_{t_0,x_0} is locally Lipschitz continuous, then the scalar-valued function $t \mapsto V(x_{t_0,x_i}(t))$ is locally Lipschitz continuous and differentiable almost everywhere. Moreover, for any $t \geq t_0$ such that $\dot{x}_{t_0,x_i}(t) \in F(t, x_{t_0,x_i}(t))$ we have

$$\frac{\left|V\left(x_{t_0,x_i}(t+h)\right) - V\left(x_{t_0,x_i}(t) + hx_{t_0,x_i}(t)\right)\right|}{h} \leq$$

$$\frac{L\left\|x_{t_0,x_i}(t+h) - x_{t_0,x_i}(t) - h\dot{x}_{t_0,x_i}(t)\right\|}{h} \to 0 \text{ as } h \to 0.$$

Hence, taking into account

$$\frac{V\left(x_{t_0,x_i}(t+h)\right) - V\left(x_{t_0,x_i}(t)\right)}{h} = \frac{V\left(x_{t_0,x_i}(t+h)\right) - V\left(x_{t_0,x_i}(t) + h\dot{x}_{t_0,x_i}(t)\right)}{h} +$$

$$\frac{V\left(x_{t_0,x_i}(t) + h\dot{x}_{t_0,x_i}(t)\right) - V\left(x_{t_0,x_i}(t)\right)}{h}$$

we derive

$$\dot{V}(x_{t_0,x_i}(t)) \overset{a.e.}{=} DV(x_{t_0,x_i}(t); \dot{x}_{t_0,x_i}(t)) \overset{a.e.}{=} D^+V(x_{t_0,x_i}(t); \dot{x}_{t_0,x_i}(t)).$$

The condition (3.4) implies $\dot{V}(x_{t_0,x_i}(t)) \overset{a.e.}{\leq} 0$ and, by Lemma 4.9 of Volume I, the function $t \mapsto V(x_{t_0,x_i}(t))$ is decreasing for $t > t_0$ as long as $x_{t_0,x_i}(t) \in \Omega\backslash\{\mathbf{0}\}$.

Repeating the proof (of sufficiency) of Theorem 3.1, we conclude Lyapunov stability for strong solutions:

for any $\varepsilon > 0$ there exists $\delta \in (0, \varepsilon)$ such that $\|x_i\| \leq \delta$ implies $\|x_{t_0,x_i}(t)\| \leq \varepsilon$ for all $t > t_0$.

Let x_{t_0,x_0} be an arbitrary solution of (3.1) with $x_0 \in \mathbb{B} : \|x_0\| \leq \delta$. Let us show that $\|x_{t_0,x_0}(t)\| \leq \varepsilon$ for all $t > t_0$. By Assumption 3.2 any solution of (3.1) is a uniform limit of strong solutions x_{t_0,x_i} with $x_i \to x_0$, i.e.,

$$\sup_{t\in[t_0,t_0+T]} \|x_{t_0,x_0}(t) - x_{t_0,x_i}(t)\| \to 0 \text{ as } i \to \infty,$$

where $T > 0$ is arbitrary instant of time such that $x_{t_0,x_0}(t) \neq \mathbf{0}$ for $t \in [t_0, t_0 + T]$. Since $V(x_{t_0,x_i}(\cdot))$ is decreasing and V is locally Lipschitz continuous then $V(x_{t_0,x_0}(\cdot))$ is also decreasing as long as $x_{t_0,x_0}(t) \neq \mathbf{0}$ and

$$\|x_{t_0,x_0}(t)\| \leq \varepsilon$$

for all $t > t_0$. The proof is complete. □

Recall that if V is Gateaux differentiable at the point x (see Appendix B) then one-sided Gateaux derivatives coincide with the Gateaux derivative, i.e., $DV(x; y) = D^+V(x; y)$. Moreover, if V is Frechét differentiable (see Appendix B) then $DV(x; y) = DV(x)y$, where $DV(x) \in \mathcal{L}(\mathbb{B}; \mathbb{R})$ is the Frechét derivative[1] of V at the point x. This means that the above Lyapunov theorem remains valid for $DV(x; y)$ (resp., $DV(x)y$) provided that V is Gateaux (Frechét) differentiable.

3.1.2 Exponential Stability

Exponential stability is the typical property of many linear systems.

Definition 3.3 *(Exponential stability)* The system (3.1) is said to be locally (globally) uniformly exponentially stable if it is locally (globally) uniformly Lyapunov stable and there exist $r > 0$, $C > 0$ and $U \subset \mathbb{B}$ (resp., $U = \mathbb{B}$) such that

$$\|x_{t_0,x_0}(t)\| \leq Ce^{-r(t-t_0)}\|x_0\|, \quad \forall t > t_0, \tag{3.6}$$

for any solution x_{t_0,x_0} of the system (3.1) with $x_0 \in \mathbb{U}$ and $t_0 \in \mathbb{R}$.

[1] Let $f : \mathbb{B} \mapsto \mathbb{V}$ be a mapping between two Banach spaces $\mathbb{B}$ and $\mathbb{V}$. A linear bounded operator $Df(x) \in \mathcal{L}(\mathbb{B}; \mathbb{V})$ is said to a Frechét derivative at the point $x \in \mathbb{B}$ if

$$\frac{\|f(x+y) - f(x) - Df(x)y\|_{\mathbb{V}}}{\|y\|_{\mathbb{B}}} \to 0 \quad \text{as} \quad \|y\|_{\mathbb{B}} \to 0.$$

Any exponentially stable system is Lyapunov stable. For $r = 0$ the condition (3.6) corresponds to a particular case of Lyapunov stability. The parameter r defines the rate of exponential convergence of all solutions of the system to zero. The constant C, in this case, characterizes overshoots (maximal deviations from the origin) of the system during transients.

Example 3.1 Let us consider the heat system

$$\frac{\partial x}{\partial t} = \frac{\partial^2 x}{\partial z^2}, \qquad x(t,0) = x(t,1) = 0, \quad x(0,z) = x_0(z).$$

Its solution is given by (see Example 2.2)

$$x_{x_0}(t,z) = 2\sum_{n=1}^{\infty} e^{-n^2\pi^2 t}\sin(\pi n z)\int_0^1 \sin(\pi n y)x_0(y)dy.$$

Let $\|\cdot\| = \sqrt{\langle\cdot,\cdot\rangle}$ denote the norm of the Lebesgue space $L^2((0,1);\mathbb{R})$. Then taking into account the orthogonality of the functions $\sin(\pi n z)$ and $\sin(\pi k z)$ for $k \neq n$ we derive

$$\|x_{x_0}(t)\|^2 = 4\sum_{n=1}^{\infty} e^{-2n^2\pi^2 t}c_n^2\int_0^1 \sin^2(\pi n z)\,dz = 2\sum_{n=1}^{\infty} e^{-2n^2\pi^2 t}c_n^2,$$

where $c_n \in \mathbb{R}$ is the Fourier coefficient given by

$$c_n = \int_0^1 \sin(\pi n z)x_0(z)dz.$$

By Parseval's identity (see Appendix A) we have $\|x_0\|_{L^2}^2 = 2\sum_{n=1}^{\infty} c_n^2$ and

$$\|x(t,x_0)\| \le \|x_0\|e^{-\pi^2 t}.$$

In the other words, the zero solution of the heat system with the homogeneous Dirichlet boundary conditions is exponentially stable.

Inspired by [6–8], we prove the following theorem.

Theorem 3.2 *Let Assumption 3.1 hold. Let the system* (3.1) *be autonomous*[2] *and have a unique solution for any* $x_0 \in \mathbb{B}$. *The system* (3.1) *is locally (globally) uniformly exponentially stable if and only if there exists a locally (globally) proper positive definite functional* $V : \mathbb{B} \mapsto \mathbb{R}_+$ *such that*

- $\exists k_1 > 0, \exists k_2 > 0, \exists \alpha > 0 \quad : \quad k_1\|x\|^\alpha \leq V(x) \leq k_2\|x\|^\alpha, \quad \forall x \in \mathbb{B};$
- *for any solution* x_{x_0} *of the autonomous system* (3.1) *with* $x_0 \in \Omega$ *and* $t_0 = 0$ *the following inequality:*

$$D_{\mathbb{K}} V(x_{x_0}(t)) \leq -\rho V(x_{x_0}(t)), \quad t \geq 0, \tag{3.7}$$

holds as long as $x_{x_0}(t) \in \Omega\backslash\{\mathbf{0}\}$, *where* $\Omega \subset \mathbb{B}$ *is a neighborhood of the origin (resp.,* $\Omega = \mathbb{B}$*) and* $\rho > 0$ *is some number.*

Proof For simplicity, let us present the proof of global exponential stability. The local case can be analyzed similarly.

Necessity. Let the system (3.1) be exponentially stable, i.e.,

$$\|x_{x_0}(t)\| \leq Ce^{-rt}\|x_0\|, \quad \forall t \geq 0.$$

Inspired by [5, 7, 9], let us define the functional $V : \mathbb{B} \mapsto \mathbb{R}_+$ as follows:

$$V(x_0) = \sup_{s\geq 0} \|x_{x_0}(s)\| + \int_0^{+\infty} \|x_{x_0}(\tau)\|d\tau. \tag{3.8}$$

This functional is single-valued (due to uniqueness of solutions). Moreover, it is finite-valued, positive definite, and proper due to

$$k_1\|x_0\| \leq \sup_{s\geq 0}\|x_{x_0}(s)\| \leq V(x_0) \leq C\left(\sup_{t\geq 0} e^{-rt} + \int_0^{+\infty} e^{-rs}ds\right)\|x_0\| = k_2\|x_0\|,$$

where $k_1 = 1$ and $k_2 = C(r+1)/r$. Moreover, for $h > 0$ we have

$$V(x_{x_0}(t+h)) - V(x_{x_0}(t)) \leq \int_0^{+\infty} \|x_{x_{x_0}(t+h)}(\tau)\|d\tau - \int_0^{+\infty} \|x_{x_{x_0}(t)}(\tau)\|d\tau.$$

Using the semigroup property of solutions, we derive

$$x_{x_{x_0}(t+h)}(\tau) = x_{x_{x_0}(t)}(\tau + h), \quad \forall \tau \geq 0, \quad \forall h \geq 0, \quad \forall t \geq 0$$

[2] The system (3.1) is autonomous if F is independent of time, i.e., $F(t, x) = F(x)$.

and

$$V(x_{x_0}(t+h)) - V(x_{x_0}(t)) \le -\int_0^h \|x_{x_{x_0}(t)}(\tau)\| d\tau.$$

Hence, dividing both sides of the above inequality by $h > 0$ and tending $h \to 0^+$, we conclude

$$\limsup_{h\to 0^+} \frac{V(x_{x_0}(t+h)) - V(x_{x_0}(t))}{h} \le -\|x_{x_0}(t)\| \le -\frac{r}{(r+1)C} V(x_{x_0}(t)), \quad t \ge 0.$$

For $t > -h$ and $h < 0$ we derive similarly

$$\begin{aligned} V(x_{x_0}(t+h)) - V(x_{x_0}(t)) &\ge \int_0^{+\infty} \|x_{x_{x_0}(t+h)}(\tau)\| d\tau - \int_0^{+\infty} \|x_{x_{x_0}(t)}(\tau)\| d\tau \\ &= \int_0^{+\infty} \|x_{x_{x_0}(t+h)}(\tau)\| d\tau - \int_0^{+\infty} \|x_{x_{x_0}(t+h)}(\tau - h)\| d\tau \\ &= \int_0^{-h} \|x_{x_{x_0}(t+h)}(\tau)\| d\tau. \end{aligned}$$

Since $h < 0$ then dividing both sides of the latter inequality by $h < 0$ and tending $h \to 0^-$, we derive

$$\limsup_{h\to 0^-} \frac{V(x_{x_0}(t+h)) - V(x_{x_0}(t))}{h} \le -\|x_{x_0}(t)\| \le -\frac{r}{(r+1)C} V(x_{x_0}(t)), \quad t \ge 0.$$

Using Corollary 4.1 of Volume I of this monograph, we conclude

$$D_{\mathbb{K}} V(x_{x_0}(t)) \le -\frac{r}{(r+1)C} V(x_{x_0}(t)), \quad \forall t \ge 0.$$

Sufficiency. The Lyapunov stability of the system follows from Theorem 3.1.

The inequality (3.7) and Lemma 4.2 of Volume I guarantee that the function $t \mapsto V(x_{x_0}(t))$ is decreasing and, by Corollary 4.3 of Volume I, differentiable almost everywhere. Hence, Lemma 4.8 of Volume I implies that

$$V(x_{x_0}(t)) - V(x_0) \le \int_0^t \dot{V}(x_{x_0}(\tau)) d\tau \le -\rho \int_0^t V(x_{x_0}(\tau)) d\tau.$$

Applying Grönwall–Bellman inequality (see Appendix E), we derive

$$V(x_{x_0}(t)) \le e^{-\rho t} V(x_0), \quad t > 0.$$

Finally, taking into account inequalities

$$k_1\|x(t)\|^\alpha \le V(x_{x_0}(t)) \le e^{-\rho t}V(x_0) \le k_2 e^{-\rho t}\|x_0\|^\alpha, \quad t>0$$

we complete the proof. □

Similar to the Lyapunov stability analysis, a sufficient condition of exponential stability can be obtained under more a restrictive assumption to the system.

Corollary 3.2 *Under Assumption 3.2, the system* (3.1) *is locally (globally) uniformly exponentially stable if there exist* $\rho > 0$ *and a locally (globally) proper locally Lipschitz continuous on* $\Omega\backslash\{\mathbf{0}\}$ *positive definite functional* $V:\mathbb{B}\mapsto\mathbb{R}_+$ *such that*

$$\exists k_1>0, \exists k_2>k_1, \exists\alpha>0 \quad : \quad k_1\|x\|^\alpha \le V(x) \le k_2\|x\|^\alpha, \quad \forall x\in\mathbb{B}$$

and

$$\sup_{y\in F(t,x)} D^+V(x;y) \le -\rho V(x), \ \forall x\in\mathcal{D}(F)\cap\Omega\backslash\{\mathbf{0}\}, \ \forall t\in\mathbb{R}, \tag{3.9}$$

where Ω *is a neighborhood of the origin (resp.,* $\Omega=\mathbb{B}$*) and* $DV(x;y)\in\mathbb{B}$ *is the right-hand upper Gateaux derivative of* V *at the point* $x\in\mathcal{D}(F)$ *in the direction* $y\in\mathbb{B}$.

Proof For any locally Lipschitz continuous strong solution x_{t_0,x_i} with $x(0)=x_i\in\mathcal{D}(F)$, we have

$$x_{t_0,x_i}(t) \overset{a.e.}{\in} \mathcal{D}(F), \dot{x}_{t_0,x_i}(t) \overset{a.e.}{\in} F(t, x_{t_0,x_i}(t)).$$

So, the scalar-valued function $t\mapsto V(x_{t_0,x_i}(t))$ is locally Lipschitz continuous and differentiable almost everywhere. Then, for any $t\ge t_0$ such that $\dot{x}_{t_0,x_i}(t)\in F(t,x_{t_0,x_i}(t))$ we have

$$\frac{\left|V\left(x_{t_0,x_i}(t+h)\right)-V\left(x_{t_0,x_i}(t)+hx_{t_0,x_i}(t)\right)\right|}{h} \le$$

$$\frac{L\left\|x_{t_0,x_i}(t+h)-x_{t_0,x_i}(t)-h\dot{x}_{t_0,x_i}(t)\right\|}{h} \to 0 \text{ as } h\to 0.$$

Hence, taking into account

$$\frac{V\left(x_{t_0,x_i}(t+h)\right)-V\left(x_{t_0,x_i}(t)\right)}{h} = \frac{V\left(x_{t_0,x_i}(t+h)\right)-V\left(x_{t_0,x_i}(t)+h\dot{x}_{t_0,x_i}(t)\right)}{h} +$$

$$\frac{V\left(x_{t_0,x_i}(t)+h\dot{x}_{t_0,x_i}(t)\right)-V\left(x_{t_0,x_i}(t)\right)}{h}$$

we derive

$$\dot{V}(x_{t_0,x_i}(t)) \overset{a.e.}{=} DV(x_{t_0,x_i}(t); \dot{x}_{t_0,x_i}(t)) \overset{a.e.}{=} D^+V(x_{t_0,x_i}(t); \dot{x}_{t_0,x_i}(t)).$$

Using the inequality (3.9) we conclude

$$\dot{V}(x_{t_0,x_i}(t)) \overset{a.e.}{\leq} -\rho V(x_{t_0,x_i}(t)).$$

Hence, using the Newton–Leibniz formula, we derive

$$V(x_{t_0,x_i}(t)) - V(x_{t_0,x_i}(t_0)) \leq \int_0^t \dot{V}(x_{t_0,x_i}(\tau))d\tau \leq -\rho \int_0^t V(x_{t_0,x_i}(\tau))d\tau,$$

and Grönwall–Bellman inequality gives

$$V(x_{t_0,x_i}(t)) \leq e^{-\rho t} V(x_{t_0,x_i}(t_0)).$$

Hence, taking into account the inequalities $k_1\|x\|^\alpha \leq V(x) \leq k_2\|x\|^\alpha$ we conclude exponential convergence of the strong solution $x_{t_0,x_i}(t)$ to $\mathbf{0}$ as time goes to $+\infty$. Being a uniform limit of strong solutions (see Assumption 3.2), any mild solution also converges to zero exponentially. The proof is complete. □

Example 3.2 Let us consider the following reaction–diffusion equation:

$$\frac{\partial x}{\partial t} = \frac{\partial^2 x}{\partial z^2} + \lambda x, \quad x(0, z) = x_0(z), \quad 0 < z < 1,$$

$$x(t, 0) = x(t, 1) = 0,$$

where $\lambda \in \mathbb{R}$ is a constant. In Example (2.2), it has been shown that the operator $A : \mathcal{D}(A) \subset \mathbb{B} \to \mathbb{B}$ with $\mathbb{B} = L^2((0, 1); \mathbb{R})$ defined as

$$Ax = \frac{\partial^2 x}{\partial z^2}, \quad x \in \mathcal{D}(A)$$

on the domain

$$\mathcal{D}(A) = H^2((0, 1); \mathbb{R}) \cap H_0^1((0, 1); \mathbb{R})$$

is a generator of a strongly continuous semigroup on $\mathbb{B}$. According to Theorem 2.4, the operator $A + \Delta A$ with $\Delta A x = \lambda x$ also generates a semigroup on $\mathbb{B}$. This means that the considered PDE has classical solutions on $\mathcal{D}(A)$, the system is well-posed, and any mild solution is a uniform limit of classical solutions with the initial conditions $x(0) = x_i \in \mathcal{D}(A)$ with $x_i \to x_0 \in \mathbb{B}$ as $i \to +\infty$.

Let us consider the Lyapunov functional $V : \mathbb{B} \mapsto [0, +\infty)$ defined as follows:

$$V(x) = \int_0^1 x^2(z)dz, \quad x \in \mathbb{B}.$$

The functional V is Frechét differentiable on $\mathcal{D}(A)$ (and, in fact, on $\mathbb{B}$):

$$DV(x)y = 2\int_0^1 x(z)y(z)dz, \quad x, y \in \mathbb{H}.$$

Then the time derivative of V along the system

$$\dot{V}(x) := DV(x)(Ax + f(x))$$

admits the representation

$$\dot{V}(x) = 2\int_0^1 x(z)\left(\frac{\partial^2 x(z)}{\partial z^2} + \lambda x(z)\right)dz = -2\int_0^1 [x'(z)]^2 dz + 2\lambda V.$$

Taking into account the Poincaré inequality (see Appendix D) we derive

$$\dot{V}(x) \leq -2(\pi^2 - \lambda)V(x).$$

Therefore, the considered PDE is exponentially stable for $\lambda < \pi^2$.

3.1.3 Asymptotic Stability

The definition of asymptotic stability of the abstract differential equations can be introduced similarly to asymptotic stability of ODEs (see Volume I).

Definition 3.4 *(Asymptotic stability)* The system (3.1) is said to be locally (globally)

- *asymptotically stable* if it is locally (globally) Lyapunov stable and for any $t_0 \in \mathbb{R}$ and any $x_0 \in U$ it holds

$$\lim_{t+\infty} \|x_{t_0,x_0}(t)\| = 0;$$

- *uniformly asymptotically stable* if it is locally (globally) uniformly Lyapunov stable and for any $\epsilon > 0$ and any $R > \epsilon$ there exists $T = T(R, \epsilon) \geq 0$ such that

$$x_0 \in U \cap B(R) \quad \Rightarrow \quad x_{t_0,x_0}(t) \in B(\epsilon), \quad \forall t \geq t_0 + T(R, \epsilon), \ \forall t_0 \in \mathbb{R},$$

where $U \subset \mathbb{B}$ is a neighborhood of the origin (resp., $U = \mathbb{B}$) and $B(r) = \{x \in \mathbb{B} : \|x\| \leq r\}$ is the ball of the radius $r > 0$ in $\mathbb{B}$. The set U is called a domain of asymptotic attraction or, shortly, *attraction domain* of the system.

Asymptotic stability of an autonomous (time-invariant) finite-dimensional system (ODE) always implies its uniform asymptotic stability (see, e.g., Chap. 3 of Volume I or [10]). In the general case, this is not true for infinite-dimensional systems.

Example 3.3 Let us consider the transport equation on $\mathbb{R}_+$

$$\frac{\partial x}{\partial t} = \frac{\partial x}{\partial z}, \quad t > 0, \quad z > 0 \tag{3.10}$$

with $\frac{\partial}{\partial z} : H^1(\mathbb{R}_+; \mathbb{R}) \subset L^2(\mathbb{R}_+, \mathbb{R}) \mapsto L^2(\mathbb{R}_+; \mathbb{R})$. The considered differential operator generates the translation semigroup on $L^2(\mathbb{R}_+; \mathbb{R})$:

$$(\Phi(t)u)(z) = u(z + t), \quad x, t \geq 0, \quad u \in L^2(\mathbb{R}_+; \mathbb{R}). \tag{3.11}$$

This means that, for any initial state $x_0 \in L^2(\mathbb{R}_+; \mathbb{R})$, the corresponding solution x_{x_0} of the transport equation is given by

$$[x_{x_0}(t)](z) = x_0(t + z), \quad \forall z, t \geq 0.$$

Since

$$\|x_{x_0}(t)\|^2_{L^2(\mathbb{R}_+;\mathbb{R})} = \int_0^{+\infty} |x_0(t + z)|^2 dz = \int_t^{+\infty} |x_0(z)|^2 dz, \quad t \geq 0,$$

then $\|x_{x_0}(t)\|_{L^2} \leq \|x_0\|_{L^2}$ and the considered system is Lyapunov stable. Moreover, since $x_0 \in L^2(\mathbb{R}_+; \mathbb{R})$ then

$$\int_t^{+\infty} |x_0(z)|^2 dz \to 0 \quad \text{as} \quad t \to +\infty.$$

This means that the system is globally asymptotically stable. However, this system is not uniformly asymptotically stable. Indeed, considering the sequence of initial conditions

$$x_0^n(z) = \begin{cases} 1 & \text{if } z \in [n, n+1], \\ 0 & \text{otherwise,} \end{cases}$$

we conclude $\|x_0^n\|_{L^2} = 1$ and

$$\|x_{x_0^n}(t)\|_{L^2} = \begin{cases} 1 & \text{if} \quad t \leq n, \\ \sqrt{n + 1 - t} & \text{if} \ n < t < n + 1, \\ 0 & \text{if} \quad t \geq n + 1. \end{cases}$$

Since $\|x_{x_0^n}(t)\|_{L^2} = 1$ for $t \leq n$, then for any $T > 0$ there exists $n \in \mathbb{N}$ such that $\|x_{x_0^n}(T)\| = 1$. Therefore, for any $\epsilon \in (0, 1)$, it is impossible to find a time T such that $\|x_{x_0}(t)\| \leq \epsilon$ for all $t \geq T$ and all $x_0 \in B(1)$. This means that the system is not uniformly asymptotically stable.

Asymptotic stability can be characterized by Lyapunov function. A sufficient condition of uniform asymptotic stability is given by the following theorem inspired by [6].

Theorem 3.3 *Under Assumption 3.2, the system (3.1) is locally (globally) uniformly asymptotically stable if there exist a locally (globally) proper locally Lipschitz continuous on $\Omega\backslash\{\mathbf{0}\}$ positive definite functional $V : \mathbb{B} \mapsto \mathbb{R}_+$ and a continuous positive function $W : (0, +\infty) \mapsto (0, +\infty)$ such that*

$$\sup_{y \in F(t,x)} D^+V(x; y) \leq -W(\|x\|), \ \forall x \in \mathcal{D}(F) \cap \Omega\backslash\{\mathbf{0}\}, \ \forall t \in \mathbb{R}, \tag{3.12}$$

where Ω is a neighborhood of the origin (resp., $\Omega = \mathbb{B}$) and $DV(x; y) \in \mathbb{B}$ is the right-hand upper Gateaux derivative of V at the point $x \in \mathcal{D}(F)$ in the direction $y \in \mathbb{B}$.

Proof By Corollary 3.1 the system is uniformly Lyapunov stable. Since Ω is a neighborhood of the origin then there exists $r^* > 0$ such that $B(\epsilon(r^*)) \in \Omega$, where $\varepsilon \in \mathcal{K}^\infty$ is introduced in Definition 3.1. Let $U = B(r^*)$.

Repeating the proof of Corollary 3.1, we conclude that any strong solution x_{t_0,x_0} of the system (3.1) satisfies the differential inequality

$$\dot{V}(x_{t_0,x_0}(t)) \overset{a.e.}{\leq} -W(\|x_{t_0,x_0}(t)\|)$$

as long as $x_{t_0,x_0}(t) \neq \mathbf{0}$. Since W is a positive function, then the locally Lipschtz continuous function $t \mapsto V(x_{t_0,x_0}(t))$ is monotone and it has a limit

$$V^* = \lim_{t \to +\infty} V(x_{t_0,x_0}(t)) \in \mathbb{R}_+.$$

Let us show that $V^* = 0$. Suppose the contrary, i.e., $V^* > 0$. Since V is locally (globally) proper then $\|x_{t_0,x_0}(t)\| \geq \overline{V}^{-1}(V^*) > 0$ for all $t \geq 0$, where $\overline{V}^{-1} \in \mathcal{K}^\infty$ is an inverse function to the function $\overline{V} \in \mathcal{K}^\infty$ is introduced in Definition 3.2. Due the Lyapunov stability of the system, we have

$$0 < \overline{V}^{-1}(V^*) \leq \|x_{t_0,x_0}(t)\| \leq \varepsilon(\|x_0\|) \quad \text{for all} \quad t \geq 0.$$

Since the positive function W is continuous then its infinitum on a compact is reached on the compact, i.e.,

$$\inf_{\overline{V}^{-1}(V^*) \leq r \leq \epsilon(\|x_0\|)} W(r) = W^* \in (0, +\infty).$$

Therefore, we have

$$\dot{V}(x_{t_0,x_0}(t)) \overset{a.e.}{\leq} -W^*$$

as long as $x_{t_0,x_0}(t) \neq \mathbf{0}$. Since the function $t \mapsto V(x_{t_0,x_0}(t))$ is monotone then

$$V(x_{t_0,x_0}(t)) - V(x_0) \leq \int_0^t \dot{V}(x_{t_0,x_0}(t))dt \leq -W^* t$$

and $V(x_{t_0,x_0}(t)) \leq 0$ for all $t \geq V(x_0)/W^*$. We obtain the contradiction to $V^* > 0$. Therefore, all strong solutions of the system converge locally (globally) to zero as $t \to +\infty$.

Let us show that the convergence is uniform. Given $\epsilon > 0$ and $R > 0$, let us define

$$W_{R,\epsilon} = \inf_{\overline{V}^{-1}(\underline{V}(\epsilon)) \leq r \leq \varepsilon(\min\{R,r^*\})} W(r) > 0$$

and

$$T = \frac{\overline{V}(\min\{R, r^*\})}{W_{R,\epsilon}},$$

where the function $\underline{V} \in \mathcal{K}^\infty$ introduced in Definition 3.2 Let us show that for any strong solution with $x_0 \in B(R) \cap U$ we have

$$V(x_{t_0,x_0}(T)) \leq \underline{V}(\epsilon).$$

Suppose the contrary, i.e., for some solution we have $V(x_{t_0,x_0}(T)) > \underline{V}(\epsilon)$. This means that $V(x_{t_0,x_0}(t)) > \underline{V}(\epsilon)$ for all $t \in [0, T]$ since the function $t \mapsto V(x_{t_0,x_0}(t))$ is monotone. In this case, one holds

$$\|x_{t_0,x_0}(t)\| \geq \overline{V}^{-1}(\underline{V}(\epsilon)), \quad \forall t \in [0, T]$$

and

$$\dot{V}(x_{t_0,x_0}(t)) \leq -W_{R,\epsilon}, \quad \forall t \in [0, T].$$

Since

$$V(x_{t_0,x_0}(t)) - V(x_0) \leq \int_0^t \dot{V}(x_{t_0,x_0}(t))dt \leq -W_{R,\epsilon} t$$

and $V(x_0) \leq \overline{V}(\min\{R, r^*\})$ (due to properness of V) then $V(x_{t_0,x_0}(T)) \leq V(x_0) - W_{R,\epsilon} T \leq 0$. We obtain the contradiction. Therefore, all strong solutions locally (globally) uniformly converge to zero.

Since any mild solution is a uniform limit of strong solutions on compact intervals of time then the same conclusion is valid for mild solutions too. Therefore, the system is locally (globally) uniformly asymptotically stable. The proof is complete. □

Converse Lyapunov function theorems for infinite-dimensional systems can be found in [6, 8, 11]. For analysis of generalized homogeneous systems in this monograph, we just use the following corollary.

Corollary 3.3 *Let*

$$K(r) = \{x \in \mathbb{H} : r^{-1} < \|x\| < r\}, \tag{3.13}$$

where $r > 0$ and $\mathbb{H}$ is a real Hilbert space. Let a linear closed densely defined operator $A : \mathcal{D}(A) \subset \mathbb{H} \mapsto \mathbb{H}$ be a generator of a strongly continuous semigroup of linear bounded operators on a real Hilbert space $\mathbb{H}$. Let an operator $f \in C(\mathbb{H}; \mathbb{H})$ be locally Lipschitz continuous on $\mathbb{H}$.

If the system

$$\dot{x} = Ax + f(x), \quad t > 0 \tag{3.14}$$

is globally uniformly asymptotically stable then for any $r > 1$ there exists a globally proper locally Lipschitz continuous (on $\mathbb{H}$) positive definite functional $V_r : \mathbb{H} \mapsto \mathbb{R}$ such that

- $\exists \epsilon_r \in \mathcal{K}_\infty \quad : \quad \|x\| \leq V_r(x) \leq \epsilon_r(\|x\|), \quad \forall x \in \mathbb{H};$
- *the inequality*

$$D^+V(x; Ax + f(x)) \leq -\frac{1}{r} \tag{3.15}$$

holds for all $x \in \mathcal{D}(A) \cap K(r)$.

Proof Since the system (3.14) is Lyapunov stable then the inclusion $x_0 \in K(r)$ implies

$$\|x_{x_0}(t)\| \leq \varepsilon(\|x_0\|) \leq \varepsilon(r) \quad \text{for all} \quad t \geq 0,$$

where $\varepsilon \in \mathcal{K}_\infty : \varepsilon(r) \geq r$ is introduced in the definition of the Lyapunov stability. Moreover, the uniform asymptotic stability means that for any $r > 0$ there exists $T_r > 0$ such that $\|x_{x_0}(t)\| \leq 1/\varepsilon(r)$ for all $t \geq T_r$ and for all $x_0 \in K(r)$. Since f is locally Lipschitz continuous then solutions of the system are unique. In this case, the functional $V : \mathbb{B} \mapsto \mathbb{R}$ defined as

$$V(x_0) = \sup_{s \in [0, T_r]} \|x_{x_0}(s)\| + \int_0^{T_r} \|x_{x_0}(\tau)\| d\tau \tag{3.16}$$

is single-valued due to uniqueness of solutions. Moreover, it is finite-valued, positive definite, and globally proper due to the estimates

$$\|x_0\| \leq \sup_{s \in [0, T_r]} \|x_{x_0}(s)\| \leq V(x_0) \leq \varepsilon(\|x_0\|)(1 + T_r).$$

Repeating the proof of Theorem 3.2, we derive that for any $x_0 \in K(r)$ it holds

$$D_{\mathbb{K}} V(x_{x_0}(t)) \leq -\|x_{x_0}(t)\| \leq -\frac{1}{r}$$

as long as $x_{x_0}(t) \in K(\varepsilon(r))$.

Applying Corollary 2.14 we conclude that all solutions depend Lipschitz continuously on initial conditions from $B(r)$ on compact intervals of time. This implies that the functional V is locally Lipschitz continuous on $\mathbb{H}$. Moreover, by construction, $\|x\| \leq V(x) \leq (1 + T_r)\varepsilon(\|x\|)$ for all $x \in \mathbb{H}$.

Let x_{x_0} be an arbitrary solution of the system with $x_0 \in \mathcal{D}(A) \cap K(r)$. By Theorem 2.16, this solution is strong and $x_{x_0}(t) \overset{a.e.}{\in} \mathcal{D}(A)$. Since

$$\frac{V(x_{x_0}(t+h))-V(x_{x_0}(t))}{h} = \frac{V\big(x_{x_0}(t+h)\big)-V(x_{x_0}(t)+h(Ax_{x_0}(t)+f(x_{x_0}(t))))}{h} +$$

$$\frac{V(x_{x_0}(t)+h(Ax_{x_0}(t)+f(x_{x_0}(t))))-V(x_{x_0}(t))}{h},$$

and V is locally Lipschitz continuous on $\mathbb{H}$, then, taking into account

$$\frac{|V\big(x_{x_0}(t+h)\big)-V(x_{x_0}(t)+h(Ax_{x_0}(t)+f(x_{x_0}(t))))|}{h} \leq$$

$$\frac{L\big\|x_{x_0}(t+h)-x_{x_0}(t)-h(Ax_{x_0}(t)+f(x_{x_0}(t)))\big\|}{h} \to 0 \text{ as } h \to 0, \quad x_{x_0}(t) \in \mathcal{D}(A) \cap B(\varepsilon(r)),$$

we derive

$$D^+ V(x_{x_0}(t)) \overset{a.e.}{=} D^+ V(x_{x_0}(t);\, Ax_{x_0}(t) + f(x_{x_0}(t))).$$

Since $D^+ V(x(t)) \in D_{\mathbb{K}} V(x(t))$ (see Chap. 4 of Volume I of this monograph) then

$$D^+ V(x_{x_0}(t);\, Ax_{x_0}(t) + f(x_{x_0}(t))) \leq -\frac{1}{r}, \quad x_{x_0}(t) \in \mathcal{D}(A) \cap K(r).$$

Taking $t - 0$ we complete the proof. □

3.2 Convergence Rates

Stabilization (see, e.g., Volume I for more details) is the classical problem of control theory. The convergence rate of the control system is an important characteristic of regulation quality. In Volume I, some concepts of fast (finite/fixed-time) convergence were introduced for nonlinear systems. The fixed-time stability can be discovered in the so-called hyper-exponentially convergent systems [12]. The convergence rate of such system is faster that any exponential. Finite-time and nearly fixed-time stable systems may demonstrate such a convergence rate locally (close to zero of close to infinity). Lyapunov characterization of infinite-dimensional finite/fixed-time stable systems is a nontrivial problem in the general case.

3.2.1 Finite-Time Stability

The definition of finite-time stability of an infinite-dimensional system can be introduced similarly to the finite-dimensional case.

Definition 3.5 *(Finite-time reaching in $\mathbb{B}$)* We say that the system (3.1) locally (globally) reaches the origin in a finite time if there exist a neighborhood $U(t_0)$ of the origin (resp., $U(t_0) = \mathbb{B}$) and a functional $\exists T : \mathbb{R} \times \mathbb{B} \mapsto \mathbb{R}_+$ such that

$$\forall x_0 \in U(t_0), \quad \exists t' \in [t_0, T(t_0, x_0)] \; : \; x_{t_0,x_0}(t') = \mathbf{0},$$

where t' depends on solution x_{t_0,x_0} of the system.

Definition 3.6 *(Finite-time stability in $\mathbb{B}$)* A locally (globally) Lyapunov stable system (3.1) is said to be locally (globally) finite-time stable if it locally (globally) reaches to the origin in a *finite time*.

By Proposition 3.1, Lyapunov stability of the system implies the uniqueness of the zero solution. This immediately implies the following corollary.

Corollary 3.4 *Any finite-time stable system converges to zero in a finite time, namely*

$$\forall x_0 \in U(t_0) \quad \Rightarrow \quad x_{t_0,x_0}(t) = \mathbf{0}, \quad \forall t \geq t_0 + T(t_0, x_0), \tag{3.17}$$

for any solution x_{t_0,x_0} of the system.

If T satisfies the definition of finite-time stability then the functional $T + T^+$ also does for any nonnegative T^+. Below we consider a minimal functional T, which is called the *settling-time functional*.

Definition 3.7 A functional $T : \mathbb{R} \times \mathbb{B} \mapsto \mathbb{R}_+$ is called the settling-time functional of the finite-time stable system (3.1) if T satisfies (3.17) and, for any $\tilde{T} : \mathbb{R} \times \mathbb{B} \mapsto \mathbb{R}_+$ such that $\tilde{T} \neq \mathbf{0}$, the functional $T_{\text{new}} = T - \tilde{T}$ does not satisfy (3.17) with the same U.

The settling-time functional T of the time-invariant finite-time stable system is independent of t_0. Finite-time stability always implies asymptotic stability. However, as shown in Volume I of this monograph, in contrast to asymptotic stability, the finite-time stability of the time-invariant system may not imply its uniform finite-time stability, which requires local boundedness of the settling-time function.

Definition 3.8 (*Uniform finite-time stability*) The system (3.1) is said to be locally (globally)

- uniformly finite-time stable (FTS) if it is finite-time stable with a time-invariant attraction domain U and the settling-time functional $T : \mathbb{R} \times U \mapsto \mathbb{R}_+$ is *locally bounded uniformly on the first argument*, i.e.,

$$\forall y \in U, \quad \exists \varepsilon > 0 \quad \text{such that} \qquad \sup_{t_0 \in \mathbb{R},\ \|x_0 - y\| < \varepsilon} T(t_0, x_0) < +\infty;$$

- 0-FTS if it is uniformly FTS and the settling-time functional is continuous at zero.

The finite-time stability, by definition, is a combination of two properties Lyapunov stability and finite-time reaching. The theorem below uses two nonnegative functionals in order to characterize this type of stability.

Theorem 3.4 *The system* (3.1) *is locally (globally) uniformly finite-time stable if and only if there exist a locally (globally) proper positive definite functional $V : \mathbb{R} \times \mathbb{B} \mapsto \mathbb{R}_+$ and a locally bounded functional $T : \mathbb{B} \mapsto \mathbb{R}_+$ such that the inequalities*

$$D_{\mathbb{K}} V(t, x_{t_0,x_0}(t)) \le 0, \qquad t > t_0,$$

$$D_{\mathbb{K}} T(x_{t_0,x_0}(t)) \le -1, \qquad t > t_0, \tag{3.18}$$

hold for any $t_0 \in \mathbb{R}$ and any solution x_{t_0,x_0} of (3.1) *as long as*

$$x_{t_0,x_0}(t) \in \Omega \backslash \{\mathbf{0}\},$$

where Ω is a neighborhood of the origin (resp., $\Omega = \mathbb{B}$).

Proof *Sufficiency.* Theorem 3.1 implies that the origin of the system (3.1) is locally (globally) Lyapunov stable. This means that any solution $x_{t_0,x_0}(t)$ remains bounded for all $t > t_0$ and $x_{t_0,x_0}(t) \in \Omega$ provided that x_0 belongs to some neighborhood U of $\mathbf{0}$ (resp., $U = \mathbb{B}$ in the case of global Lyapunov stability). Since $x_{t_0,x_0}(t) \in \Omega$ for all $t \ge 0$ then $T(x_{t_0,x_0}(t)) \in \mathbb{R}_+$, $\forall t \ge 0$.

Let us consider the interval $[t_0, t_1 + \varepsilon]$, $\varepsilon > 0$, $t_1 = t_0 + T(x_0)$ and show that there exists an instant of time $t^* \in [t_0, t_1 + \varepsilon]$ such that $x_{t_0,x_0}(t^*) = \mathbf{0}$. Suppose the contrary, i.e., $x_{t_0,x_0}(t) \ne \mathbf{0}$ for $\forall t \in [t_0, t_1 + \varepsilon]$.

The inequality (3.18) and Lemma 4.2 of Volume I imply that the function $T(x_{t_0,x_0}(\cdot))$ is decreasing on $[t_0, t_1 + \varepsilon]$ and, consequently, differentiable almost everywhere on $[t_0, t_1 + \varepsilon]$. Then, using Lemma 4.8 of Volume I, we derive

$$T(x_{t_0,x_0}(t_1+\varepsilon)) - T(x_{t_0,x_0}(t_0)) \le \int_{t_0}^{t_1+\varepsilon} \dot{T}(x_{t_0,x_0}(\tau))d\tau \le -(t_1+\varepsilon-t_0) = -T(x_0) - \varepsilon,$$

i.e., $T(x_{t_0,x_0}(t_1+\varepsilon)) \leq T(x_{t_0,x_0}(t_0)) - T(x_0) \leq -\varepsilon$. This contradicts to the inequality $T(x) \geq 0$, which holds for all $x \in U \subset \Omega$. Therefore, there exists $t^* \in [t_0, t_1+\varepsilon]$ such that $x_{t_0,x_0}(t^*) = \mathbf{0}$. Since $\varepsilon > 0$ can be selected arbitrary small then $t^* \in [t_0, t_1]$, i.e., $t^* \leq t_0 + T(x_0)$.

Necessity. Let the system (3.1) be locally (globally) uniformly finite-time stable with a finite-time attraction domain $U \subset \mathbb{B}$. Proposition 3.1 implies that $x_{t_0,\mathbf{0}} \equiv \mathbf{0}$ is the unique solution of our system with the zero initial condition. The existence of a functional V is proven by Theorem 3.1.

Let S_{t_0,x_0} denote a set of solutions of the system (3.1) with the initial condition $x(t_0) = x_0$. Let us introduce the settling-time functional T as follows:

$$T(x_0) = \sup_{t_0\in\mathbb{R}} \sup_{x\in S_{t_0,x_0}} \inf_{\tau>0\,:\,x_{t_0,x_0}(t_0+\tau)=\mathbf{0}} \tau. \tag{3.19}$$

Since the origin is uniformly finite-time stable then there exists a ball $\Omega := B(\delta)$, $\delta > 0$ such that $T : \Omega \mapsto [0, +\infty)$ is locally bounded on Ω. Notice that for any $h \in \mathbb{R}$ one holds

$$\sup_{t_0\in\mathbb{R}} \sup_{x\in S_{t_0,x_0}} \inf_{\tau>0:x(t_0+\tau)=\mathbf{0}} \tau$$

$$= \sup_{t_0\in\mathbb{R}} \sup_{\tilde{x}\in S_{t_0+h,x_0}} \inf_{\tau>0:\tilde{x}(t_0+h+\tau)=\mathbf{0}} \tau.$$

Let $x_0 \in U\backslash\{\mathbf{0}\}$, let $h > 0$ be a real number, and let $\tilde{x} \in S_{t_0,x(t_0)}$ be an arbitrary solution of the system (3.1) such that $\tilde{x}(p) \in \Omega\backslash\{\mathbf{0}\}$ for all $p \in [t_0, t_0+h]$. In this case, $\inf_{\tau>0\,:\,\tilde{x}(t_0+\tau)=\mathbf{0}} \tau \geq h$ and

$$\begin{aligned} T(\tilde{x}(t_0)) - h &= \sup_{t_0\in\mathbb{R}, x\in S_{t_0,\tilde{x}(t_0)}} \inf_{\tau>0\,:\,x(t_0+\tau)=\mathbf{0}} \tau - h \\ &= \sup_{t_0\in\mathbb{R}, x\in S_{t_0,\tilde{x}(t_0)}} \inf_{\tilde{\tau}>-h\,:\,x(t_0+h+\tilde{\tau})=\mathbf{0}} \tilde{\tau} \\ &= \sup_{t_0\in\mathbb{R}, x\in S_{t_0,\tilde{x}(t_0)}} \inf_{\tilde{\tau}>0\,:\,x(t_0+h+\tilde{\tau})=\mathbf{0}} \tilde{\tau}. \end{aligned}$$

Let us denote

$$S^h_{t_0,x_0} = \left\{ z_{[t_0+h,+\infty)} \in C([t_0+h,+\infty); \mathbb{B}) : z \in S_{t_0,x_0} \right\},$$

where $h > 0$ and $z_{[t_0+h,+\infty)}$ denotes the restriction of a function $z : [t_0, +\infty) \mapsto \mathbb{B}$ to the time interval $[t_0+h, +\infty)$. Since $S_{t_0+h,\tilde{x}(t_0+h)} \subset S^h_{t_0,\tilde{x}(t_0)}$ then

$$T(\tilde{x}(t_0+h)) = \sup_{t_0\in\mathbb{R}, x\in S_{t_0+h,\tilde{x}(t_0+h)}} \inf_{\tilde{\tau}>0\,:\,x(t_0+h+\tilde{\tau})=\mathbf{0}} \tilde{\tau}$$
$$\leq \sup_{t_0\in\mathbb{R}, x\in S^h_{t_0,\tilde{x}(t_0)}} \inf_{\tilde{\tau}>0: x(t_0+h+\tilde{\tau})=\mathbf{0}} \tilde{\tau}$$
$$= \sup_{t_0\in\mathbb{R}, x\in S_{t_0,\tilde{x}(t_0)}} \inf_{\tilde{\tau}>0: x(t_0+h+\tilde{\tau})=\mathbf{0}} \tilde{\tau}.$$

Hence, we obtain $h^{-1}(T(\tilde{x}(t_0+h)) - T(\tilde{x}(t_0))) \leq -1$ for small enough $h > 0$. Similarly, one can be shown $h^{-1}(T(\tilde{x}(t_0)) - T(\tilde{x}(t_0-h))) \leq -1$ for small enough $h > 0$. Therefore, $D_{\mathbb{K}}(T(x(t))) \leq -1$ as long as $x(t) \in \Omega\backslash\{\mathbf{0}\}$. The proof is complete. □

The sufficient condition of the uniform finite-time stability is given by the following corollary.

Corollary 3.5 *Let Assumption 3.2 hold. If there exist positive numbers $\rho > 0$, $\alpha \in [0, 1)$, and a locally (globally) proper locally Lipschitz continuous on $\mathbb{B}\backslash\{\mathbf{0}\}$ positive definite functional $V : \mathbb{B} \mapsto \mathbb{R}_+$ such that*

$$\sup_{y\in F(t,x)} D^+V(x; y) \leq -\rho V^{\alpha}(x), \quad \forall x \in \Omega \cap \mathcal{D}(F)\backslash\{\mathbf{0}\}, \quad \forall t \in \mathbb{R}, \tag{3.20}$$

then the system (3.1) *is locally (globally) 0-FTS with some finite-time attraction domain $U \subset \Omega$, and the settling time $T(x_0)$ admits the estimate*

$$T(x_0) \leq \frac{V^{1-\alpha}(x_0)}{\rho(1-\alpha)}, \quad \forall x_0 \in U, \tag{3.21}$$

where Ω and $U \subset \Omega$ are some neighborhoods of the origin (resp., $U = \Omega = \mathbb{B}$) and $D^+V(x; y)$ is the right-hand upper Gateaux derivative of V at the point $x \in \mathcal{D}(F)$ in the direction $y \in \mathbb{B}$.

Proof We provide a proof only for the case $\alpha = 0$, since the case $\alpha \in (0, 1)$ can be reduced to the case $\alpha = 0$ by using the Lyapunov function $\tilde{V} = \frac{1}{(1-\alpha)}V^{1-\alpha}$ satisfying $\sup_{y\in F(t,x)} D\tilde{V}(x)y \leq -\rho$. The Lyapunov stability of the system (3.1) follows from Corollary 3.1. Let x_{t_0,x_i} be a locally Lipschitz continuous strong solution of (3.1). In this case, $x_{t_0x_i}(t) \in \mathcal{D}(F)$ for almost all $t > t_0$ and, using chain rule for Gateaux derivative together with the inequality (3.4), we derive (see the proof of Corollary 3.2)

$$D_{\mathbb{K}}\tilde{V}(x_{t_0,x_i}(t)) \overset{a.e.}{\leq} -\rho.$$

Since V and x_{t_0,x_0} are locally Lipschitz continuous then the function $t \to V(x_{t_0,x_i}(t))$ is locally Lipschitz continuous and monotone decreasing for $t > t_0$ as long as $x_{t_0,x_i}(t) \in \Omega\backslash\{\mathbf{0}\}$. Repeating the proof of Theorem 3.4, we derive finite-time convergence for strong solutions: $x_{t_0,x_i}(t) = \mathbf{0}$ for $t \geq \frac{\tilde{V}(x_i)}{\rho}$.

We complete the proof taking into account that any mild solution is a uniform limit of strong solutions (see Assumption 3.2). Since V is locally (globally) proper and the settling-time functional admits the estimate (3.21) then T is continuous at zero. □

Example 3.4 Let us consider the evolution equation

$$\dot{x} = Ax + f(x), \quad t > 0, \quad x(0) = x_0 \in \mathbb{H}, \tag{3.22}$$

where

$$f(x) = -\frac{x}{\|x\|}, \quad x \in \mathbb{H}, \quad \|x\| = \sqrt{\langle x, x\rangle},$$

and the operator $A : \mathcal{D}(A) \subset \mathbb{H} \mapsto \mathbb{H}$ generates a strongly continuous semigroup of linear bounded operators on a real Hilbert space $\mathbb{H}$. Assume also that A is a *dissipative operator*, i.e.,

$$\langle Ax, x\rangle \le 0, \quad x \in \mathcal{D}(A).$$

The considered evolution system has classical solutions for all $x_0 \in \mathcal{D}(A)\backslash\{\mathbf{0}\}$ at least locally (see Example 2.15 for more details about the proof). Due to continuous dependence of solutions on initial conditions, any mild solution of the system with $x_0 \in \mathbb{H}\backslash\{\mathbf{0}\}$ is a uniform limit of classical solutions on compact intervals of time. Since f is discontinuous at the origin, then following the regularization scheme (2.52) we derive the evolution inclusion (3.1) with

$$F(x) = \bigcap_{\varepsilon>0} \overline{\text{co}} f(t, x \dot{+} B(\varepsilon)\backslash\{\mathbf{0}\}) = \begin{cases} \frac{x}{\|x\|}, & \text{if } x \neq \mathbf{0}, \\ B(1), & \text{if } x = \mathbf{0}. \end{cases}$$

The obtained evolution inclusion has mild solutions on whole $\mathbb{H}$ (see Example 2.18). Therefore, it satisfies all conditions of Corollary 3.5. The functional $V : \mathbb{H} \mapsto \mathbb{R}_+$ given by

$$V(x) = \|x\|$$

is Frechét differentiable on $\mathbb{H}\backslash\{\mathbf{0}\}$ and for any $x \in D(A)\backslash\{\mathbf{0}\}$ we have

$$\dot{V}(x) = \sup_{y \in F(x)} DV(x)(Ax + y) = DV(x)\left(Ax - \frac{x}{\|x\|}\right) =$$

$$\frac{1}{\|x\|}\left\langle Ax - \frac{x}{\|x\|}, x\right\rangle = \frac{\langle Ax, x\rangle}{\|x\|} - 1 \le -1, \quad \forall x \neq \mathbf{0}.$$

The origin of the considered evolution system is globally finite-time stable and the settling-time functional T admits the estimate $T(x_0) \le \|x_0\|$ for any $x_0 \in \mathbb{H}$.

If the settling-time functional T is continuous at the origin then the finite-time stability can be characterized using the Lyapunov function V only.

Corollary 3.6 *The autonomous system* (3.1) *is locally (globally) 0-FTS if and only if there exists a locally (globally) proper positive definite functional* $V : \mathbb{B} \mapsto \mathbb{R}_+$ *such that the inequality*

$$D_{\mathbb{K}} V(x_{t_0,x_0}(t)) \leq -1, \qquad \forall t > t_0 \tag{3.23}$$

holds for any $t_0 \in \mathbb{R}$ *and any solution* x_{t_0,x_0} *of* (3.1) *as long as*

$$x_{t_0,x_0}(t) \in \Omega \backslash \{\mathbf{0}\},$$

where Ω *is a neighborhood of the origin (resp.,* $\Omega = \mathbb{B}$*). Moreover, the settling-time functional of the finite-time stable system* (3.1) *admits the estimate*

$$T(x_0) \leq V(x_0) \quad \textit{for all} \quad x_0 \in U,$$

where $U \subset \Omega$ *is a neighborhood of the origin (resp.* $U = \Omega = \mathbb{B}$*).*

Proof *Sufficiency* can be proved similarly to Theorem 3.4 using V instead of T in all considerations. *Necessity.* Repeating the proof of Theorem 3.4 we derive two functionals V (Lyapunov function) and T (the settling-time functional) such that V is positive definite locally proper, T is nonnegative definite locally bounded, and $D_{\mathbb{K}} V(x_{t_0,x_0}(t)) \leq 0$, $D_{\mathbb{K}} T(x_{t_0,x_0}(t)) \leq -1$. Notice that V is independent of t in the autonomous case. Since, by assumption, T is continuous at $\mathbf{0}$, then the function $V^{new} := V + T$ is locally (globally) proper and $D_{\mathbb{K}} V^{new}(x_{t_0,x_0}(t)) \leq -1$. The proof is complete. □

In the finite-dimensional case (see Volume I), the settling-time functional T is a Lyapunov function of the finite-time stable system provided that T is continuous at zero. In the infinite-dimensional case, a setting-time function T may be continuous in $\mathbb{B}$, vanishing at zero, but not proper at $\mathbf{0}$, so we cannot take $V = T$ in the proof of Corollary 3.6. This specific feature of finite-time stable infinite-dimensional systems is illustrated by the following example.

Example 3.5 Let $\mathbb{H} = L^2((-1, 1); \mathbb{R})$ and the operator A in the system (3.22) be defined as follows:

$$Az = \frac{\partial^2}{\partial z^2}, \quad z \in \mathcal{D}(A)$$

with the domain

$$D(A) = H^2((-1, 1); \mathbb{R}) \cap H_0^1((-1, 1); \mathbb{R}).$$

It is easy to see that $\phi_i = \sin(\pi i z) \in D(A), z \in [-1, 1]$ is an eigenvector of the operator A:

$$A\phi_i = -\lambda_i x_i, \quad \lambda_i = \pi^2 i^2, \quad i = 1, 2...$$

Being an orthonormal basis in the separable Hilbert space $\mathbb{H}$, the eigenvectors ϕ_i allow us to represent any $y \in \mathcal{D}(A)$ in the form

$$y = \sum_{i=1}^{\infty} x_i \phi(x). \tag{3.24}$$

Taking into account $\langle \phi_i, \phi_j \rangle = 0$ for $i \neq j$ and $\langle \phi_i, \phi_i \rangle = 1$ we derive

$$\langle \phi_i, Ay \rangle = -\lambda_i x_i, \quad \langle \phi_i, f(y) \rangle = -\frac{x_i}{\sqrt{\langle y, y \rangle}} = -\frac{x_i}{\sqrt{\sum_{i=1}^{+\infty} x_i^2}}.$$

Therefore, any classical solution of (3.22) admits the representation (3.24), where the time-varying functions x_i satisfy the following (infinite) system of ODEs:

$$\dot{x}_i = -\left(\lambda_i + \frac{1}{\sqrt{\sum_{i=1}^{+\infty} x_i^2}}\right) x_i, \quad i = 1, 2, \ldots.$$

Since the operator A is dissipative then following Example 3.4 it can be shown that the considered system is finite-time stable, the settling-time functional T is continuous at the origin, and $T(\mathbf{0}) = 0$.

Let us consider the sequence of initial conditions $x(0) = y_0^j = \phi_j \in \mathcal{D}(A)$, $j = 1, 2, \ldots$. Obviously, $\|y_0^j\| = 1$ for all $j \geq 1$, the corresponding solution y^j has the form $y_j(t) = x_j(t)\phi_j$, and $\|y_j(t)\| = x_j(t)$, where

$$\dot{x}_j = -\left(\lambda_j + \frac{1}{|x_j|}\right) x_j, \quad x_j(0) = 1.$$

Simple computations show

$$T(y_0^j) = \frac{\ln(1 + \lambda_j)}{\lambda_j} \to 0 \quad \text{as} \quad j \to +\infty,$$

while $\|y_0^j\| = 1$ for all $j \geq 1$. This means that $\inf_{\|x_0\|=1} T(x_0) = 0$, i.e., T is not a proper function, and it cannot be considered as a Lyapunov function of the system.

3.2.2 *Fixed-Time Stability*

Nearly fixed-time stability is a property, in some sense, adjoint to finite-time stability. A system is nearly fixed-time stable if it "escapes from infinity" in a finite time.

Definition 3.9 (*Nearly fixed-time stability*) The system (3.1) is said to be locally (globally) uniformly nearly fixed-time stable (nFxTS) if it is locally (globally) uniformly Lyapunov stable in the domain $U_{t_0} \subset \mathbb{B}$ (resp., $U_{t_0} = \mathbb{B}$) and nearly fixed-time attractive:

$$\forall r > 0, \forall t_0 \in \mathbb{R}, \exists T_{t_0,r} > 0 \ : \ \|x_{t_0,x_0}(t)\| \leq r, \ \ \forall t \geq t_0 + T_{t_0,r}, \ \forall x_0 \in U_{t_0}.$$

The nearly fixed-time stability of finite-dimensional system is characterized by means of the so-called escape-time function, which defines the blow up time of the system in the backward time. The infinite-dimensional system may not have solutions in the backward time or solutions may blow up infinitely fast. The Lyapunov theorem given below provides just sufficient conditions of nearly fixed-time stability.

Theorem 3.5 *The system* (3.1) *is locally (globally) uniformly nFxTS if here exist positive numbers* $q > 0$, $\beta > 1$ *and a locally (globally) proper continuous function* $V : \mathbb{B} \mapsto \mathbb{R}_+$ *such that the inequality*

$$D_{\mathbb{K}} V(x_{t_0,x_0}(t)) \leq -q V^{\beta}(x_{t_0,x_0}(t)), \qquad \forall t > t_0 = 0, \tag{3.25}$$

holds for any solution x_{x_0} *of* (3.1) *as long as*

$$x_{x_0}(t) \in \Omega \backslash \{\mathbf{0}\},$$

where $\Omega \subset \mathbb{B}$ *is a neighborhood of the origin (resp.,* $\Omega = \mathbb{B}$*).*

Proof We provide the proof only for the case $\beta = 2$, since for $\beta > 1$ we may consider the Lyapunov function $\tilde{V} = \frac{1}{\beta-1} V^{\beta-1}$ in order to obtain

$$D_{\mathbb{K}} \tilde{V}(x_{x_0}(t)) \leq -q \tilde{V}^2(x_0(t)).$$

By Theorem 3.1, the system is Lyapunov stable, so, by Proposition 3.1, it has a unique zero solution.

Let us consider the function $\tau = \frac{1}{r} - \frac{1}{V}$, where $r \in (0, +\infty)$ is such that

$$\{x \in \mathbb{B} : V(x) \leq r\} \subset \Omega$$

(resp., $r = +\infty$ and $\tau = -\frac{1}{V}$ in the case of global stability analysis). Due to local (global) properness of V, the function $t \mapsto \tilde{\tau}(x_{t_0,x_0}(t))$ is well defined as long as $x_{t_0,x_0}(t) \neq \mathbf{0}$. By Lemma 4.3 of Volume I, we derive $D_{\mathbb{K}}\tau(x_{t_0,x_0}(t)) \leq 0$ and

$$\dot{\tau}(x_{t_0,x_0}(t)) \overset{a.e.}{=} \frac{\dot{V}(x_{t_0,x_0}(t))}{V^2(x_{t_0,x_0}(t))} \leq -q. \tag{3.26}$$

Hence, by Lemma 4.2 of Volume I, the function $t \mapsto \tau(x_{t_0,x_0}(t))$ is deceasing and differentiable almost everywhere. By Lemma 4.8 of Volume I, we have

$$\tau(x_{t_0,x_0}(t)) - \tau(x_0) \leq \int_{t_0}^{t} \dot{V}(x_{t_0,x_0}(\tau))d\tau \leq \int_{t_0}^{t} (-q)d\tau = -q(t-t_0).$$

This implies the following inequality:

$$V(x_{t_0,x_0}(t)) \leq \frac{1}{q(t-t_0) + 1/V(x_0)}, \quad t > t_0$$

as long as $x_{t_0,x_0}(t) \neq \mathbf{0}$. Taking into account the local/global properness of V we complete the proof. □

Corollary 3.7 *Let Assumption 3.2 hold. If there exist positive numbers $q > 0$, $\beta > 1$ and a locally (globally) proper locally Lipschitz continuous positive definite functional $V : \mathbb{B} \mapsto \mathbb{R}$ such that V is Frechét differentiable on $\mathcal{D}(F) \cap \Omega\backslash\{\mathbf{0}\}$ and*

$$\sup_{y \in F(t,x)} D^+V(x; y) \leq -qV^{\beta}(x), \quad \forall x \in D(F) \cap \Omega\backslash\{\mathbf{0}\}, \ \forall t \in \mathbb{R}, \qquad (3.27)$$

then the system (3.1) *is locally(globally) uniformly nFxTS, where $\Omega \subset \mathbb{B}$ is a neighborhood of the origin (resp., $\Omega = \mathbb{B}$) and $D^+V(x; y)$ is the right-hand upper Gateaux derivative of V at the point $x \in \mathcal{D}(F)$ in the direction $y \in \mathbb{B}$.*

Proof The proof repeats considerations of the proof of Theorem 3.5 for locally Lipschitz continuous strong solutions of (3.1). Next, the conclusion is extended to mild solutions using Assumption 3.2. □

The fixed-time stability assumes that the system is simultaneously nearly fixed-time stable and finite-time stable.

Definition 3.10 (*Fixed-time stability*) The system (3.1) is said to be

- locally (globally) uniformly fixed-time stable (FxTS) if it is locally (globally) uniformly finite-time stable in a domain $U \subset \mathbb{B}$ (resp., $U = \mathbb{B}$) and the settling-time functional is **bounded on** $\mathbb{R} \times U$, i.e.,

$$\exists T_{\max} > 0 : x_{t_0,x_0}(t) = 0, \quad \forall t > t_0 + T_{\max}, \quad \forall t_0 \in \mathbb{R}, \quad \forall x_0 \in U;$$

- 0-FxTS if its uniformly FxTS and the setting-time function is continuous at $\mathbf{0}$.

Exponential stability is the usual stability property of the linear infinite-dimensional systems, due to exponential estimates of the corresponding semigroups (see Chap. 2 or [8]). However, fixed-time (faster than any exponential) convergence can also be discovered for *linear PDEs*.

Example 3.6 Let us consider the wave equation

$$u_{tt} = u_{xx}, \quad t > 0, \quad 0 < x < 1, \quad u : \mathbb{R}_+ \times [0, 1] \to \mathbb{R},$$

with the so-called "transparent" boundary condition (see, e.g., [13])

$$u_x(t, 0) = u_t(t, 0), \qquad u_x(t, 1) = -u_t(t, 1), \qquad t \geq 0,$$

and the initial conditions

$$u(0, x) = \phi(x), \quad u_t(0, x) = \psi(x), \quad 0 \leq x \leq 1$$

from

$$\mathbb{Y} = \left\{(\phi, \psi) \in H^1((0, 1); \mathbb{R}) \times L^2((0, 1); \mathbb{R}) : \phi(0) + \phi(1) + \int_0^1 \psi(s)ds = 0\right\},$$

where L^2 and H^1 are Lebesgue and Sobolev spaces, respectively. The boundary conditions are transparent in the sense that any wave $u(t, x) = f_1(x - t)$ traveling to the right leaves the domain at $x = 1$ and does not generate any reflected wave. Any wave $u(t, x) = f_2(x + t)$ traveling to the left leaves the domain at $x = 0$ similarly. Since any solution of the wave system is given by

$$u(t, x) = f_1(x - t) + f_2(x + t)$$

then it vanishes after the time instant $t = 1$ independently of the initial condition.

Being a version of finite-time stability, the fixed-time stability can be characterized in a similar way.

Theorem 3.6 *The system* (3.1) *is locally (globally) uniformly fixed-time stable if and only if there exist a locally (globally) proper positive definite functional* $V : \mathbb{R} \times \mathbb{B} \mapsto \mathbb{R}$ *and a bounded functional* $T : \Omega \mapsto \mathbb{R}_+$ *such that* $T(\mathbf{0}) = 0$, $T(x) \leq T_{\max} < +\infty$ *for all* $t \in \mathbb{R}$, $x \in \Omega$ *and the inequalities*

$$D_{\mathbb{K}} V(t, x_{t_0, x_0}(t)) \leq 0, \qquad \forall t > t_0,$$

$$D_{\mathbb{K}} T(x_{t_0, x_0}(t)) \leq -1, \qquad \forall t > t_0,$$

hold for any $t_0 \in \mathbb{R}$ *and any solution* x_{t_0, x_0} *of* (3.1) *as long as*

$$x_{t_0, x_0}(t) \in \Omega \backslash \{\mathbf{0}\},$$

where Ω *is a neighborhood of the origin (resp.* $\Omega = \mathbb{B}$*).*

Proof The proof literally repeats the proof of Theorem 3.4 using the property of uniform boundedness of T (i.e., $\exists T_{\max} > 0$ such that $T(t, x) \leq T_{\max}$ for all $t \in \mathbb{R}$ and all $x \in \Omega$). □

The 0-FxTS can be analyzed using only the Lyapunov function V provided that a settling-time functional T is continuous at the origin.

Corollary 3.8 *The system* (3.1) *is locally (globally) 0-FxTS if there exist a locally (globally) proper functional* $V : \mathbb{B} \mapsto \mathbb{R}$ *and a number* $q > 0$ *such that the inequality*

$$D V_{\mathbb{K}}(x_{t_0,x_0}(t)) \leq -q(1 + V^2(x_{t_0,x_0}(t))), \qquad t > t_0, \tag{3.28}$$

hold for any $t_0 \in \mathbb{R}$ *and any solution* x_{t_0,x_0} *of* (3.1) *as long as*

$$x_{t_0,x_0}(t) \in \Omega \backslash \{\mathbf{0}\},$$

where $\Omega \subset \mathbb{B}$ *is a neighborhood of the origin. The settling-time functional admits the estimate*

$$T(x_0) \leq \frac{\pi}{2q}$$

for all $x_0 \in \mathbb{B}$ *(resp.,* $\Omega = \mathbb{B}$*).*

Proof Theorem 3.1 implies that the origin of the system (3.1) is locally (globally) Lyapunov stable. In particular this means that any solution initiated in U exists for all $t > 0$, where U is a neighborhood of the origin (resp. $U = \mathbb{B}$). Notice that $V(x_{t,x_0})$ is a monotone function and it is differentiable almost everywhere.

Let us prove the settling-time functional is bounded by $T_{\max} = 0.5\pi/q$. Suppose that there exists a solution x_{t_0,x_0} which exists on $[0, +\infty)$ and does not vanish after the time $T_{\max}$.

Let us consider the functional $T : \mathbb{B} \mapsto [0, +\infty)$ defined as follows: $T(x) = \frac{\arctan(V(x))}{q} \leq T_{\max}$. Obviously, T is locally proper. Since arctan is differentiable then, using Lemma 4.2 of Volume I, we derive

$$D_{\mathbb{K}} T(x_{t_0,x_0}(t)) \leq 0$$

for all $t \geq [t_0, t_0 + T_{\max}]$ and

$$\dot{T}(x_{t_0,x_0}(t)) \overset{a.e.}{\leq} \frac{\dot{V}(x_{t_0,x_0}(t))}{q(1+V^2(x_{t_0,x_0}(t))} \leq -1, \quad t \in [t_0, t_0 + T_{\max}].$$

Hence, using Lemma 4.8 of Volume I, we derive

$$T(x_{t_0,x_0}(t_0 + T_{\max}) - T(x_{t_0,x_0}(t_0) \leq \int_{t_0}^{t_0+T_{\max}} \dot{T}(x_{t_0,x_0}(\tau))d\tau \leq -T_{\max},$$

i.e., $T(x_{t_0,x_0}(t_0 + T_{\max})) = 0$. Proposition 3.1 implies that $x_{t_0,x_0}(t) = 0$ for all $t \geq t_0 + T_{\max}$. The proof is complete. □

Notice that the same abstract differential equation considered on different spaces may have different stability properties and convergence rates. For example, the system may be nonuniformly asymptotically stable on whole space and be fixed-time stable on its subspace being a Banach space as well.

Example 3.7 Let us consider again the transport equation (3.10) which generates the translation semigroup (3.11) on $L^2(\mathbb{R}_+; \mathbb{R})$. In Example 3.3 it is shown that this equation is globally asymptotically stable on $L^2(\mathbb{R}_+; \mathbb{R})$, but nonuniformly.

Let us consider the same equation on the subspace space of $L^2(\mathbb{R}_+; \mathbb{R})$ given by

$$\mathbb{B} = \left\{x \in L^2(\mathbb{R}_+; \mathbb{R}) : \text{supp}(x) \subset [0, 1]\right\},$$

where supp(x) denotes the support of the function $x \in L^2(\mathbb{R}_+; \mathbb{R})$. The space $\mathbb{B}$ is a Banach space of compactly supported functions from $L^2(\mathbb{R}_+; \mathbb{R})$. The transport equation (3.10) is globally fixed-time stable on $\mathbb{B}$ since for any $x_0 \in \mathbb{B}$ the solution of the transport equation satisfies

$$[x_{x_0}(t)](z) = x_0(t + z) = 0, \quad \forall t \geq 1, \forall z \geq 0.$$

A sufficient condition of FxTS is given by the following corollary.

Corollary 3.9 *Let Assumption 3.2 hold. If there exists a locally (globally) proper locally Lipschitz continuous positive definite functional $V : \mathbb{B} \mapsto \mathbb{R}$ such that V is Frechét differentiable on $\mathcal{D}(F) \cap \Omega\backslash\{\mathbf{0}\}$ and*

$$\sup_{y \in F(t,x)} DV(x)y \leq -q(1 + V^2(x)), \quad \forall x \in D(F) \cap \Omega\backslash\{\mathbf{0}\}, \ \forall t \in \mathbb{R}, \tag{3.29}$$

where $\Omega \subset \mathbb{B}$ is a neighborhood of the origin (resp., $\Omega = \mathbb{B}$), then the system (3.1) is locally (globally) 0-FxTS with an attraction domain $U \subset \Omega$ (resp., $U = \mathbb{B}$) and the settling time $T(x_0)$ admits the estimate

$$T(x_0) \leq \frac{\pi}{2q}$$

for all $x_0 \in \mathbb{B}$.

Proof The proof repeats considerations of the proof of Corollary 3.8 for locally Lipschitz continuous strong solutions of (3.1). Next, the conclusion is extended to mild solutions using Assumption 3.2. □

Example 3.8 Let us consider the evolution equation

$$\dot{x} = Ax - \tfrac{x}{\|x\|} - \|x\|x, \quad x \in \mathbb{H}$$

in a Hilbert space $\mathbb{H}$, where A is assumed to be a dissipative operator, i.e.,

$$\langle Ax, x\rangle \leq 0, \quad \forall x \in \mathcal{D}(A).$$

Obviously, selecting

$$V(x) = \|x\| = \sqrt{\langle x, x\rangle}, \quad x \in \mathbb{H}$$

we derive

$$\dot{V}(x) = \tfrac{\langle \dot{x}, x\rangle}{\|x\|} = \langle Ax, x\rangle - 1 - V^2, \quad \forall x \in \mathcal{D}(A)\backslash\{\mathbf{0}\}.$$

This means that the considered system is globally uniformly fixed-time stable and the settling-time functional admits the estimate

$$T(x_0) \leq \tfrac{\pi}{2}.$$

Concluding Remarks

The Lyapunov analysis of infinite-dimensional systems is similar to the finite-dimensional case. However, there are some specific features. For example, an asymptotic stability of an autonomous evolution equation does not imply its uniform asymptotic stability in the general case. However, even linear nonuniformly asymptotically stable autonomous evolution equations may be fixed-time stable on a subspace of the state space. Since the state of an infinite-dimensional system may converge to zero arbitrary fast dependently on an initial condition on a unit sphere, then the settling-time functional may have the zero infimum on any sphere. This means that, in the general case, the setting-time functional (being positive definite on whole space) is not even locally proper. Therefore, some converse Lyapunov theorems about finite-time stability, which, in the finite-dimensional case, are proven using the settling-time functional as a Lyapunov function, cannot be extended for the infinite-dimensional case. Being adjoint to the settling-time functional, the escape-time function of the nearly fixed-time stable system may be unbounded on any sphere due to possibility of an arbitrary fast blowup of the backward solutions (if they exist) of an infinite-dimensional system. So, the escape-time function cannot be introduced and utilized for the Lyapunov function design as well. This is the main reason why a converse Lyapunov theorem for nearly fixed-time stability is not presented in this chapter. However, converse Lyapunov theorems for finite-time and fixed-time stability formulated in terms of two Lyapunov-like functions are still valid in the general infinite-dimensional setting. Notice that all mentioned difficulties can be overcame in the case of homogeneous systems allowing the Zubov–Rosier theorem to be expanded to abstract differential equations (see Chap. 6).

Some evolution equations with unbounded nonlinear operators may have mild solutions only in a subspace $\mathbb{Y}$ of the state space $\mathbb{B}$. In this case, the stability analysis should be done in $\mathbb{Y}$.

Definition 3.11 (*Stability in a subspace*) Let $\mathbb{B}$ be a Banach space and let $\mathbb{Y} \subset \mathbb{B}$ be a Banach space with a norm $\|\cdot\|_{\mathbb{Y}}$. The system (3.1) defined in $\mathbb{B}$ is said to be

- *globally uniformly Lyapunov stable in* $\mathbb{Y}$ if there exists $\varepsilon \in \mathcal{K}^{\infty}$ such that

$$\|x_{t_0,x_0}(t)\|_{\mathbb{Y}} \leq \varepsilon(\|x_0\|_{\mathbb{Y}}), \quad \forall t \geq t_0 \tag{3.30}$$

  for any mild solution $x_{t_0,x_0} : [t_0, +\infty) \mapsto \mathbb{Y}$ of (6.3) and any $t_0 \in \mathbb{R}$;
- *globally uniformly asymptotically stable in* $\mathbb{Y}$ if it is globally uniformly Lyapunov stable in $\mathbb{Y}$ and $\forall \varepsilon > 0$ and $\forall R > \varepsilon$ there exists $\hat{T} = \hat{T}(R, \varepsilon) > 0$ such that $\|x_0\|_{\mathbb{Y}} \leq R$ implies $\|x_{t,x_0}(t)\|_{\mathbb{Y}} \leq \varepsilon$ for all $t > t_0 + \hat{T}$ and for all $t_0 \in \mathbb{R}$;
- globally uniformly exponentially stable in $\mathbb{Y}$ if there exist $C \geq 1$ and $r \geq 1$ such that

$$\|x_{t_0,x_0}\|_{\mathbb{Y}} \leq C\|x_0\|_{\mathbb{Y}} e^{-r(t-t_0)}, \quad \forall t \geq t_0, \forall t_0 \in \mathbb{R};$$

- *globally uniformly finite-time stable (FTS) in* $\mathbb{Y}$ if it is globally uniformly Lyapunov stable in $\mathbb{Y}$ and there exists a a locally bounded on $\mathbb{Y}$ settling-time function $T : \mathbb{Y} \mapsto \mathbb{R}_+$, i.e.,

$$\forall x_0 \in \mathbb{Y}, \quad x_{t_0,x_0}(t) = \mathbf{0}, \quad \forall t \geq t_0 + T(x_0);$$

- 0-FTS in $\mathbb{Y}$ if it is uniformly FTS in $\mathbb{Y}$ and the settling-time functional is continuous at zero;
- globally uniformly nearly fixed-time stable in $\mathbb{Y}$ if it is globally uniformly Lyapunov stable in $\mathbb{Y}$ and uniformly nearly fixed-time attractive in $\mathbb{Y}$:

$$\forall r > 0, \exists T_r > 0 \; : \; \|x_{t_0,x_0}(t)\|_{\mathbb{Y}} \leq r, \quad \forall t \geq t_0 + T_r, \; \forall x_0 \in \mathbb{Y}, \; \forall t_0 \in \mathbb{R};$$

- *globally uniformly fixed-time stable in* $\mathbb{Y}$ if it is finite-time stable in $\mathbb{Y}$ and there exists $T_{\max} > 0$ such that

$$T(x_0) \leq T_{\max}, \quad \forall x_0 \in \mathbb{Y}.$$

If the above conditions are fulfilled only for $x_0 \in \Omega$ then the corresponding stability is local provided that $\Omega \subset \mathbb{Y}$ is a neighborhood of the origin.

Since $\mathbb{Y}$ is assumed to be a Banach space then all results obtained in this chapter remain valid for stability analysis in $\mathbb{Y}$. Examples 3.6 and 3.7 demonstrate systems which are fixed-time stable in a subspace of the state space.

References

1. Lyapunov AM (1992) The general problem of the stability of motion. Taylor & Francis
2. Zubov VI (1964) Methods of AM Lyapunov and their applications. Noordhoff, Leiden, (Translated from Zubov VI, Metody AM Lyapunova i ih primenenie, Izdatelstvo Leningradskogo universiteta, 1957) edition
3. Bacciotti A, Rosier L (2001) Lyapunov functions and stability in control theory. Springer, Berlin
4. Mironchenko A, Wirth F (2019) Non-coercive Lyapunov functions for infinite-dimensional systems. J Differ Equ 266(11):7038–7072
5. Yoshizawa T (1955) On the stability of solutions of a system of differential equations. Mem. Coll. Sci., Univ. Kyoto, Ser. A: Math. 29(1):27–33
6. Deimling K (1992) Multivariable differential equations. Walter de Gruyter, Berlin, New York
7. Massera JL (1949) On Lyapunovff's conditions of stability. Ann Math 50:705–721
8. Curtain RF, Zwart H (2012) An introduction to infinite-dimensional linear systems theory. Springer Science & Business
9. Polyakov A (2021) Input-to-state stability of homogeneous infinite dimensional systems with locally Lipschitz nonlinearities. Automatica 129:109615
10. Clarke F, Ledyaev YS, Stern R, Wolenski P (1995) Qualitative properties of trajectories of control systems: a survey. J Dyn Control Syst 1(1):1–48
11. Karafyllis I, Krstic M (2017) Predictor feedback for delay systems: implementations and approximations. Birkhauser
12. Polyakov A (2018) Fast control systems: nonlinear approach. In: Clempner JB, Yu W (eds) New perspectives and applications of modern control theory. Springer, Berlin, pp 287–316
13. Perrollaz V, Rosier L (2014) Finite-time stabilization of 2×2 hyperbolic systems on tree-shaped networks. SIAM J Control Optim 52(1):143–163

Chapter 4
Linear Dilations in Abstract Spaces

In this chapter, linear dilations in abstract (Banach and Hilbert) spaces are introduced as a group of linear bounded operators satisfying a *limit property*. Special attention is paid to strongly continuous linear dilations in Lebesgue and Sobolev spaces. Geometric objects induced by dilations are considered. The canonical homogeneous norm and the canonical homogeneous inner product are defined for generalized homogeneous Banach and Hilbert spaces, respectively.

4.1 Basic Definitions and Properties

Let $\mathcal{L}(\mathbb{B}; \mathbb{B})$ be the Banach space of linear bounded operators $\mathbb{B} \mapsto \mathbb{B}$ with the supremum norm:

$$\|A\| = \sup_{x \in \mathbb{B}\setminus\{\mathbf{0}\}} \frac{\|Ax\|}{\|x\|},$$

where $\mathbb{B}$ is a real Banach space with a norm $\|\cdot\|$ and $A \in \mathcal{L}(\mathbb{B}; \mathbb{B})$. Let $I \in \mathcal{L}(\mathbb{B}; \mathbb{B})$ denote the identity operator.

Definition 4.1 A mapping $\mathbf{d} : \mathbb{R} \mapsto \mathcal{L}(\mathbb{B}; \mathbb{B})$ is said to be a **linear dilation** in $\mathbb{B}$ if

- **(Group property)**

$$\mathbf{d}(0) = I \quad \text{and} \quad \mathbf{d}(t+s) = \mathbf{d}(t)\mathbf{d}(s) = \mathbf{d}(s)\mathbf{d}(t) \quad \text{for} \quad t, s \in \mathbb{R};$$

- **(Limit property)**

$$\limsup_{s\to-\infty} \|\mathbf{d}(s)x\| = 0 \quad \text{and} \quad \liminf_{s\to+\infty} \|\mathbf{d}(s)x\| = +\infty, \quad \forall x \in \mathbb{B}\setminus\{\mathbf{0}\}.$$

A. Polyakov, *Generalized Homogeneity in Systems and Control Volume II*,
Communications and Control Engineering,
https://doi.org/10.1007/978-3-031-77496-6_4

The limit property specifies groups being dilations in normed spaces [1, 2].

Definition 4.2 A linear dilation $\mathbf{d}$ in $\mathbb{B}$ is

- **strongly continuous** if $\mathbf{d}(\cdot)x : \mathbb{R} \mapsto \mathbb{B}$ is continuous for any $x \in \mathbb{B}$;
- **uniformly continuous** if $\mathbf{d} : \mathbb{R} \mapsto \mathcal{L}(\mathbb{B}; \mathbb{B})$ is continuous in $\mathcal{L}(\mathbb{B}; \mathbb{B})$.

These two types of continuity are indistinguishable for linear dilations in $\mathbb{B} = \mathbb{R}^n$, since any strongly continuous group of linear bounded operators in a finite-dimensional space is uniformly continuous (see Chap. 2). The strong continuity is important for an analysis of dilations in infinite-dimensional spaces, where the uniform continuity may be a too restrictive assumption (see Chap. 2).

Example 4.1 Let $\mathbb{B}$ be the Banach space of bounded uniformly continuous functions[1] $\mathbb{R} \mapsto \mathbb{R}$ with the supremum norm: $\|x\| = \sup_{z\in\mathbb{R}} |x(z)|$ for $x \in \mathbb{B}$. A linear dilation $\mathbf{d}$ in $\mathbb{B}$ can be defined as follows:

$$(\mathbf{d}(s)x)(z) = e^{\alpha s}x(z + \beta s),$$

where $s \in \mathbb{R}$ is the group parameter, $x \in \mathbb{B}$, $z \in \mathbb{R}$ and $\alpha > 0$ and $\beta \in \mathbb{R}$ are constant parameters. Let us show that $\mathbf{d}$ is a group. Indeed, $\mathbf{d}(s)x \in \mathbb{B}$ if $x \in \mathbb{B}$, $s \in \mathbb{R}$, and for $v = \mathbf{d}(s)x$ one has

$$(\mathbf{d}(t)\mathbf{d}(s)x)(z) = (\mathbf{d}(t)v)(z) = e^{\alpha t}v(z + t) = e^{\alpha t}e^{\alpha s}x(z + \beta s + \beta t) = (\mathbf{d}(s + t)x)(z).$$

The limit property also holds since

$$\|\mathbf{d}(s)x\| = \sup_{z\in\mathbb{R}} |e^{\alpha s}x(z + s)| = e^{\alpha s} \sup_{z\in\mathbb{R}} |x(z + \beta s)| = e^{\alpha s} \|x\|.$$

Finally, the strong continuity of the dilation follows from the estimate

$$\|\mathbf{d}(s)x - x\| = \sup_{z\in\mathbb{R}} |e^{\alpha s}x(z + \beta s) - x(z)| \leq$$

$$\sup_{z\in\mathbb{R}} |e^{\alpha s}x(z + \beta s) - x(z + \beta s)| + \sup_{z\in\mathbb{R}} |x(z + \beta s) - x(z)| \leq$$

$$(e^{\alpha s} - 1)\|x\| + \sup_{z\in\mathbb{R}} |x(z + \beta s) - x(z)|.$$

Since the function $x \in \mathbb{B}$ is uniformly continuous then there exists the so-called modulus of continuity $\omega \in \mathcal{K}^{\infty}$ such that $|x(y) - x(z)| \leq \omega(|y - z|)$ for all $y, z \in \mathbb{R}$, where $\omega(|y - z|) \to 0$ as $|y - z| \to 0$. Hence, the group $\mathbf{d}$ is strongly continuous.

Let us recall the following notation from Volume I:

[1] A function $f : X \subset \mathbb{X} \mapsto \mathbb{Y}$ is said to be uniformly continuous on X if there exists a function $\omega \in \mathcal{K}_{\infty}$ (*known as modulus of continuity*) such that $\|f(x_1) - f(x_2)\|_{\mathbb{Y}} \leq \omega(\|x_1 - x_2\|_{\mathbb{X}})$ for all $x_1, x_2 \in X$, where $\mathbb{X}$ and $\mathbb{Y}$ are normed vector spaces.

$$\|\mathbf{d}(s)\| = \sup_{x\in S} \|\mathbf{d}(s)x\| = \sup_{x\neq 0} \tfrac{\|\mathbf{d}(s)x\|}{\|x\|} \quad \text{and} \quad \lfloor \mathbf{d}(s)\rfloor = \inf_{x\in S} \|\mathbf{d}(s)x\| = \inf_{x\neq 0} \tfrac{\|\mathbf{d}(s)x\|}{\|x\|},$$

where $S = \{x \in \mathbb{B} : \|x\| = 1\}$ is the unit sphere. The above quantities characterize a "rate of dilation" in $\mathbb{B}$.

Proposition 4.1

- *If* $\mathbf{d}$ *is a strongly continuous linear dilation in* $\mathbb{B}$ *then*

  (1) $\lfloor \mathbf{d}(s)\rfloor < 1$, $\|\mathbf{d}(s)\| \le 1$ *for* $s < 0$ *and* $\lfloor \mathbf{d}(s)\rfloor \ge 1$, $\|\mathbf{d}(s)\| > 1$ *for* $s > 0$;
  (2) $0 < \lfloor \mathbf{d}(s)\rfloor \le \|\mathbf{d}(s)\| < +\infty \quad s \in \mathbb{R}$;
  (3) $\mathbf{d}(s) \neq I$ *if* $s \neq 0$.

- *If* $\mathbf{d}$ *is a uniformly continuous linear dilation in* $\mathbb{B}$ *then the functions* $\|\mathbf{d}(\cdot)\| : \mathbb{R} \mapsto \mathbb{R}$ *and* $\lfloor \mathbf{d}(\cdot)\rfloor : \mathbb{R} \mapsto \mathbb{R}$ *are continuous.*

Proof (1) Let us prove the inequalities for $\lfloor \mathbf{d}(s)\rfloor$. Suppose the contrary, there exists $s_0 > 0$ such that $\|\mathbf{d}(-s_0)z\| \ge 1$ for all $z \in S = \{x \in \mathbb{B} : \|x\| = 1\}$. If $z_0 \in S$ then

$$z_n = \tfrac{\mathbf{d}(-s_0)z_{n-1}}{\|\mathbf{d}(-s_0)z_{n-1}\|} \in S, \quad n \in \mathbb{N}.$$

So, we have

$$1 \le \|\mathbf{d}(-s_0)z_n\| = \left\| \tfrac{\mathbf{d}(-2s_0)z_{n-1}}{\|\mathbf{d}(-s_0)z_{n-1}\|} \right\| \le \|\mathbf{d}(-2s_0)z_{n-1}\| \le \cdots \le \|\mathbf{d}(-(n+1)s_0)z_0\|.$$

Therefore, $\|\mathbf{d}(-(n+1)s_0)z_0\| \ge 1$ for any $n \ge 1$ and we derive the contradiction to the limit property $\limsup_{n\to\infty} \|\mathbf{d}(-(n+1)s_0)z_0\| = 0$. Similarly, if $\lfloor \mathbf{d}(s_0)\rfloor < 1$ for some $s_0 > 0$ then there exists $z_0 \in S$ such that $\|\mathbf{d}(s_0)z_0\| < 1$. So, due to the group property, we have

$$\|\mathbf{d}(ns_0)z_0\| \ge \lfloor \mathbf{d}(s_0)\rfloor \cdot \|\mathbf{d}((n-1)s_0)z_0\| \ge \ldots \ge \lfloor \mathbf{d}(s_0)\rfloor^n \to 0 \text{ as } n \to +\infty.$$

This contradicts the limit property $\liminf_{n\to+\infty} |\mathbf{d}(ns_0)z_0\| = +\infty$. Taking into account the identity $\lfloor \mathbf{d}(-s)\rfloor \cdot \|\mathbf{d}(s)\| = 1, \forall s \in \mathbb{R}$ proven by Corollary 2.1, we derive $\|\mathbf{d}(s)\| \le 1$ for $s < 0$ and $\|\mathbf{d}(s)\| > 1$ for $s > 0$.

(2) Since $\mathbf{d}$ is a strongly continuous group then there exists $\alpha \in \mathbb{R}$ and $M \ge 1$ such that $\|\mathbf{d}(s)\| < Me^{\alpha s}$ for all $s \ge 0$. Hence, using the identity (2.1) we derive $1 = \lfloor \mathbf{d}(-s)\rfloor \|\mathbf{d}(s)\| \le \lfloor \mathbf{d}(-s)\rfloor Me^{\alpha s}$ for $s \ge 0$ and

$$\lfloor \mathbf{d}(-s)\rfloor \ge \frac{e^{-\alpha s}}{M} > 0, \quad \forall s \ge 0.$$

Hence, $0 < \lfloor \mathbf{d}(s)\rfloor \le \|\mathbf{d}(s)\|$ for all $s \le 0$. Again, using the identity (2.1) we derive $\lfloor \mathbf{d}(-s)\rfloor \ge \frac{1}{\|\mathbf{d}(s)\|} > 0$ for $s \le 0$.

(3) Suppose the contrary, i.e., there exists $s' \neq 0$ such that $\mathbf{d}(s') = I$. Then the group property implies $\mathbf{d}(ns') = I$ and $\|\mathbf{d}(ns')u\| = 1$ for any $n = 1, 2, \ldots$ and any $u \in S$. This contradicts the limit property.

(4) Since

$$\|\mathbf{d}(s)u\| = \|\mathbf{d}(s)u - u + u\| \leq \|\mathbf{d}(s)u - u\| + \|u\|$$

and

$$\|u\| = \|u - \mathbf{d}(s)u + \mathbf{d}(s)u\| \leq \|\mathbf{d}(s)u - u\| + \|\mathbf{d}(s)u\|$$

then $-\|\mathbf{d}(s)u - u\| \leq \|u\| - \|\mathbf{d}(s)u\| \leq \|\mathbf{d}(s)u - u\|$, or, equivalently,

$$\left|1 - \tfrac{\|\mathbf{d}(s)u\|}{\|u\|}\right| \leq \tfrac{\|\mathbf{d}(s)u - u\|}{\|u\|} \leq \|\mathbf{d}(s) - I\|.$$

On the other hand, we have

$$\sup_{u \neq 0} \left|1 - \tfrac{\|\mathbf{d}(s)u\|}{\|u\|}\right| = \max\left\{|1 - \|\mathbf{d}(s)\||, |1 - \lfloor \mathbf{d}(s) \rfloor|\right\}$$

$$\inf_{u \neq 0} \left|1 - \tfrac{\|\mathbf{d}(s)u\|}{\|u\|}\right| = \min\left\{|1 - \|\mathbf{d}(s)\||, |1 - \lfloor \mathbf{d}(s) \rfloor|\right\}.$$

Hence, we conclude

$$|\|\mathbf{d}(s)\| - 1| \leq \|\mathbf{d}(s) - I\| \quad \text{and} \quad |\lfloor \mathbf{d}(s) \rfloor - 1| \leq \|\mathbf{d}(s) - I\|.$$

Therefore, if $\mathbf{d}$ is uniformly continuous then the functions $s \mapsto \|\mathbf{d}(s)\|$ and $s \mapsto \lfloor \mathbf{d}(s) \rfloor$ are continuous at zero and the group property implies their continuity at any $s \in \mathbb{R}$. The proof is complete. □

Any strongly continuous linear dilation introduces an alternative norm topology on $\mathbb{B}$ by means of the so-called *canonical homogeneous norm* studied in the next section. Here, we just consider some related geometric objects in $\mathbb{B}$, which are introduced similar to the finite-dimensional case (see Chap. 5 of Volume I).

- Given $x \in \mathbb{B}$ the set
$$\Gamma_{\mathbf{d}}(x) = \{\mathbf{d}(s)x : s \in \mathbb{R}\} \tag{4.1}$$
is called a **d**-*homogeneous curve.*
- The set
$$S_{\mathbf{d}}(r, z_0) = \{z \in \mathbb{B} : \|\mathbf{d}(-\ln(r))(z - z_0)\| = 1\}$$
is called a **d**-*homogeneous sphere* of the radius $r > 0$ with the center at $z_0 \in \mathbb{B}$.
- The set
$$B_{\mathbf{d}}(r, z_0) = \{z \in \mathbb{B} : \|\mathbf{d}(-\ln(r))(z - z_0)\| \leq 1\}$$

is called a **d**-*homogeneous ball* of the radius $r > 0$ with the center at $z_0 \in \mathbb{B}$.

- For spheres and balls centered at the origin, we use the notation $S_{\mathbf{d}}(r) := S_{\mathbf{d}}(r, \mathbf{0})$ and $B_{\mathbf{d}}(r) = B_{\mathbf{d}}(r, \mathbf{0})$, respectively.
- Given $z \in \mathbb{B}$, a point $z_0 \in S$ is a **d**-*homogeneous projection* of z on the unit sphere S if $\exists s_0 \in \mathbb{R} : z_0 = \mathbf{d}(s_0)z$. Recall that the homogeneous projection may be nonunique even in the Euclidean space $\mathbb{R}^2$ (see Chap. 5 of Volume I).
- A non-empty set $\mathcal{D} \subseteq \mathbb{B}$ is a **d**-*homogeneous cone* $\mathbb{B}$ if $\mathcal{D}$ is invariant with respect to the dilation **d**, i.e.,

$$x \in \mathcal{D} \quad \Rightarrow \mathbf{d}(s)x \in \mathcal{D}, \quad \forall s \in \mathbb{R}.$$

In Chap. 2 we have shown that strongly continuous (semi)groups of linear bounded operators in Banach spaces can be characterized by means of the so-called infinitesimal generators (linear closed densely defined operators). The generator of linear dilation plays an important role for analysis and design of homogeneous systems.

Definition 4.3 An operator $G_{\mathbf{d}} : \mathcal{D}(G_{\mathbf{d}}) \subset \mathbb{B} \mapsto \mathbb{B}$ defined as

$$G_{\mathbf{d}}x = \lim_{s \to 0} \frac{\mathbf{d}(s)x - x}{s}$$

is called an infinitesimal *generator* of a strongly continuous linear dilation **d**. The domain of the generator $G_{\mathbf{d}}$ is given by

$$\mathcal{D}(G_{\mathbf{d}}) = \left\{ x \in \mathbb{B} : \lim_{s \to 0} \frac{\mathbf{d}(s)x - x}{s} \text{ exists} \right\}.$$

A generator of a strongly continuous linear dilation in $\mathbb{B}$ is a linear (possibly) unbounded operator. The generator $G_{\mathbf{d}}$ is bounded if and only if **d** is a uniformly continuous group of linear bounded operators on $\mathbb{B}$ (see Sect. 2.1). Similar to the finite-dimensional case, a dilation rate of the uniformly continuous dilation can be estimated by means of the generator.

Corollary 4.1 *If* **d** *be a uniformly continuous linear dilation in a real Hilbert space* $\mathbb{H}$ *then for any* $z \in \mathbb{H}$*, we have*

$$\begin{aligned} e^{\beta s}\|z\| \le \|\mathbf{d}(s)z\| \le e^{\alpha s}\|z\| \quad \text{for } s \ge 0, \\ e^{\alpha s}\|z\| \le \|\mathbf{d}(s)z\| \le e^{\beta s}\|z\| \quad \text{for } s \le 0, \end{aligned} \tag{4.2}$$

where

$$\alpha = \sup_{z \in S}\langle G_{\mathbf{d}}z, z\rangle > 0 \quad \text{and} \quad \beta = \inf_{z \in S}\langle G_{\mathbf{d}}z, z\rangle,$$

where $S = \{x \in \mathbb{H} : \|x\| = 1\}$*, as before, is a the unit sphere.*

Proof Since for any $x \in \mathbb{H} \neq \mathbf{0}$ we have

$$\frac{d}{ds}\langle \mathbf{d}(s)x, \mathbf{d}(s)x\rangle = 2\langle G_{\mathbf{d}}\mathbf{d}(s)x, \mathbf{d}(s)x\rangle = 2\left\langle G_{\mathbf{d}}\frac{\mathbf{d}(s)x}{\|\mathbf{d}(s)x\|}, \frac{\mathbf{d}(s)x}{\|\mathbf{d}(s)x\|}\right\rangle \|\mathbf{d}(s)x\|^2,$$

the boundedness of $G_{\mathbf{d}}$ and the uniform continuity of the group $\mathbf{d}$ imply (see Corollary 2.2) that the curve $s \mapsto \|\mathbf{d}(s)x\|^2$ is continuously differentiable and

$$2\beta\|\mathbf{d}(s)x\|^2 \leq \frac{d}{ds}\|\mathbf{d}(s)x\|^2 \leq 2\alpha\|\mathbf{d}(s)x\|^2.$$

Using $\|\mathbf{d}(0)x\| = \|x\|$ we derive the estimates (4.2). Hence, taking into account the limit property we immediately conclude that $\alpha > 0$.

Example 4.2 Let $\mathbb{B}$ be the Banach space of bounded uniformly continuous functions $\mathbb{R} \mapsto \mathbb{R}$ with the supremum norm. In Example 4.1 it has been proven that the dilation in $\mathbb{B}$ can be defined as $(\mathbf{d}(s)x)(y) = e^{\alpha s}x(y + \beta s)$, $s \in \mathbb{R}$, $x \in \mathbb{B}$, $y \in \mathbb{R}$ and $\alpha > 0$, $\beta \in \mathbb{R}$ are constant parameters. The generator of this dilation is defined as follows:

$$(G_{\mathbf{d}}x)(y) = \alpha x(y) + \beta\frac{\partial x(y)}{\partial y}, \quad y \in \mathbb{R}$$

with the domain

$$\mathcal{D}(G_{\mathbf{d}}) = \left\{x \in \mathbb{B} : \frac{\partial x}{\partial y} \in \mathbb{B}\right\}.$$

Indeed, if $x \in \mathcal{D}(G_{\mathbf{d}})$ then $G_{\mathbf{d}}x \in \mathbb{B}$ and

$$\lim_{s\to 0}\left\|\tfrac{\mathbf{d}(s)x-x}{s} - G_{\mathbf{d}}x\right\| = \lim_{s\to 0}\sup_{y\in\mathbb{R}}\left|\tfrac{e^{\alpha s}x(y+\beta s)-x(y)}{s} - \beta\tfrac{\partial x(y)}{\partial y} - \alpha x(y)\right| =$$

$$\lim_{s\to 0}\sup_{y\in\mathbb{R}}\left|\left(\tfrac{e^{\alpha s}-1}{s} - \alpha\right)x(y+\beta s) + \alpha(x(y+\beta s)-x(y)) + \tfrac{x(y+\beta s)-x(y)}{s} - \beta\tfrac{\partial x(y)}{\partial y}\right| \leq$$

$$\lim_{s\to 0}\left|\tfrac{e^{s}-1}{s} - \alpha\right|\sup_{y\in\mathbb{R}}|x(y+\beta s)| + \alpha\lim_{s\to 0}\sup_{y\in\mathbb{R}}|x(y+\beta s) - x(y)| +$$

$$\lim_{s\to 0}\sup_{y\in\mathbb{R}}\left|\tfrac{x(y+\beta s)-x(y)}{s} - \beta\tfrac{\partial x(y)}{\partial y}\right| = \lim_{s\to 0}\sup_{y\in\mathbb{R}}\left|\beta\left.\tfrac{\partial x(p)}{\partial p}\right|_{p=\theta(y)} - \beta\tfrac{\partial x(y)}{\partial y}\right|$$

where $\theta(y) \in [y, y + \beta s]$ if $\beta s > 0$ or $\theta(y) \in [y + \beta s, y]$ if $\beta s < 0$ (by Mean Value theorem). Since $\frac{\partial x}{\partial y}$ is uniformly continuous then

$$\left|\left.\frac{\partial x(p)}{\partial p}\right|_{p=\theta(y)} - \frac{\partial x(y)}{\partial y}\right| \leq \omega(|\beta s|) \to 0 \text{ as } s \to 0$$

uniformly on y, where $\omega \in \mathcal{K}^\infty$ is the modulus of continuity of x. By Definition 4.3, this implies that $G_{\mathbf{d}}$ is the infinitesimal generator of $\mathbf{d}$.

Lemma 4.1 *Let $\mathbb{B}$ be a real Banach space. If a linear densely defined operator $G_0 : \mathcal{D}(G_0) \subset \mathbb{B} \mapsto \mathbb{B}$ is an infinitesimal generator of a strongly continuous group $\mathbf{d}_0$ of linear bounded operators on $\mathbb{B}$ such that*

$$\exists \omega \in \mathbb{R} \text{ and } \exists M \geq 1 \quad : \quad \|\mathbf{d}_0(s)\| \leq M e^{\omega s}, \forall s \leq 0, \tag{4.3}$$

then the operator $G_{\mathbf{d}} = I + \mu G_0$, $\mathcal{D}(G_{\mathbf{d}}) = \mathcal{D}(G_0)$ is a generator of a strongly continuous dilation $\mathbf{d}$ in $\mathbb{B}$ for any $\mu \in \mathbb{R}$ satisfying the inequality

$$1 + \mu\omega > 0. \tag{4.4}$$

Proof If G_0 is a generator of a strongly continuous group $\mathbf{d}_0$ then μG_0 with $\mu \in \mathbb{R}$ is the generator of a strongly continuous group $\mathbf{d}_\mu$. Moreover, for any $\mu \in \mathbb{R}$ the operator $G_{\mathbf{d}} = I + \mu G_0$ is the generator of the strongly continuous group $\mathbf{d}$ such that

$$\mathbf{d}(s) = e^s \mathbf{d}_\mu(s), \forall s \in \mathbb{R}.$$

Since $\mathbf{d}_0$ is a strongly continuous group then, by Corollary 2.6, there exist $M \geq 1$ and $\omega \in \mathbb{R}$ such that

$$\|\mathbf{d}_0(s)\| \leq M e^{\omega s}, \forall s \leq 0.$$

Moreover, for any $\mu \in \mathbb{R}$ we have

$$\|\mathbf{d}_\mu(s)\| \leq M e^{\mu\omega s}, \forall s \leq 0.$$

Since $\mathbf{d}_\mu$ is a group then $\mathbf{d}_\mu(-s)$ is the inverse operator to $\mathbf{d}_\mu(s)$ and

$$1 = \|I\| = \|\mathbf{d}_\mu(s)\mathbf{d}_\mu(-s)\| \leq \|\mathbf{d}_\mu(s)\| \cdot \|\mathbf{d}_\mu(-s)\|.$$

Hence, we derive

$$\tfrac{e^{\mu\omega s}}{M} \leq \|\mathbf{d}_\mu(s)\|, \forall s \geq 0,$$

and

$$\tfrac{e^{(1+\mu\omega)s}}{M} \leq \|\mathbf{d}(s)\| = \|e^s \mathbf{d}_\mu(s)\|, \forall s \geq 0,$$

$$\|\mathbf{d}(s)\| = \|e^s \mathbf{d}_\mu(s)\| \leq M e^{(1+\mu\omega)s}, \forall s \leq 0.$$

Therefore, for any $\mu \in \mathbb{R}$ such that $1 + \mu\omega > 0$, we have $\|\mathbf{d}(s)\| \to +\infty$ as $s \to +\infty$ and $\|\mathbf{d}(s)\| \to 0$ as $s \to -\infty$, i.e., the strongly continuous group $\mathbf{d}$ is a dilation in $\mathbb{B}$. The proof is complete. □

Notice that the condition (4.3) is fulfilled for any strongly continuous group of linear bounded operators on $\mathbb{B}$ (see Chap. 2), so any C_0-group in $\mathbb{B}$ can be utilized for the design of a linear dilation in $\mathbb{B}$.

4.1.1 Monotonicity

Monotonicity of dilation plays an important role for the further investigations of the homogeneous geometrical structures on Banach spaces as well as for an analysis of infinite-dimensional homogeneous dynamical systems.

Definition 4.4 A dilation $\mathbf{d}$ in $\mathbb{B}$ is **monotone** if the function $s \mapsto \|\mathbf{d}(s)x\|$ is strictly increasing for any $x \in \mathbb{B}\backslash\{\mathbf{0}\}$.

Examples of monotone and non-monotone dilation in $\mathbb{R}^n$ can be found in Volume I of this monograph. Monotone dilations in Sobolev and Lebesgue spaces are studied in the following sections of this chapter. Here we consider just some general properties of monotone dilations in abstract Banach and Hilbert spaces.

The following proposition proves the uniqueness of homogeneous projection in the case of strongly continuous monotone dilation.

Proposition 4.2 *If a linear strongly continuous dilation* $\mathbf{d}$ *in* $\mathbb{B}$ *is monotone then the* $\mathbf{d}$*-homogeneous projection of any vector* $x \in \mathbb{B}\backslash\{\mathbf{0}\}$ *on the unit sphere is unique. Namely, there exists a* ***unique*** *pair* $(s_0, u_0) \in \mathbb{R} \times S$ *such that* $u = \mathbf{d}(s_0)u_0$, *where* S *is the unit sphere in* $\mathbb{B}$.

Proof Since the function $s \mapsto \|\mathbf{d}(s)x\|$ with $x \in \mathbb{B}\backslash\{\mathbf{0}\}$ is strictly increasing then, due to the limit property of the dilation, there exists a unique $s_0 \in \mathbb{R}$ such that $\|\mathbf{d}(s_0)x\| = 1$. The corresponding vector $\mathbf{d}(s_0)x$, by definition, is the $\mathbf{d}$-homogeneous projection of x on the unit sphere S.

Strict monotonicity allows us to establish the regularity to the so-called canonical homogeneous norm studied below.

Definition 4.5 A linear dilation $\mathbf{d}$ in $\mathbb{B}$ is said to be **strictly monotone** if $\exists \beta > 0$ such that

$$\|\mathbf{d}(s)\| \leq e^{\beta s}, \quad \forall s \leq 0. \tag{4.5}$$

The strictly monotone strongly continuous dilations have properties similar to uniformly continuous dilations in the view of the following proposition.

Proposition 4.3 *If* $\mathbf{d}$ *is a strongly continuous strictly monotone linear dilation in* $\mathbb{B}$ *then there exist* $M \geq 1$, $\alpha \geq \beta > 0$ *such that*

$$\begin{array}{ll} e^{\beta s} \leq \lfloor \mathbf{d}(s) \rfloor \leq \|\mathbf{d}(s)\| \leq M e^{\alpha s} & \text{if } s \geq 0, \\ \frac{e^{\alpha s}}{M} \leq \lfloor \mathbf{d}(s) \rfloor \leq \|\mathbf{d}(s)\| \leq e^{\beta s} & \text{if } s \leq 0. \end{array} \tag{4.6}$$

Proof The result immediately follows from the identity (2.1), from the definition of strict monotonicity, and from the property of the strongly continuous group $\|\mathbf{d}(s)\| \leq M e^{\alpha s}$, $s \geq 0$ (see Corollary 2.6). The proof is complete. □

Theorem 4.1 *Let* $\mathbf{d}$ *be a strongly continuous group of linear bounded operators on a real Banach space* $\mathbb{B}$. *If* $\exists s_0 < 0 : \|\mathbf{d}(s_0)\| < 1$ *then* $\mathbf{d}$ *is a strongly continuous linear dilation in* $\mathbb{B}$, *and there exist an equivalent norm* $\|\cdot\|_*$ *in* $\mathbb{B}$ *such that the estimate* (4.6) *holds for the norm* $\|\cdot\|_*$, *a number* $M \geq 1$, *and a number* $\alpha \geq \beta$ *with*

$$\beta = \frac{\ln \|\mathbf{d}(s_0)\|}{s_0} > 0.$$

Proof Since the group $\mathbf{d}(s)$ is strongly continuous then, by Banach–Steinhaus theorem (see Appendix A), there exists $K_0 \in [1, +\infty)$ such that

$$\sup_{s \in [s_0, 0]} \|\mathbf{d}(s)\| \leq K_0.$$

Notice that any $s \leq 0$ admits the representation $s = n s_0 + \delta$, where $n \in \{0, 1, 2, ...\}$ and $\delta \in (s_0, 0]$. Denoting $\beta = \frac{\ln \|\mathbf{d}(s_0)\|}{s_0} > 0$ and using the group property of $\mathbf{d}$, we derive

$$\|\mathbf{d}(s)\| = \|\mathbf{d}(\delta)\mathbf{d}^n(s_0)\| \leq K_0 \|\mathbf{d}(s_0)\|^n = K_0 e^{s_0 n \beta} = K_0 e^{s\beta - \delta\beta} \leq K e^{\beta s}$$

where $K = K_0 e^{-s_0 \beta} \geq 1$.

Let the norm $\|\cdot\|_*$ in $\mathbb{B}$ be defined as follows:

$$\|x\|_* = \sup_{s \leq 0} e^{-\beta s} \|\mathbf{d}(s) x\|.$$

Notice that $\|x_*\| < +\infty$ for any $x \in \mathbb{B}$ due to the estimate $\|\mathbf{d}(s)\| \leq K e^{\beta s}$, $s \leq 0$. On the one hand, $\|\cdot\|_*$ satisfies all properties of the norm: positive definiteness $\|x\|_* > 0$ for $x \neq 0$, $\|\mathbf{0}\|_* = 0$; standard homogeneity $\|e^s x\|_* = e^s \|x\|_*$; reflection symmetry $\|-x\|_* = \|x\|_*$ and sub-additivity $\|x + y\|_* \leq \|x\|_* + \|y\|_*$. On the other hand, we derive

$$\|x\| = e^{-\beta \cdot 0} \|\mathbf{d}(0) x\| \leq \|x\|_* \leq \sup_{s \leq 0} e^{-\beta s} \|\mathbf{d}(s)\| \|x\| \leq K \|x\|, \quad \forall x \in \mathbb{B}.$$

Therefore, $\|\cdot\|_*$ is, indeed, an equivalent norm in $\mathbb{B}$. We complete the proof taking into account that for $t \leq 0$ the simple calculations show

$$\|\mathbf{d}(t)\|_* = \sup_{x\neq 0} \frac{\sup_{s\leq 0} e^{-\beta s}\|\mathbf{d}(t+s)x\|}{\sup_{s\leq 0} e^{-\beta s}\|\mathbf{d}(s)x\|} = e^{\beta t} \sup_{x\neq 0} \frac{\sup_{t+s\leq t} e^{-\beta(t+s)}\|\mathbf{d}(t+s)x\|}{\sup_{s\leq 0} e^{-\beta s}\|\mathbf{d}(s)x\|} \leq e^{\beta t}.$$

Using the identity (2.1) we derive $1 = \lfloor \mathbf{d}(-s)\rfloor_* \|\mathbf{d}(s)\|_* \leq \lfloor \mathbf{d}(-s)\rfloor_* e^{\beta s}$ for all $s \leq 0$. Then $e^{-\beta s} \leq \lfloor \mathbf{d}(-s)\rfloor_* \leq \|\mathbf{d}(-s)\|$ for all $s \leq 0$. Since $\mathbf{d}$ is a strongly continuous group then there exists $M \geq 1$ and $\alpha \in \mathbb{R}$ such that $\|\mathbf{d}(s)\|_* \leq Me^{\alpha s}$ for all $s \geq 0$. Hence, we conclude that $\alpha \geq \beta > 0$ and the estimate (4.6) holds for the norm $\|\cdot\|_*$.

The above theorem proves that a strongly continuous linear dilation in $\mathbb{B}$ being non-monotone with respect to the canonical norm in $\mathbb{B}$ may be strictly monotone with respect to some equivalent norm in $\mathbb{B}$.

To describe some spectral properties of generators of dilations, the Hille–Yosida theorem (see Theorem 2.3) can be utilized. However, the generator of strictly monotone dilation in a Hilbert space admits also the following simple characterization.

Proposition 4.4 *A strongly continuous linear dilation* $\mathbf{d}$ *in a real Hilbert space* $\mathbb{H}$ *is strictly monotone if and only if there exists a* $\mathbf{d}$*-homogeneous cone* $\mathcal{D} \subset \mathcal{D}(G_{\mathbf{d}})$ *dense in* $\mathbb{H}$ *such that*

$$\beta := \inf_{z\in S\cap\mathcal{D}} \langle G_{\mathbf{d}}x, x\rangle > 0, \tag{4.7}$$

where $G_{\mathbf{d}}$ *is the generator of* $\mathbf{d}$ *and* $S = \{x \in \mathbb{H} : \|x\| = 1\}$.

Proof *Necessity.* Let $\mathbf{d}$ be strictly monotone and $z \in S \cap \mathcal{D}(G_{\mathbf{d}})$. Using the Cauchy–Schwarz inequality for $s > 0$ we derive

$$1 = \langle z, z\rangle = \langle \mathbf{d}(-s)\mathbf{d}(s)z, z\rangle \leq e^{-\beta s}\|\mathbf{d}(s)z\|.$$

Hence we have

$$e^{2\beta s} - 1 \leq \langle \mathbf{d}(s)z, \mathbf{d}(s)z\rangle - \langle z, z\rangle = \langle \mathbf{d}(s)z - z, \mathbf{d}(s)z\rangle + \langle z, \mathbf{d}(s)z - z\rangle.$$

Dividing both sides of the last inequality by s and taking the limit $s \to 0^+$, we obtain

$$2\beta \leq \langle G_{\mathbf{d}}z, z\rangle + \langle z, G_{\mathbf{d}}z\rangle = 2\langle G_{\mathbf{d}}z, z\rangle.$$

Sufficiency. The inequality (4.7) implies that $\langle G_{\mathbf{d}}z, z\rangle \geq \beta\|z\|^2$ for any $z \in \mathcal{D}$. Since $\mathcal{D}$ is a $\mathbf{d}$-homogeneous cone then $\mathbf{d}(s)z \in \mathcal{D}$ for any $s \in \mathbb{R}$. Therefore, by Corollary 2.6, for $z \in \mathcal{D} \subset \mathcal{D}(G_{\mathbf{d}})$ we have

$$\frac{d}{ds}\|\mathbf{d}(s)z\|^2 = \frac{d}{ds}\langle \mathbf{d}(s)z, \mathbf{d}(s)z\rangle = 2\langle G_{\mathbf{d}}\mathbf{d}(s)z, \mathbf{d}(s)z\rangle = 2\langle \mathbf{d}(s)G_{\mathbf{d}}z, \mathbf{d}(s)z\rangle.$$

Hence, the strong continuity of the $\mathbf{d}$ implies the continuity of the derivative $\frac{d}{ds}\|\mathbf{d}(s)z\|^2$. Taking into account $\mathbf{d}(s)z \in \mathcal{D}, \forall s \in \mathbb{R}$ we derive

$$\frac{d}{ds}\|\mathbf{d}(s)z\|^2 \geq 2\beta\|\mathbf{d}(s)z\|^2, \forall s \in \mathbb{R}$$

and $\|\mathbf{d}(s_2)z\|^2 \geq e^{2\beta(s_2-s_1)}\|\mathbf{d}(s_1)z\|^2$ for any $s_1 < s_2$. Hence, for $s_2 = 0$, $s_1 = s < 0$ we have $\|\mathbf{d}(s)z\| \leq e^{\beta s}\|z\|, \forall z \in \mathcal{D}$. Finally, since D is dense in $\mathbb{H}$ and the operator $\mathbf{d}(s)$ is bounded for any $s \in \mathbb{R}$ then

$$e^{\beta s} \geq \sup_{z\in\mathcal{D}\backslash\{\mathbf{0}\}} \frac{\|\mathbf{d}(s)z\|}{\|z\|} = \sup_{z\in\mathcal{D}(G_{\mathbf{d}})\backslash\{\mathbf{0}\}} \frac{\|\mathbf{d}(s)z\|}{\|z\|} = \sup_{z\in\mathbb{H}\backslash\{\mathbf{0}\}} \frac{\|\mathbf{d}(s)z\|}{\|z\|} = \|\mathbf{d}(s)\|$$

for any $s \leq 0$. We complete the proof by applying Theorem 4.1.

The above theorem generalizes Proposition 5.3 of Volume I to real Hilbert spaces. In many concrete cases, the strict monotonicity of a dilation group can be easily checked.

Example 4.3 Let $\mathbb{B} = L^2(\mathbb{R};\mathbb{R})$ and $\mathbf{d}$ be a family of linear operators be defined as follows:

$$(\mathbf{d}(s)x)(z) = e^{\gamma s}x(e^{\theta s}z),$$

where $s \in \mathbb{R}$ is a varying parameter, $x \in L^2(\mathbb{R};\mathbb{R})$, $z \in \mathbb{R}$ is the argument of x, and $\gamma, \theta \in \mathbb{R}$ are constant parameters specifying $\mathbf{d}$. Since for $v = \mathbf{d}(s)x$ it holds

$$(\mathbf{d}(t)\mathbf{d}(s)x)(z) = (\mathbf{d}(t)v)(z) = e^{\gamma t}v(e^{st}z) = e^{\gamma t}e^{\gamma s}x(e^{\theta t}e^{\theta s}z) = (\mathbf{d}(s+t)x)(z),$$

and $\mathbf{d}(0)x = x$ then $\mathbf{d}$ is a group of linear operators on $\mathbb{B}$. Using the change of variables in the Lebesgue integral, we derive

$$\|\mathbf{d}(s)x\|_{L^2}^2 = \int_{\mathbb{R}} e^{2\gamma s}|x(e^{\theta s}z)|^2 dz = e^{(2\gamma-\theta)s}\int_{\mathbb{R}} |x(\tilde{z})|^2 d\tilde{z} = e^{(2\gamma-\theta)s}\|x\|_{L^2}.$$

Hence, we conclude that

$$\|\mathbf{d}(s)\| = \sup_{x\neq\mathbf{0}} \frac{\|\mathbf{d}(s)x\|}{\|x\|} = e^{(\gamma-0.5\theta)s},$$

so the group $\mathbf{d}$ is a linear strictly monotone dilation in $L^2(\mathbb{R};\mathbb{R})$ provided that

$$2\gamma - \theta > 0.$$

Below we show that the corresponding dilation is strongly continuous in $L^2(\mathbb{R};\mathbb{R})$.

Theorem 4.1 for Hilbert spaces can be refined as follows.

Corollary 4.2 *If conditions of Theorem 4.1 are fulfilled for a real Hilbert space* $\mathbb{H}$ *then the inequality* (4.7) *holds for the inner product*

$$\langle x, y\rangle_* = \int_{-\infty}^{0} e^{-2\gamma s}\langle \mathbf{d}(s)x, \mathbf{d}(s)y\rangle ds,$$

where $\gamma \in (0, \beta)$ *is any number.*

Proof (1) Let us show that $\langle x, y\rangle_*$ is an inner product in $\mathbb{H}$. On the one hand, since $\mathbf{d}$ is a strongly continuous group in $\mathbb{H}$ then the function $s \mapsto \langle \mathbf{d}(s)x, \mathbf{d}(s)y\rangle$ is continuous for any fixed $x, y \in \mathbb{H}$. On the other hand, since $\gamma \in (0, \beta)$ and $\|\mathbf{d}(s)z\|^2 \leq e^{2\beta s}\|z\|^2$ for all $s \leq 0$ and for all $z \in \mathbb{H}$ then

$$\langle x, y\rangle_* = \int_{-\infty}^{0} e^{-2\gamma s}|\langle \mathbf{d}(s)x, \mathbf{d}(s)y\rangle| ds \leq \|x\|\|y\| \int_{-\infty}^{0} e^{2(\beta-\gamma)s} ds = \tfrac{\|x\|\|y\|}{2(\beta-\gamma)}. \tag{4.8}$$

Hence, $\langle x, y\rangle_*$ is finite for any $x, y \in \mathbb{H}$. The linearity

$$\langle \mu x, y\rangle_* = \mu\langle x, y\rangle_*, \quad \langle x + y, z\rangle_* = \langle x, z\rangle_* + \langle y, z\rangle_*, \quad \forall x, y, z \in \mathbb{H}, \forall \mu \in \mathbb{R}$$

and the Hermitian symmetry $\langle x, y\rangle_* = \langle y, x\rangle_*, \forall x, y \in \mathbb{H}$ of $\langle\cdot,\cdot\rangle_*$ comes from the linearity and the Hermitian symmetry of $\langle\cdot,\cdot\rangle$. Finally, since the function $s \mapsto \|\mathbf{d}(s)x\|^2$ is continuous and $\|\mathbf{d}(0)x\|^2 = \|x\|^2$ then $\|x\|_* := \langle x, x\rangle_* > 0$ for any $x \neq 0$.

(2) Let us show that the inequality (4.7) holds for the inner product $\langle\cdot,\cdot\rangle_*$. Since $\frac{d}{ds}\mathbf{d}(s)x = \mathbf{d}(s)G_{\mathbf{d}}x$ for all $x \in \mathcal{D}(G_{\mathbf{d}})$ (see Corollary 2.6) then

$$\langle G_{\mathbf{d}}x, x\rangle_* = \int_{-\infty}^{0} e^{-2\gamma s}\langle \mathbf{d}(s)G_{\mathbf{d}}x, \mathbf{d}(s)x\rangle ds = \int_{-\infty}^{0} e^{-2\gamma s}\left\langle \tfrac{d}{ds}\mathbf{d}(s)x, \mathbf{d}(s)x\right\rangle ds$$

$$= \int_{-\infty}^{0} \frac{e^{-2\gamma s}}{2}\frac{d}{ds}\|\mathbf{d}(s)x\|^2 ds = \tfrac{e^{-2\gamma s}}{2}\|\mathbf{d}(s)x\|^2\Big|_{-\infty}^{0} + \gamma\int_{-\infty}^{0} e^{-2\gamma s}\|\mathbf{d}(s)x\|^2 ds.$$

Since $\gamma \in (0, \beta)$ and $\|\mathbf{d}(s)x\|^2 \leq e^{2\beta s}\|x\|^2, \forall s \leq 0$ then $e^{-2\gamma s}\|\mathbf{d}(s)x\|^2 \to 0$ as $s \to -\infty$ and

$$\langle G_{\mathbf{d}}x, x\rangle_* = \frac{\|x\|^2}{2} + \gamma\|x\|_*^2 \geq \frac{2(\beta-\gamma)\|x\|_*^2}{2} + \gamma\|x\|_* = \beta\|x\|_*$$

for all $x \in \mathcal{D}(G_{\mathbf{d}})$, where the inequality (4.8) with $x = y$ is utilized in the last step. The proof is complete. □

4.1.2 Linear Dilations in Lebesgue and Sobolev Spaces

Let $\mathbb{V}(\mathbb{R}^n;\mathbb{R}^k)$ be a vector space of functions $x : \mathbb{R}^n \mapsto \mathbb{R}^k$. If $\Theta \in \mathbb{R}^{n\times n}$ and $\Gamma \in \mathbb{R}^{k\times k}$ then the family of linear operators $\mathbf{d}(s) : \mathbb{V} \mapsto \mathbb{V}$ given by

$$(\mathbf{d}(s)x)(z) = e^{\Gamma s}x(e^{s\Theta}z), \quad z \in \mathbb{R}^n, \quad x \in \mathbb{V}, \quad s \in \mathbb{R} \tag{4.9}$$

is, obviously, a group:

$$\mathbf{d}(0)x = x \quad \text{and} \quad \mathbf{d}(s+t)x = \mathbf{d}(s) \circ \mathbf{d}(t)x, \quad \forall t, s \in \mathbb{R}.$$

To be a linear dilation in a normed vector space $\mathbb{V}$, the operator $\mathbf{d}(s)$ must be bounded for any $s \in \mathbb{R}$ and the group $\mathbf{d}$ must satisfy the limit property $\|\mathbf{d}(s)x\| \to 0$ as $s \to -\infty$ and $\|\mathbf{d}(s)x\| \to +\infty$ as $s \to +\infty$ for any $x \neq \mathbf{0}$. The group (4.9) may be a dilation in a function space $\mathbb{V}$ provided that the matrix Γ, Θ as well as a norm in $\mathbb{V}$ are properly defined. The following lemma characterizes a class of linear dilations in Lebesgue spaces studied in this monograph.

Lemma 4.2 *Let $p \geq 1$ and*

$$\tilde{\Gamma} = \Gamma - \tfrac{\mathrm{trace}(\Theta)}{p} I_k. \tag{4.10}$$

The family of operators (4.9) *is*

(1) a group of linear invertible operators on $\mathbb{V}(\mathbb{R}^n;\mathbb{R}^k)$, $C^\infty(\mathbb{R}^n;\mathbb{R}^k)$, $C_c^\infty(\mathbb{R}^n;\mathbb{R}^k)$, and $L^1_{loc}(\mathbb{R}^n;\mathbb{R}^k)$;
(2) a strongly continuous group of linear bounded operators on $L^p(\mathbb{R}^n;\mathbb{R}^k)$:

$$\lfloor e^{\tilde{\Gamma} s}\rfloor \cdot \|x\|_{L^p} \leq \|\mathbf{d}(s)x\|_{L^p} \leq |e^{\tilde{\Gamma} s}| \cdot \|x\|_{L^p}, \quad \forall s \in \mathbb{R}, \quad \forall x \in L^p; \tag{4.11}$$

moreover, for $k = 1$ the inequalities (4.11) *are fulfilled as identities;*
(3) a strongly continuous linear dilation in the Lebesgue space $L^p(\mathbb{R}^n;\mathbb{R}^k)$ if the matrix $\tilde{\Gamma}$ is anti-Hurwitz.

Proof (1) For $z \in \mathbb{R}^n$ and $x : \mathbb{R}^n \mapsto \mathbb{R}^k$ one has

$$(\mathbf{d}(t)\mathbf{d}(s)x)(z) = e^{\Gamma t}e^{\Gamma s}x(e^{\Theta t}e^{\Theta s}z) = (\mathbf{d}(s+t)x)(z),$$

i.e., $\mathbf{d}$ is a group of operators on $\mathbb{V}(\mathbb{R}^n;\mathbb{R}^k)$. Due to the group property we have $\mathbf{d}(-s)\mathbf{d}(s)x = \mathbf{d}(s)\mathbf{d}(-s)x = x$ for any $x : \mathbb{R}^n \mapsto \mathbb{R}^k$, i.e., $\mathbf{d}(s)$ is invertible and $\mathbf{d}(s) = [\mathbf{d}(-s)]^{-1}$. The operator $\mathbf{d}(s)$ is, obviously, linear

$$\mathbf{d}(s)(\mu_1 x_1 + \mu_2 x_2) = \mu_1 \mathbf{d}(s) x_1 + \mu_2 \mathbf{d}(s) x_2,$$

$$\forall \mu_1, \mu_2 \in \mathbb{R}, \forall x_1, x_2 \in \mathbb{V}(\mathbb{R}^n; \mathbb{R}^k).$$

Obviously, $x \in C^\infty(\mathbb{R}^n; \mathbb{R}^k)$ implies $\mathbf{d}(s)x \in C^\infty(\mathbb{R}^n; \mathbb{R}^k)$, so $\mathbf{d}$ is a group of linear invertible operators on $C^\infty(\mathbb{R}^n; \mathbb{R}^k)$.

Since the linear function $z \mapsto e^{\Theta s} z$ maps a compact in $\mathbb{R}^n$ to a compact in $\mathbb{R}^n$ then $\mathbf{d}(s)$ is defined on whole $C_c^\infty(\mathbb{R}^n; \mathbb{R}^k)$ and for any $x \in C_c^\infty(\mathbb{R}^n; \mathbb{R}^k)$ (i.e., for any compactly supported smooth x) we have $\mathbf{d}(s)x \in C_c^\infty(\mathbb{R}^n; \mathbb{R}^k)$ (i.e., $\mathbf{d}(s)x$ is also smooth and compactly supported in $\mathbb{R}^n$). Hence, $\mathbf{d}$ is a group of linear invertible operators on $C_c^\infty(\mathbb{R}^n; \mathbb{R}^k)$.

Since linear scaling of an argument of a function does not destroy its local integrability, then $x \in L^1_{loc}(\mathbb{R}^n; \mathbb{R}^k)$ implies $\mathbf{d}(s)x \in L^1_{loc}(\mathbb{R}^n; \mathbb{R}^k)$ and $\mathbf{d}$ is a group of linear invertible operators on $L^1_{loc}(\mathbb{R}^n; \mathbb{R}^k)$.

(2) Let a vector $x \in L^p(\mathbb{R}^n; \mathbb{R}^k)$ be chosen arbitrarily. For $\mathbf{d}(s)$ given by (4.9), using the change of the integration variable in the Lebesgue integral, we derive

$$\|\mathbf{d}(s)x\|_{L^p}^p = \int_{\mathbb{R}^n} \left|e^{s\Gamma} x(e^{\Theta s} z)\right|^p dz = \int_{\mathbb{R}^n} \det(e^{-s\Theta}) \left|e^{\Gamma s} x(\tilde{z})\right|^p d\tilde{z}.$$

Since $\det(e^{-s\Theta}) = e^{-s \,\mathrm{trace}(\Theta)}$ then

$$\|\mathbf{d}(s)x\|_{L^p}^p = \int_{\mathbb{R}^n} \left| e^{s\left(\Gamma - \frac{\mathrm{trace}(\Theta)}{p} I_k\right)} x(\tilde{z}) \right|^p d\tilde{z}. \tag{4.12}$$

Hence, we conclude that $\mathbf{d}(s)$ is a group of bounded linear operators on $L^p(\mathbb{R}^n; \mathbb{R}^k)$ and the inequalities (4.11) are fulfilled.

Let us show that for any $\varepsilon > 0$ there exists $\bar{s} > 0$ such that $\|\mathbf{d}(s)x - x\|_{L^p} \le \varepsilon$ for all $s \in [-\bar{s}, \bar{s}]$. Let $M_1 = \sup_{s\in[-1,1]} \|\mathbf{d}(s)\| \le \sup_{s\in[-1,1]} \left|e^{s\tilde{\Gamma}}\right| < +\infty$. Since the space of infinitely smooth compactly supported functions is densely embedded in L^p then there exists $x_\infty \in C_c^\infty(\mathbb{R}^n; \mathbb{R}^n)$ such that $\|x - x_\infty\|_{L^p} \le \frac{\varepsilon}{3(M_1+1)}$. This implies that

$$\|\mathbf{d}(s)x - x\|_{L^p} \le \|\mathbf{d}(s)x_\infty - x_\infty + \mathbf{d}(s)(x - x_\infty) + (x_\infty - x)\|_{L^p}$$

$$\le \|\mathbf{d}(s)x_\infty - x_\infty\|_{L^p} + (\|\mathbf{d}(s)\| + 1)\|x - x_\infty\|_{L^p} \le \|\mathbf{d}(s)x_\infty - x_\infty\|_{L^p} + \frac{\varepsilon}{3}$$

for all $s \in [-1, 1]$. Since x_∞ is compactly supported then there exists $r > 0$ such that $x_\infty(z) = \mathbf{0}$ for all $z : |z| > r$. If $M_2 = \sup_{s\in[-1,1]} |e^{\Theta s}| \ge 1$ then $x_\infty(e^{\Theta s}z) - x_\infty(z) = \mathbf{0}$ for all $z : |z| > rM_2$ and $s \in [-1, 1]$. In this case, we derive

$$\|\mathbf{d}(s)x_\infty - x_\infty\|_{L^p} = \left(\int_{\mathbb{R}^n} \left|e^{s\Gamma} x_\infty(e^{\Theta s} z) - x_\infty(z)\right|^p dz\right)^{\frac{1}{p}} \le$$

$$\left(\int_{\mathbb{R}^n} |(e^{s\Gamma}-I_k)x_\infty(e^{\Theta s}z)|^p dz\right)^{\frac{1}{p}} + \left(\int_{\mathbb{R}^n} |x_\infty(e^{\Theta s}z)-x_\infty(z)|^p dz\right)^{\frac{1}{p}} =$$

$$|e^{s\Gamma} - I_k| \cdot \|x_\infty\|_{L^p} + \left(\int_{|z|\le rM_2} |x_\infty(e^{\Theta s}z) - x_\infty(z)|^p\, dz\right)^{\frac{1}{p}} \le$$

$$|e^{s\Gamma} - I_k| \cdot \|x_\infty\|_{L^p} + \left(\int_{|z|\le rM_2} M_3^p\, |(e^{\Theta s} - I_n)z|^p\, dz\right)^{\frac{1}{p}},$$

where $M_3 = \sup\limits_{|z|\le M_2 r} |\nabla x_\infty| < +\infty$ (due to $x_\infty \in C_c^\infty$) and the identity

$$x_\infty(z+h) - x_\infty(z) = \left(\int_0^1 \nabla x_\infty(z+\lambda h)d\lambda\right) h, \quad \forall h \in \mathbb{R}^n$$

is utilized in the last step. Since

$$M_4 = \sup_{s\in[-1,1]} \tfrac{|e^{s\Gamma}-I_k|}{|s|} < +\infty, \quad M_5 = \sup_{s\in[-1,1]} \tfrac{|e^{s\Theta}-I_n|}{|s|} < +\infty,$$

$M_6 = \int_{|z|\le rM_2} dz = \frac{\pi^{n/2}}{\Gamma(n/2+1)}(rM_2)^n < +\infty$, then for

$$|s| \le \bar{s} = \min\left\{1, \tfrac{\varepsilon}{3M_4\|x_\infty\|}, \tfrac{\varepsilon}{3rM_2M_3M_5M_6^{1/p}}\right\}$$

we have

$$\begin{aligned}\|\mathbf{d}(s)x_\infty - x_\infty\|_{L^p} \le& \tfrac{|e^{s\Gamma}-I_k|}{|s|}|s|\|x_\infty\|_{L^p} + M_3\tfrac{|e^{s\Theta}-I_k|}{|s|}|s|\left(\int_{|z|\le rM_2} |z|^p\, dz\right)^{\frac{1}{p}}\\ \le& |s|M_4\|x_\infty\|_{L^p} + |s|rM_2M_3M_5M_6^{1/p} \le \frac{\varepsilon}{3} + \frac{\varepsilon}{3}.\end{aligned}$$

Therefore, $\forall \varepsilon > 0, \exists \bar{s} > 0 : \|\mathbf{d}(s)x - x\|_{L^p} \le \varepsilon$ for all $s \in [-\bar{s}, \bar{s}]$, i.e., the group $\mathbf{d}$ is strongly continuous at $s = 0$. Using the group property $\mathbf{d}(s+t) = \mathbf{d}(s)\mathbf{d}(t)$, $\forall$ $s, t \in \mathbb{R}$ we expand the strong continuity to all $s \in \mathbb{R}$.

(3) Since, by assumption, $\tilde{\Gamma} = \Gamma - \frac{\mathrm{trace}(\Theta)}{p} I_k$ is anti-Hurwitz matrix then $e^{s\tilde{\Gamma}}r \to \mathbf{0}$ as $s \to -\infty$ and $e^{s\tilde{\Gamma}}r \to \infty$ as $s \to +\infty$ for any $r \in \mathbb{R}^n\backslash\{\mathbf{0}\}$. Hence, we conclude that the limit property is fulfilled, so $\mathbf{d}$ is a linear dilation in $L^p(\mathbb{R}^n; \mathbb{R}^k)$. The proof is complete. □

To be a dilation in $\mathbb{R}^k$, the mapping $s \mapsto e^{\Gamma s}$ must have an anti-Hurwitz matrix Γ (see Chap. 5 of Volume I). However, a dilation (4.15) in L^p may have a Hurwitz matrix Γ. In other words, $e^{\Gamma s}$ may not be a dilation in $\mathbb{R}^k$, but the group $\mathbf{d}$ given by (4.15) may be a dilation in L^p.

Example 4.4 Let us consider the group (4.9) on $L^2(\mathbb{R}^n; \mathbb{R}^k)$ with $\Theta = -I_n$ and $\Gamma = -I_k$. Using the change of the integration variable in the Lebesgue integral (see Appendix C), we derive

$$\|\mathbf{d}(s)x\|_{L^2}^2 = \int_{\mathbb{R}^n} |e^{-s}x(e^{-s}z)|^2 dz = e^{s(-2+n)}\|x\|_{L^2}^2.$$

Therefore, $\mathbf{d}$ is a dilation in $L^2(\mathbb{R}^n; \mathbb{R}^k)$ if $n \geq 3$, but $\mathbf{d}$ is not a dilation in $L^2(\mathbb{R}^2; \mathbb{R}^k)$ or $L^2(\mathbb{R}; \mathbb{R}^k)$.

Corollary 4.3 *If the matrix $Q = Q^\top \in \mathbb{R}^{k\times k}$ satisfies*

$$\left(\Gamma - \tfrac{\mathrm{trace}(\Theta)}{p} I_k\right)^\top Q + Q\left(\Gamma - \tfrac{\mathrm{trace}(\Theta)}{p} I_k\right) \succ 0, \quad Q \succ 0 \tag{4.13}$$

then the dilation (4.9) *is strictly monotone in the weighted Lebesgue space $L^p_*(\mathbb{R}^n; \mathbb{R}^k)$ with the norm*

$$\|x\|_{L^p_*} = \left(\int_{\mathbb{R}^n} \left(x^\top(z)Qx(z)\right)^{\frac{p}{2}} dz\right)^{\frac{1}{p}}. \tag{4.14}$$

Moreover, it holds

$$\begin{array}{l} e^{\beta s}\|z\|_{L^p_*} \leq \|\mathbf{d}(s)z\|_{L^p_*} \leq e^{\alpha s}\|z\|_{L^p_*} \text{ for } s \geq 0, \\ e^{\alpha s}\|z\|_{L^p_*} \leq \|\mathbf{d}(s)z\|_{L^p_*} \leq e^{\beta s}\|z\|_{L^p_*} \text{ for } s \leq 0, \end{array} \tag{4.15}$$

where

$$\alpha = \frac{\lambda_{\max}\left(Q^{-1/2}\left(\Gamma - \frac{\mathrm{trace}(\Theta)}{p} I_k\right)^\top Q^{1/2} + Q^{1/2}\left(\Gamma - \frac{\mathrm{trace}(\Theta)}{p} I_k\right) Q^{-1/2}\right)}{2},$$

$$\beta = \frac{\lambda_{\min}\left(Q^{-1/2}\left(\Gamma - \frac{\mathrm{trace}(\Theta)}{p} I_k\right)^\top Q^{1/2} + Q^{1/2}\left(\Gamma - \frac{\mathrm{trace}(\Theta)}{p} I_k\right) Q^{-1/2}\right)}{2}.$$

Proof For any $x \neq \mathbf{0}$ and any $s \in \mathbb{R}$ we have

$$\begin{aligned}
\frac{d}{ds}\|\mathbf{d}(s)x\|_{L^p_*}^p =& \frac{d}{ds}\int_{\mathbb{R}^n}\left(x^\top\left(e^{\Theta s}z\right)e^{\Gamma^\top s}Qe^{\Gamma s}x\left(e^{\Theta s}z\right)\right)^{\frac{p}{2}}dz\\
=&\frac{d}{ds}\int_{\mathbb{R}^n}\det(e^{-\Theta s})\left(x^\top(\tilde{z})\,e^{\Gamma^\top s}Qe^{\Gamma s}x\,(\tilde{z})\right)^{\frac{p}{2}}d\tilde{z}\\
=&\frac{d}{ds}\int_{\mathbb{R}^n}e^{-\operatorname{trace}(\Theta)s}\left(x^\top(\tilde{z})\,e^{\Gamma^\top s}Qe^{\Gamma s}x\,(\tilde{z})\right)^{\frac{p}{2}}d\tilde{z}\\
=&\int_{\mathbb{R}^n}\frac{p}{2}\left(x^\top(\tilde{z})\,e^{\tilde{\Gamma}^\top s}Qe^{\tilde{\Gamma}s}x\,(\tilde{z})\right)^{\frac{p}{2}-1}\cdot\\
&\left(x^\top(\tilde{z})\,e^{\tilde{\Gamma}^\top s}(\tilde{\Gamma}^\top Q+Q\tilde{\Gamma})e^{\tilde{\Gamma}s}x\,(\tilde{z})\right)d\tilde{z},
\end{aligned}$$

where $\tilde{\Gamma}=\Gamma-\frac{\operatorname{trace}(\Theta)}{p}I_k$. Since

$$2\beta x^\top(\tilde{z})\,Qx\,(\tilde{z})\le x^\top(\tilde{z})\,e^{\tilde{\Gamma}^\top s}(\tilde{\Gamma}^\top Q+Q\tilde{\Gamma})e^{\tilde{\Gamma}s}x\,(\tilde{z})\le 2\alpha x^\top(\tilde{z})\,Qx\,(\tilde{z})$$

then

$$p\beta\|\mathbf{d}(s)x\|_{L^p_*}^p\le\frac{d}{ds}\|\mathbf{d}(s)x\|_{L^p_*}^p\le p\alpha\|\mathbf{d}(s)x\|_{L^p_*}^p,\quad\forall s\in\mathbb{R},\quad\forall x\in L^p_*.$$

Hence, taking into account $\|\mathbf{d}(0)x\|_{L^p_*}=\|x\|_{L^p_*}$ we derive (4.15), but the matrix inequality (4.13) implies the strict monotonicity of $\mathbf{d}$ in L^p_*. The proof is complete. □

A generator of the dilation (4.9) in L^p is characterized as follows.

Lemma 4.3 *An infinitesimal generator of the dilation* (4.9) *in the Lebesgue space $L^p(\mathbb{R}^n;\mathbb{R}^k)$ is the linear closed densely defined operator $G_{\mathbf{d}}:\mathcal{D}(G_{\mathbf{d}})\subset L^p(\mathbb{R}^n;\mathbb{R}^k)\mapsto L^p(\mathbb{R}^n;\mathbb{R}^k)$ given by*

$$G_{\mathbf{d}}x=\Gamma x+((\Theta z)\cdot\nabla)\,x,\quad z\in\mathbb{R}^n,\tag{4.16}$$

whose domain $\mathcal{D}(G_{\mathbf{d}})$ is a completion of the space $C_c^\infty(\mathbb{R}^n;\mathbb{R}^k)$ with respect to the norm $\|x\|_{G_{\mathbf{d}}}=\|x\|_{L^p}+\|G_{\mathbf{d}}x\|_{L^p}$, where

$$(\Theta z)\cdot\nabla=\sum_{i-1}^{n}(\Theta z)_i\frac{\partial}{\partial z_i}.$$

Proof Taking $x_\infty \in C_c^\infty$ we derive

$$\left\|\tfrac{\mathbf{d}(s)x-x}{s}-G_{\mathbf{d}}x\right\|_{L^p}=\left(\int_{\mathbb{R}^n}\left|\tfrac{e^{s\Gamma}x(e^{s\Theta}z)-x(z)}{s}-(\Theta z)\cdot\nabla x(z)-\Gamma x(z)\right|^p dz\right)^{\frac{1}{p}}=$$

$$\left(\int_{\mathbb{R}^n}\left|\tfrac{e^{s\Gamma}-I_k-\Gamma s}{s}x(e^{s\Theta}z)+\Gamma(x(e^{\Theta s}z)-x(z))+\tfrac{x(e^{\Theta}z)-x(z)}{s}-\Theta z\cdot\nabla x(z)\right|^p dz\right)^{\frac{1}{p}}$$

$$\leq\left|\tfrac{e^{s\Gamma}-I_k}{s}-\Gamma\right|\left(\int_{\mathbb{R}^n}|x(e^{s\Gamma}z)|^p dz\right)^{\frac{1}{p}}+|\Gamma|\left(\int_{\mathbb{R}^n}|x(e^{s\Theta}z)-x(z)|^p dz\right)^{\frac{1}{p}}+$$

$$\left(\int_{\mathbb{R}^n}|\tfrac{x(e^{s\Theta}z)-x(z)}{s}-(\Theta z)\cdot\nabla x(z)|^p dz\right)^{\frac{1}{p}},$$

where $|A|=\sup_{y\in\mathbb{R}^k}\frac{|Ay|}{|y|}$ is the matrix norm induced by the Euclidean norm $|y|=\sqrt{y^\top y}$ with $y\in\mathbb{R}^k$ and $A\in\mathbb{R}^{k\times k}$. All three terms in the above estimate tend to zero as $s\to 0$, so

$$\lim_{s\to 0}\tfrac{\mathbf{d}(s)x_\infty-x_\infty}{s}=\Gamma x_\infty+((\Theta z)\cdot\nabla)\,x_\infty.$$

Making a completion of C_c^∞ in the norm $\|x\|_{G_{\mathbf{d}}}$ we derive $\mathcal{D}(G_{\mathbf{d}})$. Since, by Lemma 4.2, $\mathbf{d}$ is a strongly continuous group in L^p then the operator $G_{\mathbf{d}}:\mathcal{D}(G_{\mathbf{d}})\subset L^p\mapsto L^p$ is closed by Corollary 2.9 of Hille–Yosida theorem (see Chap. 2). The proof is complete. □

The dilation (4.9) in L^p could also be a dilation in Sobolev spaces $W^{p,m}(\mathbb{R}^n;\mathbb{R}^k)$ under some additional restriction to parameters Γ and Θ.

Corollary 4.4 *Let a strongly continuous linear dilation* $\mathbf{d}$ *in* $L^2(\mathbb{R}^n;\mathbb{R}^k)$ *be defined by the formula* (4.9) *with a diagonal matrix*

$$\Theta=\mathrm{diag}(\theta_1,\ldots,\theta_n)\in\mathbb{R}^{n\times n}. \tag{4.17}$$

- *The Sobolev space* $W^{m,p}(\mathbb{R}^n;\mathbb{R}^k)$ *is a* $\mathbf{d}$*-homogeneous cone in* $L^p(\mathbb{R}^n;\mathbb{R}^k)$.
- *The group* $\mathbf{d}$ *is a strongly continuous linear dilation in* $W^{m,p}(\mathbb{R}^n;\mathbb{R}^k)$ *provided that the matrices*

$$\tilde{\Gamma}=\Gamma-\tfrac{\mathrm{trace}\,\Theta}{p}I_k,\quad \tilde{\Gamma}_{\overline{m}}=\Gamma+\left(\theta_{\overline{m}}-\tfrac{\mathrm{trace}\,\Theta}{p}\right)I_k,\quad 0\leq|\overline{m}|\leq m \tag{4.18}$$

are anti-Hurwitz, where

$$\overline{m} = (m_1, ..., m_n)^\top \in \mathbb{N}_*^n, \quad \mathbb{N}_* = \mathbb{N} \cup \{0\},$$

$$\theta_{\overline{m}} = \theta_1 m_1 + \cdots + \theta_n m_n.$$

Proof We prove the claim for $m = 1$ and $m = 2$. The cases $m \geq 3$ can be studied similarly by induction.

(1) Let $m = 1$. Using the linearity of the differential operator and the differentiation rule for compositions of functions, we derive

$$\left(\frac{\partial}{\partial z_i}\mathbf{d}(s)x\right)(z) = \frac{\partial}{\partial z_i} e^{s\Gamma} x(e^{s\Theta} z) = e^{s\Gamma} \frac{\partial}{\partial z_i} x(e^{s\Theta} z) = e^{s\Gamma} \frac{\partial x(\tilde{z})}{\partial \tilde{z}} e^{s\Theta} h_i \bigg|_{\tilde{z}=e^{s\Theta}z},$$

where $h_i = (0, ..., 0, 1, 0, ..., 0)^\top$ is the identity vector. Since Θ is a diagonal matrix then $e^{s\Theta} h_i = e^{\theta_i s} h_i$. Making the change of variables in the Lebesgue integral, we obtain

$$\left\|\frac{\partial}{\partial z_i}\mathbf{d}(s)x\right\|_{L^p}^p = \int_{\mathbb{R}^n} \left| e^{s\theta_i} e^{s\Gamma} \frac{\partial x(\tilde{z})}{\partial \tilde{z}} h_i \bigg|_{\tilde{z}=e^{s\Theta}z} \right|^p dz = \int_{\mathbb{R}^n} \left| e^{s\tilde{\Gamma}_i} \frac{\partial x(\tilde{z})}{\partial \tilde{z}} h_i \right|^p d\tilde{z}.$$

Since $\tilde{\Gamma}_i = \Gamma + \left(\theta_i - \frac{\text{trace}\,\Theta}{p}\right) I_k$ is anti-Hurwitz then $\left|e^{s\tilde{\Gamma}_i} \frac{\partial x(\tilde{z})}{\partial \tilde{z}} h_i\right| \to 0$ as $s \to 0$ and $\left|e^{s\tilde{\Gamma}_i} \frac{\partial x(\tilde{z})}{\partial \tilde{z}} h_i\right| \to +\infty$ as $s \to +\infty$ provided that $\frac{\partial x(\tilde{z})}{\partial \tilde{z}} h_i = \frac{\partial x(\tilde{z})}{\partial \tilde{z}_i} \neq \mathbf{0}$. Taking into account that

$$\|\mathbf{d}(s)x\|_{W^{1,p}} = \|\mathbf{d}(s)x\|_{L^p} + \sum_{i=1}^n \left\|\frac{\partial}{\partial z_i}\mathbf{d}(s)x\right\|_{L^p} \leq |e^{s\tilde{\Gamma}}| \cdot \|x\|_{L^p} + \sum_{i=1}^n |e^{s\tilde{\Gamma}_i}| \left\|\frac{\partial x}{\partial z_i}\right\|_{L^p},$$

we conclude that $W^{1,p}$ is a $\mathbf{d}$-homogeneous cone in L^2. Moreover, $\mathbf{d}$ is a dilation in $W^{1,p}$ if $\tilde{\Gamma}$ and $\tilde{\Gamma}_i$ are anti-Hurwitz matrices. The proof of strong continuity of $\mathbf{d}$ in $W^{1,p}$ repeats the corresponding proof from Lemma 4.2.

(2) Let $m = 2$. Using the diagonal structure of Θ we derive

$$\left(\frac{\partial^2}{\partial z_j \partial z_i}\mathbf{d}(s)x\right)(z) = e^{\theta_i s} e^{s\Gamma} \frac{\partial}{\partial z_j} \frac{\partial x(\tilde{z})}{\partial \tilde{z}_i}\bigg|_{\tilde{z}=e^{s\Theta}z} = e^{\theta_j s} e^{\theta_i s} e^{s\Gamma} \frac{\partial^2 x(\tilde{z})}{\partial \tilde{z}_j \partial \tilde{z}_i}\bigg|_{\tilde{z}=e^{s\Theta}z}.$$

If the matrices $\Gamma_{\overline{m}}$ are anti-Hurwitz for all $\overline{m} \in \mathbb{N}_*^n : |\overline{m}| \leq m$ then matrices $\Gamma + \theta_i I_k + \theta_j I_k$ are anti-Hurwitz as well. So, $e^{\theta_j s} e^{\theta_i s} e^{s\Gamma} y \to 0$ as $s \to -\infty$ and $e^{\theta_j s} e^{\theta_i s} e^{s\Gamma} y \to \infty$ as $s \to +\infty$ for any $y \neq \mathbf{0}$. Taking into account

$$\|\mathbf{d}(s)x\|_{W^{1,p}} = \|\mathbf{d}(s)x\|_{W^{p,1}} + \sum_{i,j=1}^n \left\|\frac{\partial^2}{\partial z_j \partial z_i}\mathbf{d}(s)x\right\|_{L^p}$$

$$\leq \|\mathbf{d}(s)x\|_{W^{p,1}} + \sum_{i,j=1}^{n} e^{\theta_j s} e^{\theta_i s} |e^{s\Gamma}| \left\| \frac{\partial^2}{\partial z_j \partial z_i} x \right\|_{L^p},$$

we conclude that $W^{1,p}$ is a **d**-homogeneous cone in L^2. Since **d** is a dilation in $W^{1,p}$ and the matrices $\Gamma + \theta_i I_k + \theta_j I_k$ are anti-Hurwitz then **d** is a dilation in $W^{2,p}$ too.

(3) The cases $m = 3, 4, \ldots$ can be analyzed similarly by induction. The proof is complete. □

The above corollary shows that a dilation **d** in a Banach space $\mathbb{B}$ may be a dilation in a Banach space $\mathbb{Y}$ embedded in $\mathbb{B}$. However, the following example shows that a dilation in $\mathbb{Y}$ may not be a dilation $\mathbb{B}$ despite that $\mathbb{Y}$ is densely embedded in $\mathbb{B}$.

Example 4.5 Let us consider a strongly continuous group of linear bounded operators on $L^2(\mathbb{R}^2; \mathbb{R})$

$$(\mathbf{d}(s)x)(z) = e^s x(e^s z), \quad s \in \mathbb{R}, \quad z \in \mathbb{R}^2, \quad x \in L^2(\mathbb{R}^2; \mathbb{R}).$$

The group **d** is not a dilation in $L^2(\mathbb{R}^2; \mathbb{R})$ since

$$\|\mathbf{d}(s)x\|^2_{L^2(\mathbb{R}^2;\mathbb{R})} = \int_{\mathbb{R}^2} \left|e^s x(e^s z)\right|^2 dz = \int_{\mathbb{R}^2} e^{-2s} \left|e^s x(\tilde{z})\right|^2 d\tilde{z} = \|x\|_{L^2(\mathbb{R}^2;\mathbb{R})}.$$

However, for $x \in H^1(\mathbb{R}^2; \mathbb{R}) \subset L^2(\mathbb{R}^2; \mathbb{R})$ we have

$$\|\mathbf{d}(s)x\|^2_{H^1(\mathbb{R}^2;\mathbb{R})} = \|x\|^2_{L^2(\mathbb{R}^2;\mathbb{R})} + \sum_{i=1}^{2} \int_{\mathbb{R}^2} \left| \frac{\partial}{\partial z_i} e^s x(e^s z) \right|^2 dz$$

$$= \|x\|^2_{L^2(\mathbb{R}^2;\mathbb{R})} + e^{2s} \sum_{i=1}^{2} \left\| \frac{\partial x}{\partial z_i} \right\|^2_{L^2(\mathbb{R}^2;\mathbb{R})}.$$

If $x \in H^1(\mathbb{R}^2; \mathbb{R})$ then the identity

$$\sum_{i=1}^{2} \left\| \frac{\partial x}{\partial z_i} \right\|_{L^2(\mathbb{R}^2;\mathbb{R})} = 0$$

may hold only for $x = \mathbf{0}$. Indeed, the constant function is the only function whose derivatives are zero. Hence, **d** is a monotone dilation in $H^1(\mathbb{R}^2; \mathbb{R})$, but **d** is not a dilation in $L^2(\mathbb{R}^2; \mathbb{R})$.

Remark 4.1 Corollary 4.4 is valid for Sobolev spaces $W^{m,p}(\mathbb{R}^n; \mathbb{R}^k)$ with a real $m \in [0, +\infty)$ and $1 < p < +\infty$.

Let us prove the remark for $\Theta = \theta I_n$ with $\theta \in \mathbb{R}$. First of all, notice that the Fourier transform and its inverse are symmetric with respect to the linear dilation $\mathbf{d}$ in the following sense:

$$(\mathcal{F}[\mathbf{d}(s)f])(\xi) = \tilde{\mathbf{d}}(s)(\mathcal{F}[f])(\xi), \quad \forall s \in \mathbb{R}, \quad \xi \in \mathbb{R}^n, \tag{4.19}$$

$$(\mathcal{F}^{-1}[\tilde{\mathbf{d}}(s)\tilde{f}])(x) = \mathbf{d}(s)(\mathcal{F}[\tilde{f}])(x), \quad \forall s \in \mathbb{R}, \quad x \in \mathbb{R}^n, \tag{4.20}$$

where $f, \tilde{f} \in L^p(\mathbb{R}^n; \mathbb{R}^k)$ and $\tilde{\mathbf{d}}(s) \in \mathcal{L}(L^p; L^p)$ is given by

$$(\tilde{\mathbf{d}}(s)\tilde{f})(\xi) = e^{(\Gamma - \mathrm{trace}(\Theta)I_k)s}\tilde{f}(e^{-\Theta^\top s}\xi).$$

Due to linearity of the Fourier transform we have

$$\left\|\mathcal{F}^{-1}\left[(1+\xi^2)^{\frac{k}{2}}\mathcal{F}(f)\right]\right\|_{L^p} < \infty \Leftrightarrow \left\|\mathcal{F}^{-1}\left[(\rho+\xi^2)^{\frac{k}{2}}\mathcal{F}(f)\right]\right\|_{L^p} < \infty, \forall \rho > 0.$$

Using the identity (4.12), for $\Theta = \theta I_n$ we derive

$$\|f\|_{W^{m,p}} = \left\|\mathcal{F}^{-1}\left[(1+\xi^2)^{\frac{m}{2}}\mathcal{F}(\mathbf{d}(s)f)\right]\right\|_{L^p} = \left\|\mathcal{F}^{-1}\left[(1+\xi^2)^{\frac{m}{2}}\tilde{\mathbf{d}}(s)\mathcal{F}(f)\right]\right\|_{L^p}$$

$$= \left\|\mathcal{F}^{-1}\left[\tilde{\mathbf{d}}(s)\{(1+e^{2\theta}\xi^2)^{\frac{m}{2}}\mathcal{F}(f)\}\right]\right\|_{L^p} = \left\|\mathbf{d}(s)\mathcal{F}^{-1}\left[(1+e^{2\theta s}\xi^2)^{\frac{m}{2}}\mathcal{F}(f)\right]\right\|_{L^p}$$

$$= \left\|e^{(\Gamma+\theta(m-n/p)I_k)s}\mathcal{F}^{-1}\left[(e^{-2\theta s}+\xi^2)^{\frac{m}{2}}\mathcal{F}(f)\right]\right\|_{L^p}.$$

Hence, for any $f \in W^{m,p}$ and for any $s \in \mathbb{R}$ we have

$$\left\lfloor e^{\left(\Gamma+\theta\left(m-\frac{n}{p}\right)I_k\right)s}\right\rfloor \left\|\mathcal{F}^{-1}\left[(e^{-2\theta s}+\xi^2)^{\frac{m}{2}}\mathcal{F}(f)\right]\right\|_{L^p} \le \|f\|_{W^{m,p}}$$

$$\le \left|e^{\left(\Gamma+\theta\left(m-\frac{n}{p}\right)I_k\right)s}\right| \left\|\mathcal{F}^{-1}\left[(e^{-2\theta s}+\xi^2)^{\frac{m}{2}}\mathcal{F}(f)\right]\right\|_{L^p} < +\infty.$$

Therefore, the group $\mathbf{d}$ with $\Theta = \theta I_n$ is a linear dilation in $W^{m,p}$ if the matrices $\Gamma + \theta\left(m - \frac{n}{p}\right)I_k$ and $\Gamma - \frac{n}{p}I_k$ are anti-Hurwitz. This corresponds to the condition given in Corollary 4.4. The proof of the remark for $\Theta = \mathrm{diag}\{\theta_1, ..., \theta_n\}$ is left for the reader.

4.2 Generalized Homogeneous Spaces

4.2.1 Canonical Homogeneous Norm

In this section we introduce the canonical homogeneous norm in $\mathbb{B}$. In Volume I, it was demonstrated that the homogeneous norm is a very efficient tool for the analysis and design of homogeneous control systems. In Volume II, we show that this tool is useful for the analysis and design of infinite-dimensional homogeneous control systems as well.

Definition 4.6 Let a linear strongly continuous dilation $\mathbf{d}$ in a real Banach space $\mathbb{B}$ be monotone. The functional $\|\cdot\|_{\mathbf{d}} : \mathbb{B} \mapsto \mathbb{R}_+$ defined as $\|\mathbf{0}\|_{\mathbf{d}} = 0$ and

$$\|x\|_{\mathbf{d}} = e^{s_x}, \quad \text{where } s_x \in \mathbb{R} : \|\mathbf{d}(-s_x)x\| = 1, \quad x \neq \mathbf{0} \tag{4.21}$$

is called the canonical $\mathbf{d}$-homogeneous norm in $\mathbb{B}$.

Below we show this functional is an norm (in the usual sense) in a Banach space $\mathbb{B}_{\mathbf{d}}$ homeomorphic to $\mathbb{B}$. In some cases, the canonical homogeneous norm may have an explicit relation with the norm $\|\cdot\|$ in $\mathbb{B}$.

Example 4.6 Let $\mathbb{B}$ be the Banach space of bounded uniformly continuous functions $\mathbb{R} \mapsto \mathbb{R}$ with the supremum norm, and let a dilation group $\mathbf{d}$ in $\mathbb{B}$ be defined as follows:

$$(\mathbf{d}(s)z)(x) = e^{\alpha s} z(x + \beta s),$$

where $s \in \mathbb{R}$ is the group parameter, $z \in \mathbb{B}$, $x \in \mathbb{R}^n$ and $\alpha > 0$ and $\beta \in \mathbb{R}$ are constant parameters. Then

$$\|\mathbf{d}(s)z\| = \sup_{x\in\mathbb{R}} |e^{\alpha s} z(x + s)| = e^{\alpha s} \sup_{x\in\mathbb{R}} |e^{\alpha s} z(x + \beta s)| = e^{\alpha s}\|z\|,$$

and $\|z\|_{\mathbf{d}} = \|z\|^{\frac{1}{\alpha}}$.

The monotonicity of the dilation group is required to guarantee that the functional $\|\cdot\|_{\mathbf{d}}$ is single-valued and continuous at the origin.

Lemma 4.4 *If a strongly continuous linear dilation $\mathbf{d}$ in $\mathbb{B}$ is monotone then*

(1) the functional $\|\cdot\|_{\mathbf{d}} : \mathbb{B} \mapsto \mathbb{R}_+$ given by (4.21) is single-valued, positive definite, and $\|x\|_{\mathbf{d}} \to 0$ as $\|x\| \to \mathbf{0}$;
(2) $\|\pm\mathbf{d}(s)x\|_{\mathbf{d}} = e^s\|x\|_{\mathbf{d}}$, $\forall s \in \mathbb{R}$, $\forall x \in \mathbb{R}^n$;
(3) $\|x\| = 1 \Leftrightarrow \|x\|_{\mathbf{d}} = 1$, $\|x\| < 1 \Leftrightarrow \|x\|_{\mathbf{d}} < 1$ and $\|x\| > 1 \Leftrightarrow \|x\|_{\mathbf{d}} > 1$.

Proof (1) Since the strongly continuous linear dilation in $\mathbb{B}$ is monotone then for any $x \in \mathbb{B}\backslash\{\mathbf{0}\}$, there exists a unique pair $s_0 \in \mathbb{R}$ and $x_0 \in S$ such that $x = \mathbf{d}(s_0)x_0$ (see

Proposition 4.2). This means that $\|\mathbf{d}(-s_0)x\| = 1$, i.e., $\|x\|_{\mathbf{d}} = e^{s_0}$ is finite, positive, and uniquely defined for any $x \in \mathbb{B}\backslash\{\mathbf{0}\}$.

Since $\|\mathbf{d}(-\ln \|x\|_{\mathbf{d}})x\| = 1$ for $x \in \mathbb{B}\backslash\{\mathbf{0}\}$ then $\|\mathbf{d}(-\ln \|x\|_{\mathbf{d}})\| \geq \frac{1}{\|x\|}$. This means that $\|\mathbf{d}(-\ln \|x\|_{\mathbf{d}})\| \to +\infty$ as $\|x\| \to 0$. Due to Proposition 4.1, this is possible only if $\ln \|x\|_{\mathbf{d}} \to -\infty$, i.e., $\|x\|_{\mathbf{d}} \to 0$ as $\|x\| \to 0$.

(2) Let $\|x\|_{\mathbf{d}} = e^{s_x}$, where $s_x \in \mathbb{R} : \|\mathbf{d}(-s_x)x\| = 1$. Since $\|\mathbf{d}(s_x)x\|_{\mathbf{d}} = e^{\tilde{s}_x}$, where $\tilde{s}_x \in \mathbb{R} : \|\mathbf{d}(-\tilde{s}_x)\mathbf{d}(s)x\| = 1$, then using the group property we derive $\|\mathbf{d}(-\tilde{s}_x + s)x\| = 1 = \|\mathbf{d}(-s_x)x\|$. Due to uniqueness of the homogeneous projection, we have $\tilde{s}_x = s + s_x$ and $\|\mathbf{d}(s)x\|_{\mathbf{d}} = e^{\tilde{s}_x} = e^s e^{s_x} = e^s \|x\|_{\mathbf{d}}$. The identity $\|x\|_{\mathbf{d}} = \|-x\|_{\mathbf{d}}$ is obvious.

(3) If $\|x\| = 1$ then for $s_x = 0$ we have $\|\mathbf{d}(-s_x)x\| = \|x\| = 1$ and $\|x\|_{\mathbf{d}} = e^{s_x} = 1$, and vice versa. If $\|x\| > 1$ and $\|\mathbf{d}(-s_x)x\| = 1$ then due to monotonicity of the dilation $-s_x < 0$. This means that $\|x\|_{\mathbf{d}} = e^{s_x} > 1$ and vice versa. The case $\|x\| < 1$ can be analyzed similarly. The proof is complete. □

The above lemma proves that the canonical homogeneous norm $\|\cdot\|_{\mathbf{d}}$ is continuous at $\mathbf{0}$. Regularity of the canonical homogeneous norm can be established for strictly monotone dilations.

Lemma 4.5 *Let a strongly continuous linear dilation* $\mathbf{d}$ *in* $\mathbb{B}$ *be strictly monotone. If* $\|\cdot\|_{\mathbf{d}}$ *is a canonical* $\mathbf{d}$*-homogeneous norm in* $\mathbb{B}$ *then*

(1) the estimates

$$\underline{\sigma}(\|x\|_{\mathbf{d}}) \leq \|x\| \leq \overline{\sigma}(\|x\|_{\mathbf{d}}), \qquad \forall x \in \mathbb{B}, \tag{4.22}$$

hold for

$$\overline{\sigma}(\rho) = \max\{M\rho^{\alpha}, \rho^{\beta}\}, \quad \underline{\sigma}(\rho) = \min\{M^{-1}\rho^{\alpha}, \rho^{\beta}\},$$

where $\alpha \geq \beta > 0$ *are defined in Proposition 4.3.*

(2) the canonical -homogeneous norm $\|\cdot\|_{\mathbf{d}}$ *is locally Lipschitz continuous on* $\mathbb{B}\backslash\{\mathbf{0}\}$, *and for any* $r > 0$ *it holds*

$$\left|\|x_1\|_{\mathbf{d}}^{\beta} - \|x_2\|_{\mathbf{d}}^{\beta}\right| \leq r^{\beta}\|\mathbf{d}(-\ln r)(x_1 - x_2)\|, \quad \forall x_1, x_2 \in \mathbb{B}\backslash B_{\mathbf{d}}(r), \tag{4.23}$$

where $\beta > 0$ *is introduced in Definition 4.5 and* $B_{\mathbf{d}}(r) = \mathbf{d}(\ln r)B$ *is the open* $\mathbf{d}$*-homogeneous ball of the radius* $r > 0$ *defined by scaling of the unit ball* $B = \{x \in \mathbb{R}^n : \|x\| < 1\}$ *in* $\mathbb{B}$;

- *if* $\alpha = \beta$ *then* $\|\cdot\|_{\mathbf{d}}$ *is uniformly continuous on* $\mathbb{B}$ *and*

$$\left|\|x_1\|_{\mathbf{d}}^{\beta} - \|x_2\|_{\mathbf{d}}^{\beta}\right| \leq \overline{\sigma}\left(\underline{\sigma}^{-1}(\|x_1 - x_2\|)\right), \quad \forall x_1, x_2 \in \mathbb{B}, \tag{4.24}$$

where $\underline{\sigma}^{-1} \in \mathcal{K}_{\infty}$ *is the inverse function to* $\underline{\sigma} \in \mathcal{K}_{\infty}$ *defined above.*

Proof (1) Since $\|\mathbf{d}(-\ln\|x\|_{\mathbf{d}})x\| = 1$ then using the estimates (4.6), we derive

$$\|x\| = \|\mathbf{d}(\ln\|x\|_{\mathbf{d}})\mathbf{d}(-\ln\|x\|_{\mathbf{d}})x\| \le \|\mathbf{d}(\ln\|x\|_{\mathbf{d}})\| \le M\|x\|_{\mathbf{d}}^{\alpha} \quad \text{for} \quad \|x\| \ge 1$$

and due to $\lfloor \mathbf{d}(s)\rfloor = \|\mathbf{d}(-s)\|^{-1}$, $s \in \mathbb{R}$ (see the formula 2.1) we derive

$$\|x\| = \|\mathbf{d}(\ln\|x\|_{\mathbf{d}})\mathbf{d}(-\ln\|x\|_{\mathbf{d}})x\| \ge \lfloor \mathbf{d}(\ln\|x\|_{\mathbf{d}})\rfloor \ge \frac{1}{Me^{-\alpha\ln\|u\|_{\mathbf{d}}}} = \frac{\|x\|_{\mathbf{d}}^{\alpha}}{M},$$

for $\|x\| \le 1$. Similarly, we derive

$$\|x\| = \|\mathbf{d}(\ln\|x\|_{\mathbf{d}})\mathbf{d}(-\ln\|u\|_{\mathbf{d}})x\| \le \|\mathbf{d}(\ln\|x\|_{\mathbf{d}})\| \le \|x\|^{\beta} \quad \text{for} \quad \|x\| \le 1$$

and $\lfloor \mathbf{d}(s)\rfloor \ge e^{\beta s}$ for $s > 0$. Hence, $\alpha \ge \beta$ and

$$\|x\| = \|\mathbf{d}(\ln\|x\|_{\mathbf{d}})\mathbf{d}(-\ln\|x\|_{\mathbf{d}})x\| \ge \lfloor \mathbf{d}(\ln\|x\|_{\mathbf{d}})\rfloor \ge \|x\|_{\mathbf{d}}^{\beta} \quad \text{for} \quad \|x\| \ge 1.$$

The inequalities (4.22) are fulfilled.

(2) By definition of the homogeneous norm one has $|\|u\|_{\mathbf{d}}^{\beta} - \|v\|_{\mathbf{d}}^{\beta}| = |e^{\beta s_u} - e^{\beta s_v}|$, where $\|\mathbf{d}(-s_u)u\| = 1$ and $\|\mathbf{d}(-s_v)v\| = 1$ and $s_u = \|u\|_{\mathbf{d}}, s_v = \|v\|_{\mathbf{d}}$. Since for strictly monotone dilations we have

$$\|\mathbf{d}(s)\| \le e^{\beta s}, \quad \forall s \le 0,$$

then for $s_u < s_v$ we derive

$$1 = \|\mathbf{d}(-s_v)(v-u) + \mathbf{d}(s_u - s_v)\mathbf{d}(-s_u)u\| \le \|\mathbf{d}(-s_v)\|\,\|u-v\| + \|\mathbf{d}(s_u - s_v)\|$$
$$\le \|\mathbf{d}(-s_v)\| \cdot \|u-v\| + e^{\beta(s_u - s_v)}$$

and for $s_u > s_v$ we obtain $1 \le \|\mathbf{d}(-s_u)\| \cdot \|u-v\| + e^{\beta(s_v - s_u)}$. Therefore,

$$\left|\|u\|_{\mathbf{d}}^{\beta} - \|v\|_{\mathbf{d}}^{\beta}\right| = \left|e^{\beta s_u} - e^{\beta s_v}\right| \le \|u-v\| \begin{cases} e^{\beta s_u}\|\mathbf{d}(-s_u)\| & \text{if } \|u\|_{\mathbf{d}} \ge \|v\|_{\mathbf{d}}, \\ e^{\beta s_v}\|\mathbf{d}(-s_v)\| & \text{if } \|u\|_{\mathbf{d}} < \|v\|_{\mathbf{d}}. \end{cases}$$

Since for the strictly monotone dilation $\mathbf{d}$ we have

$$u, v \in \mathbb{B}\backslash B \Rightarrow \|u\|_{\mathbf{d}} > 1, \|v\|_{\mathbf{d}} > 1 \Rightarrow s_v > 0, s_u > 0$$
$$\Rightarrow \max\left\{e^{\beta s_v}\|\mathbf{d}(-s_v)\|, e^{\beta s_u}\|\mathbf{d}(-s_u)\|\right\} = 1$$

then

$$\left|\|u\|_{\mathbf{d}}^{\beta} - \|v\|_{\mathbf{d}}^{\beta}\right| = \left|e^{\beta s_u} - e^{\beta s_v}\right| \le \|u-v\|, \quad \forall u, v \in \mathbb{B}\backslash B.$$

By definition of the $\mathbf{d}$-homogeneous ball, we derive $\mathbb{B}\backslash B_{\mathbf{d}}(r) = \{x \in \mathbb{B} : \|\mathbf{d}(-\ln r)x\| > 1\}$ then $u, v \in \mathbb{B}\backslash B_{\mathbf{d}}(r) \Rightarrow \mathbf{d}(-\ln r)u, \mathbf{d}(-\ln r)v \in \mathbb{B}\backslash B$ and

$$|\|\mathbf{d}(-\ln r)u\|_{\mathbf{d}}^{\beta} - \|\mathbf{d}(-\ln r)v\|_{\mathbf{d}}^{\beta}| \leq \|\mathbf{d}(-\ln r)u - \mathbf{d}(-\ln r)v\|, \quad \forall u, v \in \mathbb{B}\backslash B_{\mathbf{d}}(r).$$

Taking into account the identity $\|\mathbf{d}(s)x\|_{\mathbf{d}} = e^s \|x\|_{\mathbf{d}}, \forall s \in \mathbb{R}, \forall x \in \mathbb{B}$, we derive the inequality (4.23). Hence, using the inequality (4.22) we derive

$$\left| \|x_1\|_{\mathbf{d}}^{\beta} - \|x_2\|_{\mathbf{d}}^{\beta} \right| \leq r^{\beta}\overline{\sigma}\left(\|\mathbf{d}(-\ln r)(x_1 - x_2)\|_{\mathbf{d}}\right) = r^{\beta}\overline{\sigma}\left(r^{-1}\|x_1 - x_2\|_{\mathbf{d}}\right)$$

$$\leq \max\{Mr^{\beta-\alpha}\|x_1 - x_2\|_{\mathbf{d}}^{\alpha}, \|x_1 - x_2\|_{\mathbf{d}}^{\beta}\}$$

for all $x_1, x_2 \in B\backslash B_{\mathbf{d}}(r)$ and for all $r > 0$. The condition $\alpha = \beta$ implies (4.24).

The proof is complete. □

The canonical homogeneous norm in a real Hilbert space $\mathbb{H}$ is differentiable on the domain $\mathcal{D}(G_{\mathbf{d}})$ of the generator $G_{\mathbf{d}}$. Recall that, on the one hand, $\mathcal{D}(G_{\mathbf{d}}) \subset \mathbb{H}$ is a linear subspace dense in $\mathbb{H}$, but, on the other hand, $\mathcal{D}(G_{\mathbf{d}})$ is a $\mathbf{d}$-homogeneous cone, i.e., $\mathbf{d}(s)\mathcal{D}(G_{\mathbf{d}}) \subset \mathcal{D}(G_{\mathbf{d}}), \forall s \in \mathbb{R}$.

Lemma 4.6 *Let* $\mathbf{d}$ *be a strongly continuous strictly monotone linear dilation in a real Hilbert space* $\mathbb{H}$ *then the homogeneous norm* $\|\cdot\|_{\mathbf{d}}$ *is differentiable on* $\mathcal{D}(G_{\mathbf{d}})\backslash\{\mathbf{0}\}$ *and the Fréchet derivative of* $\|\cdot\|_{\mathbf{d}}$ *at* $u \in \mathcal{D}(G_{\mathbf{d}})\backslash\{\mathbf{0}\}$ *is given by*

$$(D\|u\|_{\mathbf{d}})(\cdot) = \frac{\langle \mathbf{d}(-\ln\|u\|_{\mathbf{d}})\,\cdot\,,\ \mathbf{d}(-\ln\|u\|_{\mathbf{d}})u\rangle}{\langle G_{\mathbf{d}}\mathbf{d}(-\ln\|u\|_{\mathbf{d}})u, \mathbf{d}(-\ln\|u\|_{\mathbf{d}})u\rangle}\|u\|_{\mathbf{d}}. \tag{4.25}$$

Proof Let us consider the function $\varphi : \mathbb{R} \times \mathbb{H} \mapsto \mathbb{R}$ defined as follows:

$$\varphi(s, u) = \langle \mathbf{d}(s)u, \mathbf{d}(s)u\rangle - 1, \quad s \in \mathbb{R}, \quad u \in \mathbb{H}.$$

The equation $\varphi(s, u) = 0$ defines an implicit function $s : \mathbb{H} \mapsto \mathbb{R}$ (i.e., $u \mapsto s(u)$) that is related with the canonical homogeneous norm as $\|u\|_{\mathbf{d}} = e^{-s(u)}$. To prove its differentiability, the Implicit Function theorem (see Appendix B or [3, Theorem E]) can be utilized. For this purpose, we need to check differentiability of the function φ at the point $(s, u) \in \mathbb{R} \times \mathcal{D}(G_{\mathbf{d}})$.

Let us show that the Fréchet derivative of the functional $\varphi : \mathbb{R} \times \mathbb{H} \mapsto \mathbb{R}$ is given by

$$D\varphi(\delta_s, \delta_u) = 2\langle G_{\mathbf{d}}\mathbf{d}(s)u, \mathbf{d}(s)u\rangle\delta_s + 2\langle \mathbf{d}(s)\delta_u, \mathbf{d}(s)u\rangle, \ \delta_s \in \mathbb{R}, \ \delta_u \in \mathbb{H}. \tag{4.26}$$

The function $D_s\varphi : \mathbb{R} \mapsto \mathbb{R}$ given by

$$D_s\varphi(\delta_s) := 2\langle G_{\mathbf{d}}\mathbf{d}(s)u, \mathbf{d}(s)u\rangle\delta_s$$

is a partial derivative of ϕ on s-variable, but the functional $D_u\varphi : \mathbb{H} \mapsto \mathbb{R}$ defined as

$$D_u\varphi(\delta_u) := 2\langle \mathbf{d}(s)\delta_u, \mathbf{d}(s)u\rangle$$

is a partial derivative of ϕ on u. Let us consider the difference

$$\begin{aligned}
&\varphi(s+\delta_s, u+\delta_u) - \varphi(s,u)\\
=&\langle \mathbf{d}(s+\delta_s)(u+\delta_u), \mathbf{d}(s+\delta_s)(u+\delta_u)\rangle - \langle \mathbf{d}(s)u, \mathbf{d}(s)u\rangle\\
=&\langle \mathbf{d}(s+\delta_s)u, \mathbf{d}(s+\delta_s)u\rangle + 2\langle \mathbf{d}(s+\delta_s)\delta u, \mathbf{d}(s+\delta_s)u\rangle\\
&+ \langle \mathbf{d}(s+\delta_s)\delta u, \mathbf{d}(s+\delta_s)\delta u\rangle - \langle \mathbf{d}(s)u, \mathbf{d}(s)u\rangle\\
=&\langle (\mathbf{d}(\delta_s) - I)\mathbf{d}(s)u, \mathbf{d}(s+\delta_s)u\rangle + \langle (\mathbf{d}(\delta_s) - I)\mathbf{d}(s)u, \mathbf{d}(s)u\rangle\\
&+ 2\langle \mathbf{d}(s+\delta_s)\delta u, \mathbf{d}(s+\delta_s)u\rangle + \langle \mathbf{d}(s+\delta_s)\delta_u, \mathbf{d}(s+\delta_s)\delta_u\rangle\\
=&\langle (\mathbf{d}(\delta_s) - I)\mathbf{d}(s)u, \mathbf{d}(s+\delta_s)u\rangle + \langle (\mathbf{d}(\delta_s) - I)\mathbf{d}(s)u, \mathbf{d}(s)u\rangle\\
&+ 2\langle \mathbf{d}(s+\delta_s)\delta_u, \mathbf{d}(s+\delta_s)u\rangle + \langle \mathbf{d}(s+\delta_s)\delta_u, \mathbf{d}(s+\delta_s)\delta_u\rangle.
\end{aligned}$$

On the one hand, since $\mathbf{d}(s)u \in \mathcal{D}(G_{\mathbf{d}})$ then we have

$$\left|\left\langle \left(\tfrac{\mathbf{d}(\delta s)-I}{\delta s} - G_{\mathbf{d}}\right)\mathbf{d}(s)u, \mathbf{d}(s)u\right\rangle\right| \to 0 \text{ as } \delta s \to 0$$

and

$$\left|\left\langle \tfrac{\mathbf{d}(\delta s)-I}{\delta s}\mathbf{d}(s)u, \mathbf{d}(s+\delta s)u\right\rangle - \langle G_{\mathbf{d}}\mathbf{d}(s)u, \mathbf{d}(s)u\rangle\right| \leq$$

$$\left|\left\langle \left(\tfrac{\mathbf{d}(\delta s)-I}{\delta s} - G_{\mathbf{d}}\right)\mathbf{d}(s)u, \mathbf{d}(s+\delta s)u\right\rangle\right|$$

$$+ \left|\left\langle \tfrac{\mathbf{d}(\delta s)-I}{\delta s}\mathbf{d}(s)u, (\mathbf{d}(\delta s) - I)\mathbf{d}(s)u\right\rangle\right| \to 0 \text{ as } \delta s \to 0.$$

On the other hand, we derive

$$\langle \mathbf{d}(s+\delta s)\delta u, \mathbf{d}(s+\delta s)u\rangle - \langle \mathbf{d}(s)\delta u, \mathbf{d}(s)u\rangle =$$

$$\langle \mathbf{d}(s+\delta s)\delta u, (\mathbf{d}(\delta s) - I)\mathbf{d}(s)u\rangle + \langle \mathbf{d}(s)(\mathbf{d}(\delta s) - I)\delta u, \mathbf{d}(s)u\rangle =$$

$$\langle \mathbf{d}(s+\delta s)\delta u, (\mathbf{d}(\delta s) - I)\mathbf{d}(s)u\rangle + \langle \delta u, (\mathbf{d}(\delta s) - I)^*\mathbf{d}^*(s)\mathbf{d}(s)u\rangle$$

where $\cdot^*$ denotes an adjoint operator (see Appendix B). For the Hilbert space $\mathbb{H}$ it can be shown (see, e.g., [4, Corollary 10.6, p. 41]) that the family of adjoint operators $\{\mathbf{d}^*(s)\}_{s\in\mathbb{R}}$ is a strongly continuous group of linear bounded operators on $\mathbb{H}$. This means that $\|(\mathbf{d}(\delta s) - I)^*\mathbf{d}^*(s)\mathbf{d}(s)u\|\| \to 0$ as $\delta s \to 0$. Taking into account $\|(\mathbf{d}(\delta s) - I)\mathbf{d}(s)u\| \to 0$ as $\delta s \to 0$, we conclude

$$|\langle \mathbf{d}(s+\delta s)\delta u, \mathbf{d}(s+\delta s)u\rangle - \langle \mathbf{d}(s)\delta u, \mathbf{d}(s)u\rangle| \leq g(\delta s)\|\delta u\|$$

where $g : \mathbb{R} \mapsto \mathbb{R}$ is a class-$\mathcal{K}$ function. Therefore, there exist a constant $c > 0$ and two class $\mathcal{K}$ functions (g_1 and g_2) such that

$$\frac{|\varphi(s+\delta s, u+\delta u)-\varphi(s,u)-D_\phi(\delta s,\delta u)|}{|\delta s|+\|\delta u\|} \leq \frac{g_1(\delta s)|\Delta s|+g_2(\delta s)\|\delta u\|+c\|\delta u\|^2}{|\delta s|+\|\delta u\|} \to 0$$

as $|\delta s| + |\delta u| \to 0$. Therefore, the Frechét derivative of ϕ is given by the formula (4.26).

Since $D_s\varphi > 0$ for $u \neq \mathbf{0}$ provided that the dilation group $\mathbf{d}$ is strictly monotone (see Proposition 4.4), then the formula (4.25) is the straightforward consequence of Implicit Function theorem (see Appendix B or [3, Theorem E]) saying that the Fréchet derivative of the function $s : \mathbb{H} \mapsto \mathbb{R}$ (implicitly defined by the equation $\varphi(s, u) = 0$) can be calculated at the point $u \in \mathcal{D}(G_{\mathbf{d}})\backslash\{\mathbf{0}\}$ as $D_u s = -\left.\frac{D_u\varphi}{D_s\varphi}\right|_{\varphi(s,u)=0}$.
The proof is complete. □

In this monograph, the canonical homogeneous norm is frequently utilized as a Lyapunov function for the analysis and design of control systems. The formula (4.25) allows us to calculate the time derivative of the canonical homogeneous norm along strong and classical solutions of evolution equations.

Example 4.7 Let us consider again the strongly continuous strictly monotone dilation (4.9) in the Sobolev space $H^p(\mathbb{R}^n; \mathbb{R}^k)$ with $\Gamma = \alpha I_k, \Theta = \beta I_n$ and $\alpha > \beta n/2 - \beta i, i = 0, 1, ..., p$. By Lemma 4.6, the canonical homogeneous norm $\|\cdot\|_{\mathbf{d},H^p}$ is Fréchet differentiable on $H^p(\mathbb{R}^n; \mathbb{R}^k)\backslash\{\mathbf{0}\}$. For example, if $v \in H^1(\mathbb{R}^n; \mathbb{R}^m)$ then using integration by parts we derive

$$\begin{aligned}\langle G_{\mathbf{d}}v, v\rangle_{H^1} &= \langle \alpha v + \beta(z\cdot\nabla)v, v\rangle_{L_2} + \langle \nabla(\alpha v + (\beta z\cdot\nabla)v), \nabla v\rangle_{L_2}\\ &= (\alpha - \beta n/2)\langle v, v\rangle_{L_2} + (\alpha + \beta(1-n/2))\langle \nabla v, \nabla v\rangle_{L_2}.\end{aligned}$$

So, the Fréchet derivative $D\|x\|_{\mathbf{d},H^1} \in \mathcal{L}(H^1(\mathbb{R}^n; \mathbb{R}^k); \mathbb{R})$ of the canonical homogeneous norm $\|\cdot\|_{\mathbf{d},H^1}$ at the point $x \in H^1(\mathbb{R}^n; \mathbb{R}^k)\backslash\{\mathbf{0}\}$ can be computed as follows:

$$\left(D\|x\|_{\mathbf{d},H^1}\right)(q) = \frac{\langle \mathbf{d}(-\ln\|x\|_{\mathbf{d},H^1})q, v\rangle_{H^1}}{(\alpha-\beta n/2)\langle v,v\rangle_{L_2}+(\alpha+\beta(1-n/2))\langle\nabla v,\nabla v\rangle_{L_2}}\|x\|_{\mathbf{d},H^1}, \tag{4.27}$$

where $q \in H^1(\mathbb{R}^n; \mathbb{R}^k)$ and $v := \mathbf{d}\left(-\ln\|x\|_{\mathbf{d},H^1}\right)x$.

4.2.2 Generalized Homogeneous Banach/Hilbert Space

Formally, the canonical homogeneous norm $\|\cdot\|_{\mathbf{d}}$ is not a norm in the usual sense, but it defines a new topology in $\mathbb{B}$. Due to monotonicity of the dilation, it is straightforwardly related with homogeneous spheres and balls. Indeed, $\|u_{\mathbf{d}}\| = r$ is equivalent to $u \in S_{\mathbf{d}}(r)$ but $\|u\|_{\mathbf{d}} < r$ is equivalent $u \in B_{\mathbf{d}}(r)$ for any $r > 0$. Moreover, below we prove that, in fact, the functional $\|\cdot\|_{\mathbf{d}}$ is the norm in a real Banach space $\mathbb{B}_{\mathbf{d}}$ homeomorphic to $\mathbb{B}$. The new space has the same set of vectors, but the rules for vector addition and for multiplication of a vector by a scalar are defined in different way, by means of the generalized homogeneous homeomorphism $\boldsymbol{\Phi}$ on $\mathbb{B}$.

Lemma 4.7 *Let a strongly continuous linear dilation* $\mathbf{d}$ *in* $\mathbb{B}$ *be strictly monotone. Let* $\|\cdot\|_{\mathbf{d}}$ *be the canonical homogeneous norm induced by the norm* $\|\cdot\|$ *in* $\mathbb{B}$*. The mapping* $\boldsymbol{\Phi} : \mathbb{B} \mapsto \mathbb{B}$ *given by*

$$\boldsymbol{\Phi}(x) = \|x\|_{\mathbf{d}}\mathbf{d}(-\ln \|x\|_{\mathbf{d}})x, \quad x \in \mathbb{B} \tag{4.28}$$

is homeomorphism on $\mathbb{B}$*, its inverse is given by*

$$\boldsymbol{\Phi}^{-1}(z) = \|z\|^{-1}\mathbf{d}(\ln \|z\|)z, \quad z \in \mathbb{B},$$

and $\boldsymbol{\Phi}(\mathbf{0}) = \boldsymbol{\Phi}^{-1}(\mathbf{0}) = \mathbf{0}$ *by continuity. Moreover, for any* $x \in \mathbb{B}$ *there exist* $\omega \in \mathcal{K}_\infty$ *and* $r > 0$ *such that*

$$\|\boldsymbol{\Phi}(x + y) - \boldsymbol{\Phi}(x)\| \le \omega(\|y\|), \quad \forall y \in \bar{B}(r) := \{z \in \mathbb{B} : \|z\| \le r\}. \tag{4.29}$$

Proof Since

$$\|\boldsymbol{\Phi}(x)\| = \|x\|_{\mathbf{d}}\|\mathbf{d}(-\ln \|x\|_{\mathbf{d}})x\| = \|x\|_{\mathbf{d}}$$

then

$$\boldsymbol{\Phi}^{-1}(\boldsymbol{\Phi}(x)) = \|\boldsymbol{\Phi}(x)\|^{-1}\mathbf{d}(\ln \|\boldsymbol{\Phi}(x)\|)\boldsymbol{\Phi}(x) = \|x\|_{\mathbf{d}}^{-1}\mathbf{d}(\ln \|x\|_{\mathbf{d}})\boldsymbol{\Phi}(x) = x.$$

On the other hand, since

$$\left\|\boldsymbol{\Phi}^{-1}(x)\right\|_{\mathbf{d}} = \left\|\mathbf{d}(\ln \|z\|)\frac{z}{\|z\|}\right\|_{\mathbf{d}} = \|z\| \left\|\frac{z}{\|z\|}\right\|_{\mathbf{d}} = \|z\|$$

then

$$\boldsymbol{\Phi}(\boldsymbol{\Phi}^{-1}(z)) = \left\|\boldsymbol{\Phi}^{-1}(z)\right\|_{\mathbf{d}}\mathbf{d}\left(-\ln \left\|\boldsymbol{\Phi}^{-1}(z)\right\|_{\mathbf{d}}\right)\boldsymbol{\Phi}^{-1}(z) = \|z\|\mathbf{d}(-\ln \|z\|)\boldsymbol{\Phi}^{-1}(z) = z.$$

Therefore, $\boldsymbol{\Phi}^{-1}$ is, indeed, the inverse of $\boldsymbol{\Phi}$.

The continuity $\boldsymbol{\Phi}$ at $\mathbf{0}$ follows from the identity $\|\boldsymbol{\Phi}(x)\| = \|x\|_{\mathbf{d}}$. The estimate (4.29) for $x = \mathbf{0}$ is a corollary of the inequality (4.22).

Let us prove continuity of $\boldsymbol{\Phi}$ at $x \in \mathbb{B}\backslash\{\mathbf{0}\}$, i.e., we need to show that

$$\forall \varepsilon > 0, \quad \exists \delta > 0 : \|y\| < \delta \Rightarrow \|\boldsymbol{\Phi}(x + y) - \boldsymbol{\Phi}(x)\| < \varepsilon.$$

Let us denote $s = \|x\|_{\mathbf{d}}$ and $\delta_s := \|x + y\|_{\mathbf{d}} - \|x\|_{\mathbf{d}}$. Then

$$\begin{aligned}\mathbf{\Phi}(x+y)-\mathbf{\Phi}(x) =&(s+\Delta s)\mathbf{d}(-\ln(s+\delta_s))(x+y)-s\mathbf{d}(-\ln s)x\\ =&(s+\delta_s)\mathbf{d}(-\ln(s+\delta_s)y+\delta_s\mathbf{d}(-\ln(s+\delta_s))x\\ &+s(\mathbf{d}(-\ln(s+\delta_s))-\mathbf{d}(-\ln s))x,\end{aligned}$$

where $s>0$ due to $\|x\|\neq 0$. Since $\|x\|\neq 0$ then, due to relation (4.22) and local Lipschitz continuity of $\|\cdot\|_{\mathbf{d}}$ on $\mathbb{B}\backslash\{\mathbf{0}\}$, there exists $r>0$ such that $\|y\|\leq r \Rightarrow |\delta_s|\leq s/2$. Since $\mathbf{d}$ is a strongly continuous group then there exists $\sigma\in\mathcal{K}_\infty$ (dependent on x) such that

$$\|(\mathbf{d}(-\ln(s+\Delta s))-\mathbf{d}(-\ln s))x\|\leq\sigma(|\delta_s|)\quad\text{for}\quad|\delta_s|\leq s/2.$$

Hence, we derive

$$\|\mathbf{\Phi}(x+y)-\mathbf{\Phi}(x)\|\leq(s+\delta_s)C\|y\|+C|\delta_s|+s\sigma(|\delta_s|),$$

where $C=\sup\limits_{\delta_s\in[-s/2,s/2]}\|\mathbf{d}(-\ln(s+\delta_s))\|$. Finally, since $|\delta_s|=|\|x+y\|_{\mathbf{d}}-\|x\|_{\mathbf{d}}|$ then, taking into account local Lipschitz continuity of $\|\cdot\|_{\mathbf{d}}$ on $\mathbb{B}\backslash\{\mathbf{0}\}$, we complete the proof.

Now we are ready to introduce a generalized homogeneous real Banach space.

Theorem 4.2 (Generalized homogeneous Banach space $\mathbb{B}_{\mathbf{d}}$) *Let a strongly continuous linear dilation* $\mathbf{d}$ *in a real Banach space* $\mathbb{B}$ *be strictly monotone. Let the canonical homogeneous norm* $\|\cdot\|_{\mathbf{d}}$ *be induced by* $\|\cdot\|$ *in* $\mathbb{B}$. *Let an addition of vectors*

$$\tilde{+}:\mathbb{B}\times\mathbb{B}\mapsto\mathbb{B}$$

and a multiplication by scalar

$$\tilde{\cdot}:\mathbb{R}\times\mathbb{B}\mapsto\mathbb{B}$$

be defined as follows:

- $x\tilde{+}y:=\mathbf{\Phi}^{-1}(\mathbf{\Phi}(x)+\mathbf{\Phi}(y))$, *where* $x,y\in\mathbb{B}$;
- $\lambda\tilde{\cdot}x:=\mathrm{sign}(\lambda)\mathbf{d}(\ln|\lambda|)x$, *where* $\lambda\in\mathbb{R}$, $x\in\mathbb{B}$,

where $\mathbf{\Phi}:\mathbb{B}\mapsto\mathbb{B}$ *is given by* (4.28). *Then the set* $\mathbb{B}$ *together with the operations* $\tilde{+}$ *and* $\tilde{\cdot}$ *is a linear normed vector space* $\mathbb{B}_{\mathbf{d}}$ *with the norm given by* $\|\cdot\|_{\mathbf{d}}$. *Moreover,* $\mathbb{B}_{\mathbf{d}}$ *is a real Banach space.*

Proof The proof, that $\mathbb{B}_{\mathbf{d}}$ is a normed vector space, literary repeats the proof of Theorem 5.3 from Volume I presented for the Euclidean space $\mathbb{R}^n$ and its generalized homogeneous counterpart $\mathbb{R}^n_{\mathbf{d}}$.

Let us prove that the normed vector space $\mathbb{B}_{\mathbf{d}}$ with the norm $\|\cdot\|_{\mathbf{d}}$ is complete. For this purpose, let us show that any Cauchy sequence in $\mathbb{B}$ is a Cauchy sequence in $\mathbb{B}_{\mathbf{d}}$. Indeed, let $\{x_n\} \in \mathbb{B}$ be a Cauchy sequence in $\mathbb{B}$ and

$$\lim_{n\to+\infty} \|x_n - x^*\| = 0.$$

Since $\mathbb{B}$ is a Banach space (a complete normed vector space) then $x^* \in \mathbb{B}$. Let us consider the case $x^* \in \mathbb{B}\backslash\{\mathbf{0}\}$. By definition of the Cauchy sequence for any $\varepsilon > 0$ there exists $N \geq 1$ such that

$$\|x_n - x_m\| < \varepsilon, \qquad \forall n, m \geq N.$$

This means that

$$\|x_n - x^*\| \leq \varepsilon \quad \text{and} \quad \|x_m - x^*\| \leq \varepsilon \qquad \forall n, m \geq N.$$

On the other hand, we have

$$\|x_n \tilde{-} x_m\|_{\mathbf{d}} = \|\boldsymbol{\Phi}^{-1}(\boldsymbol{\Phi}(x_n) - \boldsymbol{\Phi}(x_m)\|_{\mathbf{d}} = \|\boldsymbol{\Phi}(x_n) - \boldsymbol{\Phi}(x_m)\| =$$

$$\|\boldsymbol{\Phi}(x_n) - \boldsymbol{\Phi}(x^*) + \boldsymbol{\Phi}(x^*) - \boldsymbol{\Phi}(x_m)\| \leq \|\boldsymbol{\Phi}(x_n) - \boldsymbol{\Phi}(x^*)\| + \|\boldsymbol{\Phi}(x^*) - \boldsymbol{\Phi}(x_m)\|,$$

where, by definition, $u \tilde{-} v := u \tilde{+} (-1) \tilde{\cdot} v$. Using the inequality (4.29) we conclude

$$\|\boldsymbol{\Phi}(x) - \boldsymbol{\Phi}(x^*)\| \leq \omega(\|x - x^*\|) \qquad \forall x : \|x - x^*\| \leq \varepsilon,$$

where $\omega \in \mathcal{K}_\infty$ depends on x^*. Hence, we conclude that

$$\|x_n \tilde{-} x_m\|_{\mathbf{d}} \leq 2\omega(\varepsilon),$$

i.e., $\{x_n\}$ is a Cauchy sequence in $\mathbb{B}_{\mathbf{d}}$. The proof is complete. □

The above theorem justifies the name "norm" for the functional $\|\cdot\|_{\mathbf{d}}$, which now can be utilized for continuity and convergence analysis in $\mathbb{B}$ alternatively to the norm $\|\cdot\|$.

Let a linear continuous dilation $\mathbf{d}$ in a real Hilbert space $\mathbb{H}$ be monotone with respect to the norm $\|x\| = \langle x, x\rangle$, where $\langle\cdot,\cdot\rangle$ is an inner product in $\mathbb{H}$. By Theorem 4.2 the corresponding vector space $\mathbb{H}_{\mathbf{d}}$ with new operations $\tilde{+}$ and $\tilde{\cdot}$ is a Banach space. An inner product in $\mathbb{H}_{\mathbf{d}}$ can be introduced as follows:

$$\langle x, y\rangle_{\mathbf{d}} = \begin{cases} 0 & \text{if } \|x\|\|y\|=0, \\ \|x\|_{\mathbf{d}}\|y\|_{\mathbf{d}}\langle \mathbf{d}(-\ln\|x\|_{\mathbf{d}})x, \mathbf{d}(-\ln\|y\|_{\mathbf{d}})y\rangle & \text{if } \|x\|\|y\|\neq 0, \end{cases} \tag{4.30}$$

where $\|\cdot\|_{\mathbf{d}}$ is the canonical $\mathbf{d}$-homogeneous norm induced by the norm $\|\cdot\|$.

Theorem 4.3 (Generalized homogeneous Hilbert space $\mathbb{H}_{\mathbf{d}}$)
The mapping $\mathbb{H}_{\mathbf{d}} \times \mathbb{H}_{\mathbf{d}} \mapsto \mathbb{R}$ *given by* (4.30) *is an inner product on* $\mathbb{H}_{\mathbf{d}}$.

The proof literally repeats the proof of Theorem 5.4 from Volume I.

4.2.3 Generalized Homogeneous Sobolev Spaces

The Space $H^1_{\mathbf{d}}$

Let us consider the Sobolev space $H^1(\mathbb{R}^n; \mathbb{R}^k) = W^{1,2}(\mathbb{R}^n; \mathbb{R}^k)$ with the inner product

$$\langle x, y\rangle_{H^1} = \langle x, y\rangle_{L^2} + \sum_{i=1}^{n} \left\langle \frac{\partial x}{\partial z_i}, \frac{\partial y}{\partial z_i} \right\rangle_{L^2} \tag{4.31}$$

and the norm $\|\cdot\|_{H_1} = \sqrt{\langle \cdot, \cdot\rangle_{H^1}}$. Let $\mathbf{d}$ be a strongly continuous group of transformations in H^1 given by the formula (4.9) with

$$\Theta = \mathrm{diag}(\theta_1, ..., \theta_n) \quad \text{and} \quad \Gamma = I_k.$$

The generator $G_{\mathbf{d}} : \mathcal{D}(G_{\mathbf{d}}) \subset H^1(\mathbb{R}^n; \mathbb{R}^k) \mapsto H^1(\mathbb{R}^n; \mathbb{R}^k)$ of the dilation $\mathbf{d}$ in H^1 is given by

$$G_{\mathbf{d}} x = \Gamma x + ((\Theta z) \cdot \nabla) x, \quad z \in \mathbb{R}^n, \tag{4.32}$$

whose domain $\mathcal{D}(G_{\mathbf{d}})$ is a completion of the space $C_c^\infty(\mathbb{R}^n; \mathbb{R}^k)$ with respect to the norm $\|x\|_{G_{\mathbf{d}}} = \|x\|_{H^1} + \|G_{\mathbf{d}} x\|_{H^1}$, where

$$((\Theta z) \cdot \nabla) = \sum_{i=1}^{n} \theta_i z_i \tfrac{\partial}{\partial z_i}.$$

Notice that

$$\begin{aligned} \|\mathbf{d}(s)x\|_{H^1}^2 =& \|\mathbf{d}(s)x\|_{L^2}^2 + \sum_{i-1}^{n} \left\| \frac{\partial \mathbf{d}(s)x}{\partial z_i} \right\|_{L^2}^2 \\ =& e^{(2-\mathrm{trace}\,\Theta)s} \|x\|_{L^2}^2 + \sum_{i=1}^{n} e^{(2+2\theta_i - \mathrm{trace}\,\Theta)s} \left\| \frac{\partial x}{\partial z_i} \right\|_{L^2}^2. \end{aligned}$$

Obviously, if

$$2 + 2\min\{0, \theta_i\} > \mathrm{trace}\,\Theta, \quad \forall i = 1, 2, ..., n \tag{4.33}$$

then $\mathbf{d}$ is strongly continuous linear dilation in H^1, which is strictly monotone with respect to the norm $\|\cdot\|_{H^1} = \sqrt{\langle \cdot, \cdot\rangle_{H^1}}$ in H^1. So, by Lemma 4.5, the canonical homogeneous norm $\|\cdot\|_{\mathbf{d},H^1}$ is continuous on H^1 and well-defined for $x \neq \mathbf{0}$ as a unique positive root of the algebraic equation:

$$1 = a_0 \|x\|_{\mathbf{d},H^1}^{-2+\mathrm{trace}\,\Theta} + \sum_{i=1}^{n} b_i \|x\|_{\mathbf{d},H^1}^{-2+\mathrm{trace}\,\Theta - 2\theta_i}, \tag{4.34}$$

where

$$a_0 = \|x\|_{L^2}^2, \quad b_i = \left\| \frac{\partial x}{\partial z_i} \right\|_{L^2}^2.$$

Moreover, by Lemma 4.6, the canonical homogeneous norm is Frechét differentiable on $\mathcal{D}(G_{\mathbf{d}})\backslash\{\mathbf{0}\}$:

$$D\|x\|_{\mathbf{d},H^1}(z) = \|x\|_{\mathbf{d},H^1} \frac{\langle \mathbf{d}(-\ln \|x\|_{\mathbf{d},H^1})z, \mathbf{d}(-\ln \|x\|_{\mathbf{d},H^1})x\rangle_{H^1}}{\langle G_{\mathbf{d}}\mathbf{d}(-\ln \|x\|_{\mathbf{d},H^1})x, \mathbf{d}(-\ln \|x\|_{\mathbf{d},H^1})x\rangle_{H^1}}, \quad z \in H^1(\mathbb{R}^n, \mathbb{R}^k), \tag{4.35}$$

where $D\|x\|_{\mathbf{d},H^1}(\cdot) \in \mathcal{L}(H^1, \mathbb{R})$ is the Frechét derivative of $\|\cdot\|_{\mathbf{d},H^1}$ at the point $x \in \mathcal{D}(G_{\mathbf{d}})\backslash\{\mathbf{0}\}$. By Theorem 4.2, the generalized homogeneous Sobolev space $H_{\mathbf{d}}^1$ can be defined introducing the new addition and multiplication rules in H^1 as follows:

- $\lambda \tilde{\cdot} x = \mathrm{sign}(\lambda)\mathbf{d}(\ln|\lambda|)x$ for $\lambda \in \mathbb{R}$ and $x \in H^1$;
- $x \tilde{+} y = \mathbf{\Phi}^{-1}(\mathbf{\Phi}(x) + \mathbf{\Phi}(y))$ for $x, y \in H^1$,

where

$$\mathbf{\Phi}(x) = \|x\|_{\mathbf{d},H^1}\mathbf{d}(-\ln \|x\|_{\mathbf{d},H^1})x, \quad \mathbf{\Phi}^{-1}(z) = \mathbf{d}(\ln \|z\|_{H^1})\frac{z}{\|z\|_{H^1}}$$

is a homeomorphism on H^1 and a diffeomorphism on $(H^1 \cap \mathcal{D}(G_{\mathbf{d}}))\backslash\{\mathbf{0}\}$. The inner product in $H_{\mathbf{d}}^1$ is given by

$$\langle x, y\rangle_{\mathbf{d},H^1} = \begin{cases} 0 & \text{if } \|x\|\|y\|=0, \\ \|x\|_{\mathbf{d},H^1}\|y\|_{\mathbf{d}}\langle \mathbf{d}(-\ln \|x\|_{\mathbf{d},H^1})x, \mathbf{d}(-\ln \|y\|_{\mathbf{d},H^1})y\rangle & \text{if } \|x\|\|y\|\neq 0. \end{cases} \tag{4.36}$$

Example 4.8 Let us consider the dilation

$$(\mathbf{d}(s)x)(z) = e^s x(e^s z), \quad z \in \mathbb{R}$$

in the Sobolev space $H^1(\mathbb{R}, \mathbb{R})$. In this case, the canonical homogeneous norm in H^1 is defined as a unique solution of the cubic equation

$$\|x\|_{\mathbf{d},H^1}^3 = a_0\|x\|_{\mathbf{d},H^1}^2 + b_1,$$

where

$$a_0 = \|x\|_{L^2}^2, \quad b_1 = \left\| \frac{\partial x}{\partial z} \right\|_{L^2}^2.$$

Using Cordano formulas (see Appendix of Volume I) we derive

$$\|x\|_{\mathbf{d},H^1} = \frac{a_0 + \sqrt[3]{\frac{2a_0^3+27b_1+\sqrt{108a_0^3b_1+27^2b_1^2}}{2}} + \sqrt[3]{\frac{2a_0^3+27b_1-\sqrt{108a_0^3b_1+27^2b_1^2}}{2}}}{2}.$$

The Space $H^2_{\mathbf{d}}$

Let us consider the Sobolev space $H^2(\mathbb{R}^n; \mathbb{R}^k) = W^{2,2}(\mathbb{R}^n; \mathbb{R}^k)$ with the inner product

$$\langle x, y\rangle_{H^2} = \langle x, y\rangle_{L^2} + \sum_{i=1}^{n}\left\langle \frac{\partial x}{\partial z_i}, \frac{\partial y}{\partial z_i}\right\rangle_{L^2} + \sum_{i=1}^{n}\sum_{j=1}^{n}\left\langle \frac{\partial^2 x}{\partial z_i z_j}, \frac{\partial^2 y}{\partial z_i \partial z_j}\right\rangle_{L^2}. \tag{4.37}$$

Let $\mathbf{d}$ be a strongly continuous group of transformations in H^2 given by the formula (4.9) with

$$\Theta = \mathrm{diag}(\theta_1, ..., \theta_n) \quad \text{and} \quad \Gamma = I_k.$$

The generator $G_{\mathbf{d}} : \mathcal{D}(G_{\mathbf{d}}) \subset H^2(\mathbb{R}^n; \mathbb{R}^k) \mapsto H^2(\mathbb{R}^n; \mathbb{R}^k)$ of the dilation $\mathbf{d}$ in H^2 is given by

$$G_{\mathbf{d}}x = \Gamma x + ((\Theta z)\cdot\nabla)x, \quad z \in \mathbb{R}^n \tag{4.38}$$

whose domain $\mathcal{D}(G_{\mathbf{d}})$ is a completion of the space $C_c^\infty(\mathbb{R}^n; \mathbb{R}^k)$ with respect to the norm $\|x\|_{G_{\mathbf{d}}} = \|x\|_{H^2} + \|G_{\mathbf{d}}x\|_{H^2}$, where

$$((\Theta z)\cdot\nabla) = \sum_{i=1}^{n}\theta_i z_i \tfrac{\partial}{\partial z_i}.$$

Since

$$\begin{aligned}\|\mathbf{d}(s)x\|^2_{H^2} =& \|\mathbf{d}(s)x\|^2_{L^2} + \sum_{i=1}^{n}\left\|\frac{\partial \mathbf{d}(s)x}{\partial z_i}\right\|^2_{L^2} + \sum_{i=1}^{n}\sum_{j=1}^{n}\left\|\frac{\partial^2 \mathbf{d}(s)x}{\partial z_i z_j}\right\|^2_{L^2}\\ =& e^{(2-\mathrm{trace}\,\Theta)s}\|x\|^2_{L^2} + \sum_{i=1}^{n} e^{(2+2\theta_i-\mathrm{trace}\,\Theta)s}\left\|\frac{\partial x}{\partial z_i}\right\|^2_{L^2}\\ &+ \sum_{i=1}^{n}\sum_{j=1}^{n} e^{(2+2\theta_i+2\theta_j-\mathrm{trace}\,\Theta)s}\left\|\frac{\partial^2 x}{\partial z_i z_j}\right\|^2_{L^2},\end{aligned}$$

obviously, if

$$2 + 4\min\{0, \theta_i\} > \mathrm{trace}\,\Theta, \quad \forall i = 1, 2, ..., n \tag{4.39}$$

then $\mathbf{d}$ is strongly continuous linear dilation in H^2, which is strictly monotone with respect to the norm $\|\cdot\|_{H^2} = \sqrt{\langle\cdot,\cdot\rangle_{H^2}}$ in H^2. So, by Lemma 4.5, the canonical homogeneous norm $\|\cdot\|_{\mathbf{d},H^2}$ is continuous on H^2 and well-defined for $x \neq \mathbf{0}$ as a unique positive root of the algebraic equation:

$$1 = a_0\|x\|_{\mathbf{d},H^2}^{-2+\operatorname{trace}\Theta} + \sum_{i=1}^{n} b_i\|x\|_{\mathbf{d},H^2}^{-2+\operatorname{trace}\Theta-2\theta_i} + \sum_{i=1}^{n}\sum_{j=1}^{n} c_{ij}\|x\|_{\mathbf{d},H^2}^{-2+\operatorname{trace}\Theta-2\theta_i-2\theta_j}, \tag{4.40}$$

where

$$a_0 = \|x\|_{L^2}^2, \quad b_i = \left\|\frac{\partial x}{\partial z_i}\right\|_{L^2}^2, \quad c_{i,j} = \left\|\frac{\partial^2 x}{\partial z_i z_j}\right\|_{L^2}^2.$$

Moreover, by Lemma 4.6, the canonical homogeneous norm is Frechét differentiable on $\mathcal{D}(G_{\mathbf{d}})\backslash\{\mathbf{0}\}$ such that the formula

$$D\|x\|_{\mathbf{d},H^2}(z) = \|x\|_{\mathbf{d},H^2}\frac{\langle \mathbf{d}(-\ln\|x\|_{\mathbf{d},H^2})z, \mathbf{d}(-\ln\|x\|_{\mathbf{d},H^2})x\rangle_{H^2}}{\langle G_{\mathbf{d}}\mathbf{d}(-\ln\|x\|_{\mathbf{d},H^2})x, \mathbf{d}(-\ln\|x\|_{\mathbf{d},H^2})x\rangle_{H^2}}, \quad z \in H^2(\mathbb{R}^n, \mathbb{R}^k), \tag{4.41}$$

where $D\|x\|_{\mathbf{d},H^2}(\cdot) \in \mathcal{L}(H^2;\mathbb{R})$ is the Frechét derivative of $\|\cdot\|_{\mathbf{d},H^2}$ at the point $x \in \mathcal{D}(G_{\mathbf{d}})\backslash\{\mathbf{0}\}$. Therefore, by Theorem 4.2, the generalized homogeneous Sobolev space $H^2_{\mathbf{d}}$ can be defined introducing the new addition and multiplication rules in H^2 as follows:

- $\lambda\tilde{\cdot}x = \operatorname{sign}(\lambda)\mathbf{d}(\ln|\lambda|)x$ for $\lambda \in \mathbb{R}$ and $x \in H^2$;
- $x\tilde{+}y = \mathbf{\Phi}^{-1}(\mathbf{\Phi}(x) + \mathbf{\Phi}(y))$ for $x, y \in H^2$,

where

$$\mathbf{\Phi}(x) = \|x\|_{\mathbf{d},H^2}\mathbf{d}(-\ln\|x\|_{\mathbf{d},H^2})x, \quad \mathbf{\Phi}^{-1}(z) = \mathbf{d}(\ln\|z\|_{H^2})\frac{z}{\|z\|_{H^2}}$$

is a homeomorphism on H^2 and a diffeomorphism on $(H^2 \cap \mathcal{D}(G_{\mathbf{d}}))\backslash\{\mathbf{0}\}$. The inner product in $H^2_{\mathbf{d}}$ is given by

$$\langle x, y\rangle_{\mathbf{d},H^2} = \begin{cases} 0 & \text{if } \|x\|\|y\|=0, \\ \|x\|_{\mathbf{d},H^2}\|y\|_{\mathbf{d}}\langle \mathbf{d}(-\ln\|x\|_{\mathbf{d},H^2})x, \mathbf{d}(-\ln\|y\|_{\mathbf{d},H^2})y\rangle & \text{if } \|x\|\|y\|\neq 0. \end{cases} \tag{4.42}$$

Example 4.9 Let us consider the dilation

$$(\mathbf{d}(s)x)(z) = e^s x(e^s z), \quad z \in \mathbb{R}$$

in the Sobolev space $H^2(\mathbb{R}, \mathbb{R})$. In this case, the canonical homogeneous norm in H^2 is defined as a unique solution of the polynomial equation

$$\|x\|_{\mathbf{d},H^2}^5 = a_0\|x\|_{\mathbf{d},H^2}^4 + b_1\|x\|_{\mathbf{d},H^2}^2 + c_{1,1},$$

where

$$a_0 = \|x\|_{L^2}^2, \quad b_1 = \left\| \frac{\partial x}{\partial z} \right\|_{L^2}^2, \quad c_{1,1} = \left\| \frac{\partial^2 x}{\partial z^2} \right\|_{L^2}^2.$$

Concluding Remarks

A linear dilation in abstract normed vector spaces is introduced as a group of invertible transformations satisfying special limit property. Strongly and uniformly continuous linear dilations in Banach and Hilbert spaces are studied. The canonical homogeneous norm in a real Banach space is defined, and its regularity (local Lipschitz continuity and differentiability) is studied. The canonical homogeneous inner product in a real Hilbert space is proposed. The generalized homogeneous Banach and Hilbert spaces are introduced. A class of strongly continuous linear dilations in Lebesgue and Sobolev spaces is considered. Most examples studied in Volume II deal with the introduced linear dilations.

References

1. Husch LS (1970) Topological characterization of the dilation and the translation in Frechet spaces. Math Ann 190:1–5
2. Kawski M (1991) Families of dilations and asymptotic stability. In: Analysis of controlled dynamical systems, pp 285–294
3. Halkin H (1974) Implicit functions and optimization problems without continuous differentiability of the data. SIAM J Control Optim 12(2)
4. Pazy A (1983) Semigroups of linear operators and applications to partial differential equations. Springer, Berlin

Chapter 5
Infinite-Dimensional Homogeneous Mappings

This chapter studies homogeneous operators and functionals in Banach and Hilbert spaces. Euler homogeneous function theorem is extended to homogeneous mappings in abstract spaces. A characterization of homogeneous linear operator in a Banach space is presented. The concept of homogeneous extension of an infinite-dimensional operator is introduced. Examples of various homogeneous operators are presented.

5.1 Definition and Basic Properties

5.1.1 Homogeneous Functionals

The homogeneous norm studied in the previous chapter is an example of the so-called homogeneous functional.

Definition 5.1 A functional $h : \mathcal{D}(h) \subset \mathbb{B} \mapsto \mathbb{R}$ is said to be $\mathbf{d}$-homogeneous of a degree $\nu \in \mathbb{R}$ if its domain $\mathcal{D}(h)$ is a $\mathbf{d}$-homogeneous cone in $\mathbb{B}$ and

$$h(\mathbf{d}(s)x) = e^{\nu s} h(x) \quad \text{for} \quad s \in \mathbb{R}, \quad x \in \mathcal{D}(h), \tag{5.1}$$

where $\mathbf{d}$ is a linear dilation in $\mathbb{B}$.

Let us denote by $\mathcal{H}_{\mathbf{d}}(\mathcal{D})$ *a set of* $\mathbf{d}$*-homogeneous functionals* $\mathcal{D} \subset \mathbb{B} \mapsto \mathbb{B}$ *and let* $\deg_{\mathbf{d}}(h) \in \mathbb{R}$ *denote the homogeneity degree of* $h \in \mathcal{H}_{\mathbf{d}}(\mathcal{D})$. For shortness we also use the notation $\mathcal{H}_{\mathbf{d}}$ if the domain is clear from the context.

A. Polyakov, *Generalized Homogeneity in Systems and Control Volume II*, Communications and Control Engineering,
https://doi.org/10.1007/978-3-031-77496-6_5

Proposition 5.1 *Let* $\mathbf{d}$ *be a strongly continuous strictly monotone dilation in* $\mathbb{B}$. *Let* $h \in \mathcal{H}_{\mathbf{d}}(\mathbb{B})$ *be such that* $\mathcal{D}(h) = \mathbb{B}$ *and* $\sup_{u \in S} |h(u)| < +\infty$.

- *If* $\deg_{\mathbf{d}}(h) > 0$ *then* h *is bounded on any homogeneous ball* $B_{\mathbf{d}}(r)$ *with* $r > 0$ *and* $h(x) \to 0$ *as* $x \to \mathbf{0}$;
- *If* $\deg_{\mathbf{d}}(h) < 0$ *then* h *is bounded on* $\mathbb{B} \backslash B_{\mathbf{d}}(r)$ *with any* $r > 0$;
- *If* $\deg_{\mathbf{d}}(h) = 0$ *then* h *is uniformly bounded on* $\mathbb{B} \backslash \{\mathbf{0}\}$ *and, moreover,* $h \equiv$ *const provided that* h *is continuous at zero.*

Proof Since the identity (5.1) with $s = -\ln \|u\|_{\mathbf{d}}$ yields

$$h(u) = \|u\|_{\mathbf{d}}^{\deg_{\mathbf{d}}(h)} h(\mathbf{d}(-\ln \|u\|_{\mathbf{d}})u)$$

then the claim follows from the inclusion $\mathbf{d}(-\ln \|u\|_{\mathbf{d}})u \in S$ and the properties of the canonical homogeneous norm (see Lemma 4.5), where $S = \{x \in \mathbb{B} : \|x\| = 1\}$ is the unit sphere.

Some elements of homogeneous arithmetics for functionals $h, g \in \mathcal{H}_{\mathbf{d}}$ are given below

- $h + g \in \mathcal{H}_{\mathbf{d}}$ with $\mathcal{D}(h+g) = \mathcal{D}(h) \cap \mathcal{D}(g)$ provided that $\deg_{\mathbf{d}}(h) = \deg_{\mathbf{d}}(g)$;
- $hg \in \mathcal{H}_{\mathbf{d}}$ with $\mathcal{D}(hg) = \mathcal{D}(h) \cap \mathcal{D}(g)$ has the degree

$$\deg_{\mathbf{d}}(hg) = \deg_{\mathbf{d}}(h) + \deg_{\mathbf{d}}(g);$$

- $\frac{h}{g} \in \mathcal{H}_{\mathbf{d}}$ with $\mathcal{D}\left(\frac{h}{g}\right) = \{u \in \mathcal{D}(h) \cap \mathcal{D}(g) : |h(u)/g(u)| < +\infty\}$ has the degree $\deg_{\mathbf{d}}\left(\frac{h}{g}\right) = \deg_{\mathbf{d}}(h) - \deg_{\mathbf{d}}(g)$.

Corollary 5.1 *Let* $\mathbf{d}$ *be a strictly monotone strongly continuous linear dilation in* $\mathbb{B}$. *Let* $f : \mathbb{B} \mapsto \mathbb{R}$ *be a globally bounded functional* $0 \le C_0 \le |f| < C_1 < +\infty$ *and* $h_1, h_2 : \mathbb{B} \backslash \{\mathbf{0}\} \mapsto \mathbb{R}$ *be two locally bounded* $\mathbf{d}$*-homogeneous functionals of the same degree* ν. *If* $\inf_{x \in S} |h_2(x)| > 0$ *then*

$$\frac{C_0 \inf_{x \in S} |h_1(x)|}{\sup_{x \in S} |h_2(x)|} \le \left|\frac{f(x)h_1(x)}{h_2(x)}\right| \le \frac{C_1 \sup_{x \in S} |h_1(x)|}{\inf_{x \in S} |h_2(x)|} < +\infty, \quad \forall x \in \mathbb{B} \backslash \{\mathbf{0}\},$$

where S *is the unit sphere in* $\mathbb{B}$.

Proof Using $\mathbf{d}$-homogeneity the functionals $h_i, i = 1, 2$ can be rewritten as follows:

$$h_i(x) = \|x\|^{\nu} h_i(\mathbf{d}(-\ln \|x\|_{\mathbf{d}})x), \quad x \in \mathbb{B} \backslash \{\mathbf{0}\}.$$

Taking into account $\mathbf{d}(-\ln \|x\|_{\mathbf{d}})x \in S$ we derive

$$0 < \|x\|_{\mathbf{d}}^{\nu} \inf_{x\in S} |h_2(x)| \leq |h_2(x)| \leq \|x\|_{\mathbf{d}}^{\nu} \sup_{x\in S} |h_2(x)|$$

for all $x \neq \mathbf{0}$ and

$$\frac{C_0 \inf_{x\in S} |h_1(x)|}{\sup_{x\in S} |h_2(x)|} \leq \left|\frac{f(x)h_1(x)}{h_2(x)}\right| = \frac{|f(x)h_1(x)|}{h_2(x)} \leq \frac{C_1 \sup_{x\in S} |h_1(x)|}{\inf_{x\in S} |h_2(x)|} < +\infty$$

for all $x \in \mathbb{B}\backslash\{\mathbf{0}\}$. The lower bounded can be obtained similarly.

Example 5.1 Let us consider the functional $f : \mathcal{D}(f) \subset \mathbb{B} \mapsto \mathbb{R}$ with $\mathbb{B} = L^2(\mathbb{R}^n; \mathbb{R}^m)$, $\mathcal{D}(f) = L^2(\mathbb{R}^n; \mathbb{R}^m) \cap L^\infty(\mathbb{R}^n; \mathbb{R}^m)$,

$$f(x) = \|x\|_2 \|x\|_\infty, \quad x \in \mathbb{B}.$$

Let the dilation $\mathbf{d}$ in $\mathbb{B}$ be defined by (4.9) with $\Gamma \in \mathbb{R}$ and $\Theta \in \mathbb{R}$ such that $\Gamma - \Theta/2 > 0$. Using the formula (4.11) with $k = 1$, we deduce that $\mathcal{D}(f)$ is a $\mathbf{d}$-homogeneous cone and f is a $\mathbf{d}$-homogeneous functional of the degree $2\Gamma - n\Theta/2$.

A derivative of a functional $h \in \mathcal{H}_{\mathbf{d}}$ is homogeneous as well.

Proposition 5.2 *Let* $\mathbf{d}$ *be a linear strongly continuous dilation in* $\mathbb{B}$ *and* $h \in \mathcal{H}_{\mathbf{d}}(\mathbb{B})$.

- *If* h *is Frechét differentiable at* $u \in \mathbb{B}$, *then* h *is Frechét differentiable at* $\mathbf{d}(s)u \in \mathbb{B}$ *and*

$$(Dh(\mathbf{d}(s)u))(\mathbf{d}(s)y) = e^{s\,\mathrm{deg}_{\mathbf{d}}(h)} Dh(u)y, \quad s \in \mathbb{R}, \quad y \in \mathbb{B}, \tag{5.2}$$

where $Dh(u) \in \mathcal{L}(\mathbb{B}; \mathbb{R})$ *denotes the Frechét derivative of* h *at* $u \in \mathcal{D}$.
- *If* h *is Gateaux differentiable at* $u \in \mathbb{B}$ *in the direction* $g \in \mathbb{B}$, *then* h *is Gateaux differentiable at* $\mathbf{d}(s)u \in \mathbb{B}$ *in the direction* $\mathbf{d}(s)g \in \mathbb{B}$ *and*

$$Dh(\mathbf{d}(s)u; \mathbf{d}(s)g) = e^{s\,\mathrm{deg}_{\mathbf{d}}(h)} Dh(u; g), \quad s \in \mathbb{R}, \tag{5.3}$$

where $Dh(u; g) = \lim_{t\to 0} \frac{h(u+tg)-h(u)}{t}$ *is the Gateaux derivative of* $u \in \mathbb{B}$ *in the direction* $g \in \mathbb{B}$.

The claim remains valid for the left/right-hand lower/upper Gateaux derivatives too.

Proof (1) By definition, $Dh(u)$ is given by

$$\lim_{\|y\|\to 0} \frac{|h(u+y) - h(u) - (Dh(u))(y)|}{\|y\|} = 0, \quad y \in \mathbb{B}.$$

Similarly, for $Dh(\mathbf{d}(s)u)$ we have

$$\lim_{\|y_s\|\to 0} \frac{|h(\mathbf{d}(s)u + y_s) - h(\mathbf{d}(s)u) - (Dh(\mathbf{d}(s)u))(y_s)|}{\|y_s\|} = 0, \quad y_s \in \mathbb{B}.$$

Using the $\mathbf{d}$-homogeneity we derive

$$\frac{|h(\mathbf{d}(s)u+y_s)-h(\mathbf{d}(s)u)-(Dh(\mathbf{d}(s)u))(y_s)|}{\|y_s\|} = \frac{e^{\nu s}|h(u+\mathbf{d}(-s)y_s)-h(u)-e^{-\nu s}(Dh(\mathbf{d}(s)u))(y_s)|}{\|y_s\|}$$

$$\leq \frac{e^{\nu s}}{\lfloor \mathbf{d}(s) \rfloor} \frac{|h(u+y)-h(u)-e^{-\nu s}(Dh(\mathbf{d}(s)u))(\mathbf{d}(s)y)|}{\|y\|} \quad \text{with } y = \mathbf{d}(-s)y_s$$

where $\nu = \deg_{\mathbf{d}}(h)$. Since $\lfloor \mathbf{d}(s) \rfloor > 0$ and $\mathbf{d}(s)$ is a linear bounded invertible operator then $\|y\| \to 0$ is equivalent $\|y_s\| \to 0$. Hence, due to uniqueness of the Frechét derivative we derive the identity (5.2).

(2) Similarly, due to homogeneity of h we have

$$\frac{h(u+tg) - h(u)}{t} = e^{-\nu s} \frac{h(\mathbf{d}(s)u + t\mathbf{d}(s)g) - h(\mathbf{d}(s)u)}{t}$$

and taking the limit $t \to 0$ we derive (5.3). The proof is complete. □

An analog of Euler's homogeneous function theorem can be proven for functionals defined on $\mathbb{H}$.

Theorem 5.1 *Let* $\mathbf{d}$ *be a strongly continuous strictly monotone dilation group on a real Hilbert space* $\mathbb{H}$. *If* $h \in \mathcal{H}_{\mathbf{d}}(\mathbb{H})$ *is Frechét differentiable on* $\mathcal{D}(G_{\mathbf{d}})\backslash\{\mathbf{0}\} \subset \mathcal{D}(h) = \mathbb{H}$ *then*

$$(Dh(u))\,(G_{\mathbf{d}}u) = \deg_{\mathbf{d}}(h)\, h(u) \text{ for any } u \in \mathcal{D}(G_{\mathbf{d}})\backslash\{\mathbf{0}\}, \tag{5.4}$$

where $Dh(u) \in \mathcal{L}(\mathbb{H}; \mathbb{R})$ *is the Frechét derivative of* h *at the point* u.

Proof Denote $\nu = \deg_{\mathbf{d}}(h)$. Since

$$h(u) = h\left(\mathbf{d}(\ln \|u\|_{\mathbf{d}})\mathbf{d}(-\ln \|u\|_{\mathbf{d}})u\right) = \|u\|^{\nu} h(\mathbf{d}(-\ln \|u\|_{\mathbf{d}})u)$$

then for any $u \in \mathcal{D}(G_{\mathbf{d}})\backslash\{\mathbf{0}\}$ using product and chain rules for Frechét derivatives we derive

$$Dh(u) = \nu \|u\|_{\mathbf{d}}^{\nu-1} h(\mathbf{d}(-\ln \|u\|_{\mathbf{d}})u) D\|u\|_{\mathbf{d}} + \|u\|_{\mathbf{d}}^{\nu} (Dh(z))(D\mathbf{d}(-\ln \|u\|_{\mathbf{d}})u),$$

where $z = \mathbf{d}(-\ln \|u\|_{\mathbf{d}})u$, $D\|u\|_{\mathbf{d}} \in \mathcal{L}(\mathbb{H}, \mathbb{R})$ is a Frechét derivative of $\|\cdot\|_{\mathbf{d}}$ at the point u (see, the formula (4.25)) and $D\mathbf{d}(-\ln \|u\|_{\mathbf{d}})u \in \mathcal{L}(\mathbb{H}; \mathbb{H})$ is a Frechét derivative of the operator $u \mapsto \mathbf{d}(-\ln \|u\|_{\mathbf{d}})u$ at the point u. Since the mapping $s \mapsto \mathbf{d}(s)u$ is differentiable for any $u \in \mathcal{D}(G_{\mathbf{d}})$ and $\frac{d}{ds}\mathbf{d}(s)u = G_{\mathbf{d}}\mathbf{d}(s)u = \mathbf{d}(s)G_{\mathbf{d}}u$ then

$$(D\mathbf{d}(-\ln\|u\|_{\mathbf{d}})u)(y) = \mathbf{d}(-\ln\|u\|_{\mathbf{d}})y - \tfrac{(D\|u\|_{\mathbf{d}})(y)}{\|u\|_{\mathbf{d}}} G_{\mathbf{d}}\mathbf{d}(-\ln\|u\|_{\mathbf{d}})u, \quad y \in \mathbb{H}.$$

Finally, since, due to by the formula (4.25), we derive $(D\|u\|_{\mathbf{d}})(G_{\mathbf{d}}u) = \|u\|_{\mathbf{d}}$ for $u \in \mathbb{B}\backslash\{\mathbf{0}\}$, then

$$(D\mathbf{d}(-\ln\|u\|_{\mathbf{d}})u)(G_{\mathbf{d}}u) = \mathbf{0}$$

and the identity (5.4) holds. The proof is complete. □

5.1.2 Homogeneous Operators

In this section, we introduce a $\mathbf{d}$-homogeneous operator in a real Banach space $\mathbb{B}$. All linear and various nonlinear evolution models of mathematical physics are $\mathbf{d}$-homogeneous in the following sense.

Definition 5.2 An operator $f : \mathcal{D}(f) \subset \mathbb{B} \mapsto \mathbb{B}$ is said to be $\mathbf{d}$-homogeneous of a degree $\nu \in \mathbb{R}$ if its domain $\mathcal{D}(f)$ is a $\mathbf{d}$-homogeneous cone in $\mathbb{B}$ and

$$f(\mathbf{d}(s)u) = e^{\nu s}\mathbf{d}(s)f(u) \quad \text{for} \quad s \in \mathbb{R}, \quad u \in \mathcal{D}(f), \tag{5.5}$$

where $\mathbf{d}$ is a linear dilation in $\mathbb{B}$.

Let $\mathcal{F}_{\mathbf{d}}(\mathcal{D})$ denote a set of $\mathbf{d}$-homogeneous operators $\mathcal{D} \subset \mathbb{B} \mapsto \mathbb{B}$ and let $\deg_{\mathbf{d}}(f) \in \mathbb{R}$ be a homogeneity degree of $f \in \mathcal{F}_{\mathbf{d}}(\mathcal{D})$, where $\mathcal{D}$ is a $\mathbf{d}$-homogeneous cone in $\mathbb{B}$. For shortness we also use the notation $\mathcal{F}_{\mathbf{d}}$ if the domain is clear from the context.

As shown below the identity (5.5) may be understood in a weak sense.

Example 5.2 (*On $\mathbf{d}$-homogeneity of the Laplace operator*) Let us consider the Laplace operator

$$\Delta : \mathcal{D}(\Delta) \subset L^2(\mathbb{R}^n; \mathbb{R}^m) \mapsto L^2(\mathbb{R}^n; \mathbb{R}^m),$$

with the domain

$$\mathcal{D}(\Delta) = \left\{ x \in L^2(\mathbb{R}^n; \mathbb{R}^m) : \begin{array}{c} \exists f \in L^1_{loc}(\mathbb{R}^n; \mathbb{R}^m) \quad \text{such that} \\ \int_{\mathbb{R}^n} u \cdot \Delta\phi = \int_{\mathbb{R}^n} f \cdot \phi, \quad \forall \phi \in C_c^\infty(\mathbb{R}^n; \mathbb{R}^m) \end{array} \right\}.$$

Let us show that the operator Δ is $\mathbf{d}$-homogeneous of degree 2θ provided that the dilation $\mathbf{d}$ in $L^2(\mathbb{R}^n; \mathbb{R}^m)$ is given by

$$(\mathbf{d}(s)x)(z) = e^{\gamma s}x(e^{s\theta}z), \quad z \in \mathbb{R}^n, \quad x \in L^2(\mathbb{R}^n; \mathbb{R}^m), \quad s \in \mathbb{R} \tag{5.6}$$

with $\gamma \geq \frac{n\theta}{2}$. By Lemma 4.2, $\mathbf{d}$ is a group of linear invertible operators on $C^\infty(\mathbb{R}^n; \mathbb{R}^m)$, $C_c^\infty(\mathbb{R}^n; \mathbb{R}^m)$, $L^1_{loc}(\mathbb{R}^n; \mathbb{R}^m)$ and, a linear dilation in $L^2(\mathbb{R}^n; \mathbb{R}^m)$. If $\phi \in C^\infty(\mathbb{R}^n; \mathbb{R}^m)$ then, obviously,

$$(\Delta \circ \mathbf{d}(s))\phi)(x) = e^{(\gamma+2\theta)s}(\Delta\phi)(e^{\theta s}x) = e^{2\theta s}((\mathbf{d}(s) \circ \Delta)\phi)(x), \quad s \in \mathbb{R}, \quad x \in \mathbb{R}^n.$$

In other words, the Laplace operator is **d**-homogeneous as an operator $C_c^\infty(\mathbb{R}^n;\mathbb{R}^m) \mapsto C_c^\infty(\mathbb{R}^n;\mathbb{R}^m)$. Since $C_c^\infty(\mathbb{R}^n;\mathbb{R}^m)$ is dense in $L^p(\mathbb{R}^n;\mathbb{R}^m)$ and $H^p(\mathbb{R}^n;\mathbb{R}^m)$ then, it is expectable, the Δ is **d**-homogeneous as an operator $\mathcal{D}(\Delta) \subset L^2(\mathbb{R}^n;\mathbb{R}^m) \mapsto L^2(\mathbb{R}^n;\mathbb{R}^m)$ as well. Let us prove this claim rigorously using the notion of weak derivatives.

Let $x \in \mathcal{D}(\Delta)$ and $\Delta x = f \in L^1_{loc}(\mathbb{R}^n;\mathbb{R}^m)$ in the weak sense:

$$\int_{\mathbb{R}^n} x\Delta\phi = \int_{\mathbb{R}^n} f\phi, \quad \forall \phi \in C_c^\infty(\mathbb{R}^n;\mathbb{R}^m).$$

Since $\mathbf{d}(s)f \in L^1_{loc}(\mathbb{R}^n;\mathbb{R}^m)$ then $e^{2\theta s}\mathbf{d}(s)\Delta x = e^{2\theta s}\mathbf{d}(s)f$ in the weak sense and using the change of variables in the Lebesgue integral (see Appendix C or, for example, [1]) we derive

$$e^{2\theta s}\int_{\mathbb{R}^n} (\mathbf{d}(s)f)\cdot\phi = e^{(\gamma+2\theta)s}\int_{\mathbb{R}^n} f(e^{\theta s}z)\cdot\phi(z)dz$$

$$= e^{(\gamma+(2-n)\theta)s}\int_{\mathbb{R}^n} f(z)\cdot\phi(e^{-\theta s}z)dz = e^{(2\gamma+(2-n)\theta)s}\int_{\mathbb{R}^n} f\cdot\underbrace{\tilde{\phi}}_{=\mathbf{d}(-s)\phi}$$

$$= e^{(2\gamma+(2-n)\theta)s}\int_{\mathbb{R}^n} x\cdot\Delta\tilde{\phi} = e^{(2\gamma+(2-n)\theta)s}\int_{\mathbb{R}^n} x\cdot\underbrace{\Delta\circ\mathbf{d}(-s))\phi}_{=e^{-2\theta}(\mathbf{d}(-s)\circ\Delta)\phi} =$$

$$e^{(\gamma-n\theta)s}\int_{\mathbb{R}^n} x(z)\cdot(\Delta\phi)(e^{-\theta s}z)dz = e^{\gamma s}\int_{\mathbb{R}^n} x(e^{\theta s}z)\cdot(\Delta\phi)(z)dx$$

$$= \int_{\mathbb{R}^n} (\mathbf{d}(s)u)\cdot\Delta\phi.$$

Hence, $\mathbf{d}(s)u \in \mathcal{D}(\Delta)$ and $(\Delta\circ\mathbf{d}(s))u = e^{2\theta s}\mathbf{d}(s)f = e^{2\theta s}(\mathbf{d}(s)\circ\Delta)u$ in the weak sense for any $s \in \mathbb{R}$ and any $u \in \mathcal{D}(\Delta)$.

Example 5.3 *Saint-Venant equation* is an example of a system of conservation laws (see, [2]). In the field of hydraulics, it models a flow in open-channels

$$\begin{aligned} \frac{\partial H}{\partial t} &= -\frac{\partial}{\partial x}(HV), \\ \frac{\partial V}{\partial t} &= -\frac{\partial}{\partial x}\left(\tfrac{1}{2}V^2 + gH\right), \end{aligned} \tag{5.7}$$

where H and V are scalar functions of time and space variables. The quantity $H(t,x)$ is the water level at the instant of time $t \in \mathbb{R}_+$ in the point $x \in \mathbb{R}$, and $V(t,x)$ is the water velocity in the same position. The parameter g denotes the gravitation constant.

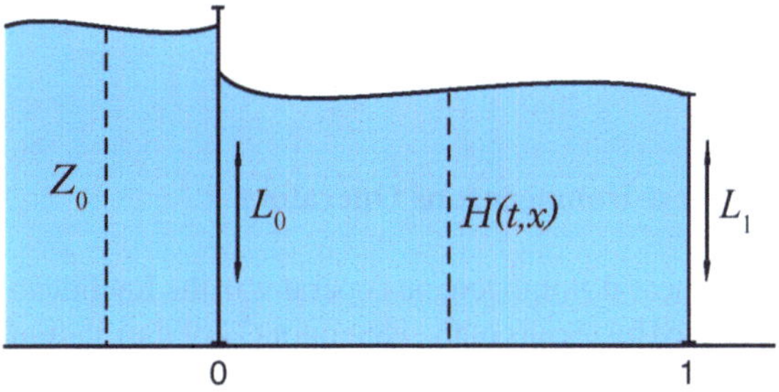

Fig. 5.1 Water channel with two spillways

Let us consider the case when the water channel is supported by two overflow spillways (Fig. 5.1), which adjust an input and output flows in a pool (between spillways). The space argument x is restricted on the segment $[0, 1]$, where $x=0$ and $x=1$ are positions of spillways.

Inspired by [2], let us consider Eq. (5.7) is supported with the following homogeneous nonlinear boundary conditions:

$$H(t,0)V(t,0) - (Z_0 - L_0)^{3/2} = 0,$$
$$H(t,1)V(t,1) - (H(t,1) - L_1)^{3/2} = 0,$$

where Z_0 is the water level above the pool and L_0, L_1 are spillways heights.

Let us show that for $L_0 = Z_0$ and $L_1 = 0$ the corresponding evolution equation is **d**-homogeneous. Let us consider the operator $f : \mathcal{D}(f) \subset \mathbb{B} \mapsto \mathbb{B} := C([0,1];\mathbb{R}) \times C([0,1];\mathbb{R})$ defined on the domain

$$\mathcal{D}(f) = \left\{ (u_1, u_2) \in C^1([0,1];\mathbb{R}_+) \times C^1([0,1];\mathbb{R}) : \begin{array}{c} u_1(0)u_2(0) = 0; \\ u_1(1)u_2(1) = u_1^{3/2}(1) \end{array} \right\}$$

as follows:

$$f(u) = \begin{pmatrix} -\frac{\partial}{\partial x}(u_1 u_2) \\ -\frac{\partial}{\partial x}\left(g u_1 + \frac{1}{2}u_2^2\right) \end{pmatrix},$$

where $u = (u_1, u_2) \in \mathcal{D}(f)$. The operator f is **d**-homogeneous of degree $\nu = 1$ with respect to the weighted (uniformly continuous) dilation $\mathbf{d}(s)u = (e^{2s}u_1, e^s u_2)$, where $u = (u_1, u_2) \in C([0,1];\mathbb{R}) \times C([0,1];\mathbb{R})$ and $s \in \mathbb{R}$. Indeed,

$$f(\mathbf{d}(s)u) = \begin{pmatrix} -\frac{\partial}{\partial x}(e^{2s}u_1 e^s u_2) \\ -\frac{\partial}{\partial x}\left(g e^{2s} u_1 + \frac{1}{2}[e^s u_2]^2\right) \end{pmatrix} =$$

$$\begin{pmatrix} -e^{3s}\frac{\partial}{\partial x}(u_1 u_2) \\ -e^{2s}\frac{\partial}{\partial x}\left(g u_1 + \frac{1}{2}u_2^2\right) \end{pmatrix} = e^s \mathbf{d}(s) f(u).$$

Finally, the equality $(\mathbf{d}(s)u)(x) = (e^{2s}u_1(x), e^s u_2(x))$, $x \in \mathbb{R}$ implies that for any $u \in \mathcal{D}(f)$ one has $\mathbf{d}(s)u \in \mathcal{D}(f)$, i.e., $\mathcal{D}(f)$ is a **d**-homogeneous cone.

Other **d**-homogeneous models of fluid dynamics (such as Burgers and Navier–Stokes equations) are considered below.

Linear d-Homogeneous Operators

Any linear **d**-homogeneous operator in the Euclidean space $\mathbb{R}^n$ can be uniquely characterized by means of the generator $G_{\mathbf{d}}$ of the dilation **d**. A similar characterization of linear **d**-homogeneous operator in a real Banach $\mathbb{B}$ is given by the following lemma.

Lemma 5.1 (On homogeneous linear operator in $\mathbb{B}$) *Let* **d** *be a linear strongly continuous dilation in* $\mathbb{B}$ *with a generator* $G_{\mathbf{d}} : \mathcal{D}(G_{\mathbf{d}}) \subset \mathbb{B} \mapsto \mathbb{B}$ *and let* $\mathcal{D}_{\exp}(G_{\mathbf{d}}^{\infty})$ *be a set of all entire vectors of* **d**. *Let an operator* $A : \mathcal{D}(A) \subset \mathbb{B} \mapsto \mathbb{B}$ *be linear, closed, and densely defined. Then the following two claims are equivalent*

i. operator A *is* **d**-*homogeneous of a degree* $\mu \in \mathbb{R}$;
ii. there exists a **d**-*homogeneous cone* $\mathcal{D} \subset \mathcal{D}_{\exp}(G_{\mathbf{d}}^{\infty})$ *dense in* $\mathbb{B}$ *such that* $\mathcal{D} \subset \mathcal{D}(A)$, $A\mathcal{D} \subset \mathcal{D}_{\exp}(G_{\mathbf{d}}^{\infty})$ *and*

$$AG_{\mathbf{d}}x = (\mu I + G_{\mathbf{d}})Ax, \quad \forall x \in \mathcal{D}. \tag{5.8}$$

Proof First of all notice that due to Corollary 2.10, $\mu I + G_{\mathbf{d}}$ is the generator of the group $\Psi(s) = e^{\mu s}\mathbf{d}(s)$, $s \in \mathbb{R}$. Moreover, it has the same set of entire vectors, i.e., $\mathcal{D}_{\exp}((\mu I + G_{\mathbf{d}})^{\infty}) = \mathcal{D}_{\exp}(G_{\mathbf{d}}^{\infty})$. Indeed, since the following operator identity

$$(\mu I + G_{\mathbf{d}})^n = \sum_{i=1}^{n} \frac{n!}{i!(n-i)!}\mu^{n-i}G_{\mathbf{d}}^i, \quad \forall n \in \mathbb{N},$$

holds then for any $y \in \mathcal{D}_{\exp}((\mu I + G_{\mathbf{d}})^{\infty})$ we have

$$\sum_{n=0}^{\infty} \frac{s^n(\mu I + G_{\mathbf{d}})^n y}{n!} = \sum_{n=0}^{\infty} \frac{s^n \sum_{i=1}^{n} \frac{n!}{i!(n-i)!}\mu^{n-i}G_{\mathbf{d}}^i y}{n!}$$

$$= \sum_{n=0}^{\infty}\sum_{i=1}^{n} \frac{s^{(n-i)}\mu^{n-i}}{i!}\frac{s^i G_{\mathbf{d}}^i y}{i!} = \sum_{j=0}^{+\infty}\frac{s^j\mu^j}{j!}\sum_{k=0}^{\infty}\frac{s^k G_{\mathbf{d}}^k y}{k!} = e^{\mu s}\sum_{k=0}^{\infty}\frac{s^k G_{\mathbf{d}}^k y}{k!},$$

i.e., $y \in \mathcal{D}_{\exp}(G_{\mathbf{d}}^{\infty})$ and vice versa.

ii $\Rightarrow$ **i.** By assumption, $\mathcal{D} \subset \mathcal{D}(A)$ and $A\mathcal{D} \subset \mathcal{D}_{\exp}(G_{\mathbf{d}}^{\infty})$. Then $G_{\mathbf{d}}^i x \in \mathcal{D}$ and $AG_{\mathbf{d}}^i x \in \mathcal{D}_{\exp}(G_{\mathbf{d}}^{\infty})$, $\forall x \in \mathcal{D}$, $\forall i \in \mathbb{N}$. Hence, using the identity (5.8) we derive

$$\sum_{i=0}^{\infty} \frac{s^i(\mu I + G_{\mathbf{d}})^i Ax}{i!} = \sum_{i=0}^{\infty} \frac{s^i(\mu I + G_{\mathbf{d}})^{i-1}AG_{\mathbf{d}}x}{i!} = \cdots = A\sum_{i=0}^{\infty}\frac{s^i G_{\mathbf{d}}^i x}{i!}.$$

This means

$$e^{\mu s}\mathbf{d}(s)Ax = A\mathbf{d}(s)x, \quad \forall x \in D.$$

Since $\mathcal{D}$ is dense in $\mathbb{B}$ then for any $x \in \mathcal{D}(A)$ there exists a sequence $x_n \in \mathcal{D} \subset \mathcal{D}(A)$ such that $x_n \to x$ as $n \to \infty$. Since the operator A is closed then $Ax_n \to Ax$ as $n \to \infty$. Similarly, taking into account that $\mathcal{D}$ is a $\mathbf{d}$-homogeneous cone, we derive $A\mathbf{d}(s)x_n \to A\mathbf{d}(s)x$ as $n \to \infty$. Therefore, using the boundedness of the operator $e^{\mu s}\mathbf{d}(s)$ we conclude

$$\underbrace{A\mathbf{d}(s)x_n}_{\substack{\downarrow \\ A\mathbf{d}(s)x}} = \underbrace{e^{\mu s}\mathbf{d}(s)Ax_n}_{\substack{\downarrow \\ e^{\mu s}\mathbf{d}(s)Ax}} \quad \text{as} \quad n \to \infty.$$

Therefore, the identity $A\mathbf{d}(s)x = e^{\mu s}\mathbf{d}(s)Ax \in \mathbb{B}$ holds for any $x \in \mathcal{D}(A)$ and any $s \in \mathbb{R}$. Notice that the obtained identity implies that $A\mathbf{d}(s)x \in \mathbb{B}$ for any $x \in \mathcal{D}(A)$ and any $s \in \mathbb{R}$, i.e., $\mathcal{D}(A)$ is a $\mathbf{d}$-homogeneous cone. So, the operator A is $\mathbf{d}$-homogeneous of degree μ.

i $\Rightarrow$ **ii**. Let us introduce the sets

$$\mathcal{D} = \left\{ y = \sqrt{\tfrac{1}{\pi}} \int_{-\infty}^{+\infty} e^{-s^2}\mathbf{d}\left(\tfrac{s}{\sqrt{n}}\right) x\, ds : x \in \mathcal{D}(A), n \in \mathbb{N} \right\} \subset \mathbb{B}.$$

$$\mathcal{D}_\mu = \left\{ y = \sqrt{\tfrac{1}{\pi}} \int_{-\infty}^{+\infty} e^{-s^2} e^{\mu\frac{s}{\sqrt{n}}}\mathbf{d}\left(\tfrac{s}{\sqrt{n}}\right) x\, ds : x \in \mathcal{D}(A), n \in \mathbb{N} \right\} \subset \mathbb{B}.$$

In the proof of Corollary 2.8 it is shown that $\mathcal{D} \subset \mathcal{D}_{\exp}(G_{\mathbf{d}}^{\infty})$ and $D_\mu \subset \mathcal{D}_{\exp}(\mu I + G_{\mathbf{d}}^{\infty})$. As shown above $\mathcal{D}_{\exp}(\mu I + G_{\mathbf{d}}^{\infty}) = \mathcal{D}_{\exp}(G_{\mathbf{d}}^{\infty})$.

Let us show that $\mathcal{D} \subset \mathcal{D}(A)$. Since the dilation $\mathbf{d}$ is strongly continuous then the curve $\tau \mapsto e^{\mu\tau}\mathbf{d}(\tau)Ax$ is continuous for any $x \in \mathcal{D}(A)$. Hence, due to $\mathbf{d}$-homogeneity of A, the curve $\tau \mapsto A\mathbf{d}(\tau)x$ is continuous too. Recall (see, Appendix B) that if A is the closed linear operator, $f(s) \in \mathcal{D}(A)$, $\forall s \in \mathbb{R}$ and the curve $\tau \mapsto Af(\tau)$ is continuous then $A\int f(\tau)d\tau = \int Af(\tau)d\tau$. Since $\mathcal{D}(A)$ is $\mathbf{d}$-homogeneous cone then $\mathbf{d}(s)x \in \mathcal{D}(A)$. Taking into account the linearity of A, for any $x \in \mathcal{D}(A)$ and any $n \in \mathbb{N}$ we derive

$$y_\mu = \sqrt{\frac{1}{\pi}} \int_{-\infty}^{+\infty} e^{s^2} e^{\frac{\mu}{\sqrt{n}}s}\mathbf{d}\left(\frac{s}{\sqrt{n}}\right) Ax\, ds = \int_{-\infty}^{+\infty} \sqrt{\frac{1}{\pi}} e^{-\frac{s^2}{2}} A\mathbf{d}\left(\frac{s}{\sqrt{n}}\right) x\, ds$$

$$= A\underbrace{\int_{-\infty}^{+\infty} \sqrt{\frac{1}{\pi}} e^{-\frac{s^2}{2}}\mathbf{d}\left(\frac{s}{\sqrt{n}}\right) x\, ds}_{x_n} = Ax_n.$$

This means that the vector $y_\mu = Ax_n \subset \mathcal{D}_{\exp}(\mu I + G_{\mathbf{d}}^{\infty}) = \mathcal{D}_{\exp}(G_{\mathbf{d}}^{\infty})$ is an entire vector of $\mathbf{d}$ and $x_n \in \mathcal{D}(A)$, i.e., we have proven that

$$A\mathcal{D} \subset \mathcal{D}_{\exp}(G_{\mathbf{d}}^{\infty}) \quad \text{and} \quad \mathcal{D} \subset \mathcal{D}(A).$$

In the proof of Corollary 2.8, it has been shown that $x_n \to x$ as $n \to +\infty$. This means that the set $\mathcal{D}$ is dense in $\mathcal{D}(A)$. Since $\mathcal{D}(A)$ is dense in $\mathbb{B}$ then $\mathcal{D}$ is dense in $\mathbb{B}$. Finally, by definition of $\mathbf{d}$-homogeneity, for any $x \in \mathcal{D}$ and any $s \neq 0$ we have

$$A\sum_{i=0}^{\infty}\frac{s^i G_{\mathbf{d}}^i}{i!}x = \sum_{i=0}^{\infty}\frac{s^i(\mu + G_{\mathbf{d}}))^i}{i!}Ax,$$

and

$$A\frac{\sum_{i=0}^{\infty}\frac{s^i G_{\mathbf{d}}^i}{i!} - I}{s}x = \frac{\sum_{i=0}^{\infty}\frac{s^i(\mu+G_{\mathbf{d}}))^i}{i!} - I}{s}Ax,$$

or, equivalently,

$$AG_{\mathbf{d}}\sum_{i=0}^{\infty}\frac{s^i G_{\mathbf{d}}^i}{(i+1)!}x = (\mu I + G_{\mathbf{d}})\sum_{i=0}^{\infty}\frac{s^i(\mu + G_{\mathbf{d}}))^i}{(i+1)!}Ax.$$

Since x and Ax are entire vectors then $\sum_{i=0}^{\infty}\frac{s^i G_{\mathbf{d}}^i}{(i+1)!}x \to x$ as $s \to 0$ and $\sum_{i=0}^{\infty}\frac{s^i(\mu+G_{\mathbf{d}})^i}{(i+1)!}Ax \to Ax$ as $s \to 0$. This means that

$$AG_{\mathbf{d}}x = (\mu I + G_{\mathbf{d}})Ax, \quad \forall x \in \mathcal{D}.$$

The proof is complete. □

In the finite-dimensional case (see Chap. 6 of Volume I), it was discovered any linear vector field with nilpotent matrix may be homogeneous of both positive and negative degree of dependently of the dilation. Example 5.2 shows that the same property takes the place for some differential operators such as Laplace operator. The example given below introduces a class of differential operators, which are homogeneous with respect to a linear dilation in the Lebesgue space (see Chap. 4).

Example 5.4 *(A class of linear homogeneous differential operators in L^2)* Let us consider the differential operator $A : \mathcal{D}(A) \subset L^2(\mathbb{R}^n; \mathbb{R}^k) \mapsto L^2(\mathbb{R}^n; \mathbb{R}^k)$ defined as follows:

$$(Ax)(z) = \sum_{|\alpha|\leq q} \Xi_\alpha(z)\frac{\partial^{\alpha_1}\ldots\partial^{\alpha_n}x(z)}{\partial z_1^{\alpha_1}\ldots z_n^{\alpha_n}}, \tag{5.9}$$

where $q \in \mathbb{N}$, $\alpha = (\alpha_1, \ldots, \alpha_n)^\top \in \mathbb{N}^n$, $z = (z_1, \ldots, z_n)^\top \in \mathbb{R}^n$, and $\Xi_\alpha \in C^\infty(\mathbb{R}^n; \mathbb{R}^{k\times k})$.

Let a dilation $\mathbf{d}$ in $L^2(\mathbb{R}^n;\mathbb{R}^k)$ be defined by the formula (4.9). As a straightforward corollary of Lemma 5.1, we conclude that the operator A is $\mathbf{d}$-homogeneous of the degree $\mu \in \mathbb{R}$ if

$$AG_{\mathbf{d}}x_\infty = (\mu I + G_{\mathbf{d}})Ax_\infty, \quad \forall x_\infty \in C^\infty(\mathbb{R}^n;\mathbb{R}^k),$$

where $G_{\mathbf{d}}$ is given by (4.16), or equivalently,

$$\mathbf{0} = \sum_{|\alpha|\le q} (\Xi_\alpha(z)\Gamma - (\Gamma+\mu I_k)\Xi_\alpha(z)) \tfrac{\partial^{\alpha_1}...\partial^{\alpha_n}}{\partial z_1^{\alpha_1}...z_n^{\alpha_n}} +$$

$$\left(\Xi_\alpha(z)\tfrac{\partial^{\alpha_1}...\partial^{\alpha_n}}{\partial z_1^{\alpha_1}...z_n^{\alpha_n}}\right)((\Theta z)\cdot\nabla) - ((\Theta z)\cdot\nabla)\left(\Xi_\alpha(z)\tfrac{\partial^{\alpha_1}...\partial^{\alpha_n}}{\partial z_1^{\alpha_1}...z_n^{\alpha_n}}\right).$$

This identity is *algebraic* and can be easily checked in many cases.

For example, if $n=k=q=1$, $\Theta=\Gamma=1$ then the dilation and its generator are defined as follows:

$$(\mathbf{d}(s)x)(z) = e^s x(e^s z), \quad G_{\mathbf{d}}x = x + z\frac{\partial x}{\partial z}, \quad z \in \mathbb{R}.$$

In this case, for $m=1$ the operator A becomes

$$Ax = \Xi_0(z)x + \Xi_1(z)\frac{\partial x}{\partial z}.$$

This operator is $\mathbf{d}$-homogeneous of degree $\mu \in \mathbb{R}$ if

$$\mathbf{0} = -\mu\Xi_0(z) - \mu\Xi_1(z)\tfrac{\partial}{\partial z} + \Xi_0(z)z\tfrac{\partial}{\partial z} - z\tfrac{\partial \Xi_0(z)}{\partial z} - z\Xi_0(z)\tfrac{\partial}{\partial z} +$$

$$\Xi_1(z)\tfrac{\partial}{\partial z}\left(z\tfrac{\partial}{\partial z}\right) - z\tfrac{\partial}{\partial z}\left(\Xi_1(z)\tfrac{\partial}{\partial z}\right)$$

or, equivalently, if

$$\mu\Xi_0(z) + z\frac{\partial \Xi_0(z)}{\partial z} = \mathbf{0}, \quad -\mu\Xi_1(z) + \Xi_1(z) - z\tfrac{\partial \Xi_1(z)}{\partial z} = \mathbf{0}.$$

For example, taking $\Xi_0(z) = z^{-\mu}$ and $\Xi_1(z) = z^{1-\mu}$ with $\mu=-1$ we derive a $\mathbf{d}$-homogeneous operator A of negative degree $\mu=-1$ with $\Xi_0, \Xi_1 \in C^\infty$.

Elements of Homogeneous Arithmetics

Obviously, if $f_1, f_2 \in \mathcal{F}_{\mathbf{d}}$ and $h_1, h_2 \in \mathcal{H}_{\mathbf{d}}$ then

- $h_i \cdot f_i \in \mathcal{F}_{\mathbf{d}}$ and $\deg_{\mathbf{d}}(h_i \cdot f_i) = \deg_{\mathbf{d}}(h_i) + \deg_{\mathbf{d}}(f_i)$, where $\mathcal{D}(h_i \cdot f_i) = \mathcal{D}(h_i) \cap \mathcal{D}(f_i)$ and $i = 1, 2$;
- $h_1 \cdot f_1 + h_2 \cdot f_2 \in \mathcal{F}_{\mathbf{d}}$ provided that $\deg_{\mathbf{d}}(h_1) + \deg_{\mathbf{d}}(f_1) = \deg_{\mathbf{d}} h_2 + \deg_{\mathbf{d}}(f_2)$, where $\mathcal{D}(h_1 \cdot f_1 + h_2 \cdot f_2) = \mathcal{D}(h_1) \cap \mathcal{D}(f_1) \cap \mathcal{D}_{h_2} \cap \mathcal{D}(f_2)$.

It is expectable that an inverse of a homogeneous operator is also a homogeneous operator.[1]

Theorem 5.2 (On homogeneity of the inverse operator) *Let* $\mathbf{d}$ *and* $\mathbf{d}_\nu(s) = e^{\nu s}\mathbf{d}(s), s \in \mathbb{R}$ *be dilations in* $\mathbb{B}$. *If an invertible operator* $f : \mathcal{D}(f) \subset \mathbb{B} \mapsto \mathbb{B}$ *is* $\mathbf{d}$*-homogeneous of a degree* $\nu \in \mathbb{R}$ *then the operator* f^{-1} *is* $\mathbf{d}_\nu$*-homogeneous of the degree* $-\nu$ *with respect to* $\mathbf{d}_\nu(s) = e^{\nu s}\mathbf{d}(s), s \in \mathbb{R}$. *Moreover, if, additionally,* $f(e^{\nu s}x) = e^{\nu s} f(x)$ *for* $x \in \mathcal{D}(f), s \in \mathbb{R}$ *then* f^{-1} *is* $\mathbf{d}$*-homogeneous of the degree* $-\nu$.

Proof Let $x \in \mathcal{D}(f), s \in \mathbb{R}$ be arbitrary and

$$y_1 = e^{\nu s}\mathbf{d}(s) f(x) \in \mathbb{B}, \quad y_2 = f(\mathbf{d}(s)x) \in \mathbb{B}.$$

Due to $\mathbf{d}$-homogeneity of f, we have $\mathcal{D}(f)$ is a $\mathbf{d}$-homogeneous cone and $e^{\nu s}\mathbf{d}(s) f(x) = f(\mathbf{d}(s)x)$. This means that $\mathbf{d}(s)x \in \mathcal{D}(f)$ and $y_1 = y_2 \in \mathcal{D}(f^{-1}) = f(\mathcal{D}(f))$. Since $x \in \mathcal{D}(f)$ is an arbitrary vector, $f(x) \in D(f^{-1})$ and $e^{\nu s}\mathbf{d}(s) f(x) = y_1 = y_2 \in \mathcal{D}(f^{-1})$ for any $s \in \mathbb{R}$, then $\mathcal{D}(f^{-1})$ is a $\mathbf{d}_\nu$-homogeneous cone and $\mathbf{d}_\nu(-s)y_1 \in \mathcal{D}(f^{-1})$ for any $s \in \mathbb{R}$. In this case, we have

$$x = f^{-1}(e^{-\nu s}\mathbf{d}(-s)y_1) \quad \text{and} \quad x = \mathbf{d}(-s) f^{-1}(y_2)$$

or, equivalently,

$$f^{-1}(\mathbf{d}_\nu(-s)y_1) = e^{-\nu(-s)}\mathbf{d}_\nu(-s) f^{-1}(y_2), \quad s \in \mathbb{R}.$$

Taking into account $y_1 = y_2 \in \mathcal{D}(f^{-1})$ we complete the proof of $\mathbf{d}_\nu$-homogeneity of f^{-1}. Moreover, if, additionally, $f(e^{\nu s}x) = e^{\nu s} f(x)$ for $x \in \mathcal{D}(f), s \in \mathbb{R}$ then $f^{-1}(e^{-\nu s}y) = e^{-\nu s} f^{-1}(y)$ for any $y \in \mathcal{D}(f^{-1})$ and any $s \in \mathbb{R}$. This implies that f^{-1} is $\mathbf{d}$-homogeneous of degree $-\nu$. The proof is complete. □

[1] An operator $f : \mathcal{D}(f) \subset \mathbb{B} \mapsto \mathbb{B}$ is said to me invertible if $\exists f^{-1} : \mathcal{D}(f^{-1}) \subset \mathbb{B} \mapsto \mathbb{B}$ such that $\mathcal{D}(f^{-1}) = f(\mathcal{D}(f)), f^{-1}(f(x)) = x$ for any $x \in \mathcal{D}(f)$ and $\mathcal{D}(f) = f^{-1}(\mathcal{D}(f)), f(f^{-1}(y)) = y$ for any $y \in \mathcal{D}(f^{-1})$.

Example 5.5 *(Homogeneity of the inverse Laplacian)* The operator given by

$$(\Delta^{-1} f)(x) = -\int_{\mathbb{R}^3} \frac{f(y)}{4\pi\|x-y\|_{\mathbb{R}^3}} dy, \quad f \in L^2(\mathbb{R}^3;\mathbb{R}^3), \quad x \in \mathbb{R}^3$$

is an inverse to Laplace operator Δ, since for any $f \in L^2(\mathbb{R}^3;\mathbb{R}^3)$ the function $u = \Delta^{-1} f$ satisfies the identity $\Delta u = f$ in the weak sense. In Example 5.2, it has been shown that the Laplace operator **d**-homogeneous of the degree 2β provided that the dilation **d** is given by (4.9) with $\Gamma = \alpha I_3$ and $\Theta = \beta I_3$. Since Δ is a linear operator then the according to Theorem 5.2 the operator Δ^{-1} is **d**-homogeneous of the degree -2β. This fact can also be proven using the change of variable in the Lebesgue integral:

$$(\Delta^{-1} f)(x) = -\int_{\mathbb{R}^3} \frac{f(y)}{4\pi\|x-y\|_{\mathbb{R}^3}} dy = -e^{3\beta s}\int_{\mathbb{R}^3} \frac{f(e^{\beta s}\tilde{y})}{4\pi\|x-e^{\beta s}\tilde{y}\|_{\mathbb{R}^3}} d\tilde{y} =$$

$$-e^{2\beta s}e^{-\alpha s}\int_{\mathbb{R}^3} \frac{(\mathbf{d}(s)f)(\tilde{y})}{4\pi\|e^{-\beta s}x-\tilde{y}\|_{\mathbb{R}^3}} d\tilde{y} = e^{2\beta s}\mathbf{d}(-s)[\Delta^{-1}\mathbf{d}(s)f](x),$$

or, equivalently,

$$\Delta^{-1}\mathbf{d}(s)f = e^{-2\beta s}\mathbf{d}(s)\Delta^{-1} f$$

due to invertibility $\mathbf{d}(-s) = [\mathbf{d}(s)]^{-1}$ and boundedness of the operators $\mathbf{d}(s)$.

Theorem 5.3 (On composition of homogeneous operators)
*Let $f_1 : \mathcal{D}(f_1) \subset \mathbb{B} \mapsto \mathbb{B}$ and $f_2 : \mathcal{D}(f_2) \subset \mathbb{B} \mapsto \mathbb{B}$ be **d**-homogeneous operators. Let $\mathcal{R}(f_2) \subset \mathcal{D}(f_1)$, where $\mathcal{R}(f_2)$ is the range of the operator f_2. The composition $f_1(f_2) : \mathcal{D}(f_2) \subset \mathbb{B} \mapsto \mathbb{B}$ is a **d**-homogeneous operator provided that* $\deg_{\mathbf{d}}(f_2) = 0$ *or f_1 is standard homogeneous:*

$$\exists \nu \in \mathbb{R} \quad : \quad f_1(e^s x) = e^{\nu s} f_1(x), \quad \forall s \in \mathbb{R}, \forall x \in \mathcal{D}(f_1).$$

Proof Since $f_2(\mathbf{d}(s)x) = e^{s\deg_{\mathbf{d}}(f_2)}\mathbf{d}(s)f_2(x)$ for all $x \in \mathcal{D}(f_2)$ then

$$f_1(f_2(\mathbf{d}(s)x)) = f_1(e^{s\deg_{\mathbf{d}}(f_2)}\mathbf{d}(s)f_2(x))$$

and the **d**-homogeneity of the composition of the operators is preserved if f_1 is a standard homogeneous operator or $\deg_{\mathbf{d}}(f_2) = 0$.

Example 5.6 *(Homogeneity of the Navier–Stokes equations)* Let us consider the system of Navier–Stokes equations (2.50). In Example 2.17, it has been shown that selecting $p(t) = \mathrm{div}((-\Delta)^{-1}(u(t)\cdot\nabla)u(t)$ the analysis can be reduced to the system of the form

$$\dot{u}(t) = \Delta u(t) + f(u(t)), \quad t > 0,$$

where $f : \mathcal{D}(f) \subset L^2(\mathbb{R}^3;\mathbb{R}^3) \mapsto L^2(\mathbb{R}^3;\mathbb{R}^3)$ is given by

$$f(u) := \nabla\mathrm{div}(\Delta^{-1}(u\cdot\nabla)u) - (u\cdot\nabla)u, \quad u \in \mathcal{D}(f)$$

with the domain

$$\mathcal{D}(f) = \left\{u \in H^{\frac{7}{4}}(\mathbb{R}^3;\mathbb{R}^3) : \mathrm{div}u = \mathbf{0}\right\}$$

being a fractional Sobolev space of divergent-free vector fields. The inverse of the Laplace operator is given by

$$(\Delta^{-1}g)(x) = -\int_{\mathbb{R}^3} \frac{g(y)}{4\pi\|x-y\|_{\mathbb{R}^3}}dy, \quad g \in L^2(\mathbb{R}^3;\mathbb{R}^3), \quad x \in \mathbb{R}^3.$$

Notice that (see Appendix D)

$$\mathcal{D}(\Delta) = H^2(\mathbb{R}^3\ \mathbb{R}^3) \subset H^{\frac{7}{4}}(\mathbb{R}^3;\mathbb{R}^3) \subset H^1(\mathbb{R}^3;\mathbb{R}^3)$$

and, according to Examples 2.14 and 2.17, the operator $u \mapsto f(\Delta^{-\frac{7}{8}}u)$ is locally Lipschitz continuous on $L^2(\mathbb{R}^3;\mathbb{R}^3)$.

Let us consider the dilation given by (4.9) with $\Gamma = I_3$ and $\Theta = I_3$. Notice that **d**-homogeneity of the Laplace operator Δ with the degree 2 has been proven in Example 5.2. Let us show that f is **d**-homogeneous of the degree 2 as well.

On the one hand, it is obvious that for any $u \in H^1(\mathbb{R}^3;\mathbb{R}^3)$ the identity $\mathrm{div}(u) = \mathbf{0}$ always implies that $\mathrm{div}(\mathbf{d}(s)u) = \mathbf{0}$. Hence, taking into account, $\mathbf{d}(s)H^{\frac{7}{4}}(\mathbb{R}^3;\mathbb{R}^3) \subset H^{\frac{7}{4}}(\mathbb{R}^3;\mathbb{R}^3)$ (see Remark 4.1) we conclude $\mathbf{d}(s)\mathcal{D}(f) \subset \mathcal{D}(f)$, i.e., $\mathcal{D}(f)$ is a **d**-homogeneous cone. On the other hand,

$$\nabla\mathbf{d}(s)u = e^s\mathbf{d}(s)\nabla u,$$

$$(\mathbf{d}(s)u\cdot\nabla)(\mathbf{d}(s)u) = e^{2s}\mathbf{d}(s)(u\cdot\nabla)u$$

and

$$\nabla\mathrm{div}(\mathbf{d}(s)u) = e^{2s}\mathbf{d}(s)\nabla\mathrm{div}(u).$$

In Example 5.5 it has been shown that

$$(-\Delta)^{-1}\mathbf{d}(s) = e^{-2s}\mathbf{d}(s)(-\Delta)^{-1}.$$

Hence, according to Theorem 5.3 f is a **d**-homogeneous operator. Moreover, for any $u \in H^{\frac{7}{4}}(\mathbb{R}^3;\mathbb{R}^3)$ we derive

$$f(\mathbf{d}(s)u) = \nabla \mathrm{div}((-\Delta)^{-1}(\mathbf{d}(s)u \cdot \nabla)(\mathbf{d}(s)u) - (\mathbf{d}(s)u \cdot \nabla)(\mathbf{d}(s)u) = e^{2s}\mathbf{d}(s)f(u),$$

i.e., the nonlinear operator $\Delta + f : \mathcal{D}(\Delta) \subset L^2(\mathbb{R}^3;\mathbb{R}^3) \mapsto L^2(\mathbb{R}^3;\mathbb{R}^3)$ is **d**-homogeneous of the degree 2.

We complete this section with the following, obvious, corollary showing that the properties of homogeneous operators and systems should depend mainly on the sign of the homogeneity degree but not on its value.

Corollary 5.2 *Let an operator f be* **d***-homogeneous of a degree $\nu \in \mathbb{R}$. Let $\mathbf{d}_\alpha(s) := \mathbf{d}(\alpha s), s \in \mathbb{R}$ for some $\alpha > 0$. Then f is $\mathbf{d}_\alpha$-homogeneous of the degree $\alpha\nu$.*

The above corollary holds for **d**-homogeneous functionals as well as for **d**-homogeneous set-valued mappings studied below.

On Regularity of Homogeneous Operators

Homogeneity allows any local property (e.g., local Lipschitz continuity or differentiability) of a finite-dimensional operator to be extended globally. For the infinite-dimensional case, similar results can be obtained.

Theorem 5.4 *Let* **d** *be a strictly monotone strongly continuous linear dilation in $\mathbb{B}$. A* **d***-homogeneous operator $f : \mathbb{B} \mapsto \mathbb{B}$ satisfies the Lipschitz condition on any* **d***-homogeneous sphere $S_{\mathbf{d}}(r)$ (resp., on any* **d***-homogeneous ball $B_{\mathbf{d}}(r)$) with $r > 0$, if and only if it satisfies the Lipshitz condition on the unit sphere S (resp., on the unit ball $B(1)$).*

Proof The *Necessity* is straightforward since $S = S_{\mathbf{d}}(1)$ (resp, $B(1) = B_{\mathbf{d}}(1)$). *Sufficiency.* If $x_i \in S_{\mathbf{d}}(r)$, $i = 1, 2$ then

$$z_i := \mathbf{d}(-\ln(\|x_i\|_{\mathbf{d}}))x_i = \mathbf{d}(-\ln(r))x_i \in S.$$

Due to monotonicity of the dilation **d** (see Proposition (4.2)) the homogeneous projections z_i are uniquely defined. Hence, if $L > 0$ is a Lipschitz constant of f on S then

$$\|f(x_1) - f(x_2)\| = \|r^\nu \mathbf{d}(-\ln r)(f(z_1) - f(z_2)\| \le r^\nu \|\mathbf{d}(-\ln r)\| L \|z_1 - z_2\| =$$

$$r^\nu \|\mathbf{d}(-\ln r)\| \; L \; \|\mathbf{d}(\ln r)(x_1 - x_2)\| \le L_r \|x_1 - x_2\|,$$

where $L_r := r^\nu \|\mathbf{d}(-\ln r)\| \, L \, \|\mathbf{d}(\ln r)\| < +\infty$ is the Lipschitz constant for f on $S_{\mathbf{d}}(r)$, $r > 0$. The case of the $\mathbf{d}$-homogeneous balls can be treated similarly. The proof is complete. □

Therefore, $\mathbf{d}$-homogeneity always allows a globally expansion of a local regularity. For example, Lipschitz continuity on a unit ball implies Lipschitz continuity on any $\mathbf{d}$-homogeneous ball in $\mathbb{B}$. The same holds for $\mathbf{d}$-homogeneous spheres. However, in the general case, an operator f may satisfy the Lipshitz condition on any sphere in $\mathbb{B}$ and be nowhere continuous (see, Chap. 6 of Volume I).

An operator $f : \mathbb{B} \mapsto \mathbb{B}$ is said to be *uniformly continuous* on $\Omega \subset \mathbb{B}$ if there exists $\omega \in \mathcal{K}$ such that $\|f(x_1) - f(x_2)\| \le \omega(\|x_1 - x_2\|)$ for all $x_1, x_2 \in \Omega$.

Theorem 5.5 *Let* $\mathbf{d}$ *be a strictly monotone uniformly continuous dilation in* $\mathbb{B}$.

- *A* $\mathbf{d}$*-homogeneous operator* $f : \mathbb{B} \mapsto \mathbb{B}$ *is locally Lipschitz continuous on* $\mathbb{B}\backslash\{\mathbf{0}\}$ *or, more precisely, it satisfies the Lipschitz condition on any set*

$$K(r_1, r_2) = \{u \in \mathbb{B} : r_1 \le \|u\| \le r_2\}$$

with $0 < r_1 < r_2$ *if and only if it satisfies the Lipschitz condition on the unit sphere* S.

- *A* $\mathbf{d}$*-homogeneous operator* $f : \mathbb{B} \mapsto \mathbb{B}$ *is locally uniformly continuous on* $\mathbb{B}\backslash\{\mathbf{0}\}$ *(i.e., it is uniformly continuous on any set* $K(r_1, r_2)$ *with* $0 < r_1 < r_2$*), if and only if it is uniformly continuous on the unit sphere* S.

Proof Let us study the case of Lipschitz continuity. The uniform continuity of f can be treated analogously. Since the *Necessity* is straightforward, we prove only the *Sufficiency*. Let us denote

$$K_{\mathbf{d}}(r_1, r_2) = \{x \in \mathbb{B} : r_1 \le \|x\|_{\mathbf{d}} \le r_2\}.$$

It is sufficient to show that f satisfy the Lipschitz condition on $K_{\mathbf{d}}(r_1, r_2)$ for any $0 < r_1 < r_2$. Indeed, Lemma 4.5 implies that for any $0 < \tilde{r}_1 < \tilde{r}_2$ there exist $0 < r_1 < r_2$ such that $K(\tilde{r}_1, \tilde{r}_2) \subset K_{\mathbf{d}}(r_1, r_2)$.

From Theorem 5.4, we derive that f satisfies Lipschitz condition on any $\mathbf{d}$-homogeneous sphere $S_{\mathbf{d}}(r)$, $r > 0$.

Let $x_i \in K_{\mathbf{d}}(r_1, r_2)$, $i = 1, 2$ and let $r_0 \ge \max\{r_2, r_1^{-1}\}$. In this case,

$$\frac{1}{r_0} \le \|x_i\| \le r_0 \quad \text{and} \quad x_i = \mathbf{d}\left(\ln \tfrac{\|x_i\|_{\mathbf{d}}}{r_0}\right) z_i$$

for some $z_i \in S_{\mathbf{d}}(r_0)$. Hence, we derive

$$f(x_1) - f(x_2) = f(\mathbf{d}(\ln \|x_1\|_{\mathbf{d}}/r_0)z_1) - f(\mathbf{d}(\ln \|x_2\|_{\mathbf{d}}/r_0)z_2)$$

$$= \tfrac{\|x_1\|_{\mathbf{d}}^{\nu}}{r_0^{\nu}}\mathbf{d}(\ln \|x_1\|_{\mathbf{d}}/r_0)f(z_1) - \tfrac{\|x_2\|_{\mathbf{d}}^{\nu}}{r_0^{\nu}}\mathbf{d}(\ln \|x_2\|_{\mathbf{d}}/r_0)f(z_2)$$

$$= \tfrac{\|x_1\|_{\mathbf{d}}^{\nu}}{r_0^{\nu}}\mathbf{d}(\ln \|x_1\|_{\mathbf{d}}/r_0)(f(z_1) - f(z_2)) +$$

$$\tfrac{\|x_1\|_{\mathbf{d}}^{\nu} - \|x_2\|_{\mathbf{d}}^{\nu}}{r_0^{\nu}}\mathbf{d}(\ln \|x_1\|_{\mathbf{d}}/r_0)f(z_2) +$$

$$\tfrac{\|x_2\|^{\nu}}{r_0^{\nu}}(\mathbf{d}(\ln \|x_1\|_{\mathbf{d}}/r_0) - \mathbf{d}(\ln \|x_2\|_{\mathbf{d}}/r_0)f(z_2).$$

(a) Let $L_{r_0} > 0$ be a Lipschitz constant for f on the $\mathbf{d}$-homogeneous sphere $S_{\mathbf{d}}(r_0)$. Since $\|x_i\|_{\mathbf{d}} < r_0$ and due to monotonicity of the dilation $\mathbf{d}$ we have $\|\mathbf{d}(\ln \|x_i\|_{\mathbf{d}}/r_0)\| < 1$ then

$$\tfrac{\|x_1\|_{\mathbf{d}}^{\nu}}{r_0^{\nu}}\mathbf{d}(\ln \|x_1\|_{\mathbf{d}}/r_0)(f(z_1) - f(z_2)) \leq L_{r_0}\|z_1 - z_2\| =$$

$$L_{r_0}\|\mathbf{d}(\ln r_0/\|x_1\|_{\mathbf{d}})x_1 - \mathbf{d}(\ln r_0/\|x_2\|_{\mathbf{d}})x_2\| \leq$$

$$L_{r_0}\|\mathbf{d}(\ln r_0/\|x_1\|_{\mathbf{d}})\| \; \|x_1 - x_2\| +$$

$$L_{r_0}\|(\mathbf{d}(\ln r_0/\|x_1\|_{\mathbf{d}}) - \mathbf{d}(\ln r_0/\|x_2\|_{\mathbf{d}}))x_2\| \leq$$

$$L_{r_0}\|\mathbf{d}\,(2\ln r_0)\| \; \|x_1 - x_2\| +$$

$$\|x_2\|L_{r_0}\|\mathbf{d}(\ln r_0/\|x_1\|_{\mathbf{d}}) - \mathbf{d}(\ln r_0/\|x_2\|_{\mathbf{d}})\|.$$

Since $\mathbf{d}$ is a uniformly continuous dilation then using Theorem 2.1 for any $y \in \mathcal{D}(G_{\mathbf{d}})$ we derive $\mathbf{d}(s_1) - \mathbf{d}(s_2) = \int_{s_2}^{s_1} G_{\mathbf{d}}\mathbf{d}(s)ds$ and, consequently,

$$\|\mathbf{d}\left(\ln \tfrac{r_0}{\|x_1\|_{\mathbf{d}}}\right) - \mathbf{d}\left(\ln \tfrac{r_0}{\|x_2\|_{\mathbf{d}}}\right)\| \leq \|G_{\mathbf{d}}\| \sup_{r \in (r_1, r_2)} \left\|\mathbf{d}\left(\ln \tfrac{r_0}{r_1}\right)\right\| \left|\ln \tfrac{r_0}{\|x_1\|_{\mathbf{d}}} - \ln \tfrac{r_0}{\|x_2\|_{\mathbf{d}}}\right|$$

$$\leq \|G_{\mathbf{d}}\|\|\mathbf{d}(2\ln r_0)\| \; |\ln \|x_2\|_{\mathbf{d}} - \ln \|x_1\|_{\mathbf{d}}| \,.$$

By Lemma 4.5, the canonical homogeneous norm $\|\cdot\|_{\mathbf{d}}$ satisfies the Lipschitz condition on any set $K_{\mathbf{d}}(r_1, r_2)$. Hence, from the smoothness of the logarithm function on $(0, +\infty)$ we conclude $\exists \tilde{L}_1 > 0$ such that

$$\|\mathbf{d}\left(\ln \tfrac{r_0}{\|x_1\|_{\mathbf{d}}}\right) - \mathbf{d}\left(\ln \tfrac{r_0}{\|x_2\|_{\mathbf{d}}}\right)\| \leq \|G_{\mathbf{d}}\|\|\mathbf{d}(2\ln r_0)\| \; \tilde{L}_1\|x_1 - x_2\|,$$

for all $x_1, x_2 \in K(r_1, r_2)$. Therefore, we derive

$$\left\| \tfrac{\|x_1\|_{\mathbf{d}}^{\nu}}{r_0^{\nu}} \mathbf{d}(\ln \|x_1\|_{\mathbf{d}}/r_0)(f(z_1) - f(z_2)) \right\| \leq L_{r_0}^1 \|x_1 - x_2\|,$$

where $L_{r_0}^1 = L_{r_0} \|\mathbf{d}(2 \ln r_0)\| (1 + \|G_{\mathbf{d}}\| \tilde{L}_1)$.

(b) Since the canonical homogeneous norm $\|\cdot\|_{\mathbf{d}}$ satisfies the Lipschitz condition on any set $K_{\mathbf{d}}(r_1, r_2)$ then there exists $\tilde{L}_2 > 0$ such that

$$\left\| \tfrac{\|x_1\|^{\nu} - \|x_2\|^{\nu}}{r_0^{\nu}} \mathbf{d}(\ln \|x_1\|_{\mathbf{d}}/r_0) f(z_2) \right\| \leq$$

$$\tilde{L}_2 \|x_1 - x_2\| \|\mathbf{d}(\ln \|x_1\|_{\mathbf{d}}/r_0)\| \, \|f(z_2)\| \leq L_{r_0}^2 \|x_1 - x_2\|,$$

where $L_{r_0}^2 = \tilde{L}_2 \sup_{z \in S_{\mathbf{d}}(r_0)} \|f(z)\|$ with $r_0 > 0$.

(c) Finally, repeating considerations given in the part (a) of this proof, we conclude

$$\left\| \tfrac{\|x_2\|^{\nu}}{r_0^{\nu}} (\mathbf{d}(\ln \|x_1\|_{\mathbf{d}}/r_0) - \mathbf{d}(\ln \|x_2\|_{\mathbf{d}}/r_0) f(z_2) \right\| \leq L_{r_0}^3 \|x_1 - x_2\|,$$

where $L_{r_0}^3 = \|G\|_{\mathbf{d}} \tilde{L}_1 \sup_{z \in S_{\mathbf{d}}(r_0)} \|f(s)\|$.

Therefore, f satisfy the Lipschitz condition on $K_{\mathbf{d}}(r_1, r_2)$ for any $0 < r_1 < r_2$ and on $K(r_1, r_2)$ for any $0 < r_1 < r_2$. The proof is complete. □

The above theorem proves that the regularity (Lipschitz continuity) on a sphere of a $\mathbf{d}$-homogeneous operator can always be expanded to $\mathbb{B}\backslash\{\mathbf{0}\}$ in the case of uniformly continuous dilations. *Does the symmetry with respect to a strongly continuous dilation allow a regularity on a sphere to be expanded to* $\mathbb{B}\backslash\{\mathbf{0}\}$? The following corollary partially answers this question.

Corollary 5.3 *Let* $\mathbf{d}$ *be a strictly monotone strongly continuous linear dilation in* $\mathbb{B}$. *Let an operator* $f : \mathbb{B} \mapsto \mathbb{B}$ *be* $\mathbf{d}$*-homogeneous of degree* $\nu \in \mathbb{R}$.

- *The operator* f *is continuous on* $\mathbb{B}\backslash\{\mathbf{0}\}$ *if and only if it is continuous on the unit sphere* S.
- *Let the Banach space* $\mathbb{B}_{\mathbf{d}}$ *be defined as in Theorem 4.2. Let* $\tilde{+}$ *and* $\tilde{\cdot}$ *denote an addition operation and a multiplication by a scalar in* $\mathbb{B}_{\mathbf{d}}$, *respectively. The operator* $\tilde{f} : \mathbb{B}\backslash\{\mathbf{0}\} \mapsto \mathbb{B}$ *given by*

$$\tilde{f}(x) = \|x\|_{\mathbf{d}}^{-\nu} f(x), \quad x \in \mathbb{B}\backslash\{\mathbf{0}\}$$

is locally Lipschitz continuous on $\mathbb{B}_{\mathbf{d}}\backslash\{\mathbf{0}\}$ *if and only if* $\exists L > 0$ *such that*

$$\left\| \tilde{f}(x) \tilde{-} \tilde{f}(y) \right\|_{\mathbf{d}} \leq L \|x \tilde{-} y\|_{\mathbf{d}}, \quad \forall x, y \in S, \tag{5.10}$$

where $u \tilde{-} v = u \tilde{+} (-1) \tilde{\cdot} v$ *denotes the vector subtraction in* $\mathbb{B}_{\mathbf{d}}$.

Proof Since f is $\mathbf{d}$-homogeneous of the degree ν then, obviously, $\tilde{f}$ is $\mathbf{d}$-homogeneous of the degree 0. In this case, we have

$$\tilde{f}(\lambda \tilde{\cdot} x) = \lambda \tilde{\cdot} \tilde{f}(x), \quad \forall \lambda > 0, \forall x \in \tilde{\mathbb{B}} \tag{5.11}$$

i.e., $\tilde{f}$ is standard homogeneous with respect to the standard dilation in $\mathbb{B}_{\mathbf{d}}$. The standard dilation is uniformly continuous in $\mathbb{B}_{\mathbf{d}}$. Using Theorem 5.5 we conclude that $\tilde{f}$ is locally Lipschitz continuous on $\mathbb{B}_{\mathbf{d}}\backslash\{\mathbf{0}\}$.

Similarly, we conclude that $\tilde{f}$ is continuous on $\mathbb{B}_{\mathbf{d}}\backslash\{\mathbf{0}\}$ if and only if $\tilde{f}$ is continuous on S. Taking into account continuity of $\|\cdot\|_{\mathbf{d}}$ and Φ on $\mathbb{B}$ (see, Lemmas 4.4, 4.5 and Corollary 4.7) we derive continuity of f on $\mathbb{B}\backslash\{\mathbf{0}\}$. The proof is complete.□

In Chap. 2, it has been discovered that the local Lipschitz continuity is the usual restriction to nonlinearity in order to guarantee a well-posedness of the abstract differential equation with unbounded linear and locally bounded nonlinear operators. According to the above corollary, this property can be discovered for homogeneous operators satisfying Lipschitz condition on the sphere. However, the corresponding analysis of the abstract differential equation has to be transferred to the homeomorphic space $\mathbb{B}_{\mathbf{d}}$. In this case, the evolution can be rewritten in terms of the so-called polar homogeneous homogeneous coordinates (see Chap. 7 for more details).

Notice that the continuity of a homogeneous operator at the origin can be obtained under less restrictive conditions.

Proposition 5.3 *Let* $\mathbf{d}$ *be a strongly continuous strictly monotone dilation in* $\mathbb{B}$ *such that* $\|\mathbf{d}(s)\| \leq e^{\beta s}$ *for all* $s < 0$ *and some* $\beta > 0$. *Let* $f : \mathbb{B} \mapsto \mathbb{B}$ *be* $\mathbf{d}$-*homogeneous operator of a degree* $\mu \in \mathbb{R}$, $f(\mathbf{0}) = \mathbf{0}$, $M := \sup_{x\in S} \|f(x)\| < +\infty$ *and* $\mu + \beta > 0$ *then* f *is continuous at the origin and* $f(x) \to \mathbf{0}$ *as* $x \to \mathbf{0}$.

Proof From homogeneity of f we derive

$$f(x) = \|x\|_{\mathbf{d}}^{\mu}\mathbf{d}(\ln \|x\|_{\mathbf{d}}) f(\mathbf{d}(\ln - \|x\|_{\mathbf{d}})x).$$

Taking into account $\|\mathbf{d}(\ln - \|x\|_{\mathbf{d}})x\| = 1$ we derive

$$\|f(x)\| \leq M\|x\|_{\mathbf{d}}^{\beta+\mu}$$

for $\|x\| \leq 1$. Using the continuity of the canonical homogeneous norm $\|\cdot\|_{\mathbf{d}}$ at the origin we complete the proof.

Frechét derivatives of homogeneous operators inherit dilation symmetry.

Proposition 5.4 *Let* $\mathbf{d}$ *be a strictly monotone strongly continuous linear dilation in* $\mathbb{B}$. *Let* $f : \mathbb{H} \mapsto \mathbb{H}$ *be a* $\mathbf{d}$-*homogeneous operator of a degree* $\nu \in \mathbb{R}$. *If* f *is Frechét differentiable on a* $\mathbf{d}$-*homogeneous cone* $\mathcal{D} \subset \mathbb{H}$, *then*

$$(Df(\mathbf{d}(s)x))(\mathbf{d}(s)y) = e^{\nu s}\mathbf{d}(s)(Df(x))(y), \quad \forall s \in \mathbb{R}, \forall x \in \mathcal{D}, \forall y \in \mathbb{H}, \tag{5.12}$$

where $Df(z) \in \mathcal{L}(\mathbb{H}; \mathbb{H})$ *denotes the Frechét derivative of* f *at a point* z.

Proof By definition, $Df(x)$ is given by

$$\lim_{\|y\|\to 0} \frac{\|f(x+y)-f(x)-(Df(x))(y)\|}{\|y\|} = 0, \quad y \in \mathbb{H}.$$

Similarly, for $Df(\mathbf{d}(s)x)$ we have

$$\lim_{\|y_s\|\to 0} \frac{\|f(\mathbf{d}(s)x+y_s)-f(\mathbf{d}(s)x)-(Df(\mathbf{d}(s)x))(y_s)\|}{\|y_s\|} = 0, \quad y_s \in \mathbb{H}.$$

Using homogeneity we derive

$$\frac{\|f(\mathbf{d}(s)x+y_s)-f(\mathbf{d}(s)x)-(Df(\mathbf{d}(s)x))(y_s)\|}{\|y_s\|} \leq$$

$$\frac{e^{\nu s}\|\mathbf{d}(s)\|\cdot\|f(x+y)-f(x)-e^{-\nu s}\mathbf{d}(-s)(Df(\mathbf{d}(s)x))(y_s)\|}{\|y_s\|} \leq$$

$$\frac{e^{\nu s}\|\mathbf{d}(s)\|}{\lfloor\mathbf{d}(s)\rfloor} \frac{\|f(x+y)-f(u)-e^{-\nu s}\mathbf{d}(-s)(Df(\mathbf{d}(s)x))(\mathbf{d}(s)y)\|}{\|y\|} \quad \text{with } y=\mathbf{d}(-s)y_s.$$

Since $\lfloor\mathbf{d}(s)\rfloor > 0$ and $\mathbf{d}(s)$ is a linear bounded invertible operator then the limit $\|y\| \to 0$ is equivalent to the limit $\|y_s\| \to 0$. Hence, due to uniqueness of the Frechét derivative we derive the identity (5.12).

The above proposition has the following straightforward corollary.

Corollary 5.4 *Let* $\mathbf{d}$ *be a strictly monotone strongly continuous linear dilation in* $\mathbb{H}$. *A* $\mathbf{d}$*-homogeneous operator* $f : \mathbb{H} \mapsto \mathbb{H}$ *is Frechet differentiable on* $\mathbb{H}\backslash\{\mathbf{0}\}$ *if and only if it is Frechet differentiable on the unit sphere* $S = \{x \in \mathbb{H} : \|x\| = 1\}$.

An analog of Euler's homogeneous function theorem can be proven for $\mathbf{d}$-homogeneous operators in $\mathbb{H}$.

Theorem 5.6 *Let* $\mathbf{d}$ *be a strictly monotone strongly continuous dilation in* $\mathbb{H}$ *and* $f : \mathbb{H} \mapsto \mathbb{H}$ *be a* $\mathbf{d}$*-homogeneous operator. If* f *is Frechét differentiable on* $\mathcal{D}\backslash\{\mathbf{0}\}$*, where* $\mathcal{D} \subset D(G_\mathbf{d}) \subset \mathbb{H}$ *is a* $\mathbf{d}$*-homogeneous cone, then*

$$(Df(x))(G_\mathbf{d}x) = ((\deg_\mathbf{d}(f)I + G_\mathbf{d}) \circ f)(x) \text{ for any } x \in \mathcal{D}, \tag{5.13}$$

where $Df(x) \in \mathcal{L}(\mathbb{H},\mathbb{H})$ *denotes the Frechét derivative of* f *at* x.

Proof Denote $\nu = \deg_{\mathcal{F}_\mathbf{d}}(h)$. Since

$$f(x) = f(\mathbf{d}(\ln\|x\|_\mathbf{d})\mathbf{d}(-\ln\|x\|_\mathbf{d})x) = \|x\|^\nu \mathbf{d}(\ln\|x\|_\mathbf{d}) f(\mathbf{d}(-\ln\|x\|_\mathbf{d})x)$$

then for any $x \in \mathcal{D}(G_{\mathbf{d}})\backslash\{\mathbf{0}\}$ using product and the chain rule for Frechét derivatives we derive

$$(Df(x))(y) = \nu\|x\|_{\mathbf{d}}^{\nu-1}(D\|x\|_{\mathbf{d}})(y)\mathbf{d}(\ln\|x\|_{\mathbf{d}})f(\mathbf{d}(-\ln\|x\|_{\mathbf{d}})x) +$$

$$\|x\|_{\mathbf{d}}^{\nu-1}(D\|x\|_{\mathbf{d}})(y)G_{\mathbf{d}}\mathbf{d}(\ln\|x\|_{\mathbf{d}})f(\mathbf{d}(-\ln\|x\|_{\mathbf{d}})x) +$$

$$\|x\|_{\mathbf{d}}^{\nu}\mathbf{d}(\ln\|x\|_{\mathbf{d}})(Df(z))((D\mathbf{d}(-\ln\|x\|_{\mathbf{d}})x)(y)),$$

where $z = \mathbf{d}(-\ln\|u\|_{\mathbf{d}})x$, $D\|x\|_{\mathbf{d}} \in \mathcal{L}(\mathbb{H};\mathbb{R})$ is a Frechét derivative of $\|\cdot\|_{\mathbf{d}}$ at the point x (see formula (4.25)) and $D\mathbf{d}(-\ln\|x\|_{\mathbf{d}})x \in \mathcal{L}(\mathbb{H};\mathbb{H})$ is a Frechét derivative of the mapping $x \mapsto \mathbf{d}(-\ln\|x\|_{\mathbf{d}})x$ at the point$x$. Since the mapping $s \mapsto \mathbf{d}(s)x$ is differentiable for any $x \in \mathcal{D}(G_{\mathbf{d}})$ and $\frac{d}{ds}\mathbf{d}(s)x = G_{\mathbf{d}}\mathbf{d}(s)x = \mathbf{d}(s)G_{\mathbf{d}}x$ then

$$(D\mathbf{d}(-\ln\|x\|_{\mathbf{d}}x)(y) = \mathbf{d}(-\ln\|x\|_{\mathbf{d}})y - \frac{(D\|x\|_{\mathbf{d}})(y)}{\|x\|_{\mathbf{d}}}G_{\mathbf{d}}\mathbf{d}(-\ln\|x\|_{\mathbf{d}})x, \quad y \in \mathbb{H}.$$

Since $(D\|x\|_{\mathbf{d}})(G_{\mathbf{d}}x) = \|x\|_{\mathbf{d}}$ (see (4.25)), then $(D\mathbf{d}(-\ln\|x\|_{\mathbf{d}})x)(G_{\mathbf{d}}x) = 0$ and the identity (5.13) holds.

5.2 Homogeneous Approximations

5.2.1 Homogeneous Approximation on a Sphere

Homogeneous approximations of non homogeneous operators in an infinite-dimensional space can be introduced similarly to the finite-dimensional case (see Chap. 6 of Volume I).

Definition 5.3 Let a real Banach space $\mathbb{B}$ be equipped with a strictly monotone strongly continuous linear dilation $\mathbf{d}$. Let $f : \mathbb{B} \mapsto \mathbb{B}$ be a locally bounded operator. An operator $f_r : \mathbb{B} \mapsto \mathbb{B}$ defined as follows:

$$f_r(x) = \frac{\|x\|_{\mathbf{d}}^{\nu}}{r^{\nu}}\mathbf{d}\left(\ln\frac{\|x\|_{\mathbf{d}}}{r}\right)f\left(\mathbf{d}\left(\ln\frac{r}{\|x\|_{\mathbf{d}}}\right)x\right), \qquad x \in \mathbb{B},$$

is said to be the $\mathbf{d}$-homogeneous approximation of f with degree $\nu \in \mathbb{R}$ on the $\mathbf{d}$-homogeneous sphere $S_{\mathbf{d}}(r)$, where $r > 0$.

Since $\mathbf{d}\left(\ln\frac{r}{\|x\|_{\mathbf{d}}}\right)x \in S_{\mathbf{d}}(r)$ then $f_r(x) = f(x)$ on if $x \in S_{\mathbf{d}}(r)$, i.e., the $\mathbf{d}$-homogeneous approximation f_r of the operator f on the sphere $S_{\mathbf{d}}(r)$ is a homogeneous expansion of f from the sphere $S_{\mathbf{d}}(r)$ to the whole space $\mathbb{B}$.

Theorem 5.7 *Let* $\mathbf{d}$ *be a strictly monotone strongly continuous linear dilation in* $\mathbb{B}$. *Let* f_r *be a* $\mathbf{d}$*-homogeneous approximation of the operator* $f : \mathbb{B} \mapsto \mathbb{B}$ *on a* $\mathbf{d}$*-homogeneous sphere* $S_{\mathbf{d}}(r)$, *where* $r > 0$. *Then the following holds.*

- *The operator* f_r *is* $\mathbf{d}$*-homogeneous of a degree* $\nu \in \mathbb{R}$.
- *If* f *is Lipschitz continuous (uniformly continuous) on* $S_{\mathbf{d}}(r)$ *and* $\mathbf{d}$ *is uniformly continuous then* f_r *is locally Lipschitz continuous (locally uniformly continuous) on* $\mathbb{B}\backslash\{\mathbf{0}\}$.

Proof (1) Homogeneity of the operator f_r is obvious. Indeed,

$$f_r(\mathbf{d}(s)x) = \frac{\|\mathbf{d}(s)x\|_{\mathbf{d}}^{\nu}}{r^{\nu}}\mathbf{d}\left(\ln\frac{\|\mathbf{d}(s)x\|_{\mathbf{d}}}{r}\right)f\left(\mathbf{d}\left(\ln\frac{r}{\|\mathbf{d}(s)x\|_{\mathbf{d}}}\right)\mathbf{d}(s)x\right) =$$

$$e^{\nu s}\tfrac{\|x\|_{\mathbf{d}}^{\nu}}{r^{\nu}}\mathbf{d}\left(s+\ln\tfrac{\|x\|_{\mathbf{d}}}{r}\right)f\left(\mathbf{d}\left(\ln\tfrac{r}{e^{s}\|x\|_{\mathbf{d}}}\right)\mathbf{d}(s)x\right)=e^{\nu s}\mathbf{d}(s)f_r(x),$$

where $x \in \mathcal{D}(f_r)\backslash\{\mathbf{0}\}$.

(2) The local Lipschit (uniform) continuity of f_r follows from Theorem 5.5.

5.2.2 *Homogeneity at Zero and at Infinity Limits*

Local homogeneity (close to zero and close to infinity) can be studied using the homogeneous approximations defined below.

Definition 5.4 Let $\mathbf{d}_L$ be a strongly continuous strictly monotone linear dilation on a Banach space $\mathbb{B}$, where $L = 0$ or $L = \infty$. A locally bounded operator $f : \mathbb{B} \mapsto \mathbb{B}$ (resp., functional $h : \mathbb{B} \mapsto \mathbb{R}$) is said to be $\mathbf{d}_L$-homogeneous in the L-limit if there exists a $\mathbf{d}_L$-homogeneous operator $f_L : \mathbb{B} \mapsto \mathbb{B}$ (resp., $\mathbf{d}_L$-homogeneous functional $h_L : \mathbb{B} \mapsto \mathbb{R}$) of degree $\nu_L \in \mathbb{R}$ such that

$$\lim_{r\to L}\ \sup_{x\in S}\|r^{-\nu}\mathbf{d}_L(-\ln r)f(\mathbf{d}_L(\ln r)x) - f_L(x)\|_{\mathbb{B}} = 0,$$
$$\left(\text{resp., } \lim_{r\to L}\ \sup_{x\in S}|r^{-\nu}h(\mathbf{d}_L(\ln r)x) - h_L(x)| = 0\right) \tag{5.14}$$

where S is the unit sphere in $\mathbb{B}$. The corresponding operator f_L (resp., the functional h_L) is called a $\mathbf{d}_L$-homogeneous approximation of the operator f (resp., the functional h).

If f satisfies the above definition for both $L = 0$ and $L = \infty$ then the operator f (resp., the functional h) is said to be *homogeneous in the bi-limit*.

Notice that the limit (5.14) can be rewritten as follows:

$$\lim_{r\to L} \sup_{x\in S_{\mathbf{d}_L}(r)} \|r^{-\nu}\mathbf{d}_L(-\ln r)(f(x)-f_L(x))\|=0,$$
$$\left(\text{resp., } \lim_{r\to L} \sup_{x\in S_{\mathbf{d}_L}(r)} r^{-\nu}|h(x)-h_L(x)|=0\right)$$

where $S_{\mathbf{d}_L}(r)=\mathbf{d}(\ln r)S$ is the $\mathbf{d}_L$-homogeneous sphere of the radius $r>0$.

Example 5.7 Let $\mathbb{B}=L^2(\mathbb{R},\mathbb{R})$. The operator $f:\mathbb{B}\mapsto\mathbb{B}$ be defined as

$$f(x)=\sin(\|x\|_{L^2})x+x$$

with $x\in\mathbb{B}$ does not have homogeneous approximation at ∞-limit, while its homogeneous approximation at the 0-limit is $f_0(x)=x$.

A precision of a $\mathbf{d}$-homogeneous approximation is characterized below.

Theorem 5.8 *Let $\mathbf{d}_L$ be a strongly continuous strictly monotone linear dilation in $\mathbb{B}$, where $L=0$ or $L=\infty$. If an operator $f:\mathbb{B}\mapsto\mathbb{B}$ (resp., a functional $h:\mathbb{X}\mapsto\mathbb{R}$) is $\mathbf{d}_L$-homogeneous of degree $\nu_L\in\mathbb{R}$ in the L-limit, then there exists $\sigma_L\in\mathcal{K}$ such that*

- $\|f(x)-f_0(x)\|\le\sigma_0(\|x\|_{\mathbf{d}})\|x\|_{\mathbf{d}_0}^{\nu_L+\beta_L}$, $\forall x\in B(1)\backslash\{\mathbf{0}\}$, *(resp,. $|h(x)-h_0(x)|\le\sigma_0(\|x\|_{\mathbf{d}})\|x\|_{\mathbf{d}_0}^{\nu_L}$, $\forall x\in B(1)\backslash\{\mathbf{0}\}$)*
- $\|f(x)-f_\infty(x)\|\le M_L\sigma_\infty\left(\|x\|_{\mathbf{d}_\infty}^{-1}\right)\|x\|_{\mathbf{d}_\infty}^{\nu_L+\alpha_L}$, $\forall x\in\mathbb{B}\backslash B(1)$, *(resp., $|h(x)-h_\infty(x)|\le\sigma_\infty\left(\|x\|_{\mathbf{d}_\infty}^{-1}\right)\|x\|_{\mathbf{d}_\infty}^{\nu_L}$, $\forall x\in\mathbb{B}\backslash B(1)$)*

where $B(1)$ is the unit ball in $\mathbb{B}$, $\|\cdot\|_{\mathbf{d}_L}$ is the canonical $\mathbf{d}_L$-homogeneous norm in $\mathbb{B}$ and the positive parameters $\alpha_L\ge\beta_L>0$, $M_{\mathbf{d}_L}\ge1$ are defined in Proposition 4.3.

Proof Let us prove the estimates for the operator f. The estimates for functional h can be derived similarly.

The case $L=0$. If f_0 is a $\mathbf{d}_0$-homogeneous approximation of f in the 0-limit then the function $\tilde{\sigma}:[0,1]\mapsto\mathbb{R}$ defined as $\tilde{\sigma}_0(0)=0$ and

$$\tilde{\sigma}_0(\rho)=\sup_{x\in B_{\mathbf{d}_0}(r)\backslash\{\mathbf{0}\}}\|r^{-\nu_0}\mathbf{d}_0(-\ln r)(f(x)-f_0(x))\|,\quad \rho\in(0,1]$$

is continuous at 0 and increasing, where $B_{\mathbf{d}_0}(r)$ is the $\mathbf{d}_0$-homogeneous ball in $\mathbb{B}$ of the radius $r>0$. Since $\tilde{\sigma}_0$ is continuous at 0 and non-decreasing then there exist $\sigma_0\in\mathcal{K}$ such that $\tilde{\sigma}_0(\rho)\le\sigma_0(\rho)$ for all $\rho\in[0,1]$. For example, we can define $\sigma_0(\rho)=(1+\rho)\hat{\sigma}_0(\rho)$, where

$$\hat{\sigma}_0(\rho) = \begin{cases} \tilde{\sigma}_0(1) & \text{if } \rho > \frac{1}{2}, \\ \tilde{\sigma}_0\left(\frac{1}{n+1}\right) + \frac{\tilde{\sigma}_0\left(\frac{1}{n}\right) - \tilde{\sigma}_0\left(\frac{1}{n+1}\right)}{\frac{1}{n+1} - \frac{1}{n+2}}\left(\rho - \frac{1}{n+2}\right) & \text{if } \rho \in \left(\frac{1}{n+2}, \frac{1}{n+1}\right], n \in \mathbb{N}, \\ 0 & \text{if } \rho = 0. \end{cases}$$

On the other hand, we have

$$\sup_{x \in S_{\mathbf{d}_0}(r)} \left\| \frac{\mathbf{d}_0(-\ln(r))(f(x) - f_L(x))}{r^{\nu_0}} \right\| \geq \frac{\inf_{y \neq 0} \frac{\|\mathbf{d}_0(-\ln r)y\|}{\|y\|} \sup_{x \in S_{\mathbf{d}_0}(r)} \|f(x) - f_L(x)\|}{r^{\nu_0}}$$

$$\geq \frac{\sup_{x \in S_{\mathbf{d}_0}(r)} \|f(x) - f_L(x)\|}{r^{\nu_0} \|\mathbf{d}_0(\ln r)\|},$$

where the inequality $\|\mathbf{d}_L(\ln r)\| \|\mathbf{d}_L(-\ln r)y\| \geq \|\mathbf{d}_L(\ln r)\mathbf{d}_L(-\ln r)y\| = \|y\|$ is utilized on the last step. Using the strict monotonicity of $\mathbf{d}_0$ (i.e., $\beta_L > 0$: $\|\mathbf{d}_L(s)\|_{\mathbb{Y}} \leq e^{\beta_L s}, \forall s \leq 0$) we derive

$$\frac{\sup_{x \in S_{\mathbf{d}_0}(r)} \|f(x) - f_L(x)\|}{r^{\nu_0 + \omega_0}} \leq \sup_{x \in S_{\mathbf{d}_0}(r)} \|r^{-\nu_0} \mathbf{d}_0(-\ln(r))(f(x) - f_L(x))\| \leq \sigma_0(r),$$

$$\|f(x) - f_0(x)\| \leq \sup_{y \in S_{\mathbf{d}_0}(\|x\|_{\mathbf{d}_0})} \|f(y) - f_0(y)\| \leq \sigma_0(\|x\|_{\mathbf{d}_0}) \|x\|_{\mathbf{d}_0}^{\nu_0 + \beta_0}.$$

In **the case** $L = \infty$, the function $\tilde{\sigma}_\infty$ can be defined as follows: $\tilde{\sigma}_\infty(0) = 0$ and

$$\tilde{\sigma}_\infty\left(r^{-1}\right) = \sup_{x \in \mathbb{B} \backslash B_{\mathbf{d}_\infty}(r)} \|r^{-\nu_\infty} \mathbf{d}_\infty(-\ln r)(f(x) - f_L(x))\|, \quad r > 1.$$

A strictly increasing continuous function σ_∞ can be obtained similarly to σ_0.

For $r > 1$ we have

$$\sup_{x \in S_{\mathbf{d}_\infty}(r)} \left\| \frac{\mathbf{d}_\infty(\ln(r))(f(x) - f_L(x))}{r^{\nu_\infty}} \right\| \geq \frac{\inf_{y \neq 0} \frac{\|\mathbf{d}_\infty(-\ln r)y\|}{\|y\|} \sup_{x \in S_{\mathbf{d}_\infty}(r)} \|f(x) - f_L(x)\|}{r^{\nu_\infty}}$$

$$\geq \frac{\sup_{x \in S_{\mathbf{d}_\infty}(r)} \|f(x) - f_L(x)\|}{r^{\nu_\infty} \|\mathbf{d}_\infty(\ln r)\|}$$

$$\geq \frac{r^{-(\nu_\infty + \alpha_\infty)}}{M_{\mathbf{d}_\infty}} \sup_{x \in S_{\mathbf{d}_\infty}(r)} \|f(x) - f_L(x)\|,$$

where the inequality $\|\mathbf{d}_L(s)\| \leq M_{\mathbf{d}_L} e^{\alpha_L s}$, $s > 0$ is utilized on the last step. We complete the proof by obtaining the following inequalities

$$\frac{\sup_{x \in S_{\mathbf{d}_\infty}(r)} \|f(x) - f_L(x)\|}{M_{\mathbf{d}_\infty} r^{\nu_\infty + \alpha_\infty}} \leq \sup_{x \in S_{\mathbf{d}_\infty}(r)} \left\| \frac{\mathbf{d}_\infty(-\ln(r))(f(x) - f_L(x))}{r^{\nu_\infty}} \right\| \leq \sigma_\infty(r^{-1}),$$

$$\|f(x)-f_0(x)\| \leq \sup_{y\in S_{\mathbf{d}_\infty}(\|x\|_{\mathbf{d}_\infty})} \|f(y)-f_0(y)\| \leq M_{\mathbf{d}_\infty}\sigma_\infty(\|x\|_{\mathbf{d}_\infty}^{-1})\|x\|_{\mathbf{d}_\infty}^{\nu_\infty+\alpha_\infty}.$$

Let $K_{\mathbb{B}}(r)=\{x\in\mathbb{B}: r^{-1}<\|x\|<r\}$ with $r>1$, $B(r)=\{x\in\mathbb{B}:\|x\|<1\}$ and $S=\{x\in\mathbb{B}:\|x\|=1\}$ are as before. The following corollary generalizes the result of [3] to functionals in Banach spaces.

Corollary 5.5 *Let $\mathbf{d}_0$ and $\mathbf{d}_\infty$ be strictly monotone dilations in $\mathbb{B}$. Let a functional $h:\mathbb{B}\mapsto\mathbb{R}_+$ be homogeneous in the bi-limit and h_0,h_∞ be its homogeneous approximations at zero and at infinity, respectively. If*

$$\begin{array}{ll} \inf\limits_{x\in S} h_0(x)=\underline{h}_0>0, \quad \inf\limits_{x\in S} h_\infty(x)=\underline{h}_\infty>0, & \\ & \forall r>0, \\ 0<\inf\limits_{x\in K(r)} h(x)\leq \sup\limits_{x\in K(r)} h(x)<+\infty, & \end{array} \tag{5.15}$$

then there exist positive real numbers $\underline{\rho}_0$, $\overline{\rho}_0$, $\underline{\rho}_\infty$, $\overline{\rho}_\infty$ such that

$$\underline{\rho}_0 h_0(x)\leq h(x)\leq \overline{\rho}_0 h_0(x), \ \forall x\in B(1)\backslash\{\mathbf{0}\},$$

$$\underline{\rho}_\infty h_\infty(x)\leq h(x)\leq \overline{\rho}_\infty h_\infty(x), \ \forall x\in\mathbb{B}\backslash B(1).$$

Moreover, if $\nu_0\geq 0$ and $\nu_\infty\geq 0$ then there exist $\underline{c}>0$, $\overline{c}>0$ such that

$$\underline{c}\,\mathfrak{H}\left(h_0(x),h_\infty(x)\right)\leq h(x)\leq \overline{c}\,\mathfrak{H}\left(h_0(x),h_\infty(x)\right), \quad \forall x\in\mathbb{Y}\backslash\{\mathbf{0}\}, \tag{5.16}$$

where

$$\mathfrak{H}(a,b)-a\frac{1+b}{1+a}, \quad a,b\in\mathbb{R}_+. \tag{5.17}$$

Proof Notice that $\|x\|_{\mathbf{d}}\leq 1$ (resp., ≥ 1) is equivalent $\|x\|\leq 1$ (resp., ≥ 1) for any strictly monotone strongly continuous dilation $\mathbf{d}$ in $\mathbb{B}$. Using the homogeneity we derive

$$h_0(x)=\|x\|_{\mathbf{d}_0}^{\nu_0} h_0(-\ln\|x\|_{\mathbf{d}_0}x), \quad h_\infty(x)=\|x\|_{\mathbf{d}_\infty}^{\nu_\infty} h_\infty(-\ln\|x\|_{\mathbf{d}_\infty}x),$$

$$\underline{h}_0\|x\|_{\mathbf{d}_0}^{\nu_0}\leq h_0(x)\leq \|x\|_{\mathbf{d}_0}^{\nu_0}h_0, \quad \underline{h}_\infty\|x\|_{\mathbf{d}_\infty}^{\nu_\infty}\leq h_\infty(x)\leq \|x\|_{\mathbf{d}_0}^{\nu_0}\bar{h}_\infty,$$

where $x\in\mathbb{B}\backslash\{\mathbf{0}\}$,

$$\underline{h}_0=\inf_{x\in S}h_0(x), \quad \bar{h}_0=\sup_{x\in S}h_0(x), \quad \underline{h}_\infty=\inf_{x\in S}h_\infty(x), \quad \bar{h}_\infty=\sup_{x\in S}h_\infty(x)$$

and $\bar{h}_0<+\infty$ and $\bar{h}_\infty<\infty$ due to the condition (5.15) and Theorem 5.8.

I. On the one hand, Theorem 5.8 gives

$$h(x) \leq \delta_0(1)\|x\|_{\mathbf{d}_0}^{\nu_0} + h_0(x), \quad \forall x \in B(1)\backslash\{\mathbf{0}\},$$

$$h(x) \leq \delta_\infty(1)\|x\|_{\mathbf{d}_\infty}^{\nu_\infty} + h_\infty(x), \quad \forall x \in \mathbb{B}\backslash B(1)$$

so using the homogeneity-based estimates for h_0 and h_∞ we derive

$$\overline{\rho}_0 = 1 + \tfrac{\delta_0(1)}{\underline{h}_0}, \quad \text{and} \quad \overline{\rho}_\infty = 1 + \tfrac{\delta_\infty(1)}{\underline{h}_\infty}.$$

On the other hand, Theorem 5.8 implies that

$$\left(1 - \tfrac{\delta_0(\|x\|_{\mathbf{d}_0})}{\underline{h}_0}\right) h_0(x) \leq h_0(x) - \delta_0(\|x\|_{\mathbf{d}_0})\|x\|_{\mathbf{d}_0}^{\nu_0} \leq h(x), \quad \forall x \in B(1)\backslash\{\mathbf{0}\},$$

$$\left(1 - \frac{\delta_\infty\left(\frac{1}{\|x\|_{\mathbf{d}_\infty}}\right)}{\underline{h}_\infty}\right) h_\infty(x) \leq h_\infty(x) - \delta_\infty\left(\tfrac{1}{\|x\|_{\mathbf{d}_\infty}}\right)\|x\|_{\mathbf{d}_\infty}^{\nu_\infty} \leq h(x), \quad \forall x \in \mathbb{B}\backslash B(1).$$

Since $\delta_0 \in \mathcal{K}$ then there exists $r_0 \in (0, 1)$ such that

$$1 - \delta_0(r_0)\underline{h}_0^{-1} > 0 \quad \text{and} \quad 1 - \delta_0(r)\underline{h}_0^{-1} > 1 - \delta_0(r_0)\underline{h}_0^{-1}, \ \forall r \in [0, r_0].$$

Notice that $\|x\|_{\mathbf{d}_0}^{\nu_0} \leq \max\{1, r_0^{\nu_0}\}$ if $r_0 \leq \|x\|_{\mathbf{d}_0} \leq 1$. By assumption, for

$$h_{r_0} = \inf_{r_0 \leq \|x\|_{\mathbf{d}_0} \leq 1} h(x)$$

we have $h_{r_0} > 0$, so we derive the estimates

$$\frac{h_{r_0} h_0(x)}{\overline{h}_0 \max\{1, r_0^{\nu_0}\}} \leq \frac{h_{r_0} h_0(x)}{\|x\|_{\mathbf{d}_0}^{\nu_0} \overline{h}_0} \leq h_{r_0} \leq h(x), \quad \forall x \in \mathbb{B} : r_0 \leq \|x\|_{\mathbf{d}_0} \leq 1,$$

$$\underline{\rho}_0 h_0(x) \leq h(x), \quad \forall x \in B(1)\backslash\{\mathbf{0}\}, \quad \underline{\rho}_0 = \min\left\{1 - \tfrac{\delta_0(r_0)}{\underline{h}_0}, \tfrac{h_{r_0}}{\overline{h}_0 \max\{1, r_0^{\nu_0}\}}\right\}.$$

Since $\delta_\infty \in \mathcal{K}$ then there exists $r_\infty > 1$ such that

$$1 - \delta_\infty\left(r_\infty^{-1}\right)\underline{h}_\infty^{-1} > 0 \quad \text{and} \quad 1 - \delta_\infty\left(r^{-1}\right)\underline{h}_\infty^{-1} \geq 1 - \delta_\infty\left(r_\infty^{-1}\right)\underline{h}_\infty^{-1}, \ \forall r \geq r_\infty.$$

Notice that $\|x\|_{\mathbf{d}_\infty}^{\nu_\infty} \leq \max\{1, r_\infty^{\nu_\infty}\}$ if $1 \leq \|x\|_{\mathbf{d}_\infty} \leq r_\infty$. By assumption, for

$$h_{r_\infty} = \inf_{1 \leq \|x\|_{\mathbf{d}_\infty} \leq r_\infty} h(x)$$

we have $h_{r_\infty} > 0$, so we derive the estimates

$$\frac{h_{r_\infty} h_\infty(x)}{\bar{h}_\infty \max\{1, r_\infty^{\nu_\infty}\}} \le \frac{h_{r_\infty} h_\infty(x)}{\|x\|_{\mathbf{d}_\infty}^{\nu_\infty} \bar{h}_\infty} \le h_{r_\infty} \le h(x), \quad \forall x \in \mathbb{B} : 1 \le \|x\|_{\mathbf{d}_\infty} \le r_\infty,$$

$$\underline{\rho}_\infty h_\infty(x) \le h(x), \ \forall x \in B(1)\backslash\{\mathbf{0}\}, \ \underline{\rho}_\infty = \min\left\{1 - \frac{\delta_\infty(r_\infty^{-1})}{\underline{h}_\infty}, \frac{h_{r_\infty}}{\bar{h}_\infty \max\{1, r_\infty^{\nu_\infty}\}}\right\}.$$

II. For $\nu_0 \ge 0$ we have $0 \le h_0(x) \le \bar{h}_0$ for all $x \in B(1)\backslash\{\mathbf{0}\}$, so

$$\mathfrak{H}\,(h_0(x), h_\infty(x)) = \tfrac{1+h_\infty(x)}{1+h_0(x)} h_0(x) \ge \tfrac{h_0(x)}{1+\bar{h}_0}, \quad \forall x \in B(1)\backslash\{\mathbf{0}\}.$$

Taking into account that $\forall x \in \mathbb{B}\backslash B(1)$ it holds

$$\mathfrak{H}\,(h_0(x), h_\infty(x)) \ge h_\infty(x) \tfrac{h_0(x)}{1+h_0(x)} \ge h_\infty(x) \tfrac{\|x\|_{\mathbf{d}_0} \underline{h}_0}{1+\|x\|_{\mathbf{d}_0} \bar{h}_0} \ge h_\infty(x) \tfrac{\underline{h}_0}{\bar{h}_0},$$

we derive (5.16) for $\bar{c} = \bar{\max}\left\{\bar{\rho}_0(1+\bar{h}_0), \frac{\bar{\rho}_\infty \bar{h}_0}{\underline{h}_0}\right\}$.

On the other hand, for $\nu_\infty \ge 0$ we have $0 \le h_\infty(x) \le \bar{h}_\infty$, so

$$\mathfrak{H}\,(h_0(x), h_\infty(x)) = \tfrac{1+h_\infty(x)}{1+h_0(x)} h_0(x) \le (1+\bar{h}_0) h_0(x), \quad \forall x \in B(1)\backslash\{\mathbf{0}\}.$$

Since

$$\mathfrak{H}\,(h_0(x), h_\infty(x)) \le 1 + h_\infty(x) \le \tfrac{h_\infty(x)}{\underline{h}_\infty \|x\|_{\mathbf{d}_\infty}^{\nu_\infty}} + h_\infty(x), \quad \forall x \in \mathbb{Y}\backslash B_{\mathbb{Y}}(1)$$

then $\mathfrak{H}\,(h_0(x), h_\infty(x)) \le \left(\underline{h}_\infty^{-1} + 1\right) h_\infty(x)$ for all $x \in \mathbb{B}\backslash B(1)$. Therefore, the inequality (5.16) holds for $\underline{c} = \min\left\{\frac{\underline{\rho}_0}{1+\bar{h}_0}, \frac{\underline{\rho}_\infty}{\underline{h}_\infty^{-1}+1}\right\}$.

5.3 Homogeneous Set-Valued Mappings

The **d**-homogeneous approximations at 0 and at ∞ limits are introduced by means of a uniform limit of f on a **d**-homogeneous sphere. This uniformity is crucial for an analysis of non homogeneous systems by means of their **d**-homogeneous approximations. Unfortunately, this does not allow us to deal with **d**-homogeneous approximations of unbounded operators like $x \to (1 + \|x\|)\Delta x$, where Δ is a Laplace operator. Indeed, in the case of the uniform dilation $\mathbf{d}(s) = e^s I$, the functional $g_0(x) = 1$ is, obviously, **d**-homogeneous approximation of $g(x) = (1 + \|x\|)$ at 0-limit. Hence, $f_0(x) = \Delta x$ is expected to be a **d**-homogeneous approximation of $f(x) = (1 + \|x\|)\Delta x$ at $\mathbf{0}$-limit with the homogeneity degree $\nu = 0$. However, considering the expression under the limit in Definition 5.4 we derive

$$\sup_{x\in\mathcal{D}(\Delta)\cap S} \|r^{-\nu_0}\mathbf{d}(-\ln(r))f(\mathbf{d}(\ln r)x) - f_0(x)\| =$$

$$r \sup_{x\in\mathcal{D}(\Delta)\cap S} \|x\| \cdot \|\Delta x\| = r \sup_{x\in\mathcal{D}(\Delta)\cap S} \|\Delta x\| = +\infty.$$

To overcome this difficulty some set-valued **d**-homogeneous (*sub*homogeneous and *sup*homogeneous) extensions can be utilized.

5.3.1 Sub-Homogeneity

Below we deal with analysis of set-valued mappings

$$F : \mathcal{D}(F) \subset \mathbb{B} \mapsto 2^{\mathbb{B}},$$

which may define differential inclusions in infinite-dimensional spaces [4].

Definition 5.5 Let **d** be a dilation in $\mathbb{B}$. A set-valued mapping $F : \mathcal{D}(F) \subset \mathbb{B} \mapsto 2^{\mathbb{B}}$ is said to be

- **d**-homogeneous of a degree $\nu \in \mathbb{R}$ if domain $\mathcal{D}(F)$ is a **d**-homogeneous cone and
$$F(\mathbf{d}(s)u) \subset e^{\nu s}\mathbf{d}(s)F(u) \quad \text{for} \quad s \in \mathbb{R}, \quad u \in \mathcal{D}(F); \tag{5.18}$$
- **d**-*sub*homogeneous of a degree $\nu \in \mathbb{R}$ if $\mathbf{d}(s)\mathcal{D}(F) \subset \mathcal{D}(F)$ for $s < 0$,
$$F(\mathbf{d}(s)u) \subset e^{\nu s}\mathbf{d}(s)F(u) \quad \text{for} \quad s < 0, \quad u \in \mathcal{D}(F); \tag{5.19}$$
- **d**-*sup*homogeneous of a degree $\nu \in \mathbb{R}$ if $\mathbf{d}(s)\mathcal{D}(F) \subset \mathcal{D}(F)$ for $s > 0$,
$$F(\mathbf{d}(s)u) \subset e^{\nu s}\mathbf{d}(s)F(u) \quad \text{for} \quad s > 0, \quad u \in \mathcal{D}(F). \tag{5.20}$$

Since the bounded operator $\mathbf{d}(s)$ is invertible with the bounded inverse $\mathbf{d}(s)^{-1} = \mathbf{d}(-s)$ then denoting $y = \mathbf{d}(s)u$ for any **d**-homogeneous operator we derive

$$e^{-\mu s}\mathbf{d}(-s)F(y) \subset F(\mathbf{d}(-s)y), \forall s \in \mathbb{R}.$$

Since $\mathbf{d}(s)$ maps any **d**-homogeneous cone onto itself then the latter inclusion holds for all $y \in \mathcal{D}(F)$, i.e., *for any* **d**-*homogeneous operator we have*

$$F(\mathbf{d}(s)u) = e^{\nu s}\mathbf{d}(s)F(u) \quad \text{for} \quad s \in \mathbb{R}, \quad u \in \mathcal{D}(F).$$

Set-valued homogeneous operators have similar properties to single-valued homogeneous operators (see Sect. 5.1.2), for example,

- a (geometric) sum of homogeneous set-valued mappings is a homogeneous set-valued mapping provided their homogeneity degree is identical;
- an inverse of a homogeneous mapping of a degree ν is a homogeneous mapping of degree $-\nu$;
- a composition of homogeneous set-valued mappings is a **d**-homogeneous mapping provided that they satisfy some conditions like in Theorem 5.3;
- local regularity (e.g., Lipschitz continuity) of a homogeneous set-valued mapping can be expanded globally provided that the dilation is uniformly continuous.

5.3.2 Homogeneous Extensions

The homogeneous approximation introduced above is a useful tool for the homogeneity-based analysis of non homogeneous systems. However, homogeneous approximations may not exist in many cases even if $\mathbb{B} = \mathbb{R}^n$ (see Chap. 6 of Volume I). The set-valued homogeneous extension introduced below can be considered as a "robust" alternative of the homogeneous approximation. The homogeneous extensions may be constructed for unbounded non homogeneous operators.

Definition 5.6 A (possibly) set-valued **d**-homogeneous operator $F : \mathcal{D}(F) \subset \mathbb{B} \mapsto 2^{\mathbb{B}}$ is said to be a **d**-*homogeneous extension* of an operator $f : \mathcal{D}(f) \subset \mathbb{B} \mapsto \mathbb{B}$ if $f(x) \subset F(x)$ for any $x \in \mathcal{D}(f) \subset \mathcal{D}(F)$.

In this book, we deal mainly with the homogeneous extension

$$F(x) = \bigcup_{s\in\mathbb{R}:\mathbf{d}(s)x\in\mathcal{D}(f)} \{e^{-\nu s}\mathbf{d}(-s) f(\mathbf{d}(s)x)\}, \tag{5.21}$$

having the domain $\mathcal{D}(F) = \bigcup_{s\in\mathbb{R}} \mathbf{d}(s)\mathcal{D}(f)$.

The operator F given by (5.21) is **d**-homogeneous of degree ν.

If, in the above identities, we restrict $s \leq 0$ (or $s \geq 0$) then the set-valued operator F becomes *sub*homogeneous (resp., *sup*homogeneous).

A set-valued homogeneous extensions in $\mathbb{R}^n$ appears, for example, as a result of Filippov's regularization procedure applied to a discontinuous homogeneous vector field [5, 6].

In Volume II, most examples of infinite-dimensional **d**-homogeneous operators are given for Lebesgue and Sobolev spaces of functions with an unbounded support. The homogeneous extension allows a (non homogeneous) operator defined on a space of a compactly supported functions to be extended to a **d**-homogeneous operator on a space of functions with an unbounded support.

Example 5.8 *(On* **d***-homogeneous extension of the Laplace operator)* Let us consider the Laplace operator $\Delta : \mathcal{D}(\Delta) \subset L^2(\Omega; \mathbb{R}^m) \mapsto L^2(\Omega; \mathbb{R}^m)$ with $\mathcal{D}(\Delta) = H^2(\Omega; \mathbb{R}^m) \cap H_0^1(\Omega; \mathbb{R}^m)$. Let Ω be the unit ball in $\mathbb{R}^n$:

$$\Omega = \{z \in \mathbb{R}^n : |z| < 1\}.$$

Let $\mathbb{B} = L^2(\mathbb{R}^n; \mathbb{R}^m)$ and a linear dilation $\mathbf{d}$ in $\mathbb{B}$ be defined as

$$(\mathbf{d}(s)x)(z) = e^{\alpha s}x(e^{\beta s}z), \quad s \in \mathbb{R}, \quad z \in \mathbb{R}^n, \quad x \in \mathbb{B},$$

where $\alpha > n\beta/2$. The dilation $\mathbf{d}$ on $\mathbb{B}$ is not a dilation in $L^2(\Omega; \mathbb{R}^m)$, so the Laplace operator Δ is not $\mathbf{d}$-homogeneous on space of compactly supported functions $\Omega \mapsto \mathbb{R}^m$. However, this operator can be extended to an operator on $\mathbb{B} = L^2(\mathbb{R}^n; \mathbb{R}^m)$ in the following way

$$(Au)(x) = \begin{cases} (\Delta u_\Omega)(x) & \text{if } x \in \Omega, \\ 0 & \text{if } x \notin \Omega, \end{cases}$$

where $u_\Omega \in L^2(\Omega; \mathbb{R}^m)$ is a restriction of $u \in L^2(\mathbb{R}^n; \mathbb{R}^m)$ to the domain Ω. The domain of A is given by

$$\mathcal{D}(A) = \{u \in \mathbb{B} : u_\Omega \in H^2(\Omega; \mathbb{R}^m) \cap H_0^1(\Omega; \mathbb{R}^m), u_{\mathbb{R}^n\backslash\Omega} \in L^2(\mathbb{R}^n\backslash\{\Omega\}; \mathbb{R}^m)\},$$

where u_Ω is, as before, and $u_{\mathbb{R}^n\backslash\Omega}$ is a restriction of u to the domain $\mathbb{R}^n\backslash\Omega$. The operator A is not $\mathbf{d}$-homogeneous but it admits the $\mathbf{d}$-homogeneous extension F of the degree $2\beta \in \mathbb{R}$ given by the formula (5.21).

The above example presents a possible way for application of homogeneity-based methods to evolution systems with a compactly supported state variable. In Chap. 8, we use this trick in order to design a control for a heat system with a distributed state variable on the segment $[0, 1]$ by means of a homogeneous control designed for the system with a state variable supported on $\mathbb{R}_+$.

Concluding Remarks

Homogeneous functionals and operators in Banach and Hilbert spaces are introduced in this chapter. It is shown that most properties of homogeneous functions and vector fields in $\mathbb{R}^n$ can be rediscovered for homogeneous functional and operators in a Banach space $\mathbb{B}$ or in a Hilbert space $\mathbb{H}$. However, some particular features of infinite-dimensional homogeneous systems are noticed. For example, strong continuity of a linear dilation in a Banach/Hilbert space does not allow a global expansion of regularity of homogeneous mapping based on its "local" regularity on a unit sphere. The desired result can be guaranteed in the topology a generalized homogeneous (homeomorphic) Banach (resp., Hilbert) space $\mathbb{B}_\mathbf{d}$ (resp., $\mathbb{H}_\mathbf{d}$) or in the case of uniformly continuous dilation in $\mathbb{B}$. Various examples of homogeneous differential operators are presented. Some of them are inspired by models of fluid dynamics. Elements of the theory of homogeneous approximations and extensions of non homogeneous operators are considered. The developed tools are utilized in the following chapters for homogeneous control systems design and analysis.

References

1. Netuka I (2011) The change-of-variables theorem for the lebesgue integral. Acta Universitatis Matthiae Belii, Ser Math 19:37–42
2. Coron J-M, d'Andrea Novel B, Bastin G (2007) A strict Lyapunov function for boundary control of hyperbolic systems of conservation laws. IEEE Trans Autom Control 52(1):2–11
3. Andrieu V, Praly L, Astolfi A (2008) Homogeneous approximation, recursive observer design, and output feedback. SIAM J Control Optim 47(4):1814–1850
4. Krein SG, Khazan MI (1985) Differential equations in a banach space. J Soviet Math 30(3):2154–2239
5. Orlov Y (2005) Finite time stability and robust control synthesis of uncertain switched systems. SIAM J Control Optim 43(4):1253–1271
6. Levant A (2005) Homogeneity approach to high-order sliding mode design. Automatica 41(5):823–830

Chapter 6
Analysis of Infinite-Dimensional Homogeneous Systems

This chapter introduces a homogeneity-based approach to the stability, regularity, and robustness analysis of infinite-dimensional dynamical systems. In particular,

- a dilation symmetry of solutions of homogeneous evolution equations/inclusions is established;
- a global expansion of regularity of solutions is discussed;
- a homogeneity-based finite-time and fixed-time stability analysis is developed;
- the existence of a homogeneous Lyapunov function for any stable homogeneous infinite-dimensional system is proven;
- a robustness (input-to-state stability) of homogeneous evolution equation is studied.

6.1 Homogeneity of Evolution Equations

6.1.1 Dilation Symmetry of Solutions

In Volume I of this monograph it is shown that the standard homogeneity is inherited by solutions of differential equations with standard homogeneous right-hand sides. Namely, if there exists $\nu \in \mathbb{R}$ such that

$$f(e^s x) = e^{(\nu+1)s} f(x), \quad x \in \mathbb{R}^n, \quad s \in \mathbb{R}$$

then for any solution $x(t, x_0)$ of the following ODE

$$\dot{x}(t) = f(x(t)), \quad t > 0, \quad x(0) = x_0$$

A. Polyakov, *Generalized Homogeneity in Systems and Control Volume II*,
Communications and Control Engineering,
https://doi.org/10.1007/978-3-031-77496-6_6

we have

$$x(t, e^s x_0) = e^s x(e^{\nu s} t, x_0), \quad t > 0.$$

This property is crucial for homogeneity-based analysis of evolution systems. In particular, it implies that the existence (uniqueness, stability, etc.) of solutions for small initial states always imply the existence (resp., uniqueness, stability, etc.) of solutions for large initial states. Below we extend this result to mild, strong, and classical solutions of homogeneous dynamical systems modeled by abstract differential equations (inclusions) studied in Chap. 2.

The Case of Linear Evolution Equation

First, let us show that a semigroup generated by a closed densely defined linear homogeneous operator in $\mathbb{B}$ is homogeneous as well.

Lemma 6.1 *Let a linear closed densely defined operator $A : \mathcal{D}(A) \subset \mathbb{B} \mapsto \mathbb{B}$ generate a strongly continuous semigroup Φ of linear bounded operators on $\mathbb{B}$ and let $\mathbf{d}$ be a linear dilation in $\mathbb{B}$. If the operator A is $\mathbf{d}$-homogeneous of degree $\mu \in \mathbb{R}$ then*

$$\Phi(t)\mathbf{d}(s) = \mathbf{d}(s)\Phi(e^{\mu s} t), \quad \forall t \geq 0, \quad \forall s \in \mathbb{R}. \tag{6.1}$$

Proof Since Φ is generated by A then $\Phi(e^{\mu s} t)u \in \mathcal{D}(A)$ for any $u \in \mathcal{D}(A)$ (see Chap. 2, Theorem 2.1).

Let $s \in \mathbb{R}$ and $u \in \mathcal{D}(A)$ be selected arbitrary. Since the operator A is $\mathbf{d}$-homogeneous then, by definition of the $\mathbf{d}$-homogeneous cone, the domain $\mathcal{D}(A)$ is invariant with respect to the transformation $\mathbf{d}(s)$, i.e., $\mathbf{d}(s)z \in \mathcal{D}(A), \forall z \in \mathcal{D}(A)$, and, consequently,

$$y_1(t) := (\Phi(t)\mathbf{d}(s))u \in \mathcal{D}(A), \quad y_2(t) := (\mathbf{d}(s)\Phi(e^{\mu s} t))u \in \mathcal{D}(A), \quad t \geq 0.$$

Being generated by A, the semigroup Φ satisfies (see Chap. 2, Theorem 2.1)

$$\frac{d}{dt}\Phi(t)z = A\Phi(t)z = \Phi(t)Az, \quad \forall t > 0, \quad \forall z \in \mathcal{D}(A).$$

Taking into account that A is $\mathbf{d}$-homogeneous of the degree μ and $\mathbf{d}(s)$ is a linear bounded operator on $\mathbb{B}$, we derive

$$\tfrac{d}{dt} y_2(t) = e^{\mu s}\mathbf{d}(s)A\Phi(e^{\mu s} t)u = A\mathbf{d}(s)\Phi(e^{\mu s} t)u = Ay_2(t), \quad \forall t > 0.$$

On the other hand, we have

$$\tfrac{d}{dt} y_1(t) = A\Phi(t)\mathbf{d}(s)u = Ay_1(t), \quad \forall t > 0.$$

Since $y_1(0) = y_2(0) = \mathbf{d}(s)u$ then due to uniqueness of the semigroup Φ generated by A (see Chap. 2, Theorem 2.2) we derive $y_1(t) = y_2(t)$ for all $t \geq 0$ and

$$\Phi(t)\mathbf{d}(s)u = \mathbf{d}(s)\Phi(e^{\mu s}t)u, \quad \forall t \geq 0, \quad \forall u \in \mathcal{D}(A).$$

Since $\Phi(t)\mathbf{d}(s)$ and $\mathbf{d}(s)\Phi(e^{\mu s}t)$ are bounded linear operators and the set $\mathcal{D}(A)$ is dense in $\mathbb{B}$ then the above identity holds for all $u \in \mathbb{B}$ and all $t \geq 0$. The proof is complete. □

If $A \in \mathbb{R}^{n\times n}$ and $x \mapsto Ax \in \mathbb{R}^n$ is a $\mathbf{d}$-homogeneous vector field with $\mathbf{d}(s) = e^{G_{\mathbf{d}}s}$ and $G_{\mathbf{d}} \in \mathbb{R}^{n\times n}$ then the lemma gives the identity:

$$e^{At}e^{G_{\mathbf{d}}s} = e^{G_{\mathbf{d}}s}e^{Ae^{\mu s}t}, \qquad t \geq 0, \quad s \in \mathbb{R}$$

proven in Chap. 6 of Volume I.

Example 6.1 The Laplace operator $\Delta : H^2(\mathbb{R}^n;\mathbb{R}^m) \subset L^2(\mathbb{R}^n;\mathbb{R}^m) \mapsto L^2(\mathbb{R}^n;\mathbb{R}^m)$ is proven to be $\mathbf{d}$-homogeneous of degree 2β with respect to the linear strongly continuous dilation $(\mathbf{d}(s)x)(z) = e^{\alpha s}x(e^{\beta s}z)$, $z \in \mathbb{R}^n$, $x \in L^2(\mathbb{R}^n;\mathbb{R}^m)$. The strongly continuous semigroup Φ generated by Δ is given by

$$(\Phi(t)x)(z) = \frac{1}{\sqrt{(4\pi t)^n}}\int_{\mathbb{R}^n} e^{-\frac{\|z-y\|^2}{4t}}x(y)dy, \quad z \in \mathbb{R}^n.$$

Lemma 6.1 guarantees that the dilation symmetry of Δ implies the dilation symmetry of Φ (see the formula (6.1)). Alternatively, this can also be proven by making a change of variable in the above integral.

It is well-known (see Chap. 2, Sect. 2.2.1) that

$$x_{x_0}(t) = \Phi(t)x_0, \quad t \geq 0$$

is a unique solution of the linear evolution equation

$$\dot{x} = Ax, \quad t > 0,$$

with the initial condition $x(0) = x_0 \in \mathbb{B}$. Lemma 6.1 proves the symmetry of these solutions:

$$x_{\mathbf{d}(s)x_0}(t) = \mathbf{d}(s)x_{x_0}(e^{\mu s}t), \quad s \in \mathbb{R}, \quad t \geq 0, \tag{6.2}$$

where x_z denotes a solution with the initial condition $x(0) = z \in \mathbb{B}$.

Below we show that this identity holds for all solutions of nonlinear $\mathbf{d}$-homogeneous evolution equations and inclusions as well.

The Case of Nonlinear Evolution Equation

Let us consider the nonlinear evolution system

$$\dot{x} = Ax + f(x), \quad t > 0, \tag{6.3}$$

where a densely defined closed linear operator

$$A : \mathcal{D}(A) \subset \mathbb{B} \mapsto \mathbb{B}$$

generates a strongly continuous semigroup Φ of linear bounded operators on $\mathbb{B}$, and

$$f : \mathcal{D}(f) \subset \mathbb{B} \mapsto \mathbb{B}$$

is a (possibly) unbounded nonlinear operator.

Recall definitions of solutions of (6.3) (see Chap. 2 for more details): a continuous function $x : [0, T) \mapsto \mathbb{B}$ is said to be

- a *mild* solution of the system (6.3) if $f(x(\cdot)) \in L^1((0, T); \mathbb{B})$ and

$$x(t) = \Phi(t)x(0) + \int_0^t \Phi(t-\tau)f(x(\tau))\,d\tau, \qquad t \in (0, T); \tag{6.4}$$

- a *strong* solution of the evolution equation (6.3) if $x \in C([0, T); \mathbb{B})$, x is differentiable almost everywhere on $(0, T)$ such that $\dot{x} \in L^1((0, T); \mathbb{B})$, $f(x(\cdot)) \in L^1((0, T); \mathbb{B})$ and x satisfies (6.3) almost everywhere on $(0, T)$;
- a *classical* solution of the evolution equation (6.3) if $x \in C([0, T); \mathbb{B})$, $\dot{x} \in C((0, T); \mathbb{B})$, $x(t) \in \mathcal{D}(A) \cap \mathcal{D}(f)$ for all $t \in (0, T)$ and x satisfies (6.3) everywhere on $(0, T)$.

Theorem 6.1 *Let* $\mathbf{d}$ *be a linear dilation in* $\mathbb{B}$. *Let* A *and* f *be* $\mathbf{d}$*-homogeneous operators of degree* $\mu \in \mathbb{R}$. *If* $x : [0, T) \mapsto \mathbb{B}$ *is a mild (or strong) solution of the evolution equation* (6.3) *and*

$$x(t) \overset{a.e.}{\in} \mathcal{D}(f), \quad t \in (0, T),$$

then the function $x^s : [0, e^{-\mu s}T) \mapsto \mathbb{B}$ *defined as*

$$x^s(t) := \mathbf{d}(s)x(e^{\mu s}t), \quad t \in [0, e^{-\mu s}T)$$

is also a mild (resp., strong) solution of the evolution equation (6.3) *and*

$$x^s(t) \overset{a.e.}{\in} \mathcal{D}(f), \quad t \in (0, e^{-\mu s}T),$$

for any $s \in \mathbb{R}$. The claim remains valid for classical solutions allowing the above inclusions to hold everywhere on the intervals $(0, T)$ and $(0, e^{-\mu s}T)$, respectively.

Proof If $x(t) \in \mathcal{D}(f)$ then due to $\mathbf{d}$-homogeneity of the operator f we have $\mathbf{d}(s)\mathcal{D}(f) \subset \mathcal{D}(f)$ and $x^s(e^{-\mu s}t) \in \mathcal{D}(f)$.

(1) *The case of mild solutions.*

Since the $\mathbf{d}$-homogeneous operator A generates a strongly continuous semigroup Φ, then according to Lemma 6.1 we have

$$\Phi(t)\mathbf{d}(s) = \mathbf{d}(s)\Phi(e^{\mu s}t), \quad \forall t \geq 0, \quad \forall s \in \mathbb{R}.$$

It is well known (see Appendix C) that $K \int_0^t \xi(s)ds = \int_0^t K\xi(s)ds$ for any bounded linear operator $K : \mathbb{B} \mapsto \mathbb{B}$ and any Bochner integrable function $\xi \in L^1((0, T); \mathbb{B})$. Hence, using $f(x(\cdot)) \in L^1((0, T); \mathbb{B})$ we derive

$$\mathbf{d}(s) \int_0^t f(x(\tau))d\tau = e^{\mu s}\mathbf{d}(s) \int_0^{e^{-\mu s}t} f(x(e^{\mu s}\tau))d\tau =$$

$$\int_0^{e^{-\mu s}t} e^{\mu s}\mathbf{d}(s) f(x(e^{\mu s}\tau))d\tau = \int_0^{e^{-\mu s}t} f(\mathbf{d}(s)x(e^{\mu s}\tau))d\tau = \int_0^{e^{-\mu s}t} f(x^s(\tau))d\tau,$$

i.e., $f(x^s(\cdot)) \in L^1((0, e^{-\mu s}T); \mathbb{B})$. Similarly, we derive

$$\begin{aligned}
\mathbf{d}(s)x(e^{\mu s}t) &= (\mathbf{d}(s)\Phi(e^{\mu s}t))x(0) + \int_0^{e^{\mu s}t} (\mathbf{d}(s)\Phi(e^{\mu s}t - \tau)f)x(\tau)\, d\tau \\
&= (\Phi(t)\mathbf{d}(s))x(0) + \int_0^{e^{\mu s}t} (\mathbf{d}(s)\Phi(e^{\mu s}t - \tau)f)x(\tau)\, d\tau \\
&= \Phi(t)\mathbf{d}(s)x(0) + e^{\mu s} \int_0^t \mathbf{d}(s)\Phi(e^{\mu s}(t - \sigma))f(x(e^{\mu s}\sigma))\, d\sigma \\
&= \Phi(t)\mathbf{d}(s)x(0) + e^{\mu s} \int_0^t \Phi(t - \sigma)\mathbf{d}(s)f(x(e^{\mu s}\sigma))\, d\sigma \\
&= \Phi(t)\mathbf{d}(s)x(0) + \int_0^t \Phi(t - \sigma)f(\mathbf{d}(s)x(e^{\mu s}\sigma))\, d\sigma,
\end{aligned}$$

where the linearity of the operator $\Phi(t - \sigma)$ and the $\mathbf{d}$-homogeneity of Φ and f are utilized on the last two steps. Therefore, we have shown that

$$x^s(t) = \Phi(t)x^s(0) + \int_0^t \Phi(t - \sigma)f(x^s(\sigma))\, d\sigma,$$

i.e., x^s is a mild solution of (6.3).

(2) *The case of strong and classical solutions.* Let x be a strong solution. Since $x(t)$ is differentiable almost everywhere and $\mathbf{d}(s)$ is a linear bounded operator then for $\tau = e^{\mu s}t$ we have

$$\dot{x}^s(t) \stackrel{a.e.}{=} e^{\mu s}\mathbf{d}(s)\dot{x}(\tau) \stackrel{a.e.}{=} e^{\mu s}\mathbf{d}(s)(Ax(\tau)+f(x(\tau))) \stackrel{a.e.}{=} Ax^s(t)+f(x^s(t)).$$

In the case of the classical solution, the above identity holds everywhere. Moreover, if $x(t) \in \mathcal{D}(A)$ then due to $\mathbf{d}$-homogeneity of the operator A we always have $x^s(t) \in \mathcal{D}(A)$. Finally, taking into account that the operator $\mathbf{d}(s)$ is bounded we conclude $x^s \in C([0,T);\mathbb{B})$ provided that $x \in C([0,T);\mathbb{B})$ as well as $\dot{x}^s \in C((0,T);\mathbb{B})$ provided that $\dot{x} \in C((0,T);\mathbb{B})$, i.e., x^s is also a classical solution.

The Case of Nonlinear Implicit Evolution Equation

Let $\tilde{\mathbb{X}} = \mathbb{B} \times \mathbb{X}$, where $\mathbb{B}$ is a Banach space and $\mathbb{X}$ is a linear vector space. Let us consider the nonlinear implicit evolution equation

$$\begin{aligned} \dot{x} &= Ax + G(x,p), \\ \mathbf{0} &= Q(x,p), \end{aligned} \qquad t > 0, \tag{6.5}$$

where a densely defined closed linear operator

$$A : \mathcal{D}(A) \subset \mathbb{B} \mapsto \mathbb{B}$$

generates a strongly continuous semigroup Φ of linear bounded operators on $\mathbb{B}$, and

$$G : \mathcal{D}(G) \subset \tilde{\mathbb{X}} \mapsto \mathbb{B} \quad \text{and} \quad Q : \mathcal{D}(Q) \subset \tilde{\mathbb{X}} \mapsto \mathbb{X}$$

are a (possibly) nonlinear operators.

Recall definitions of solutions of the implicit evolution equation:
a pair (x,p) with $x : [0,T) \mapsto \mathbb{B}$ and $p : [0,T) \mapsto \mathbb{X}$ is said to be

- a *mild* solution of the implicit evolution equation (6.5) if $x \in C([0,T);\mathbb{B})$, $G(x,p) \in L^1((0,T);\mathbb{B})$ and

$$\begin{aligned} x(t) &= \Phi(t)x(0) + \int_0^t \Phi(t-\tau)G(x(\tau),p(\tau))\,d\tau, \\ \mathbf{0} &\stackrel{a.e.}{=} Q(x(t),p(t)), \end{aligned} \qquad t \in (0,T); \tag{6.6}$$

- a *strong* solution of the implicit evolution equation (6.5) if $x \in C([0,T);\mathbb{B})$, x is differentiable almost everywhere on $(0,T)$, $\dot{x}$, $G(x,p) \in L^1((0,T);\mathbb{B})$ and (6.5) is satisfied almost everywhere on $(0,T)$;

- a *classical* solution of the evolution equation (6.5) if $x \in C([0,T);\mathbb{B})$, $\dot{x} \in C((0,T);\mathbb{B})$, $x(t) \in \mathcal{D}(A)$ and $(x(t), p(t)) \in \mathcal{D}(G) \cap \mathcal{D}(Q)$ for $t \in (0,T)$ and (6.5) is satisfied on $(0,T)$.

A symmetry of solutions of the implicit evolution equation (6.5) is also preserved provided that the operators A, Q, G are $\mathbf{d}$-homogeneous.

Theorem 6.2 *Let* $\mathbf{d}$ *be a linear dilation in a Banach space* $\tilde{\mathbb{X}} = \mathbb{B} \times \mathbb{X}$ *be defined as follows:*

$$\mathbf{d}(s)(x, p) = (\mathbf{d}_1(s)x, \mathbf{d}_2(s)p), \quad s \in \mathbb{R}, u \in \mathbb{B}, p \in \mathbb{X},$$

where $\mathbf{d}_1$ *is a linear dilation in a Banach space* $\mathbb{B}$, $\mathbf{d}_2$ *is a linear dilation in a vector space* $\mathbb{X}$.

Let us denote $F := (G, Q)$ *and* $\mathcal{D}(F) := \mathcal{D}(G) \cap \mathcal{D}(Q)$,

$$F : \mathcal{D}(F) \subset \tilde{\mathbb{X}} \mapsto \tilde{\mathbb{X}}.$$

Let A *be a* $\mathbf{d}_1$*-homogeneous operator of degree* $\mu \in \mathbb{R}$ *and* F *be a* $\mathbf{d}$*-homogeneous operator of the same degree* μ.

If the pair (x, p) *with* $x : [0,T) \mapsto \mathbb{B}$ *and* $p : [0,T) \mapsto \mathbb{X}$ *is a mild (or strong) solution of the implicit evolution equation* (6.5) *such that*

$$(x(t), p(t)) \overset{a.e}{\in} \mathcal{D}(F), \quad t \in (0,T),$$

then the pair (x^s, p^s) *with* $x^s : [0, e^{-\mu s}T) \mapsto \mathbb{B}$ *and* $p^s : [0, e^{-\mu s}T) \mapsto \mathbb{X}$ *defined as follows:*

$$(x^s(t), p^s(t)) := \mathbf{d}(s)(x(e^{\mu s}t), p(e^{\mu s}t)) \quad \text{with} \quad t \in [0, e^{-\mu s}T)$$

is also a mild (resp., strong) solution of the evolution equation (6.3) *and*

$$(x^s(t), p^s(t)) \overset{a.e}{\in} \mathcal{D}(F), \quad t \in (0, e^{-\mu s}T)$$

for any $s \in \mathbb{R}$. *Moreover, the claim remains valid for classical solutions and the above inclusions hold everywhere on* $(0,T)$ *and* $(0, e^{-\mu s}T)$, *respectively.*

Proof Since the operator F is $\mathbf{d}$ homogeneous then

$$\mathbf{d}(s)\mathcal{D}(F) \subset \mathcal{D}(F)$$

and

$$G(\mathbf{d}_1(s)x, \mathbf{d}_2(s)p) = e^{\mu s}\mathbf{d}_1(s)G(x, p), \quad \forall (x, p) \in \mathcal{D}(F), \forall s \in \mathbb{R},$$

$$Q(\mathbf{d}_1(s)x, \mathbf{d}_2(s)p) = e^{\mu s}\mathbf{d}_2(s)Q(x, p), \quad \forall (x, p) \in \mathcal{D}(F), \forall s \in \mathbb{R}.$$

Hence, for $(x(t), p(t)) \in \mathcal{D}(F)$ we conclude $(x^s(e^{-\mu^s}t), p^s(e^{-\mu^s}t)) \in \mathcal{D}(F)$.

(1) *The case of mild solutions.*

Since the $\mathbf{d}_1$-homogeneous operator A generates a strongly continuous semigroup Φ, then according Lemma 6.1 we have

$$\Phi(t)\mathbf{d}_1(s) = \mathbf{d}_1(s)\Phi(e^{\mu s}t), \quad \forall t \geq 0, \quad \forall s \in \mathbb{R}.$$

It is well known (see Appendix C) that $K\int_0^t \xi(s)ds = \int_0^t K\xi(s)ds$ for any bounded linear operator $K : \mathbb{B} \mapsto \mathbb{B}$ and any Bochner integrable function ξ. Hence, using $G(x, p) \in L^1((0, T); \mathbb{B})$ we derive

$$\mathbf{d}_1(s)\int_0^t G(x(\tau), p(\tau))d\tau = \int_0^t \mathbf{d}_1(s)G(x(\tau), p(\tau))d\tau =$$

$$e^{-\mu s}\int_0^t G(\mathbf{d}_1(s)x(\tau), \mathbf{d}_2(s)p(\tau))d\tau = \int_0^{e^{-\mu s}t} G(\mathbf{d}_1(s)x(e^{\mu s}\tau), \mathbf{d}_2(s)p(e^{\mu s}\tau))d\tau$$

$$= \int_0^{e^{-\mu s}t} G(x^s(\tau), p^s(\tau))d\tau,$$

i.e., $G(x^s, p^s) \in L^1((0, e^{-\mu s}T); \mathbb{B})$. Similarly, we derive

$$\mathbf{d}_1(s)x(e^{\mu s}t) = \mathbf{d}_1(s)\Phi(e^{\mu s}t)x(0) + \mathbf{d}_1(s)\int_0^{e^{\mu s}t} \Phi(e^{\mu s}t - \tau)G(x(\tau), p(\tau))\, d\tau$$

$$= \Phi(t)\mathbf{d}_1(s)x(0) + \int_0^{e^{\mu s}t} \mathbf{d}_1(s)\Phi(e^{\mu s}t - \tau)G(x(\tau), p(\tau))\, d\tau =$$

$$\Phi(t)\mathbf{d}_1(s)x(0) + e^{\mu s}\int_0^t \mathbf{d}_1(s)\Phi(e^{\mu s}(t - \sigma))G(x(e^{\mu s}\sigma), p(e^{\mu s}\sigma))\, d\sigma =$$

$$\Phi(t)\mathbf{d}_1(s)x(0) + e^{\mu s}\int_0^t \Phi(t - \sigma)\mathbf{d}_1(s)G(x(e^{\mu s}\sigma), p(e^{\mu s}\sigma))\, d\sigma =$$

$$\Phi(t)\mathbf{d}_1(s)x(0) + \int_0^t \Phi(t - \sigma)G(\mathbf{d}_1(s)x(e^{\mu s}\sigma), \mathbf{d}_2(s)p(e^{\mu s}\sigma))\, d\sigma,$$

where the linearity of $\Phi(t - \sigma)$ and a homogeneity of Φ and G are utilized in the last two steps. Therefore, we have shown that

$$x^s(t) = \Phi(t)x^s(0) + \int_0^t \Phi(t-\sigma)G(x^s(\sigma), p^s(\sigma))\,d\sigma.$$

Finally, since (x, p) is a mild solution on $[0, T)$ then

$$Q(x(t), p(t)) \overset{a.e.}{=} \mathbf{0}, \quad \forall t \in [0, T)$$

and using the identity $\mathbf{d}_2(s)\,\mathbf{0} = \mathbf{0}$ we derive

$$Q(x^s(\tau), p^s(\tau)) = Q((\mathbf{d}_1(s)x)(\tau), (\mathbf{d}_2(s)p)(\tau)) =$$

$$e^{\mu s}\mathbf{d}_2(s)Q(x(e^{\mu s}\tau), p(e^{\mu s}\tau)) \overset{a.e.}{=} \mathbf{0}, \qquad \tau \in [0, e^{-\mu s}T).$$

(2) *The case of strong and classical solutions.* Let (x, p) be a strong solution. Since $x(t)$ is differentiable almost everywhere and $\mathbf{d}(s)$ is a linear bounded operator then for $\tau = e^{\mu s}t$ we have

$$\tfrac{dx^s(t)}{dt} \overset{a.e.}{=} e^{\mu s}\mathbf{d}(s)\tfrac{dx(\tau)}{d\tau} \overset{a.e.}{=} e^{\mu s}\mathbf{d}(s)(Ax(\tau)+G(x(\tau), p(\tau)) \overset{a.e.}{=}$$

$$Ax^s(t) + G(x^s(t), p^s(t)).$$

In the case of the classical solution the above identity holds everywhere. Moreover, if $x(t) \in \mathcal{D}(A)$ then due to $\mathbf{d}$-homogeneity of the operator A we always have $x^s(t) \in D(A)$, i.e., x^s is a classical solution.

The Case of Nonlinear Evolution Inclusion

Let us consider the nonlinear evolution inclusion

$$\dot{x} - Ax \in F(x), \quad t > 0, \tag{6.7}$$

where a densely defined closed linear operator

$$A : \mathcal{D}(A) \subset \mathbb{B} \mapsto \mathbb{B}$$

generates a strongly continuous semigroup Φ of linear bounded operators on $\mathbb{B}$, and

$$F : \mathcal{D}(F) \subset \mathbb{B} \mapsto 2^{\mathbb{B}}$$

is a (possibly) set-valued nonlinear operator.

Recall the definitions of solutions of the evolution inclusion (6.7):

- A continuous function $x : [t_0, t_0 + T) \mapsto \mathbb{B}$ is said to be a *mild solution* to (6.7) if there exists a selector (i.e., a single-valued mapping) $f \in L^1((t_0, t_0 + T); \mathbb{B})$ such that

$$f(s) \in F(x(s)) \quad \text{for almost all} \quad s \in (0, T)$$

and

$$x(t) = \Phi(t)x_0 + \int_{t_0}^{t} \Phi(t-s) f(s) ds, \quad \forall t \in [t_0, t_0 + T).$$

- If a mild solution x given above satisfies (2.51) for (almost) all $t \in (0, T)$ then x is called *classical* (*strong*) solution of (2.51).
- A continuous function $x : [0, T) \mapsto \mathbb{B}$ is said to be a *generalized solution* of the evolution inclusion (6.7) if x is a uniform limit on compact intervals of time in $[0, T)$ of approximate solutions x_ε as $\varepsilon \to 0$, where $x_\varepsilon : [t_0, t_N] \mapsto \mathbb{B}$ is given by

$$x_\varepsilon(t) := \tfrac{t-t_{i-1}}{t_i-t_{i-1}} x_i + \tfrac{t_i-t}{t_i-t_{i-1}} x_{i-1}, \quad \text{for} \quad t \in [t_i, t_{i-1}),$$

with

$$t_i : 0 < t_i - t_{i-1} < \varepsilon, \quad i = 1, 2, ..., N,$$

and $(t_i, x_i) \in \mathcal{D}(A) \cap \mathcal{D}(F)$ satisfying

$$\sum_{t_0 < t_i \leq t_0 + T'} \inf_{f_i \in F(t_i, x_i)} \|x_i - x_{i-1} - (t_i - t_{i-1}) f_i\| \leq \varepsilon.$$

Theorem 6.3 *Let* $\mathbf{d}$ *be a linear dilation in* $\mathbb{B}$. *Let a linear closed densely defined operator* $A : \mathcal{D}(A) \subset \mathbb{B} \mapsto \mathbb{B}$ *generate a strongly continuous semigroup* Φ *of linear bounded operators on* $\mathbb{B}$. *Let* A *and* F *be* $\mathbf{d}$*-homogeneous operators of degree* $\mu \in \mathbb{R}$.

If $x : [0, T) \mapsto \mathbb{B}$ *is a mild (or strong) solution of the evolution inclusion* (6.7) *and*

$$x(t) \overset{a.e.}{\in} \mathcal{D}(F), \quad t \in (0, T),$$

then the function $x^s : [0, e^{-\mu s}T) \mapsto \mathbb{B}$ *defined as*

$$x^s(t) := \mathbf{d}(s) x(e^{\mu s} t), \quad t \in [0, e^{-\mu s} T)$$

is also a mild (resp., strong) solution of the evolution inclusion (6.7) *and*

$$x^s(t) \overset{a.e.}{\in} \mathcal{D}(F), \quad t \in (0, e^{-\mu s} T),$$

for any $s \in \mathbb{R}$.

The claim remains true for classical solutions and the above inclusions hold everywhere on the intervals $(0, T)$ *and* $(0, e^{-\mu s} T)$, *respectively.*

Proof If x is a mild solution then

$$x(t) = \Phi(t)x(0) + \int_0^t \Phi(t-\tau)f(\tau)d\tau$$

for some $f \in L^1([0,T);\mathbb{B})$ satisfying

$$f(\tau) \overset{a.e.}{\in} F(x(\tau, x_0)).$$

Hence (see the proof of Theorem 6.1 for details) we derive

$$x^s(t) = \mathbf{d}(s)x(e^{\mu s}t) = \mathbf{d}(s)\Phi(e^{\mu s}t)x(0) + \int_0^{e^{\mu s}t} \mathbf{d}(s)\Phi(e^{\mu s}t-\tau)f(\tau)d\tau =$$

$$\Phi(t)\mathbf{d}(s)x(0) + \int_0^t \Phi(t-\tau)e^{\mu s}\mathbf{d}(s)f(e^{\mu s}\tau)d\tau.$$

Since $f(\tau) \overset{a.e.}{\in} F(x(\tau))$ then due to homogeneity of F we have

$$e^{\mu s}\mathbf{d}(s)f(e^{\mu s}\tau) \overset{a.e.}{\in} e^{\mu s}\mathbf{d}(s)F(x(e^{\mu s}\tau)) \subset F(\mathbf{d}(s)x(e^{\mu s}\tau)) = F(x^s(\tau)).$$

Therefore, x^s is also a solution of the evolution inclusion (6.7) for any $s \in \mathbb{R}$.

Example 6.2 Let $\mathbb{B} = L^2(\mathbb{R}^n;\mathbb{R}^m)$ and $\Delta : H^2(\mathbb{R}^n;\mathbb{R}^m) \subset \mathbb{B} \mapsto \mathbb{B}$ be the Laplace operator. In Example 5.2, it is shown that the Laplace operator is $\mathbf{d}$-homogeneous of degree 2β with respect to the dilation (4.9) given by $(\mathbf{d}(s)x(z) = e^{\alpha s}x(e^{\beta s}z)$, $x \in L^2(\mathbb{R}^n;\mathbb{R}^m)$, $z \in \mathbb{R}^n$.

Since $\|\mathbf{d}(s)x\|_{L_2} = e^{(\alpha-n\beta/2)s}\|x\|_{L_2}$ then for $\alpha - n\beta/2 \neq 0$ the evolution equation

$$\dot{x} = \Delta x + f(x), \quad f(x) := -\|x\|_{L^2}^{\frac{2\beta}{\alpha-n\beta/2}}x, \quad x \in \mathbb{B} \tag{6.8}$$

is $\mathbf{d}$-homogeneous of degree 2β.

For $1 + \frac{2\beta}{\alpha-n\beta/2} > 0$ we have $\|f(x)\|_{L^2} = \|x\|_{L^2}^{1+\frac{2\beta}{\alpha-n\beta/2}} \to 0$ as $x \to \mathbf{0}$ and the existence of strong solutions of the above evolution equation can be established using Corollary 2.13 and Theorem 2.16.

For $\frac{2\beta}{\alpha-n\beta/2} = -1$ we have the case studied in Example 2.18, which proves the existence of mild Filippov solutions for this system and considering the evolution inclusion

$$\dot{x} - \Delta x \in F(x), \quad x \in \mathbb{B},$$

$$F(x) = \bigcap_{\varepsilon>0} \overline{\text{co}} f(t, x\dot{+}B(\varepsilon)\backslash\{\mathbf{0}\}) = \alpha \cdot \begin{cases} \frac{x}{\|x\|}, & \text{if } x \neq \mathbf{0}, \\ B(1), & \text{if } x = \mathbf{0}. \end{cases}$$

Theorem 6.1 implies that all mentioned solutions are symmetric and satisfy the relation (6.2) with $\mu = 2\beta$.

The result similar to Theorem 6.3 can be proven for the generalized solutions as well [1].

Theorem 6.4 *Let* $\mathbf{d}$ *be a linear dilation in* $\mathbb{B}$ *and the operators* A *and* F *be* $\mathbf{d}$*-homogeneous of a degree* μ*. If* $x : [0, T) \mapsto \mathbb{B}$ *is a generalized solution of the evolution inclusion* (6.7) *then the function* $x^s : [0, e^{-\mu s}T) \mapsto \mathbb{B}$ *defined as*

$$x^s(t) := \mathbf{d}(s)x(e^{\mu s}t), \quad t \in [0, e^{-\mu s}T)$$

is also a generalized solution of the evolution inclusion (6.7) *for any* $s \in \mathbb{R}$.

Proof Let $T' < T$ and x^ε be an arbitrary approximate solution of (6.7) with the sequences t_i and x_i (see Definition (2.14)). Let us consider the function $\tau \to x^{\varepsilon,s}(\tau)$ given by

$$x^{\varepsilon,s}(\tau) = \mathbf{d}(s)x^\varepsilon(e^{\nu s}\tau),$$

where $s \in \mathbb{R}$ is an arbitrary number.

Let us denote $\tau_i = e^{-\nu s}t_i$ and $x_i^s = \mathbf{d}(s)x_i$. Since $\mathcal{D}(A)$ and $\mathcal{D}(F)$ are $\mathbf{d}$-homogeneous cones and $x_i \in \mathcal{D}(A) \cap \mathcal{D}(F)$ then $x_i^s \in \mathcal{D}(A) \cap \mathcal{D}(F)$.

The homogeneity of the operators A and F implies

$$(A + F)(\mathbf{d}(s)x_i) = F(x_i^s) = e^{\nu s}\mathbf{d}(s)(A + F)(x_i).$$

Hence, we derive

$$\sum_{t_0 < t_i < T'} \inf_{\tilde{f}_i \in (A+F)(x_i^s)} \left\| x_i^s - x_{i-1}^s - (\tau_i - \tau_{i-1})\tilde{f}_i \right\| =$$

$$\sum_{t_0 < t_i < T'} \inf_{\tilde{f}_i \in e^{\nu s}\mathbf{d}(s)(A+F)(x_i)} \left\| \mathbf{d}(s)x_i - \mathbf{d}(s)x_{i-1} - e^{-\nu s}(t_i - t_{i-1})\tilde{f}_i \right\| =$$

$$\sum_{t_0 < t_i < T'} \inf_{f_i \in (A+F)(x_i)} \|\mathbf{d}(s)(x_i - x_{i-1} - (t_i - t_{i-1})f_i)\| < \|\mathbf{d}(s)\|\varepsilon,$$

i.e., the function $x_{\varepsilon,s}$ is an approximate solution of the evolution equation (6.7) with the initial condition $x(0) = \mathbf{d}(s)x_0$.

Finally, since $\mathbf{d}(s)$ is a linear bounded operator for any $s \in \mathbb{R}$, then the inequality $\|x^{\varepsilon,s}(\tau) - x^s(\tau)\| \leq \|\mathbf{d}(s)\| \|(x^\varepsilon(t) - x(t))\|$ implies that $x^s(\tau) = \lim_{\varepsilon \to 0} x^{\varepsilon,s}(\tau)$ uniformly on $\tau \in [0, T']$. The proof is complete. □

Below we say that an evolution equation/inclusion is **d**-homogeneous of degree $\nu \in \mathbb{R}$ in $\mathbb{B}$ if for any $x_0 \in \mathbb{B}$ it has a solution x_{x_0} defined, at least, locally in time and all solution satisfy the identity (6.2).

6.1.2 On Uniqueness and Forward Completeness

In Chap. 5, we have discovered that local properties of homogeneous operators can be expanded globally. The same conclusions are valid for solutions of homogeneous systems. Namely, if solutions of the system (6.3) are well defined for small initial data $x(0) = x_0 \in \mathbb{B} : \|x_0\| < \varepsilon$ then this system has well-defined solutions for large initial data. Such a property can be utilized for global expansion of regularity of homogeneous evolution systems. Below we say an *evolution system* if there is no need to distinguish evolution inclusions and (explicit/implicit) evolution equations. The results presented in this section hold for all types of solutions considered in this monograph.

Initially we show that uniqueness of solutions can be expanded using dilation symmetry

Lemma 6.2 (On uniqueness of solutions) *Let* **d** *be a linear dilation in* $\mathbb{B}$. *If a* **d***-homogeneous evolution system of degree* $\mu \in \mathbb{R}$ *has a unique solution* x_{x_0} *with an initial value* $x(0) = x_0 \in \mathcal{D}$*, where* $\mathcal{D} \subset \mathbb{B}$ *is a* **d***-homogeneous cone belonging a set of all admissible initial conditions of the system, then, for any* $s \in \mathbb{R}$,

$$x_{\mathbf{d}(s)x_0}(t) = \mathbf{d}(s)x_{x_0}(e^{\mu s}t)$$

is a unique solution of the evolution system with the scaled initial data $x(0) = \mathbf{d}(s)x_0 \in \mathcal{D}$.

Proof Assume that x_{x_0} is a unique solution with $x(0) = x_0$. In the previous section, we have shown (for various types of solution) that $x_{\mathbf{d}(s)x_0}$ defined above is a solution with $x_{\mathbf{d}(s)x_0}(0) = \mathbf{d}(s)x_0$. Let us prove that this solution is unique. Suppose that there exists another solution $y_{\mathbf{d}(s)x_0}(t)$ such that

$$\mathbf{d}(s)x_{x_0}(e^{\mu s}t) \neq y_{\mathbf{d}(s)x_0}(t)$$

on some time interval. Since **d** is a group of linear bounded invertible operators then $\mathbf{d}(-s)$ is one to one mapping and $\mathbf{d}(-s)y_{\mathbf{d}(s)x_0}(e^{-\mu s}t) \neq x_{x_0}(t)$. On the other hand, as shown in the previous section, $\mathbf{d}(-s)y_{\mathbf{d}(s)x_0}(e^{-\mu s}t)$ is a solution with the initial data $x(0) = x_0$. This contradicts the assumption that the solution with the initial data $x(0) = x_0$ is unique. The proof is complete. □

Corollary 6.1 *Let* $\mathbf{d}$ *be a linear strongly continuous strictly monotone dilation in a Hilbert space* $\mathbb{H}$. *Let* $A : \mathcal{D}(A) \subset \mathbb{H} \mapsto \mathbb{H}$ *be a generator of a strongly continuous semigroup* Φ *of linear bounded operators on* $\mathbb{H}$ *and*

$$\langle Ax, x\rangle \le \gamma\|x\|^2, \quad \forall x \in \mathcal{D}(A) \tag{6.9}$$

with some $\gamma \in \mathbb{R}$. *Let* $f : \mathbb{H} \mapsto \mathbb{H}$ *be locally Lipschitz continuous on* $\mathbb{H}\backslash\{\mathbf{0}\}$ *and* $f(\mathbf{0}) = \mathbf{0}$. *If the operators* A *and* f *are* $\mathbf{d}$*-homogeneous of degree* $\mu \le 0$ *then the system* (6.3) *is forward complete.*

Proof By Corollary 2.13, for any $x_0 \in \mathbb{H}\backslash\{\mathbf{0}\}$ the system (6.3) has a local-in-time mild solution.[1] For $x_0 = \mathbf{0}$ the system, at least, has the zero solution. Due to Theorem 2.16 any mild solution of (6.3) with $x_0 \in \mathcal{D}(A)$ is strong. Let us show that all solutions are defined globally in the forward time. Let $\|\cdot\|_{\mathbf{d}}$ be the canonical homogeneous norm induced by the norm $\|\cdot\| = \sqrt{\langle\cdot,\cdot\rangle}$. In this case, by the formula (4.25) we derive

$$\begin{aligned} D\|x\|_{\mathbf{d}}(Ax + f(x)) &= \|x\|_{\mathbf{d}} \frac{\langle \mathbf{d}(-\ln\|x\|_{\mathbf{d}})x, \mathbf{d}(-\ln\|x\|_{\mathbf{d}})(Ax+f(x))\rangle}{\langle G_{\mathbf{d}}\mathbf{d}(-\ln\|x\|_{\mathbf{d}})x, \mathbf{d}(-\ln\|x\|_{\mathbf{d}})x\rangle} \\ &= \|x\|_{\mathbf{d}}^{1+\mu} \frac{\langle \mathbf{d}(-\ln\|x\|_{\mathbf{d}})x, A\mathbf{d}(-\ln\|x\|_{\mathbf{d}})x + f(\mathbf{d}(-\ln\|x\|_{\mathbf{d}})x)\rangle}{\langle G_{\mathbf{d}}\mathbf{d}(-\ln\|x\|_{\mathbf{d}})x, \mathbf{d}(-\ln\|x\|_{\mathbf{d}})x\rangle} \end{aligned}$$

for all $x \in \mathcal{D}$, where the set $\mathcal{D} \subset \big(\mathcal{D}(A) \cap \mathcal{D}_{\exp}(G_{\mathbf{d}}^{\infty})\big)$ is dense in $\mathbb{H}$ and defined in Lemma 5.1. By definition of the canonical homogeneous norm, we have $\|\mathbf{d}(-\ln\|x\|_{\mathbf{d}})x\| = 1$. The condition (6.9) means

$$\langle A\mathbf{d}(-\ln\|x\|_{\mathbf{d}})x, \mathbf{d}(-\ln\|x\|_{\mathbf{d}})x\rangle \le \gamma, \quad \forall x \in \mathcal{D}.$$

Due to strict monotonicity of the the dilation $\mathbf{d}$, there exists $\beta > 0$ (see Proposition 4.4) such that

$$\langle G_{\mathbf{d}}\mathbf{d}(-\ln\|x\|_{\mathbf{d}})x, \mathbf{d}(-\ln\|x\|_{\mathbf{d}})x\rangle \ge \beta.$$

Hence, we derive

$$D\|x\|_{\mathbf{d}}(Ax + f(x)) \le \frac{C+\gamma}{\beta}\|x\|_{\mathbf{d}}^{1+\mu}, \quad \forall x \in \mathcal{D},$$

where $C = \sup_{\|y\|=1}\|f(y)\| < +\infty$. Since $\mathcal{D}$ is dense in $\mathbb{H}$ (and, in particular, in $\mathcal{D}(A)$), then the canonical homogeneous norm of any strong solution x_{x_0} of the system with $x(0) = x_0 \in \mathcal{D}(A)$ satisfies the inequality

$$\frac{d}{dt}\|x_{x_0}(t)\|_{\mathbf{d}} \overset{a.e.}{=} D\|x\|_{\mathbf{d}}(Ax + f(x))|_{x=x_0(t)} \overset{a.e.}{\le} \frac{C}{\beta}\|x_{x_0}(t)\|_{\mathbf{d}}^{1+\mu}.$$

[1] The system (6.3) is forward complete it has global-in-time mild solutions for all initial conditions in $\mathbb{B}$.

For $\mu \leq 0$ the obtained differential inequality implies that $\|x_{x_0}(t)\|_{\mathbf{d}}$ remains finite for any finite $t \geq 0$. Taking into account that any mild solution is a uniform limit of strong solutions (see Theorem 2.16), we complete the proof. □

Theorem 6.5 (On global expansion of regularity) *Let an evolution equation/inclusion be* **d***-homogeneous with respect to a linear dilation* **d** *in* $\mathbb{B}$. *Let* $h : \mathcal{D}(h) \subset \mathbb{B} \mapsto \mathbb{R}$ *be a densely defined* **d***-homogeneous functional of a degree* $\nu \neq 0$ *and there exist* $\varepsilon > 0$ *such that the evolution system has a global-in-time solution for any*

$$x(0) = x_0 \in \mathcal{D}_\varepsilon := \{x \in \mathcal{D}(h) : |h(x)| \leq \varepsilon\}$$

defined on $[0, +\infty)$. *Then it has a global-in-time solution for any* $x_0 \in \mathcal{D}(h)$. *Moreover, each solution for* $x_0 \in \mathcal{D}(h)$ *is unique if and only if each solution with the initial data* $x_0 \in \mathcal{D}_\varepsilon$ *is unique.*

Proof If $x_0 \in \mathcal{D}(h)$ is such that $h(x_0) > \varepsilon$ then due to **d**-homogeneity of h we have

$$h(\mathbf{d}(s)x) = e^{\nu s}h(x), \quad \forall s \in \mathbb{R}, \quad \forall x \in \mathcal{D}(h)$$

and, obviously, there exists $s_0 \in \mathbb{R}$ such that

$$|h(\mathbf{d}(s_0)x_0)| \leq \varepsilon.$$

By assumption, the evolution system with $x(0) = \mathbf{d}(s_0)x_0$ has a unique solution x^s defined on $[0, +\infty)$. From Theorems 6.3 and 6.4, we derive that the system (6.3) with $x(0) = \mathbf{d}(s)\mathbf{d}(s_0)x_0$ has a solution defined on $[0, +\infty)$ for any $s \in \mathbb{R}$. Lemma 6.2 guarantees that the uniqueness is preserved as well. Taking $s = -s_0$ we complete the proof.

If, in the above theorem, we have $\mathcal{D}(h) = \mathbb{B}$ then the existence and uniqueness of solutions can be established for all $x_0 \in \mathbb{B}$.

Since $\mathbf{d}(s)$ *is a linear bounded invertible operator then continuous dependence on initial conditions can be expanded similarly.*

Example 6.3 *(On global-in-time solutions of Navier–Stokes equations)* As a potential application of the above theorem, the problem of the expansion of regularity of Navier–Stokes equations can be mentioned. It is well known [2, 3] that the existence of global-in-time classical solutions of Navier–Stokes equations can be proven for small (in a certain sense) initial data. The system of Navier–Stokes equations is **d**-homogeneous (see Example 5.6) with respect to the group of linear bounded operators $(\mathbf{d}(s)z)(x) = e^s z(e^s x)$, $s \in \mathbb{R}$, $z \in L^\infty(\mathbb{R}^3; \mathbb{R}^3)$. Consequently, the regularity can be expanded globally if it is possible to find an appropriate **d**-homogeneous functional h satisfying Theorem 6.5. This is a nontrivial problem for $n = 3$. For example, selecting

$$\tilde{h}(x_0) := \|x_0\|_2^2 \|x_0\|_\infty$$

with $x_0 \in \mathcal{D}(h) = L^2 \cap L^\infty$ one can be shown (see [3]) that there exists $\varepsilon > 0$ such that for any $x_0 \in \mathcal{D}(h) : |\tilde{h}(x_0)| < \varepsilon$ the system of Navier–Stokes equations has a unique global-in-time classical solution belonging to $L^2 \cap L^\infty$. Moreover, it is easy to check that the functional $\tilde{h}(\mathbf{d}(s)x)$ is **d**-homogeneous. However, its degree is zero, i.e., the selected $\tilde{h}$ does not satisfy the conditions of Theorem 6.5. A global expansion of regularity of Navier–Stokes equations for $n = 3$ is still an open interesting problem.

6.2 Stability Analysis

A dilation symmetry also simplifies global expansion of local stability.

Corollary 6.2 *Let* **d** *be a linear dilation in* $\mathbb{B}$. *If an evolution system is* **d**-*homogeneous and locally uniformly Lyapunov (asymptotically) stable then it is globally uniformly Lyapunov (resp., asymptotically) stable.*

Proof The stability implies that all solutions of an evolution system are global-in-time for small initial data: $\|x_0\| \leq \varepsilon$ with $x_0 \in \mathbb{B}$. Applying Theorem 6.5 we conclude that all solutions are global-in-time for all initial conditions in $\mathbb{B}$. The dilation symmetry of an evolution system means that solutions satisfy the identity

$$x_{\mathbf{d}(s)x_0}(t) = \mathbf{d}(s)x_{x_0}(e^{\mu s}t), \quad \forall t \geq 0,$$

where $\mu \in \mathbb{R}$ is a homogeneity degree of the system and $s \in \mathbb{R}$ is the group parameter. Since $\mathbf{d}(s)$ is a linear bounded invertible operator then

- uniform (on t) boundedness of $x_{x_0}(t)$ implies uniform boundedness of $x_{\mathbf{d}(s)x_0}(t)$;
- asymptotic convergence of $x_{x_0}(t)$ to $\mathbf{0}$ as $t \to T_{x_0}$, where $T_{x_0} \leq +\infty$ implies asymptotic convergence of $x_{\mathbf{d}(s)x_0}(t)$ to $\mathbf{0}$ as $t \to T_{\mathbf{d}(s)x_0}$, where $T_{\mathbf{d}(s)x_0} = e^{-\mu s}T_{x_0}$

and vice versa. Since **d** is a dilation then a global behavior of solutions (i.e., for $\|x_0\| < +\infty$) repeats their local behavior (i.e., for $\|x_0\| < \varepsilon$). Indeed, if $\|x_0\| < +\infty$ then from Limit property of the dilation we derive that there exists $s_0 \in \mathbb{R}$ such that $\|\mathbf{d}(s_0)x_0\| \leq \varepsilon$. Therefore, local uniform Lyapunov (asymptotic) stability of a **d**-homogeneous system is equivalent to its global uniform Lyapunov (asymptotic) stability. The proof is complete. □

6.2.1 Convergence Rates of Homogeneous Systems

The homogeneity degree of a **d**-homogeneous uniformly asymptotically stable system specifies its convergence rate [4, p. 110], [1, 5].

Theorem 6.6 *Let* **d** *be a strongly continuous strictly monotone linear dilation in* $\mathbb{B}$. *If an evolution equation/inclusion is* **d**-*homogeneous of degree* $\nu \in \mathbb{R}$ *in* $\mathbb{B}$ *and locally uniformly asymptotically stable then*

- *for* $\nu < 0$ *it is globally uniformly finite-time stable and there exists a* **d**-*homogeneous settling-time functional* $T : \mathbb{B} \mapsto [0, +\infty)$ *of the degree* $-\nu$ *that is locally bounded and continuous at* **0**;
- *for* $\nu = 0$ *it is globally uniformly exponentially stable;*
- *for* $\nu > 0$ *it is globally uniformly nearly fixed-time stable, i.e.,*

$$\forall r > 0, \quad \exists T_r > 0 \quad : \quad \|x_{x_0}(t)\| < r, \quad \forall t \geq T_r, \quad \forall x_0 \in \mathbb{B}.$$

Proof **I.** In the view of Proposition 4.3, for any $s > 0$ we have $c := \|\mathbf{d}(s)\| > 1$. By Corollary 6.2, the local uniform asymptotic stability of a homogeneous system always implies the global one. Consequently, there exists $T' > 0$ such that $\|x_{x_0}(t)\| \leq 1/c$ for $t \geq T'$ for all $x_0 \in \{x \in \mathbb{B} : \|x_0\| \leq 1\}$.

Let us introduce the following notation:

- $\Delta T_0 = e^{\nu s} T'$ and $\Delta T_i = e^{\nu s} \Delta T_{i-1}$ for $i = 1, 2, ...$;
- $T_0 = 0$ and $T_i = T_{i-1} + \Delta T_{i-1}$ for $i = 1, 2, ...$;
- $x_i = x_{x_0}(T_i)$ for $i = 1, 2, ...$

Since $x_{x_0}(T_1) = x_1$ then $x_{x_1}(\tau) = x_{x_0}(T_1 + \tau), \tau > 0$ due Remark 2.1 and, by construction, $\|x_1\| \leq 1/c$. Using homogeneity of the evolution system we conclude that

$$x_{\mathbf{d}(s)x_1}(t) = \mathbf{d}(s)x_{x_1}(e^{\nu s}t) = \mathbf{d}(s)x_{x_0}(T_1 + e^{\nu s}t), \quad t > 0$$

is also a solution. Moreover, $x_{\mathbf{d}(s)x_1}(0) = \mathbf{d}(s)x_1$ and

$$\|x_{\mathbf{d}(s)x_1}(0)\| \leq \|\mathbf{d}(s)\| \, \|x_1\| \leq 1.$$

In this case, the uniform asymptotic stability guarantees

$$\|x_{\mathbf{d}(s)x_1}(T')\| = \|\mathbf{d}(s)x_{x_0}(T_1 + e^{\nu s}T')\| =$$

$$\|\mathbf{d}(s)x_{x_0}(T_1 + \Delta T_1)\| = \|\mathbf{d}(s)x_{x_0}(T_2)\| = \|\mathbf{d}(s)x_2\| \leq 1/c$$

and

$$\|\mathbf{d}(2\,s)x_2\| \leq \|\mathbf{d}(s)\| \cdot \|\mathbf{d}(s)x_2\| \leq 1.$$

Repeating the same consideration by induction we derive

$$\|\mathbf{d}(is)x_i\| \leq 1$$

and

$$\|x_{x_0}(T_i)\| = \|\mathbf{d}(-is)\mathbf{d}(is)x_i\| \leq \|\mathbf{d}(-is)\| \to 0 \quad \text{as} \quad i \to \infty$$

due to the limit property of the dilation $\mathbf{d}$. Evidently, $\Delta T_i = T'e^{i\nu s}$ and for $\nu < 0$ we obtain

$$T_i = T' \sum_{n=0}^{i-1} e^{n\nu s} \to \frac{T'}{1-e^{\nu s}} \quad \text{as} \quad i \to \infty.$$

In other words, $x_{x_0}(t) \to 0$ as $t \to \frac{T'}{1-e^{\nu s}}$ provided that $\|x_0\| \leq 1$. This means that the system is locally uniformly finite-time stable, but the homogeneity implies the globally uniform finite-time stability.

Show now that the system has a $\mathbf{d}$-homogeneous settling-time functional of the degree $-\nu$. Let $S(x_0)$ be a set of solutions of our $\mathbf{d}$-homogeneous system with the initial condition $x(0) = x_0 \in \mathbb{B}$. Let us introduce the following functional

$$T(x_0) = \sup_{x_{x_0} \in S(x_0)} \inf_{\tau \geq 0 : x_{x_0}(\tau) = \mathbf{0}} \tau.$$

Since the origin is uniformly finite-time attractive then the functional T is well-defined and locally bounded on $\mathbb{B}$. Lyapunov stability implies that $x(t) = \mathbf{0}$ is the unique solution of the evolution system with the zero initial condition. Hence, by construction, T is the settling-time functional. Since $\mathbf{d}$-homogeneity of the system implies $x_{\mathbf{d}(s)x_0}(\tau) = \mathbf{d}(s)x_{x_0}(e^{\nu s}\tau)$ then

$$T(\mathbf{d}(s)x_0) = \sup_{x_{\mathbf{d}(s)x_0} \in S(\mathbf{d}(s)x_0)} \inf_{\tau \geq 0 : x_{\mathbf{d}(s)x_0}(\tau) = \mathbf{0}} \tau =$$

$$\sup_{x_{x_0} \in S(x_0)} \inf_{\tau \geq 0 : x_{x_0}(e^{\nu s}\tau) = \mathbf{0}} \tau = e^{-\nu s} \sup_{x_{x_0} \in S(x_0)} \inf_{\tau \geq 0 : x_{x_0}(\tau) = \mathbf{0}} \tau = e^{-\nu s} T(x_0),$$

i.e., the functional T is $\mathbf{d}$-homogeneous of degree $-\nu$. Finally, since T is locally bounded then

$$\sup_{\|x_0\|_{\mathbf{d}} \leq e^s} T(x_0) = \sup_{\|\mathbf{d}(-s)x_0\|_{\mathbf{d}} \leq 1} T(x_0) = \sup_{\|y_0\|_{\mathbf{d}} \leq 1} T(\mathbf{d}(s)y_0) = e^{-\nu s} \sup_{\|y_0\|_{\mathbf{d}} \leq 1} T(y_0),$$

where $y_0 = \mathbf{d}(-s)x_0$. Taking into account $\nu < 0$ we derive $e^{-\nu s} \to 0$ as $s \to -\infty$ and

$$\sup_{\|x_0\| \leq e^s} T(x_0) \to 0 \quad \text{as} \quad s \to -\infty,$$

i.e., T is continuous at $\mathbf{0}$.

II. Repeating the above considerations for $\nu = 0$ we derive

$$\|x(T_i)| \leq \|\mathbf{d}(-is)\| \leq \|\mathbf{d}(-s)\|^i, \quad T_i = iT'.$$

In the view of Proposition 4.3, the parameter s can always be selected such that $\|\mathbf{d}(s)\| > c$ and $\|\mathbf{d}(-s)\| < T'$ for some fixed $c > 1$. In this case, we have

$$\|x(T_i)\| \leq e^{T_i \ln \frac{\|\mathbf{d}(-s)\|}{T'}}$$

and taking into account uniform asymptotic stability we conclude that the system is locally exponentially stable, but the homogeneity implies global exponential stability.

III. Let S_{x_0} denote the solution set of the evolution system with the initial condition $x(0) = x_0$. Let us consider the functional

$$T(x_0, r) = \sup_{x \in S_{x_0}} \inf_{\tau : \|x(t)\|_{\mathbf{d}} \leq r, \forall t \geq \tau} \tau,$$

which estimates the convergence time of solutions (initiated at x_0) into the $\mathbf{d}$-homogeneous ball $B_{\mathbf{d}}(r)$ of the radius $r > 0$. This functional is well-defined and locally bounded since the evolution system is assumed to be uniformly asymptotically stable and the estimates (4.22) hold.

The $\mathbf{d}$-homogeneity of solutions (see the formula (6.2)) implies that

$$\begin{aligned} T(\mathbf{d}(s)x_0, r) &= \sup_{x \in S_{\mathbf{d}(s)x_0}} \inf_{\tau : \|x(t)\|_{\mathbf{d}} \leq r, \forall t \geq \tau} \tau \\ &= \sup_{x \in S_{x_0}} \inf_{\tau : \|\mathbf{d}(s)x(e^{\nu s}t)\|_{\mathbf{d}} \leq r, \forall t \geq \tau} \tau \\ &= \sup_{x \in S_{x_0}} \inf_{\tau : e^s \|x(e^{\nu s}t)\|_{\mathbf{d}} \leq r, \forall t \geq \tau} \tau \\ &= \sup_{x \in S_{x_0}} \inf_{\tilde{\tau} : e^s \|x(\tilde{t})\|_{\mathbf{d}} \leq r, \forall \tilde{t} \geq \tilde{\tau}} e^{-\nu s} \tilde{\tau} \\ &= e^{-\nu s} T(x_0, e^{-s} r) \end{aligned}$$

for any $r > 0$, any $s \in \mathbb{R}$ and any $x_0 \in \mathbb{B}$.

Let us denote $r_i = e^{-is}$ and

$$T_i = \sup_{x_0 : \|x_0\|_{\mathbf{d}} \leq r_{i+1}} T(x_0, r_i),$$

where $s < 0$ and $i = 0, 1, 2, \ldots$. The number $T_i > 0$ estimates a time of convergence of any solutions of the evolution system from the $\mathbf{d}$-homogeneous ball of the radius r_i into the $\mathbf{d}$-homogeneous ball of the radius $r_{i-1} < r_i$. In this case, we have

$$\begin{aligned}
T_i = & \sup_{x_0:\|x_0\|_{\mathbf{d}}\le r_{i+1}} T(x_0, r_i)\\
= & \sup_{\tilde{x}_0:\|\mathbf{d}(s)\tilde{x}_0\|_{\mathbf{d}}\le r_{i+1}} T(\mathbf{d}(s)\tilde{x}_0, r_i)\\
= & \sup_{\tilde{x}_0:\|\mathbf{d}(s)\tilde{x}_0\|_{\mathbf{d}}\le r_{i+1}} e^{-\nu s}T(\tilde{x}_0, e^{-s}r_i)\\
= & e^{-\nu s} \sup_{\tilde{x}_0:e^s\|\tilde{x}_0\|_{\mathbf{d}}\le r_{i+1}} T(\tilde{x}_0, r_{i+1})\\
= & e^{-\nu s} \sup_{\tilde{x}_0:\|\tilde{x}_0\|_{\mathbf{d}}\le r_{i+2}} T(\tilde{x}_0, r_{i+1})\\
= & e^{-\nu s}T_{i+1}.
\end{aligned}$$

Hence, we derive

$$T_i = e^{\nu s}T_{i-1} = e^{2\nu s}T_{i-2} = \ldots = e^{is}T_0.$$

Notice that $T_0 > 0$ is bounded due to uniform asymptotic stability of the system.

Since $s < 0$ then $r_i \to +\infty$ as $i \to +\infty$. In this case for $\nu > 0$ we have

$$T(x_0, 1) \le \sum_{i=0}^{\infty} T_i = T_0 \sum_{i=0}^{\infty} e^{\nu is} = \frac{T_0}{1-e^{\nu s}} < +\infty, \quad \forall x_0 \in \mathbb{B}.$$

This means that any trajectory of the system converges into the unit ball in a fixed time. In this case, the uniform asymptotic stability implies the nearly fixed-time stability. The proof is complete. □

The above theorem does not specify the type of solution (mild, strong, etc.) and the form of a evolution equation/inclusion, since its proof uses just a semigroup property of solutions, **d**-homogeneity and the uniform asymptotic stability of the system. Therefore, the obtained result is valid for any kind of dynamical system in a Banach space.

Example 6.4 *(Super-twisting control for heat system)* Let us consider the controlled diffusion process modeled by

$$\frac{\partial y}{\partial t} = \nu \frac{\partial^2 y}{\partial z^2} + u, \quad z \in \mathbb{R}, \quad t > 0 \tag{6.10}$$

with a distributed nonlinear PI (Proportional-Integral) feedback control given by [6]

$$u(t, z) = -\lambda_1 \frac{y(t,z)}{\|y(t,z\cdot)\|_{L^2}^{1/2}} + w(t, z), \tag{6.11}$$

$$\frac{\partial w}{\partial t} = -\lambda_2 \frac{y(t,z)}{\|y(t,\cdot)\|_{L^2}} \quad \text{for} \quad \|y(t, \cdot)\|_{L^2} \neq 0, \tag{6.12}$$

where $y(t, \cdot), w(t, \cdot) \in L^2(\mathbb{R}; \mathbb{R})$, $\nu > 0$ is a diffusion coefficient and $\lambda_1, \lambda_2 > 0$ are control parameters. The state vector of the closed-loop system is

$$x = (y, w)^\top,$$

so it can be represented in the form (6.7) with the state space

$$\mathbb{B} = L^2(\mathbb{R}; \mathbb{R}) \times L^2(\mathbb{R}; \mathbb{R}),$$

with the linear operator

$$A : \mathcal{D}(A) \subset \mathbb{B} \mapsto \mathbb{B}, \quad \mathcal{D}(A) = H^2(\mathbb{R}; \mathbb{R}) \times L^2(\mathbb{R}; \mathbb{R})$$

given by

$$A\begin{pmatrix} y \\ w \end{pmatrix} = \begin{pmatrix} \frac{\partial^2}{\partial z^2} & I \\ \mathbf{0} & \mathbf{0} \end{pmatrix}\begin{pmatrix} y \\ w \end{pmatrix},$$

and with the nonlinear operator $f : \mathbb{B} \mapsto \mathbb{B}$ given by

$$f(x) = \begin{cases} \begin{pmatrix} -\lambda_1 \frac{y}{\|y\|_{L^2}^{1/2}} \\ \\ -\lambda_2 \frac{y}{\|y\|_{L^2}} \end{pmatrix}, & y \neq \mathbf{0}, \\ \\ \begin{pmatrix} \mathbf{0} \\ \\ \mathbf{0} \end{pmatrix}, & y = \mathbf{0}, \end{cases} \qquad x = (y, w)^\top.$$

The system has mild solutions on $\mathbb{B}_0 = (L^2(\mathbb{R}; \mathbb{R})\backslash\{\mathbf{0}\}) \times L^2(\mathbb{R}; \mathbb{R})$ (see Chap. 2) for $x(0) \in \mathbb{B}_0$ and classical solutions for $x(0) \in \mathcal{D}(A) \cap \mathbb{B}_0$. The right-hand side of the equation (6.12) is discontinuous at $y = \mathbf{0}$. It may be regularized using Filippov-like method (see Chap. 2). For more details about solutions existence and uniqueness problem for such a system we refer the reader to [6, 7].

Let us consider the Lyapunov functional candidate

$$V(x) = \lambda_2 \|y\|_{L^2} + \frac{\|w\|_{L^2}^2}{2}, \quad x = (y, w)^\top.$$

For $x = (y, w)^\top \in \mathbb{B}_0 \cap \mathcal{D}(A)$, we have

$$\begin{aligned} DV(x)(Ax + f(x)) =& \lambda_2 \frac{\left\langle y, \frac{\partial y}{\partial t} \right\rangle}{\|y\|_{L^2}} + \left\langle w, \frac{\partial w}{\partial t} \right\rangle \\ =& \lambda_2 \frac{\left\langle y, \nu \frac{\partial^2 y}{\partial z^2} \right\rangle}{\|y\|_{L^2}} - \lambda_1 \lambda_2 \sqrt{\|y\|_{L^2}} \\ =& -\nu \lambda_2 \frac{\left\langle \frac{\partial y}{\partial z}, \frac{\partial y}{\partial z} \right\rangle}{\|y\|_{L^2}} - \lambda_1 \lambda_2 \sqrt{\|y\|_{L^2}} \leq -\lambda_1 \lambda_2 \sqrt{\|y\|_{L^2}}, \end{aligned}$$

where the integration by parts is utilized in the last step. Therefore, V is a Lyapunov functional of the system (see Chap. 3) and the system is globally asymptotically stable in the view of LaSalle invariance principle (see Chap. 4 of Volume I or [8]), which is valid for abstract differential equations (see, e.g., [9]).

Let us show that the system is **d**-homogeneous of negative degree. Indeed, the linear strongly continuous dilation **d** in $\mathbb{B}$ can be defined as follows:

$$(\mathbf{d}(s)x)(z) = \begin{pmatrix} e^{1.75\,s}y(e^{-0.5\,s}z) \\ e^{0.75\,s}w(e^{-0.5\,s}z) \end{pmatrix}, \quad x = (y, w)^\top \in \mathbb{B}$$

(please see Lemma 4.2). Moreover, simple computations show

$$A\mathbf{d}(s)x = e^{-s}\mathbf{d}(s)Ax$$

and

$$[f(\mathbf{d}(s)x)](z) = \begin{pmatrix} -\lambda_1 \dfrac{e^{1.75\,s}y(e^{-0.5\,s}z)}{\left(\int_{\mathbb{R}}\left(e^{1.75\,s}y(e^{-0.5\,s}z)\right)^2 dz\right)^{1/4}} \\ -\lambda_2 \dfrac{e^{1.75\,s}y(e^{-0.5\,s}z)}{\left(\int_{\mathbb{R}}\left(e^{1.75\,s}y(e^{-0.5\,s}z)\right)^2 dz\right)^{1/2}} \end{pmatrix}$$

$$= \begin{pmatrix} -\lambda_1 \dfrac{e^{0.75\,s}y(e^{-0.5\,s}z)}{\|y\|_{L^2}^{1/2}} \\ -\lambda_2 \dfrac{e^{-0.25\,s}y(e^{-0.5\,s}z)}{\|y\|_{L^2}} \end{pmatrix} = [e^{-s}\mathbf{d}(s)f(x)](z), \qquad z \in \mathbb{R}.$$

Therefore, the system is **d**-homogeneous of the degree -1, so, by Theorem 6.6, the uniform asymptotic stability of the system would imply its finite-time stability.

According to Theorem 6.6, the characterization of the convergence rate of a homogeneous system by means of the homogeneity degree is possible if the system is uniformly asymptotically stable of the system in $\mathbb{B}$. In many cases this condition may be too conservative. Indeed, Examples 3.6 and 3.7 show that an infinite-dimensional system may not be uniformly asymptotically stable in $\mathbb{B}$ but it may be uniformly asymptotically stable in a subspace of $\mathbb{Y} \subset \mathbb{B}$ (see Definition 3.11). If $\mathbb{Y} \subset \mathbb{B}$ is a Banach space and **d** is a dilation in $\mathbb{Y}$ then Theorem 6.6 is applicable in $\mathbb{Y}$. In other cases, the following corollary may be useful.

Corollary 6.3 *Let $\mathbb{B}$ be a Banach space and $\mathbb{Y} \subset \mathbb{B}$ be a Banach space with the norm $\|\cdot\|_{\mathbb{Y}} = \|\cdot\|_{\mathbb{B}}$. Let* **d** *be a strongly continuous linear dilation $\mathbb{B}$. Let an evolution equation/inclusion be* **d***-homogeneous of degree ν in $\mathbb{B}$ and locally uniformly asymptotically stable in $\mathbb{Y}$.*

- *If $\nu < 0$ and $\exists s > 0 : \mathbf{d}(s)\mathbb{Y} \subset \mathbb{Y}$ then the evolution equation/inclusion is globally uniformly finite-time stable in $\mathbb{Y}$.*
- *If $\nu > 0$ and $\exists s < 0 : \mathbf{d}(s)\mathbb{Y} \subset \mathbb{Y}$ then the evolution equation/inclusion is globally uniformly nearly fixed-time stable in $\mathbb{Y}$.*

Proof The uniform asymptotically stability a system on $\mathbb{Y}$, in particular, implies that then $\mathbb{Y}$ is a positively invariant set of the system. For any finite $s \in \mathbb{R}$, the system is uniformly asymptotically stable on $\mathbf{d}(s)\mathbb{Y}$ due to $\mathbf{d}$-homogeneity of the system.

Let S_{x_0} denote the solution set of the evolution system with the initial condition $x(0) = x_0 \in \mathbb{B}$. Let us consider the functional

$$T(x_0, r) = \sup_{x \in S_{x_0}} \inf_{\tau : \|x_{x_0}(t)\|_{\mathbf{d}} \le r, \forall t \ge \tau} \tau,$$

which estimates the convergence time of solutions (initiated at x_0) into the $\mathbf{d}$-homogeneous ball $B_{\mathbf{d}}(r)$ of the radius $r > 0$. This functional is well-defined and locally bounded since the evolution system is assumed to be uniformly asymptotically stable and the estimates (4.22) hold.

The $\mathbf{d}$-homogeneity of solutions (see the formula (6.2)) implies that

$$\begin{aligned}
T(\mathbf{d}(s)x_0, r) &= \sup_{x \in S_{\mathbf{d}(s)x_0}} \inf_{\tau : \|x_{\mathbf{d}(s)x_0}(t)\|_{\mathbf{d}} \le r, \forall t \ge \tau} \tau \\
&= \sup_{x \in S_{x_0}} \inf_{\tau : \|\mathbf{d}(s)x_{x_0}(e^{\nu s}t)\|_{\mathbf{d}} \le r, \forall t \ge \tau} \tau \\
&= \sup_{x \in S_{x_0}} \inf_{\tau : e^{s}\|x_{x_0}(e^{\nu s}t)\|_{\mathbf{d}} \le r, \forall t \ge \tau} \tau \\
&= \sup_{x \in S_{x_0}} \inf_{\tilde{\tau} : e^{s}\|x_{x_0}(\tilde{t})\|_{\mathbf{d}} \le r, \forall \tilde{t} \ge \tilde{\tau}} e^{-\nu s}\tilde{\tau} \\
&= e^{-\nu s} T(x_0, e^{-s} r)
\end{aligned}$$

for any $r > 0$, any $s \in \mathbb{R}$ and any $x_0 \in \mathbb{B}$.

Case $\nu < 0$ Let us denote $r_i = e^{-is}$ and

$$T_i = \sup_{x_0 \in \mathbb{Y} : \|x_0\|_{\mathbf{d}} \le r_i} T(x_0, r_{i+1}),$$

where $s > 0$ is some fixed number and $i = 0, 1, 2, \ldots$. The number $T_i > 0$ estimates a time of convergence of any solutions of the evolution system from the $\mathbf{d}$-homogeneous ball of the radius r_i into the $\mathbf{d}$-homogeneous ball of the radius $r_{i+1} < r_i$. This number is well-defined since the system is uniformly asymptotically stable in $\mathbb{Y}$.

Since $\mathbf{d}(s)\mathbb{Y} \subset \mathbb{Y}$ then the $\mathbf{d}$-homogeneity of solutions (see the formula (6.2)) implies that

$$\begin{aligned}
T_i = & \sup_{x_0\in\mathbb{Y}:\|x_0\|_{\mathbf{d}}\leq r_i} T(x_0, r_{i+1}) \\
\geq & \sup_{x_0\in\mathbf{d}(s)\mathbb{Y}:\|x_0\|_{\mathbf{d}}\leq r_i} T(x_0, r_{i+1}) \\
= & \sup_{\tilde{x}_0\in\mathbb{Y}:\|\mathbf{d}(s)\tilde{x}_0\|_{\mathbf{d}}\leq r_i} T(\mathbf{d}(s)\tilde{x}_0, r_{i+1}) \\
= & \sup_{\tilde{x}_0\in\mathbb{Y}:\|\mathbf{d}(s)\tilde{x}_0\|_{\mathbf{d}}\leq r_i} e^{-\nu s} T(\tilde{x}_0, e^{-s} r_{i+1}) \\
=& e^{-\nu s} \sup_{\tilde{x}_0\in\mathbb{Y}:e^s\|\tilde{x}_0\|_{\mathbf{d}}\leq r_i} T(\tilde{x}_0, r_{i+2}) \\
=& e^{-\nu s} \sup_{\tilde{x}_0\in\mathbb{Y}:\|\tilde{x}_0\|_{\mathbf{d}}\leq r_{i+1}} T(\tilde{x}_0, r_{i+2}) \\
=& e^{-\nu s} T_{i+1}.
\end{aligned}$$

Hence, we derive

$$T_i \leq e^{\nu s} T_{i-1} \leq e^{2\nu s} T_{i-2} \leq \ldots \leq e^{i\nu s} T_0.$$

Notice that $T_0 > 0$ is bounded due to uniform asymptotic stability of the system in $\mathbb{Y}$.

Since $s > 0$ then $r_i \to 0$ as $i \to +\infty$. In this case for $\nu < 0$ we have

$$T(x_0, 0) \leq \sum_{i=0}^{\infty} T_i = T_0 \sum_{i=0}^{\infty} e^{\nu i s} = \frac{T_0}{1 - e^{\nu s}} < +\infty, \quad \forall x_0 \in B(1) \cap \mathbb{Y}.$$

This means that any trajectory initiated in $B(1) \cap \mathbb{Y}$ of the system converges to zero in a fixed time. In this case, the uniform asymptotic stability in $\mathbb{Y}$ implies the uniform finite-time stability.

Case $\nu > 0$ Let us denote $r_i = e^{-is}$ and

$$T_i = \sup_{x_0\in\mathbb{Y}:\|x_0\|_{\mathbf{d}}\leq r_{i-1}} T(x_0, r_i),$$

where $s < 0$ is some fixed number and $i = 0, 1, 2, \ldots$. The number $T_i > 0$ estimates a time of convergence of any solutions of the evolution system from the **d**-homogeneous ball of the radius r_i into the **d**-homogeneous ball of the radius $r_{i-1} < r_i$. This number is well-defined since the system is uniformly asymptotically stable in $\mathbb{Y}$.

Since $\mathbf{d}(s)\mathbb{Y} \subset \mathbb{Y}$ then the **d**-homogeneity of solutions (see the formula (6.2)) implies that

$$\begin{aligned}
T_i &= \sup_{x_0 \in \mathbb{Y}: \|x_0\|_{\mathbf{d}} \le r_{i-1}} T(x_0, r_i) \\
&\ge \sup_{x_0 \in \mathbf{d}(s)\mathbb{Y}: \|x_0\|_{\mathbf{d}} \le r_{i-1}} T(x_0, r_i) \\
&= \sup_{\tilde{x}_0 \in \mathbb{Y}: \|\mathbf{d}(s)\tilde{x}_0\|_{\mathbf{d}} \le r_{i-1}} T(\mathbf{d}(s)\tilde{x}_0, r_i) \\
&= \sup_{\tilde{x}_0 \in \mathbb{Y}: \|\mathbf{d}(s)\tilde{x}_0\|_{\mathbf{d}} \le r_{i-1}} e^{-\nu s} T(\tilde{x}_0, e^{-s} r_i) \\
&= e^{-\nu s} \sup_{\tilde{x}_0 \in \mathbb{Y}: e^s \|\tilde{x}_0\|_{\mathbf{d}} \le r_{i-1}} T(\tilde{x}_0, r_{i+1}) \\
&= e^{-\nu s} \sup_{\tilde{x}_0 \in \mathbb{Y}: \|\tilde{x}_0\|_{\mathbf{d}} \le r_i} T(\tilde{x}_0, r_{i+1}) \\
&= e^{-\nu s} T_{i+1}
\end{aligned}$$

for any $r > 0$, any $s \in \mathbb{R}$ and any $x_0 \in \mathbb{Y}$. Hence, we derive

$$T_i \le e^{\nu s} T_{i-1} \le e^{2\nu s} T_{i-2} \le \ldots \le e^{is} T_0.$$

Notice that $T_0 > 0$ is bounded due to uniform asymptotic stability of the system in $\mathbb{Y}$.

Since $s < 0$ then $r_i \to +\infty$ as $i \to +\infty$. In this case for $\nu > 0$ we have

$$T(x_0, 1) \le \sum_{i=0}^{\infty} T_i = T_0 \sum_{i=0}^{\infty} e^{\nu i s} = \frac{T_0}{1 - e^{\nu s}} < +\infty, \quad \forall x_0 \in B(1) \cap \mathbb{Y}.$$

This means that any trajectory of the system converges into the set $B(1) \cap \mathbb{Y}$ in a fixed time. In this case, the uniform asymptotic stability in $\mathbb{Y}$ implies the uniform nearly fixed-time stability. The proof is complete. □

Corollary 6.4 *Let $\mathbb{B}$ be a Banach space and $\mathbb{Y} \subset \mathbb{B}$ be a Banach space with the norm $\|\cdot\|_{\mathbb{Y}} = \|\cdot\|_{\mathbb{B}}$. Let $\mathbf{d}_1$ and $\mathbf{d}_2$ be strongly continuous strictly monotone linear dilations in $\mathbb{B}$. If all conditions of Corollary 6.3 are fulfilled for $\mathbf{d} = \mathbf{d}_1$, $\nu = \nu_1 < 0$ and for $\mathbf{d} = \mathbf{d}_2$, $\nu = \nu_2 > 0$, simultaneously, then the evolution equation/inclusion is globally uniformly fixed-time stable in $\mathbb{Y}$.*

Proof By Corollary 6.3 the system is both nearly fixed-time stable in $\mathbb{Y}$ and finite-time stable in $\mathbb{Y}$. This implies the fixed-time stability of the system on $\mathbb{Y}$ (see Definition 3.11). The proof is complete. □

Example 6.5 Let us consider the transport equation on $\mathbb{R}_+$

$$\frac{\partial x}{\partial t} = \frac{\partial x}{\partial z}, \quad t > 0, \quad z > 0 \tag{6.13}$$

with $\frac{\partial}{\partial z} : H^1(\mathbb{R}_+; \mathbb{R}) \subset L^2(\mathbb{R}_+; \mathbb{R}) \mapsto L^2(\mathbb{R}_+; \mathbb{R})$. The differential operator is $\mathbf{d}_1$-homogeneous of the degree -1 with respect to the dilation

$$[\mathbf{d}_1(s)x](z) = e^s x(e^{-s}z), \quad s \in \mathbb{R}, \quad z \in \mathbb{R}_+, \quad x \in L^2(\mathbb{R}_+; \mathbb{R})$$

and $\mathbf{d}_2$-homogeneous of the degree 1 with respect to the dilation

$$[\mathbf{d}_2(s)x](z) = e^s x(e^s z), \quad s \in \mathbb{R}, \quad z \in \mathbb{R}_+, \quad x \in L^2(\mathbb{R}_+; \mathbb{R}).$$

The considered system is non uniformly asymptotically stable on $L^2(\mathbb{R}_+; \mathbb{R})$ (see Example 3.3), so Theorem 6.6 is not applicable. However, the system is uniformly asymptotically stable in the subspace

$$\mathbb{Y} = \{x \in L^2(\mathbb{R}_+; \mathbb{R}) : \text{supp}(x) \subset [0, 1]\},$$

where supp(x) denotes the support of the function $x \in L^2(\mathbb{R}_+; \mathbb{R})$ and $\|\cdot\|_{\mathbb{Y}} := \|\cdot\|_{L^2(\mathbb{R}_+;\mathbb{R})}$. In Example 3.7, it is shown that the considered system is fixed-time stable. Moreover, the system satisfies almost all conditions of Corollary 6.4 except the conditions $\mathbf{d}_1(s_1)\mathbb{Y} \subset \mathbb{Y}$ for some $s_1 > 0$ and $\mathbf{d}_2(s_2)\mathbb{Y} \subset \mathbb{Y}$ for some $s_2 < 0$. Therefore, the mentioned conditions are too conservative. Their relaxation is an interesting open problem.

6.2.2 Homogeneous Lyapunov Function Theorem

The **d**-homogeneity of an evolution system is inherited by Lyapunov functions.

Theorem 6.7 *Let* **d** *be a linear strongly continuous strictly monotone dilation in* $\mathbb{B}$. *A* **d***-homogeneous evolution inclusion* (6.7) *is uniformly Lyapunov stable if and only if there exists a globally proper positive definite* **d***-homogeneous functional* $V : \mathbb{B} \mapsto \mathbb{R}$ *of the degree* 1 *such that the inequality*

$$D_{\mathbb{K}} V(x_{x_0}(t)) \le 0, \quad \forall t \ge 0$$

holds for any solution x_{x_0} *with* $x_0 \in \mathbb{B}\backslash\{\mathbf{0}\}$ *as long as* $x_{x_0}(t) \ne \mathbf{0}$, *where* $D_{\mathbb{K}}$ *denotes the contingent derivative (see Chap. 4 of Volume I).*

Proof We prove the *necessity* only, since the *sufficiency* immediately follows from Theorem 3.1.

Let S_{x_0} denote a set of solutions of an evolution system with the initial condition $x(t_0) = x_0$. In this case, the functional $V : \mathbb{B} \mapsto \mathbb{B}$ given by

$$V(x_0) := \sup_{x_{x_0} \in S_{x_0}} \sup_{t \ge 0} \|x_{x_0}(t)\|_{\mathbf{d}}$$

is globally proper. Indeed, due to the properties of $\|\cdot\|_{\mathbf{d}}$ (see Lemma 4.5) there exists $\underline{\sigma}, \bar{\sigma} \in \mathcal{K}^{\infty}$

$$\underline{\sigma}(\|x\|) \leq \|x\|_{\mathbf{d}} \leq \bar{\sigma}(\|x\|), \quad \forall x \in \mathbb{B}.$$

Hence, using Lyapunov stability of the origin we derive

$$\underline{\sigma}(\|x\|) \leq \|x\|_{\mathbf{d}} \leq V(x) \leq \bar{\sigma}(\varepsilon(x)), \quad \forall x \in \mathbb{B},$$

where $\varepsilon \in \mathcal{K}_{\infty}$ is introduced in the definition of the Lyapunov stability (see Chap. 3).

Due to uniform Lyapunov stability of the system, $V(x_{x_0}(t))$ is decreasing and, by Lemma 4.2 of Volume I, $D_{\mathbb{K}} V(x_{x_0}(t)) \leq 0$ as long as $x_{x_0}(t) \in \mathbb{B} \backslash \{\mathbf{0}\}$.

Finally, since for any $\mathbf{d}$-homogeneous evolution system we have

$$x_{\mathbf{d}(s)x_0}(t) = \mathbf{d}(s) x_{x_0}(e^{\mu s} t), \quad t \geq 0,$$

where μ is a homogeneity degree of the evolution system, then

$$V(\mathbf{d}(s)x_0) := \sup_{x_{\mathbf{d}(s)x_0} \in S_{\mathbf{d}(s)x_0}} \sup_{t \geq 0} \|x_{\mathbf{d}(s)x_0}(t)\|_{\mathbf{d}} =$$

$$\sup_{x_{x_0} \in S_{x_0}} \sup_{t \geq 0} \|\mathbf{d}(s) x_{x_0}(e^{\mu s} t)\|_{\mathbf{d}} = e^{s} \sup_{x_{x_0} \in S_{x_0}} \sup_{t \geq 0} \|x_{x_0}(e^{\mu s} t)\|_{\mathbf{d}} = e^{s} V(x_0),$$

i.e., V is a $\mathbf{d}$-homogeneous functional of the degree 1. The proof is complete. □

The similar result can be proven for asymptotically stable homogeneous evolution equations.

Theorem 6.8 *Let* $\mathbf{d}$ *be a dilation in* $\mathbb{B}$ *and the evolution inclusion* (6.7) *satisfying Assumption 3.1 be* $\mathbf{d}$*-homogeneous of a degree* $\mu \in \mathbb{R}$.

The system (6.7) *is uniformly asymptotically stable if and only if there exists a globally proper positive definite* $\mathbf{d}$*-homogeneous functional* $V : \mathbb{B} \mapsto \mathbb{R}$ *of the degree* 1 *such that the inequality*

$$D_{\mathbb{K}} V(x_{x_0}(t)) \leq -\rho V^{\mu+1}(x_{x_0}(t)), \quad \forall t \geq 0, \quad \forall x_0 \in \mathbb{B} \backslash \{\mathbf{0}\} \tag{6.14}$$

holds for any solution x_{x_0} *as long as* $x_{x_0}(t) \neq \mathbf{0}$*, where* $\rho > 0$ *is some positive number.*

Proof *Necessity.* If $\mu = -1$ then, by Theorem 6.6, the asymptotically stable $\mathbf{d}$-homogeneous evolution system is globally uniformly finite-time stable and there exists a locally bounded and continuous at zero $\mathbf{d}$-homogeneous settling-time function T of the degree 1. On the one hand, Theorem 6.7 implies that there exists a globally proper positive definite $\mathbf{d}$-homogeneous functional $V : \mathbb{B} \mapsto \mathbb{R}$ of the degree 1 such that

$$\mathcal{D}_{\mathbb{K}} V(x_{x_0}(t)) \leq 0$$

as long as $x_{x_0}(t) \neq \mathbf{0}$. On the other hand, the settling-time functional satisfies the inequality

$$D_{\mathbb{K}} T(x_{x_0}(t)) \leq -1, \quad \forall t \geq 0$$

as long as $x_{x_0}(t) \neq \mathbf{0}$ (see the proof of Theorem 3.4). Hence, considering $V_0 = V + T$ we derive

$$D_{\mathbb{K}} V_0(x_{x_0}(t)) \leq -1, \quad \forall t \geq 0$$

as long as $x_{x_0}(t) \neq 0$. Since V is globally proper and T is locally bounded and continuous at $\mathbf{0}$ then V_0 is globally proper. Taking into account that V_0 is a positive definite functional we conclude that the theorem is proven for $\mu = -1$.

Let $\mu \neq -1$. Let us consider the evolution inclusion

$$\dot{z} \in \|z\|_{\mathbf{d}}^{-\mu-1}(Az + F(z)), \quad z(0) = x_0 \in \mathbb{B}\backslash\{\mathbf{0}\} \tag{6.15}$$

obtained using the operators A and F from the evolution equation (6.7). Let the right-hand side of the above system be assigned to $\mathbf{0}$ at $z = \mathbf{0}$.

The considered system is, obviously, $\mathbf{d}$-homogeneous of the degree -1 and Lemma 2.4 proves that it has a solution z_{x_0} for any initial condition $z(0) = x_0 \in \mathbb{B}\backslash\{\mathbf{0}\}$ and any solution of (6.7) with $x(0) = x_0 \neq \mathbf{0}$ satisfies

$$x_{x_0}(t) = z_{x_0}\left(\int_0^t \|x_{x_0}(\sigma)\|_{\mathbf{d}}^{\mu+1} d\sigma\right)$$

as long as $x_{x_0}(t) \neq \mathbf{0}$, and, inversely, any solution of (6.15) satisfies

$$z_{x_0}(\tau) = x_{x_0}\left(\int_0^\tau \frac{d\sigma}{\|z_{x_0}(\sigma)\|_{\mathbf{d}}^{\mu+1}}\right)$$

as long as $z_{x_0}(\tau) \neq \mathbf{0}$. The latter also implies that if $x_{x_0}(t) \to \mathbf{0}$ as $t \to +\infty$ then $z_{x_0}(\tau)$ converges to $\mathbf{0}$ as $\tau \to \tau' = \int_0^{+\infty} \|x_{x_0}(\sigma)\|_{\mathbf{d}}^{\mu+1} d\sigma \leq +\infty$ and vice versa. Hence, we conclude that the system (6.15) is globally asymptotically stable if and only if the system (6.7) is globally uniformly asymptotically stable.

For the system with the homogeneity degree -1 it has been already proven the existence of a globally proper positive definite functional $V_0 : \mathbb{B} \mapsto \mathbb{R}$ of degree 1 such that

$$D_{\mathbb{K}} V_0(z_{x_0}(\tau)) \leq -1, \quad \forall \tau \geq 0$$

as long as $z_{x_0}(\tau) \neq 0$. Denoting $\tau(t) = \int_0^t \|x_{x_0}(\sigma)\|_{\mathbf{d}}^{\mu+1} d\sigma$ and using Lemma 4.3 of Volume I we derive

$$D_K V_0(x_{x_0}(t)) = D_K V_0\left(z_{x_0}(\tau(t))\right) \subset D_K V_0(z_{x_0}(\tau))\dot{\tau} \leq$$

$$-\|x_{x_0}(t)\|_{\mathbf{d}}^{\mu+1} \leq -\rho V_0^{\mu+1}(x_{x_0}(t)),$$

where $\rho = \frac{1}{\sup_{x\in S} V^{1+\mu}(x)} > 0$ due to Corollary 5.1.

Sufficiency. The inequality (6.14) and Theorem 3.1 guarantee that the system (6.7) is globally uniformly Lyapunov stable. The differential inequality (6.14) and the global properness of V also imply that $x_{x_0}(t) \to \mathbf{0}$ as $t \to +\infty$ (see the proof of Theorem 3.2 for more details). The proof is complete. □

The above theorem is an extension of Zubov's homogeneous Lyapunov function theorem [10] to evolution inclusions in Banach spaces. Zubov's theorem does not guarantee a smoothness of the Lyapunov function even for ODEs (i.e., for $\mathbb{B} = \mathbb{R}^n$). However, the robustness (input-to-state stability) analysis requires, at least, a local Lipschitz continuity of the Lyapunov function function (see, [11–13]). Rosier's theorem [14] generalizes Zubov's result and guarantees the existence of a smooth homogeneous Lyapunov function for homogeneous asymptotically stable ODE. An analog of the Rosier theorem in an infinite-dimensional space is proven in [15] and in [16] for generalized homogeneous evolution equations with locally Lipschitz continuous nonlinearities and with unbounded (but M-regular) nonlinear operators, respectively. Notice that a continuously differentiability may be a too conservative condition for Lypunov functions in abstract spaces. The theorem given below proves the existence of a locally Lipschitz continuous homogeneous Lyapunov function for homogeneous abstract differential equation in a Hilbert space.

Theorem 6.9 *Let* $\mathbf{d}$ *be a strictly monotone strongly continuous dilation in a Hilbert space* $\mathbb{H}$, *a linear closed densely defined operator* $A : \mathcal{D}(A) \subset \mathbb{H} \mapsto \mathbb{H}$ *be a generator of a strongly continuous semigroup* Φ *of linear bounded operators on* $\mathbb{H}$, $f : \mathbb{H} \mapsto \mathbb{H}$ *be locally Lipschitz continuous on* $\mathbb{H}\backslash\{\mathbf{0}\}$ *and the evolution equation* (6.3) *be* $\mathbf{d}$*-homogeneous of degree* $\mu \in \mathbb{R}$. *Let* $m > 0$ *be an arbitrary real number. The system* (6.3) *is uniformly asymptotically stable if and only if there exists a continuous positive definite functional* $V : \mathbb{H} \mapsto \mathbb{R}_+$ *such that*

(1) V *is* $\mathbf{d}$*-homogeneous of degree* m, *locally Lipschitz continuous on* $\mathbb{H}\backslash\{\mathbf{0}\}$ *and there exist* $0 < \underline{k} \leq k < +\infty$ *satisfying*

$$\underline{k}\|x\|_{\mathbf{d}}^m \leq V(x) \leq \overline{k}\|x\|_{\mathbf{d}}^m, \quad \forall x \in \mathbb{H}, \tag{6.16}$$

where $\|\cdot\|_{\mathbf{d}}$ *is the canonical homogeneous norm in* $\mathbb{H}$;

(2) the inequality

$$D^+V(x; Ax + f(x)) \leq -\rho V^{1+\frac{\mu}{m}}(x), \quad \forall x \in \mathcal{D}(A)\backslash\{\mathbf{0}\} \tag{6.17}$$

holds for some $\rho > 0$, *where* $D^+V(x; g) = \limsup_{h\to 0^+} \frac{V(x+hg)-V(x)}{h}$ *is the right-hand upper Gateaux derivative in the direction* $g \in \mathbb{H}$.

Proof [2] *Sufficiency*. Since by Theorem 2.16 any solution x_{x_0} of (6.3) with $x_0 \in \mathcal{D}(A)$ is a locally Lipschitz continuous strong solution then the inequality (6.17) implies

$$D^+V(x_{x_0}(t)) \overset{a.e.}{=} D^+V(x_{x_0}(t); Ax_{x_0}(t) + f(x_{x_0}(t))) \le -W(x_{x_0}(t)).$$

The obtained inequality implies asymptotic stability of strong solutions. Taking into account that any mild solution is the uniform limit of strong solutions (see Theorem 2.16) we complete the proof of sufficiency.

Necessity. We design the Lyapunov function in several steps: first, we construct a function $V_r : \mathbb{H} \mapsto \mathbb{R}_+$ such that $t \mapsto V_r(x_{x_0}(t))$ is strictly decreasing as long as $x_{x_0}(t) \in K(r)$ for some $r > 1$; next, using V_r we derive a $\mathbf{d}$-homogeneous Lyapunov function $V : \mathbb{H} \mapsto \mathbb{R}_+$ by means of a special integral transformation inspired by [14]. In the final step, we show that the condition (6.17) holds for the constructed V.

If the system is asymptotically stable then it is, in particular, Lyapunov stable and there exists $\varepsilon \in \mathcal{K}_\infty$ such that

$$\|x_{x_0}(t)\| \le \varepsilon(\|x_0\|), \quad \forall t \ge 0.$$

I. Let $s_0 > 0$,

$$r_0 := \sup_{s\in[-s_0,s_0]} \|\mathbf{d}(s)\|$$

and $r > r_0^2 > 1$. Let $\delta > \max\{r, \varepsilon(r)\}$ and $a_\delta \in C_c^\infty$ be such that $a_\delta(p) = 0$ for $p \in [0, 0.5/\delta] \cap [2\delta, +\infty)$ and $a_\delta(p) = 1$ for $p \in [1/\delta, \delta]$. Then for any $x_0 \in \mathbb{H}$ the system

$$\dot{x} = Ax + f_\delta(x), \quad t > 0, \quad x(0) = x_0 \in \mathbb{H}, \quad f_\delta(x) = a_\delta(\|x\|)f(x) \qquad (6.18)$$

has the unique mild solution $x_{x_0}^\delta : [0, +\infty) \mapsto \mathbb{H}$ which, for $x_0 \in K(r)$, coincides with the unique solution x_{x_0} of (6.3) as long as $x_{x_0}^\delta(t) \in K(\delta)$. Moreover, for any $T > 0$ there exists $L_{\delta,T} > 0$ such that

$$\|x_{x_1}^\delta(t) - x_{x_2}^\delta(t)\| \le L_{\delta,T}\|x_1 - x_2\|,$$

for all $t \in [0, T]$.

Since $\delta > \epsilon(r)$, then the Lyapunov stability of the system (6.3) implies that $x_0 \in K(r) \Rightarrow \|x_{x_0}(t)\| < \delta, \forall t \ge 0$. The global uniformly asymptotic stability also guarantees that there exists $T_r > 0$ such that

$$\|x_0\| < r \quad \Rightarrow \quad \|x_{x_0}(t)\| \le 1/r, \quad \forall t \ge T_r.$$

[2] The proof is reprinted from *Automatica*, 129/7, A. Polyakov, Input-to-State Stability of homogeneous infinite-dimensional systems with locally Lipschitz nonlinearities, 109615, https://doi.org/10.1016/j.automatica.2021.109615. ©2021 Elsevier, with permission from Elsevier.

Therefore, we have $x^\delta_{x_0}(t) = x_{x_0}(t)$ as long as $\|x_{x_0}(t)\| \geq 1/\delta$ provided that $x_0 \in K(r)$. In this case, for the functional $V_0 : \mathbb{H} \mapsto \mathbb{R}_+$ given by

$$V_0(x_0) := \sup_{t\geq 0} \|x_{x_0}(t)\|$$

we have

$$V_0(x_0) = \sup_{t\in[0,T_r]} \|x^\delta_{x_0}(t)\|$$

for all $x_0 \in K(r)$. Hence, using the triangle inequality we derive

$$|V_0(x_1) - V_0(x_2)| \leq \sup_{t\in[0,T_r]} \|x^\delta_{x_1}(t) - x^\delta_{x_2}(t)\| \leq$$

$$L_{\delta,T_r}(\|x_1 - x_2\|), \forall x_1, x_2 \in K(r).$$

The latter means that the functional V_0 satisfy the Lipschitz condition on $K(r)$.

The global uniform Lyapunov stability of (6.3) implies that $\|x\| \leq V_0(x) \leq \varepsilon(\|x\|)$. Moreover, the function $t \mapsto V_0(x_{x_0}(t))$ is decreasing as long as $x_{x_0}(t) \in K(r)$.

Let us consider the functional

$$V_1(x_0) := \int_0^{+\infty} \tilde{a}(\|x_{x_0}(s)\|)\|x_{x_0}(s)\|ds,$$

where $\tilde{a} \in C^\infty_c$ is such that $\tilde{a}(\rho) = 0$ for $\rho \notin (1/r, r)$, $0 < \tilde{a}(\rho) \leq 1$ for $\rho \in (1/r, r)$ and $\tilde{a}(\rho) = 1$ for $\rho \in [1/r_0, r_0]$.

Notice that $\tilde{a}(\|x_{x_0}(s)\|)\|x_{x_0}(s)\| = 0$ if $x_{x_0}(s) \notin K(r)$. Repeating the above considerations we conclude that

$$V_1(x_0) = \int_0^{T_r} \tilde{a}(\|x^\delta_{x_0}(s)\|)\|x^\delta_{x_0}(s)\|ds$$

for all $x_0 \in K(r)$. Since $x \mapsto \tilde{a}(\|x\|)\|x\|$ satisfy the Lipschitz condition on $\mathbb{H}$ with some Lipschitz constant $\tilde{L} > 0$ then

$$\begin{aligned} |V_1(x_1) - V_1(x_2)| \leq& \tilde{L} \int_0^{T_r} \|x^\delta_{x_1}(s) - x^\delta_{x_2}(s)\|ds \\ \leq& T_r L_{\delta,T_r} \tilde{L}\|x_1 - x_2\|, \quad \forall x_1, x_2 \in K(r), \end{aligned}$$

i.e., V_1 also satisfies the Lipschitz condition on $K(r)$.

Moreover, if $h > 0$ and $t \geq 0$ are such that $x_{x_0}(t + \theta) \in K(r)$, $\forall \theta \in [0, h]$ then $x_{x_0}(t + \theta) = x^\delta_{x_{x_0}(t)}(\theta)$ for all $\theta \in [0, h]$ and

$$V_1(x_{x_0}(t + h)) - V_1(x_{x_0}(t))$$

$$
\begin{aligned}
&= \int_0^{T_r} \tilde{a}(\|x^\delta_{x(t+h)}(s)\|)\|x^\delta_{x(t+h)}(s)\|ds \\
&\quad - \int_0^{T_r} \tilde{a}(\|x^\delta_{x(t)}(s)\|)\|x^\delta_{x(t)}(s)\|ds \\
&= \int_0^{T_r} \tilde{a}(\|x^\delta_{x(t+h)}(s)\|)\|x^\delta_{x(t+h)}(s)\|ds \\
&\quad - \int_0^{h} \tilde{a}(\|x^\delta_{x(t)}(s)\|)\|x^\delta_{x(t)}(s)\|ds \\
&\quad - \int_h^{T_r+h} \tilde{a}(\|x^\delta_{x(t)}(s)\|)\|x^\delta_{x(t)}(s)\|ds \\
&= \int_0^{T_r} \tilde{a}(\|x^\delta_{x(t+h)}(s)\|)\|x^\delta_{x(t+h)}(s)\|ds \\
&\quad - \int_0^{h} \tilde{a}(\|x^\delta_{x(t)}(s)\|)\|x^\delta_{x(t)}(s)\|ds \\
&\quad - \int_0^{T_r} \tilde{a}(\|x^\delta_{x(t)}(s+h)\|)\|x^\delta_{x(t)}(s+h)\|ds \\
&= - \int_0^{h} \tilde{a}(\|x^\delta_{x(t)}(s)\|)\|x^\delta_{x(t)}(s)\|ds \\
&= - \int_0^{h} \tilde{a}(\|x_{x(t)}(s)\|)\|x_{x(t)}(s)\|ds \\
&= - \tilde{a}(\|x_{x_0}(t+\theta_h)\|)\|x_{x_0}(t+\theta_h)\|h, \quad \theta_h \in [0, h],
\end{aligned}
$$

where the mean value theorem for integrals (see Appendix of Volume I) is utilized in the last step. Hence, we derive

$$
\limsup_{h \to 0^+} \frac{V_1(x_{x_0}(t+h)) - V_1(x_{x_0}(t))}{h} \le -\tilde{a}(\|x_{x_0}(t)\|)\|x_{x_0}(t)\|
$$

as long as $x_{x_0}(t) \in K(r)$. Notice that $0 \le V_1(x_0) \le T_r \varepsilon(\|x_0\|)$ for all $x_0 \in \mathbb{H}$.

In this case, the functional $V_r : \mathbb{H} \mapsto \mathbb{R}_+$ given by

$$
V_r(x) := V_0(x) + V_1(x)
$$

satisfies the Lipschitz condition on $K(r)$ and

$$
\|x\| \le V_r(x) \le (1 + T_r)\varepsilon(\|x\|)
$$

for all $x \in \mathbb{H}$. Moreover, for any $x_0 \in K(r)$ the function $t \mapsto V_r(x_{x_0}(t))$ is strictly decreasing,

$$\limsup_{h\to 0^+} \tfrac{V_r(x_{x_0}(t+h))-V_r(x_{x_0}(t))}{h} \le -W_r(\|x_{x_0}(t)\|),$$

as long as $x_{x_0}(t) \in K(r)$, where $W_r(\rho) := \tilde{a}(\rho)\rho$ with $\rho \in \mathbb{R}_+$.

Due to the homogeneity of the system (7.22) the solutions x_{x_0} are symmetric (see Theorem 6.1). Hence, if $\mathbf{d}(-s)x_{x_0}(t+\theta) \in K(r)$ for all $\theta \in [0, h]$ then using the identity $\mathbf{d}(-s)x_{x_0}(t) = x_{\mathbf{d}(-s)x_0}(e^{\mu s}t)$, $\forall s \in \mathbb{R}$ we derive

$$\begin{aligned} V_r(\mathbf{d}(-s)x_{x_0}(t+h)) - & V_r(\mathbf{d}(-s)x_{x_0}(t)) \\ & = V_r(x_{\mathbf{d}(-s)x_0}(e^{\mu s}(t+h))) - V_r(x_{\mathbf{d}(-s)x_0}(e^{\mu s}t)) \\ & \le -\tilde{a}(\|x_{\mathbf{d}(-s)x_0}(e^{\mu s}t+\theta_h)\|)\|x_{\mathbf{d}(-s)x_0}(e^{\mu s}t+\theta_h)\|e^{\mu s}h, \end{aligned}$$

where $\theta_h \in [0, e^{\mu s}h]$, and $\mu \in \mathbb{R}$ is the homogeneity degree of the system (7.22), i.e., the function $t \mapsto V_r(\mathbf{d}(-s)x_{x_0}(t))$ is strictly decreasing,

$$\begin{aligned} \limsup_{h\to 0^+} & \frac{V_r(\mathbf{d}(-s)x_{x_0}(t+h)) - V_r(\mathbf{d}(-s)x_{x_0}(t))}{h} \\ & \le -\tilde{a}(\|\mathbf{d}(-s)x_{x_0}(t)\|)\|\mathbf{d}(-s)x_{x_0}(t)\|e^{\mu s} \end{aligned}$$

as long as $\mathbf{d}(-s)x_{x_0}(t) \in K(r)$.

II. Inspired by [14] let us consider the functional $V : \mathbb{H} \mapsto \mathbb{R}_+$ given by

$$V(x) := \int_{-\infty}^{+\infty} e^{ms}\hat{a}(V_r(\mathbf{d}(-s)x))ds$$

where $\hat{a} \in C^\infty$ is a function such that $\hat{a}(\rho) = 1$ for $\rho \ge (1+T_r)\varepsilon(r)$, $\hat{a}(\rho) = 0$ for $\rho \le 1/r$ and $\hat{a}'(\rho) > 0$ for $1/r < \rho < (1+T_r)\varepsilon(r)$. By construction, the functional $V : \mathbb{B} \mapsto \mathbb{R}_+$ is $\mathbf{d}$-homogeneous of degree $m > 0$. Moreover,

$$\begin{aligned} V(x) & \le \int_{-\infty}^{+\infty} e^{ms}\hat{a}((1+T_r)\varepsilon(\|\mathbf{d}(-s)x)\|))ds \\ & \le \int_{-\infty}^{\bar{s}(x)} e^{ms}\hat{a}((1+T_r)\varepsilon(\|\mathbf{d}(-s)x)\|))ds \le e^{m\bar{s}(x)} \end{aligned}$$

where $\bar{s}(x) := \ln\left(\|x\|_{\mathbf{d}}/\underline{\sigma}\left(\varepsilon^{-1}\left(\frac{1}{r(1+T_r)}\right)\right)\right)$ and $\underline{\sigma} \in \mathcal{K}^\infty$ is defined in Lemma 4.5. Indeed, if $s \ge s(x)$ then

$$\underline{\sigma}(\|\mathbf{d}(-s)x\|) \le e^{-s}\|x\|_{\mathbf{d}} \le \underline{\sigma}\left(\varepsilon^{-1}\left(\tfrac{1}{r(1+T_r)}\right)\right)$$

and taking into account $\underline{\sigma}, \varepsilon \in \mathcal{K}^\infty$ we conclude $\|\mathbf{d}(-s)x)\| \le \varepsilon^{-1}\left(\frac{1}{r(1+T_r)}\right)$ or, equivalently, $(1+T_r)\varepsilon(\|\mathbf{d}(-s)x)\|) \le 1/r$ for all $s \ge \bar{s}(x)$. This means $V_r(\mathbf{d}(-s)x) \le 1/r$ and $\hat{a}(V_r(\mathbf{d}(-s)x))) = 0$ for all $s \ge \bar{s}(x)$. Similarly, we derive

$$V(x) \geq \int_{-\infty}^{+\infty} e^{ms}\hat{a}(\|\mathbf{d}(-s)x)\|)ds \geq \int_{-\infty}^{\underline{s}(x)} e^{ms}ds = e^{m\underline{s}(x)},$$

where $\underline{s}(x) := \ln\left(\|x\|_{\mathbf{d}}/\bar{\sigma}\left((1+T_r)\varepsilon(r)\right)\right)$. Indeed, if $s \leq s(x)$ then from Lemma 4.5 we have

$$\bar{\sigma}(\|\mathbf{d}(-s)x\|) \geq e^{-s}\|x\|_{\mathbf{d}} \geq \bar{\sigma}((1+T_r)\varepsilon(r)).$$

Since $\bar{\sigma} \in \mathcal{K}^{\infty}$ then $\|\mathbf{d}(-s)x\| \geq (1+T_r)\varepsilon(r)$ and $\hat{a}(\|\mathbf{d}(-s)x\|)) = 1$ for all $s \leq \underline{s}(x)$. Therefore, we have proven that the inequality (6.16) holds for

$$\underline{k} := \frac{1}{\bar{\sigma}^m\left((1+T_r)\varepsilon(r)\right)}, \quad \bar{k} := \frac{1}{\underline{\sigma}^m\left(\varepsilon^{-1}\left(\frac{1}{r(1+T_r)}\right)\right)}.$$

If $x(t+\theta) \in K(r_0)$ for all $\theta \in [0, h]$ then $\mathbf{d}(-s)x(t+\theta) \in K(r)$ for all $\theta \in [0, h]$ and for all $s \in [-s_0, s_0]$. Indeed,

$$\|\mathbf{d}(-s)x(t+\theta)\| \leq \|\mathbf{d}(-s)\| \cdot \|x(t+\theta)\| \leq r_0^2 = r$$

and

$$1/r_0 \leq \|x(t+\theta)\| = \|\mathbf{d}(s)\| \cdot \|\mathbf{d}(-s)x(t+\theta)\|.$$

This means that

$$1/r \leq V_r(\mathbf{d}(-s)x_{x_0}(t+\theta)) \leq (1+T_r)\varepsilon(r)$$

for all $s \in [-s_0, s_0]$ and all $\theta \in [0, h]$. Using

$$V_r(\mathbf{d}(-s)x_{x_0}(t+h)) - V_r(\mathbf{d}(-s)x_{x_0}(t)) \leq$$
$$- W_r(x_{\mathbf{d}(-s)x_0}(e^{\mu s}t+\theta_h))e^{\mu s}h =$$
$$- W_r(\mathbf{d}(-s)x_{x_0}(t+e^{-\mu s}\theta_h))e^{\mu s}h$$

we derive

$$\begin{aligned}
&V(x_{x_0}(t+h)) - V(x_{x_0}(t)) \\
&= \int_{-\infty}^{+\infty} e^{ms}[\hat{a}(V_r(\mathbf{d}(-s)x_{x_0}(t+h))) - \hat{a}(V_r(\mathbf{d}(-s)x_{x_0}(t)))]ds \\
&= \int_{-s_0}^{s_0} e^{ms}[\hat{a}(V_r(\mathbf{d}(-s)x_{x_0}(t+h))) - \hat{a}(V_r(\mathbf{d}(-s)x_{x_0}(t)))]ds \\
&= \int_{-s_0}^{s_0} e^{ms}\hat{a}'(\theta)[V_r(\mathbf{d}(-s)x_{x_0}(t+h)) - V_r(\mathbf{d}(-s)x_{x_0}(t))]ds \\
&\leq -h\int_{-s_0}^{s_0} e^{ms}\hat{a}'(\theta)W_r(\mathbf{d}(-s)x_{x_0}(t+e^{-\mu s}\theta_h))e^{\mu s}ds,
\end{aligned}$$

where the mean value theorem (see Appendix of Volume I) is utilized with $\theta \in [V_r(\mathbf{d}(-s)x_{x_0}(t+h)), V_r(\mathbf{d}(-s)x_{x_0}(t))]$. Therefore, the function $t \mapsto V(x_{x_0}(t))$ is strictly decreasing,

$$\begin{aligned}\limsup_{h\to 0^+} \frac{V(x_{x_0}(t+h)) - V(x_{x_0}(t))}{h} &= D^+V(x(t)) \\ \leq -\int_{-\infty}^{+\infty} &e^{(m+\mu)s}\hat{a}'(V_r(\mathbf{d}(-s)x_{x_0}(t)))W_r(\|\mathbf{d}(-s)x_{x_0}(t)\|)ds \\ = -\, &W(x_{x_0}(t)),\end{aligned}$$

as long as $x_{x_0}(t) \in K(r_0)$, where D^+ is the right-hand upper Dini derivative (see Volume I) and

$$W(x) := \int_{-\infty}^{+\infty} e^{(m+\mu)s}\hat{a}'(V_r(\mathbf{d}(-s)x))W_r(\|\mathbf{d}(-s)x\|)ds.$$

The functional W is nonnegative and $\mathbf{d}$-homogeneous of degree $\mu+1$. If $\|x\| = 1$ then $\mathbf{d}(-s)x \in K(r_0)$ and $W_r(\|\mathbf{d}(-s)x\|) = \|\mathbf{d}(-s)x\|$ for all $s \in [-s_0, s_0]$. Using the mean value theorem for integrals we derive

$$W(x) \geq \int_{-s_0}^{s_0} e^{(m+\mu)s-q}\hat{a}'(V_r(\mathbf{d}(-s)x))\|\mathbf{d}(-s)x\|ds \geq$$

$$\frac{1}{r_0}e^{(m+\mu)s^*}\hat{a}'(V_r(\mathbf{d}(-s^*)x)),$$

where $s^* \in [-s_0, s_0]$. Since $\mathbf{d}(-s^*)x \in K(r_0)$ then

$$\frac{1}{r} < \frac{1}{r_0} \leq V_r(\mathbf{d}(-s^*)x) \leq (1+T_r)\varepsilon(r_0) < (1+T_r)\varepsilon(r).$$

Hence, we derive $\inf_{\|x\|=1} W(x) \geq \bar{c}$, where

$$\bar{c} = \frac{\min\{e^{(m+\mu)s_0}, e^{-(m+\mu)s_0}\} \min_{1/r_0 \leq \rho \leq (1+T_r)\varepsilon(r_0)} \hat{a}'(\rho)}{r_0} > 0$$

and using $\mathbf{d}$-homogeneity of W we conclude

$$\begin{aligned}W(x) =& W(\mathbf{d}(\ln \|x\|_{\mathbf{d}})\mathbf{d}(-\ln \|x\|_{\mathbf{d}})x) \\ =& \|x\|_{\mathbf{d}}^{m+\mu} W(\mathbf{d}(-\ln \|x\|_{\mathbf{d}})x) \geq \underline{c}\|x\|_{\mathbf{d}}^{m+\mu}.\end{aligned}$$

We have proven that

$$D^+V(x(t)) \leq -W(x_{x_0}(t)) \leq -\underline{c}\|x_{x_0}(t)\|_{\mathbf{d}}^{m+\mu}$$

as long as $x_{x_0}(t) \in K(r_0)$. If $x_{x_0}(t) \in \mathbf{d}(\tau)K(r_0)$, where $\tau \in \mathbb{R}$, then $\mathbf{d}(-\tau)x_{x_0}(t) \in K(r_0)$. Using homogeneity of V we derive

$$V(x_{x_0}(t)) = V(\mathbf{d}(-\tau)\mathbf{d}(\tau)x_{x_0}(t)) = e^{-\tau}V(\mathbf{d}(\tau)x_{x_0}(t)).$$

Since by Theorem 6.1 we have $x_{\mathbf{d}(\tau)x_0}(e^{-\mu\tau}t) = \mathbf{d}(\tau)x_{x_0}(t)$ then the function $t \mapsto V(x_{x_0}(t))$ is strictly decreasing as well and

$$\begin{aligned}
\limsup_{h\to 0^+} \tfrac{V(x_{x_0}(t+h))-V(x_{x_0}(t))}{h} &= D^+V(x_{x_0}(t)) \\
=&e^{-\tau s} \limsup_{h\to 0^+} \tfrac{V(x_{\mathbf{d}(\tau)x_0}(e^{-\mu\tau}(t+h)))-V(x_{\mathbf{d}(\tau)x_0}(e^{-\mu\tau}t))}{h} \\
\leq& - e^{-\tau(1+\mu)}W(x_{\mathbf{d}(\tau)x_0}(e^{-\mu\tau}t)) \\
=& - W(\mathbf{d}(-\tau)x_{\mathbf{d}(\tau)x_0}(e^{-\mu\tau}t)) = -W(x_{x_0}(t))
\end{aligned}$$

as long as $x_{x_0}(t) \in \mathbf{d}(\tau)K(r_0)$, where $\tau \in \mathbb{R}$ is any real number. Taking into account $\bigcup_{\tau\in\mathbb{R}} \mathbf{d}(\tau)K(r_0) = \mathbb{H}\backslash\{\mathbf{0}\}$ we conclude that the inequality

$$D^+V(x_{x_0}(t)) \leq -W(x_{x_0}(t)) \leq -\underline{c}\|x_{x_0}(t)\|_{\mathbf{d}}^{m+\mu} \leq -\rho V^{1+\frac{\mu}{m}}(x_{x_0}(t)) \tag{6.19}$$

holds for any solution of the system as long as $x_{x_0}(t) \neq \mathbf{0}$, where $\rho = \underline{c}/\overline{k}^{1+\mu/m}$.

III. Let us show that the inequality (6.17) holds. Let x be an arbitrary solution of the system with $x_0 \in \mathcal{D}(A)$. By Theorem 2.16, this solution is strong and $x(t) \overset{a.e.}{\in} \mathcal{D}(A)$. Since

$$\tfrac{V(x(t+h))-V(x(t))}{h} = \tfrac{V(x(t+h))-V(x(t)+h(Ax(t)+f(x(t))))}{h} + \\ \tfrac{V(x(t)+h(Ax(t)+f(x(t))))-V(x(t))}{h}$$

and V is locally Lipschitz continuous then, taking into account

$$\tfrac{|V(x(t+h))-V(x(t)+h(Ax(t)+f(x(t))))|}{h} \leq \\ \tfrac{L\|x(t+h)-x(t)-h(Ax(t)+f(x(t)))\|}{h} \to 0 \text{ as } h\to 0, \quad x(t)\in\mathcal{D}(A),$$

we derive

$$D^+V(x(t)) \overset{a.e.}{=} D^+V(x(t);\, Ax(t)+f(x(t))).$$

Since $x_0 \in \mathcal{D}(A)$ is arbitrary, then, taking $t = 0$ in the above identity and in (6.19), we come up with (6.17). The proof is complete. □

Remark 6.1 Since Gateaux derivative of a **d**-homogeneous functional is **d**-homogeneous too (see Proposition 5.2) then

$$D^+V(x; Ax + f(x)) = e^{-ms} D^+V(\mathbf{d}(s)x; \mathbf{d}(s)Ax + \mathbf{d}(s)f(x))$$

$$= e^{-ms} D^+V(\mathbf{d}(s)x; e^{-\mu s}(A\mathbf{d}(s)x + f(\mathbf{d}(s)x))), \quad \forall s \in \mathbb{R}, \quad \forall x \neq \mathbf{0},$$

where m is the homogeneity degree of V and μ is the homogeneity degree of the system $\dot{x} = Ax + f(x)$. Notice that $D^+V(x, \lambda d) = \lambda D^+V(x, d), \forall \lambda > 0$ and $d \in \mathbb{H}$. Then taking $s = -\frac{1}{m} \ln \|x\|_{\mathbf{d}}$ we derive

$$D^+V(x; Ax + f(x)) = \|x\|_{\mathbf{d}}^{1+\frac{\mu}{m}} D^+V(y, Ay + f(y)), \quad y = \mathbf{d}(-\ln \|x\|_{\mathbf{d}})x.$$

This means that the inequality (6.17) in Theorem 6.9 can be replaced with

$$\sup_{y \in \mathcal{D}(A) \cap S} D^+V(y; Ay + f(y)) \leq -\rho, \tag{6.20}$$

where $S = \{x \in \mathbb{H} : \|x\| = 1\}$. Therefore, the analysis of the asymptotic stability of the **d**-homogeneous system on $\mathbb{H}$ can always be reduced to the analysis of a Lyapunov function on a unit sphere.

Despite that the above theorem proves the existence of strict locally Lipschitz continuous Lyapunov $V : \mathbb{B} \mapsto \mathbb{R}_+$ for any uniformly stable homogeneous system, the design of such as function for a concrete system may be a very challenging task.

Example 6.6 Let us consider the system (6.10)–(6.12) from Example 6.4 on the space $\mathbb{H} = H^1(\mathbb{R}; \mathbb{R}) \times L^2(\mathbb{R}; \mathbb{R})$ with $\mathcal{D}(A) = H^3(\mathbb{R}; \mathbb{R}) \times H^1(\mathbb{R}; \mathbb{R})$. Let the Lyapunov function candidate $V : \mathbb{H} \mapsto \mathbb{H}$ be given by

$$V(x) = a\left(\lambda_2 \|y\|_{L^2} + \frac{\|w\|_{L^2}^2}{2}\right)^{\frac{3}{2}} - b\langle y, w\rangle_{L^2} + c\left\|\frac{\partial y}{\partial z}\right\|_{L^2}^2, \quad x = (y, w)^\top, \tag{6.21}$$

where $b > 0, c > 0$ satisfy

$$b > \frac{(2c + b)^2}{8\nu c}$$

and $a > 0$ is large enough to guarantee the positive definiteness and global properness of the functional V. Notice that the above inequality is feasible for $\nu > 1$. Using integration by part we conclude that the identities $\langle \frac{\partial^2 y}{\partial z^2}, y\rangle_{L^2} = -\langle \frac{\partial y}{\partial z}, \frac{\partial y}{\partial z}\rangle_{L^2}$, $\langle \frac{\partial^3 y}{\partial z^3}, \frac{\partial y}{\partial z}\rangle_{L^2} = -\langle \frac{\partial^2 y}{\partial z^2}, \frac{\partial^2 y}{\partial z^2}\rangle_{L^2}$ and $\langle \frac{\partial y}{\partial z}, \frac{\partial w}{\partial z}\rangle_{L^2} = -\langle \frac{\partial^2 y}{\partial z^2}, w\rangle_{L^2}$ hold for any $x = (y, w)^\top \in D(A)$. Hence, we derive

$$DV(x)(Ax+f(x)) = \tfrac{3a}{2}\sqrt{\lambda_2\|y\|_{L^2}+\tfrac{\|w\|_{L^2}^2}{2}}\left(-\nu\lambda_2\tfrac{\left\langle\frac{\partial y}{\partial z},\frac{\partial y}{\partial z}\right\rangle}{\|y\|_{L^2}}-\lambda_1\lambda_2\sqrt{\|y\|_{L^2}}\right)$$
$$-b\left\langle\tfrac{\partial^2 y}{\partial z^2},w\right\rangle+b\lambda_1\tfrac{\langle y,w\rangle_{L^2}}{\sqrt{\|y\|_{L^2}}}-b\|w\|_{L^2}^2+b\lambda_2\|y\|_{L^2}$$
$$-c\nu\left\|\tfrac{\partial^2 y}{\partial z^2}\right\|_{L^2}-\tfrac{2c\lambda_1}{\sqrt{\|y\|_{L^2}}}\left\|\tfrac{\partial y}{\partial z}\right\|_{L^2}-2c\left\langle\tfrac{\partial^2 y}{\partial z^2},w\right\rangle_{L^2}.$$

Since

$$2c\nu\left\|\frac{\partial^2 y}{\partial z^2}\right\|_{L^2}^2-(2c+b)\left\|\frac{\partial^2 y}{\partial z^2}\right\|_{L^2}\|w\|_{L^2}+\frac{(2c+b)^2}{8\nu c}\|w\|_{L^2}^2\geq 0$$

then, using the Cauchy–Schwarz inequality, we obtain

$$DV(x)(Ax+f(x))\leq-\left(\frac{3}{2}a\lambda_1\lambda_2^{\frac{3}{2}}-b\lambda_2\right)\|y\|_{L^2}$$
$$+\left(b\lambda_1-\frac{3}{2\sqrt{2}}a\lambda_1\lambda_2\right)\sqrt{\|y\|_{L^2}}\|w\|_{L^2}$$
$$-\left(b-\frac{(2c+b)^2}{8\nu c}\right)\|w\|^2-\tfrac{2c\lambda_1}{\sqrt{\|y\|_{L^2}}}\left\|\tfrac{\partial y}{\partial z}\right\|_{L^2}.$$

Therefore, taking $b=\frac{3}{2\sqrt{2}}a\lambda_2$, $c=b/2$ and

$$\lambda_1>\sqrt{\frac{\lambda_2}{2}},\quad \lambda_2>0,$$

we conclude that the inequality (6.20) holds, but, for a sufficiently large $a>0$, the functional V is positive definite and globally proper. The function V is locally Lipschitz continuous on $\mathbb{H}$ and $\mathbf{d}$-homogeneous of degree 3 with respect to the dilation $\mathbf{d}$ introduced in Example 6.4. Therefore, a strict $\mathbf{d}$-homogeneous Lyapunov function for the system (6.10)–(6.12) admits a rather simple representation (6.21) if this system is considered in the space $\mathbb{H}=H^1(\mathbb{R};\mathbb{R})\times L^2(\mathbb{R};\mathbb{R})$.

Notice that we cannot obtain a Lyapunov function for the system (6.10)–(6.12) in $\mathbb{B}=L^2(\mathbb{R};\mathbb{R})\times L^2(\mathbb{R};\mathbb{R})$ (see Example 6.4) just by taking $c=0$, since, in this case, there is no clear way to compensate the term $\left\langle\frac{\partial^2 y}{\partial z^2},w\right\rangle$ in the representation of $DV(x)(Ax+f(x))$.

Theorem 6.9 can be extended to homogeneous systems with unbounded nonlinearities. However, the functional V in this case is defined and Lipschitz continuous only on a subspace $\mathbb{Y}$ of $\mathbb{H}$.

Let us consider the system (6.3) with (locally) M-regular operator (see Definition 2.9). This system may not have mild solutions for initial conditions in $\mathbb{H}$, but, according to Theorem 2.15, it may have mild solutions for initial conditions in $\mathcal{D}(M)$. The stability of such a system should be studied in the corresponding subspace of $\mathbb{H}$ (see

Definition 3.11). Let us consider the Banach space $\mathbb{Y} = \mathcal{D}(M)$ with the graph norm

$$\|x\|_{\mathbb{Y}} = \|x\| + \|Mx\|, \quad x \in \mathbb{Y}.$$

Since the operator M is closed then $\mathbb{Y}$ is a real Banach space. The Zubov–Rosier theorem for homogeneous systems stable in $\mathbb{Y}$ is given by the following corollary.

Corollary 6.5 *Let a linear closed densely defined operator $A : \mathcal{D}(A) \subset \mathbb{H} \mapsto \mathbb{H}$ be a generator of a strongly continuous semigroup Φ of linear bounded operators on $\mathbb{H}$ and an operator $f : \mathcal{D}(f) \subset \mathbb{H} \mapsto \mathbb{H}$ be M-regular with $M : \mathbb{Y} \mapsto \mathbb{H}$ satisfying all conditions of Theorem 2.15 and*

$$\sup_{t\in[t_1,t_2]} \|M\Phi(t)\| < +\infty, \quad 0 < t_1 < t_2 < +\infty. \tag{6.22}$$

Let **d** *be a strictly monotone strongly continuous dilation in a Hilbert space $\mathbb{H}$ and in the Banach space $\mathbb{Y}$, the evolution equation* (6.3) *be* **d**-*homogeneous of degree $\mu \in \mathbb{R}$ and $M^{-1}\mathcal{D}(A)$ be* **d**-*homogeneous cone. Let $m > 0$ be an arbitrary real number. The system* (6.3) *is uniformly asymptotically stable in $\mathbb{Y}$ if and only if there exists a continuous positive definite functional $V : \mathbb{Y} \mapsto \mathbb{R}$ such that*

(1) V is **d**-*homogeneous of the degree m and locally Lipschitz continuous on $\mathbb{Y}\backslash\{\mathbf{0}\}$;*
(2) there exist $\underline{k}, \overline{k} > 0$ satisfying

$$\underline{k}\|x\|^m_{\mathbf{d},\mathbb{Y}} \le V(x) \le \overline{k}\|x\|^m_{\mathbf{d},\mathbb{Y}}, \quad \forall x \in \mathbb{Y}, \tag{6.23}$$

where $\|\cdot\|_{\mathbf{d},\mathbb{Y}}$ is the canonical homogeneous norm induced by $\|\cdot\|_{\mathbb{Y}}$;
(3) there exists $\rho > 0$ such that

$$D^+V(x; Ax + f(x)) \le -\rho V^{1+\frac{\mu}{m}}(x), \quad \forall x \in M^{-1}\mathcal{D}(A). \tag{6.24}$$

Proof[3] (1) Let $\delta > 1$ be an arbitrary real number and

$$f_\delta(t, x) = a_\delta(\|x\|_{\mathbb{Y}}) f(t, x), \quad x \in \mathbb{Y},$$

where $a_\delta \in C^\infty_c(\mathbb{R}_+; \mathbb{R})$ such that $a_\delta(s) = 0$ for $s \notin (1/(2\delta), 2\delta)$ and $a_\delta(s) = 1$ for $s \in [1/\delta, \delta]$. Let us shown that for any $x_0 \in \mathbb{Y}$ the initial value problem

$$\dot{x} = Ax + f_\delta(x), \quad t \ge 0 \quad x(0) = x_0 \in \mathbb{Y} \tag{6.25}$$

[3] The proof is reprinted from *Systems & Control Letters*, 148/2, A. Polyakov, Homogeneous Lyapunov functions for homogeneous infinite-dimensional systems with unbounded nonlinear operators, 104854, https://doi.org/10.1016/j.sysconle.2020.104854. ©2020 Elsevier, with permission from Elsevier.

has a unique mild solution $x^{\delta} : [0, +\infty) \mapsto \mathbb{Y}$, which is continuous in $\mathbb{Y}$ and coincides with the mild solution x_{x_0} of (7.22) as long as $x_{x_0}(t) \in K_{\mathbb{Y}}(\delta)$, where

$$K_{\mathbb{Y}}(\delta) = \{x \in \mathbb{Y} : 1/\delta \leq \|x\|_{\mathbb{Y}} \leq \delta\}. \tag{6.26}$$

Let us show that for any $x_0 \in \mathbb{Y}\backslash\{\mathbf{0}\}$ the system (6.3) has a mild solution x_{x_0}, which is uniquely defined as long as $0 < \|x_{x_0}(t)\|_{\mathbb{Y}} < +\infty$.

Indeed, for $x_1 \in \mathbb{B}$, $x_2 \in \mathbb{B}$ we have $M^{-1}x_1, M^{-1}x_2 \in \mathbb{Y}$ and

$$\|f_\delta(M^{-1}x_1) - f_\delta(M^{-1}x_2)\| =$$

$$\|a_\delta(\|M^{-1}x_1\|_{\mathbb{Y}})f(M^{-1}x_1) - a_\delta(\|M^{-1}x_2\|_{\mathbb{Y}})f(M^{-1}x_2)\| \leq$$

$$\|f(M^{-1}x_1)\|_{\mathbb{X}}|a_\delta(\|M^{-1}x_1\|_{\mathbb{Y}}) - a_\delta(\|M^{-1}x_2\|_{\mathbb{Y}})| +$$

$$|a_\delta(\|M^{-1}x_2\|_{\mathbb{Y}})|\|f(M^{-1}x_1) - f(M^{-1}x_2)\| =$$

$$\|f(M^{-1}x_1)\| \cdot |a'_\delta(\theta)| \cdot \left|\|M^{-1}x_1\|_{\mathbb{Y}} - \|M^{-1}x_2\|_{\mathbb{Y}}\right| +$$

$$|a_\delta(\|M^{-1}x_2\|_{\mathbb{Y}})| \cdot \|f(M^{-1}x_1) - f(M^{-1}x_2)\|$$

where the Mean Value theorem is utilized for

$$\theta \in [\min \|M^{-1}x_i\|_{\mathbb{Y}}, \max \|M^{-1}x_i\|_{\mathbb{Y}}], \quad i = 1, 2.$$

Notice that $M^{-1}x_i \in K_{\mathbb{Y}}(2\delta)$ implies $x_i \in K(2\xi\delta)$, where $\xi = 1 + \|M^{-1}\| < +\infty$ and

$$K(r) = \{x \in \mathbb{B} : 1/r \leq \|x\| \leq r\}, \quad r > 1. \tag{6.27}$$

Local Lipschitz continuity of $f \circ M^{-1}$ yields

$$\sup_{x \in K_{\mathbb{X}}(2\xi\delta)} \|f(M^{-1}x)\| < +\infty,$$

so $\exists \tilde{L}_\delta > 0$ such that

$$\|f(M^{-1}x_1) - f(M^{-1}x_2)\| \leq \tilde{L}_\delta\|x_1 - x_2\| \quad \forall x_1, x_2 \in K(2\xi\delta).$$

Since $a_\delta \in C_c^\infty$ then $\sup_{s\in\mathbb{R}} |a'_\delta(s)| < +\infty$. Notice also

$$M^{-1}x_2 \notin K_{\mathbb{Y}}(2\delta) \quad \Rightarrow \quad a_\delta(\|M^{-1}x_2\|_{\mathbb{Y}}) = 0.$$

Therefore, taking into account

$$\left|\|M^{-1}x_1\|_{\mathbb{Y}} - \|M^{-1}x_2\|_{\mathbb{Y}}\right| \leq \|M^{-1}(x_1 - x_2)\|_{\mathbb{Y}} \leq (\|M^{-1}\| + 1)\|x_1 - x_2\|,$$

we conclude that $f_\delta \circ M^{-1}$ satisfy the Lipschitz condition on $\mathbb{B}$ with the Lipschitz constant

$$L_\delta := \tilde{L}_\delta + (\|M^{-1}\| + 1) \sup_{\theta \in \mathbb{R}} |a'_\delta(\theta)| \sup_{x \in K(2\xi\delta)} \|f(M^{-1}x)\| < +\infty.$$

Given $x_0 \in \mathbb{Y}$ let us consider the metric space

$$\mathbf{Z} = \left\{ z \in C([0, T], \mathbb{B}) : z(0) = Mx_0, \sup_{t \in [0,T]} \|z(t)\| < +\infty \right\},$$

with $T > 0$ and the metric $\rho(z_1, z_2) = \sup_{t \in [0,T]} \|z_1(t) - z_2(t)\|$. Let the operator F be defined on Z as follows:

$$(Fz)(t) = \Phi(t)Mx_0 + \int_0^t M\Phi(t-\tau) f_\delta(M^{-1}z(\tau))d\tau.$$

The function $\tau \mapsto f_\delta(\tau, M^{-1}z(\tau))$ is Bochner integrable in $\mathbb{H}$ if and only if $\tau \mapsto \|f_\delta(\tau, M^{-1}z(\tau))\|$ is Lebesgue integrable. Since $z \in C([0, T]; \mathbb{H})$ then M-regularity of f implies $f(M^{-1}x(\cdot)) \in L^1((0, T); \mathbb{B})$ and $F : \mathbf{Z} \mapsto \mathbf{Z}$. Indeed, one has

$$\begin{aligned}\|(Fz)(t)\| \le& \|\Phi(t)\| \cdot \|Mx_0\| + \int_0^t \|M\Phi(t-\tau)\| \cdot \|f_\delta(\tau, M^{-1}z(\tau)) - f_\delta(\tau, \mathbf{0})\| \, d\tau \\ \le& k\|Mx_0\| + L_\delta \int_0^t \|M\Phi(t-\tau)\| \cdot \|z(\tau)\| \, d\tau,\end{aligned}$$

where $k = \sup_{t \in [0,T]} \|\Phi(t)\| < +\infty$. Taking into account integrability of $\|M\Phi(\cdot)\|$ on $[0, T]$, we derive

$$\sup_{t \in [0,T]} \|(Fz)(t)\| < +\infty$$

for any $z \in C([0, T]; \mathbb{B})$. Let us show that the operator $F : \mathbf{Z} \mapsto \mathbf{Z}$ has the unique fixed point $F(z^*) = z^*$ with $z^* \in \mathbf{Z}$. For any $z_1, z_2 \in \mathbf{Z}$ we have

$$\begin{gathered}\|F(z_1)(t) - F(z_2)(t)\| \le \\ \int_0^t \|M\Phi(t-\tau)\| \cdot \|f_\delta(M^{-1}z_1(\tau), q(\tau)) - f_\delta(M^{-1}z_2(\tau), q(\tau))\| d\tau \le \\ L_\delta \int_0^t \|M\Phi(\sigma)\| d\sigma \sup_{t \in [0,T]} \|z_1(\tau)) - z_2(\tau))\| \le L_\delta \int_0^T \|M\Phi(\sigma)\| d\sigma \cdot \rho(z_1, z_2).\end{gathered}$$

Since, by assumption, $\int_0^T \|M\Phi(\sigma)\| d\sigma \to 0$ as $T \to 0$ then for a sufficiently small $T > 0$ we have $\eta = L_\delta \int_0^T \omega(\sigma) d\sigma < 1$ and

$$\|F^k(z_1)(t) - F^k(z_2)(t)\| \le \eta^k \rho(x_1, x_2), \quad k = 1, 2, \ldots .$$

Using the contraction principle (Banach Fixed-Point theorem, see Appendix B) we deduce the existence of a unique fixed point z^* of the operator F in $\mathbf{Z}$, which corresponds to a mild solution $x^\delta := M^{-1}z^*$ of the nonlinear system (6.25). Indeed, using the identity $M^{-1}\Phi(t)Mx_0 = \Phi(t)x_0, \forall x_0 \in \mathcal{D}(M)$ we derive

$$\begin{aligned} x^\delta(t) =& M^{-1}z^*(t) \\ =& M^{-1}\Phi(t)Mx_0 + M^{-1}\int_0^t M\Phi(t-\tau)f_\delta(\tau, M^{-1}z^*(\tau))d\tau \\ =& \Phi(t)x_0 + \int_0^t \Phi(t-\tau)f_\delta(\tau, x^\delta(\tau))d\tau. \end{aligned}$$

Notice also $x^\delta(t) = M^{-1}z^*(t) \in \mathbb{Y}$ for all $t \in [0, T]$.

The solution x^δ can be prolonged for all $t \geq 0$. Suppose the converse, i.e., exists $t_1 \in \mathbb{R}_+$ such that $\|x^\delta(t)\|_{\mathbb{Y}} \to +\infty$ as $t \to t_1$. In this case, $\|M^{-1}z^*(t)\|_{\mathbb{Y}} \to +\infty$ as $t \to t_1$ and there exists $t' \in (0, t_1)$ such that $\|M^{-1}z^*(t)\|_{\mathbb{Y}} \geq 2\delta$ for all $t \in (t', t_1)$. The latter implies that $f_\delta(\mathbf{0}, M^{-1}z^*(t)) = \mathbf{0}$ for all $t \in (t', t_1)$,

$$z^*(t) = \Phi(t-t')Mx^\delta(t') \quad \text{and} \quad x^\delta(t) = \Phi(t-t')x^\delta(t'), \quad \forall t \geq [t', t_1).$$

Since Φ is a strongly continuous semigroup of linear bounded operators on $\mathbb{B}$ then the inequality $t_1 - t' < +\infty$ implies $\sup_{s\in[0,t_1-t']}\|\Phi(s)\| < +\infty$. Therefore, we derive

$$\begin{aligned} \|x^\delta(t)\|_{\mathbb{Y}} =& \|\Phi(t-t')x^\delta(t')\| + \|M\Phi(t-t')x^\delta(t')\| \\ =& \|\Phi(t-t')x^\delta(t')\| + \|\Phi(t-t')Mx^\delta(t')\| \\ \leq& \|\Phi(t-t')\|\|x^\delta(t')\|_{\mathbb{Y}} \end{aligned}$$

and $\limsup_{t\to t_1}\|x^\delta(t)\|_{\mathbb{Y}} < +\infty$.

If $x_0 \in K_{\mathbb{Y}}(\delta)$ then the mild solution x^δ is the unique mild solution of (7.22) as long as $x^\delta(t) = M^{-1}z^*(t) \in K_{\mathbb{Y}}(\delta)$. Since $\|x^\delta(t+h) - x(t)\|_{\mathbb{Y}} = \|M^{-1}(z^*(t+h) - z^*(t))\| + \|z^*(t+h) - z^*(t)\|, t \geq t_0, h > 0$ then the continuity of z in $\mathbb{B}$ implies the continuity of x^δ in $\mathbb{Y}$.

(2) Let us show for any $T > 0$ there exists $L_{\delta,T} > 0$ such that for any $x_1, x_2 \in \mathbb{Y}$ one has

$$\|x^\delta_{x_1}(t) - x^\delta_{x_2}(t)\|_{\mathbb{Y}} \leq L_{\delta,T}\|x_1 - x_2\|_{\mathbb{Y}}, \quad \forall t \in [0, T]$$

where $x^\delta_{x_i}(t)$ is a mild solution of (6.25) with $x(t_0) = x_i, i = 1, 2$.

Indeed, since

$$\|x^\delta_{x_1}(t) - x^\delta_{x_2}(t)\| \leq \|\Phi(t)\| \cdot \|x_1 - x_2\| +$$

$$\int_0^t \|\Phi(t-\tau)\| \cdot \|f_\delta(x^\delta_{x_1}(\tau)) - f_\delta(x^\delta_{x_2}(\tau))\|d\tau$$

then taking into account Lipschitz continuity of $f_\delta \circ M^{-1}$ we derive

$$\|x_{x_1}^\delta(t) - x_{x_2}^\delta(t)\| \le k\|x_1 - x_2\| + kL_\delta \int_0^t \|Mx_{x_1}^\delta(\tau) - Mx_{x_2}^\delta(\tau)\|d\tau,$$

where $k := \sup_{s\in[0,T]} \|\Phi(s)\| < +\infty$. On the other hand, we have

$$Mx_{x_i}^\delta(t) = z_{x_i}^*(t) = \Phi(t-t_0)Mx_0 + \int_0^t M\Phi(t-\tau)f_\delta(M^{-1}z_{x_i}^*(\tau))d\tau,$$

where $i = 1, 2$ and $z_{x_i}^*$ is defined above. Hence, we derive

$$\begin{aligned}\|M(x_{x_1}^\delta(t) - x_{x_2}^\delta(t))\| \le & h\|Mx_1 - Mx_2\| + \\ & \int_0^t \|M\Phi(t-\tau)\| \cdot \|f_\delta(M^{-1}z_{x_1}^*(\tau)) - f_\delta(M^{-1}z_{x_2}^*(\tau))\|d\tau \\ \le & k\|Mx_1 - Mx_2\| + \\ & L_\delta \int_0^t \|M\Phi(t-\tau)\| \cdot \|M^{-1}z_{x_1}^*(\tau) - M^{-1}z_{x_2}^*(\tau)\|d\tau \\ \le & k\|Mx_1 - Mx_2\| + \\ & L_\delta \int_0^t \|M\Phi(t-\tau)\| \cdot \|x_{x_1}^\delta(\tau) - x_{x_2}^\delta(\tau)\|d\tau,\end{aligned}$$

where $L_\delta \ge 0$ is defined above. Combining the two obtained estimates we derive

$$\|x_{x_1}^\delta(t) - x_{x_2}^\delta(t)\|_{\mathbb{Y}} \le k\|x_1 - x_2\|_{\mathbb{Y}} + L_\delta \int_0^t (k + \|M\Phi(t-\tau)\|)\|x_{x_1}^\delta(\tau) - x_{x_2}^\delta(\tau)\|_{\mathbb{Y}} d\tau.$$

Hence, using the Grönwall–Bellman inequality we come up with the estimate

$$\|x_{x_1}^\delta(t) - x_{x_2}^\delta(t)\|_{\mathbb{Y}} \le ke^{L_\delta \int_0^t (k+\|M\Phi(t-\tau)\|)d\tau}\|x_1 - x_2\|_{\mathbb{Y}} \le L_{\delta,T}\|x_1 - x_2\|_{\mathbb{Y}}$$

for all $t \in [0, T]$, where $L_{\delta,T} = \sup_{t\in[0,T]} ke^{L_\delta \int_0^t (k+\|M\Phi(t-\tau)\|)d\tau} < +\infty$.

(3) Let us show that for $x_0 \in M^{-1}\mathcal{D}(A)$ the mild solution $x_{x_0}^\delta$ of (6.25) is

(a) locally Lipschitz continuous as a function $[0, T) \mapsto \mathbb{H}$;
(b) locally uniformly continuous as a function $[0, T) \mapsto \mathbb{Y}$;
(c) a strong solution of (6.25), $x(t) \overset{a.e.}{\in} M^{-1}\mathcal{D}(A)$ and

$$M\dot{x}^\delta(t) \overset{a.e.}{=} MAx^\delta(t) + Mf(x^\delta(t)), \quad t \in (0, T),$$

where $T \in \mathbb{R}_+$ or $T = +\infty$ defines the maximal interval of existence of the solution $x_{x_0}^\delta$.

First of all notice that $x_0 \in M^{-1}\mathcal{D}(A)$ implies $x_0 \in \mathcal{D}(A)$. Indeed, $x_0 \in M^{-1}\mathcal{D}(A)$ implies $Mx_0 \in \mathcal{D}(A)$ and $x_0 \in \mathbb{Y}$ (since $M^{-1} : \mathbb{H} \mapsto \mathbb{Y}$). In this case, we have

$$\lim_{h\to 0^+} \frac{\Phi(h)Mx_0 - Mx_0}{h} = AMx_0.$$

On the other hand,the identity $M^{-1}\Phi(t)Mx = \Phi(t)x, \forall t \geq 0, \forall x \in \mathcal{D}(M)$ implies that

$$\lim_{h\to 0^+} \frac{\Phi(h)Mx_0 - Mx_0}{h} = \lim_{h\to 0^+} M\frac{\Phi(h)x_0 - x_0}{h}.$$

Since M by assumption is a closed linear operator and $\mathbb{Y}$ is its domain then the existence of the above limits yields $\frac{\Phi(h)x_0 - x_0}{h} \to y^* \in \mathbb{Y}$ as $h \to 0$. Since A is a closed linear operator as well then the existence of the limit $\lim_{h\to 0^+} \frac{\Phi(h)x_0 - x_0}{h} \in \mathbb{H}$ guarantees $x_0 \in \mathcal{D}(A)$ and

$$MAM^{-1}z = Az, \quad \forall z \in \mathcal{D}(A).$$

(a) Recall (see Chap. 2) that if $z \in \mathcal{D}(A)$ then $A\Phi(s)z = \Phi(s)Az$ and

$$\Phi(s+h)z - \Phi(s)z = A\int_s^{s+h} \Phi(\tau)zd\tau = \int_s^{s+h} \Phi(\tau)Azd\tau$$

for all $s \geq 0$ and all $h > 0$. Let us denote

$$C(T) := \operatorname{ess}\sup_{s\in[0,T]} \|Mx_{x_0}^{\delta}(s)\| < +\infty, \quad \text{if} \quad x_0 \in \mathbb{Y}.$$

Since for $x_0 \in M^{-1}\mathcal{D}(A)$ we have $x_0 \in \mathcal{D}(A)$ and

$$\|f_\delta(x_{x_0}^{\delta}(\tau))\| = \|f_\delta(M^{-1}Mx_{x_0}^{\delta}(\tau)) - f_\delta(0)\| \leq L_\delta\|Mx_{x_0}^{\delta}(\tau)\|$$

then

$$\|x_{x_0}^{\delta}(t+h) - x_{x_0}^{\delta}(t)\| \leq \|\Phi(t+h)x_0 - \Phi(t)x_0\| + \int_t^{t+h} \|\Phi(t-\tau)f_\delta(x_{x_0}^{\delta}(\tau))\|d\tau \leq$$

$$\int_t^{t+h} \|\Phi(\tau)\| \cdot \|Ax_0\|d\tau + \int_t^{t+h} \|\Phi(s)\| \cdot \|f_\delta(x_{x_0}^{\delta}(\tau))\|\, d\tau \leq hk(\|Ax_0\| + L_\delta C(T))$$

for all $h \in [0, T]$.

(b) Now let us prove the uniform continuity of the solution in $\mathbb{Y}$. We have

$$\|Mx^{\delta}_{x_0}(t+h) - Mx^{\delta}_{x_0}(t)\| \leq \|\Phi(t+h)Mx_0 - \Phi(t)Mx_0\| +$$

$$\int_t^{t+h} \omega(t-\tau)\|f_\delta(x^{\delta}_{x_0}(\tau))\|\,d\tau \leq$$

$$\int_t^{t+h} \|\Phi(\tau)\|_{\mathbb{X}} \cdot \|AMx_0\|_{\mathbb{X}} d\tau + L_\delta \int_t^{t+h} \|M\Phi(t-\tau)\|\|Mx^{\delta}_{x_0}(\tau)\| \leq$$

$$hk\|AMx_0\|_{\mathbb{X}} + L_\delta C(T)\int_0^h \|M\Phi(s)\|ds$$

for all $h \in [0, T]$. Combining the above inequality with the estimate obtained in the case a), we derive

$$\|x^{\delta}_{x_0}(t+h) - x^{\delta}_{x_0}(t)\|_{\mathbb{Y}} \leq h(k\|Ax_0\|_{\mathbb{Y}} + L_\delta C(T)k) + L_\delta C(T)\int_0^h \|M\Phi(s)\|ds.$$

Taking into account $\int_0^h \|M\Phi(s)\|ds \to 0$ as $h \to 0$, we complete the proof.

(c) Let $x^{\delta}_{x_0}$ be a mild solution of (6.25) with $x_0 \in M^{-1}\mathcal{D}(A)$. Recall that $x^{\delta}_{x_0}(t) = Mz_{x_0}(t)$, where $t \geq t_0$ and

$$z_{x_0}(t) = \Phi(t)Mx_0 + \int_0^t M\Phi(t-s)f_\delta(M^{-1}z(s))ds.$$

Let us show that $z_{x_0} : [p, T] \mapsto \mathbb{H}$ satisfies a Lipschitz condition for each $p \in (0, T)$. Indeed, since

$$z_{x_0}(t+h) - z_{x_0}(t) = (\Phi(t+h) - \Phi(t))Mx_0 + \int_0^{t+h} M\Phi(t+h-s)f_\delta(M^{-1}z_{x_0}(s))ds -$$

$$\int_{t_0}^{t} M\Phi(t-s)f_\delta(M^{-1}z_{x_0}(s))ds = \int_t^{t+h} \Phi(s)AMx_0 +$$

$$\int_0^h M\Phi(t+h-s)f_\delta(M^{-1}z_{x_0}(s))ds +$$

$$\int_0^t M\Phi(t-s)(f_\delta(M^{-1}z_{x_0}(s+h)) - f_\delta(M^{-1}z_{x_0}(s)))ds.$$

Notice that

$$\left\| \int_0^{t+h} M\Phi(t+h-s) f_\delta(M^{-1} z_{x_0}(s)) ds \right\| \leq$$

$$\int_0^h \|M\Phi(t+\sigma)\| \cdot \|f_\delta(M^{-1} z_{x_0}(h-\sigma))\| d\sigma \leq hC(T)\bar{\omega}(T,p),$$

where $\bar{\omega}(T,p) := \sup_{t\in[p,T]} \|M\Phi(t)\| < +\infty$ due to the condition (6.22). Hence, using Lipschitz continuity of f_δ we derive

$$\|z_{x_0}(t+h) - z_{x_0}(t)\| \leq (k\|AMx_0\| + C(T)\bar{\omega}(T,p))\, h +$$

$$L_\delta \int_0^t \|M\Phi(t-s)\| \|z_{x_0}(s+h) - z_{x_0}(s)\| ds.$$

Applying the Grönwall–Bellman inequality we conclude

$$\|z_{x_0}(t+h) - z_{x_0}(t)\| \leq (k\|AMx_0\| + C(T)\bar{\omega}(T,p))\, e^{L_\delta \int_0^t \|M\Phi(t-s)\| ds} h,$$

i.e., $z_{x_0} : [p,T] \mapsto \mathbb{H}$ satisfies Lipschitz condition for each $p \in (0,T)$. Since $z \mapsto f_\delta(M^{-1}z)$ is Lipschitz continuous and $f_\delta(x^\delta_{x_0}(t)) = f_\delta(M^{-1}z_{x_0}(t)), t \in [0,T]$ then the function $t \mapsto f(x^\delta_{x_0}(t))$ is locally Lipschitz continuous. Repeating the proof of Theorem 2.16 we conclude that $x^\delta_{x_0}$ is a strong solution, i.e., $\dot{x}^\delta_{x_0}(t) \overset{a.e.}{=} Ax^\delta_{x_0}(t) + f_\delta(x^\delta_{x_0}(t)), t > 0$ provided that $x_0 \in M^{-1}\mathcal{D}(A)$. Since the Hilbert space $\mathbb{H}$ is a reflexive Banach space then it satisfies the Radon–Nikodym property, i.e., locally Lipschitz continuity of z_{x_0} implies its differentiability almost everywhere. Since M is a closed linear operator then, taking into account $Mx^\delta_{x_0}(t) = z_{x_0}(t)$, we derive $\dot{z}_{x_0}(t) \overset{a.e.}{=} M\dot{x}^\delta_{x_0}(t) \overset{a.e.}{=} MAx^\delta_{x_0}(t) + Mf(x^\delta_{x_0}(t)), t \geq 0$ and

$$\dot{z}_{x_0}(t) \overset{a.e.}{=} MAM^{-1} z_{x_0}(t) + Mf(M^{-1} z_{x_0}(t)), \quad t \geq 0.$$

Since

$$\tilde{\Phi}(t)v := M\Phi(t)M^{-1}v = \Phi(t)v, \quad \forall t \geq 0, \forall v \in \mathbb{H},$$

then $\tilde{\Phi}$ is a strongly continuous semigroup of linear bounded operators on $\mathbb{H}$. The operator $MAM^{-1} : \mathcal{D}(A) \subset \mathbb{X} \mapsto \mathbb{X}$ is an infinitesimal generator of $\tilde{\Phi}$. Since $\tilde{\Phi} = \Phi$ then $MAM^{-1} = A$ due to the uniqueness of an infinitesimal generator. Using Theorem 2.9 we derive $z(t) \overset{a.e.}{\in} \mathcal{D}(A)$ and, consequently, $x^\delta(t) = M^{-1}z(t) \overset{a.e.}{\in} M^{-1}\mathcal{D}(A)$.

(4) Let us show that if the origin of (7.22) is globally uniformly Lyapunov stable in $\mathbb{Y}$, then, for any $r > 1$ and for

$$\delta > \max\left\{\varepsilon(r), \frac{1}{\varepsilon^{-1}(1/r)}\right\}, \tag{6.28}$$

where $\varepsilon \in \mathcal{K}^{\infty}$ is given by (3.30) and ε^{-1} is the inverse function to ε, one holds

(i) for any $x_0 \in \mathbb{Y} : 1/r \leq \|x_0\|_{\mathbb{Y}} \leq r$ we have

$$\|x^{\delta}_{x_0}(t)\|_{\mathbb{Y}} \leq \varepsilon(\|x_0\|_{\mathbb{Y}}) \leq \delta, \quad \forall t \geq 0$$

(ii) if $\exists t_1 \geq 0$ such that $\|x^{\delta}(t_1)\|_{\mathbb{Y}} \leq 1/\delta$ then

$$\|x^{\delta}_{x_0}(t)\|_{\mathbb{Y}} \leq 1/r, \quad \forall t \geq t_1,$$

where $x^{\delta}_{x_0}$ is the unique mild solution of (6.25).

The proof of (i). Notice that the inequality (3.30) and $x_{x_0}(0) = x_0$ yield $\varepsilon(r) \geq r$ and $1/\delta < 1/r < r < \delta$. Since $x^{\delta}_{x_0}$ coincide with the solution x_{x_0} of (7.22) as long as $x^{\delta}_{x_0}(t) \in K_{\mathbb{Y}}(\delta)$ then the inequality (3.30) implies $\|x^{\delta}_{x_0}(t)\|_{\mathbb{Y}} \leq \varepsilon(\|x_0\|_{\mathbb{Y}}) \leq \delta$ for all $t \geq 0$ and any $x_0 : \|x_0\|_{\mathbb{Y}} \leq [1/r, r]$.

The poof of (ii) Suppose the contrary, i.e., $\exists t_2 > t_1$ such that $\|x^{\delta}_{x_0}(t_2)\| > 1/r$ and $\|x^{\delta}_{x_0}(t)\| > 1/\delta$ for $t \in (t_1, t_2]$. In this case, we have $x^{\delta}_{x_0}(t) = x_{x_0}(t)$ for all $t \in (t_1, t_2]$ and using the inequality (3.30) we conclude

$$\|x^{\delta}_{x_0}(t)\|_{\mathbb{Y}} \leq \varepsilon(\|x^{\delta}_{x_0}(t_1)\|_{\mathbb{Y}}) = \varepsilon(1/\delta) \leq \varepsilon(\varepsilon^{-1}(1/r)) = 1/r$$

for all $t \geq t_1$.

(5) Tending δ to 0 we conclude that for any $x_0 \in M^{-1}\mathcal{D}(A)\backslash\{\mathbf{0}\}$ we have a unique mild solution $x_{x_0} : [0, T) \mapsto M^{-1}\mathcal{D}(A)$ of the homogeneous system (6.3) satisfying

$$\dot{x}_{x_0}(t) \overset{a.e.}{=} Ax_{x_0}(t) + f(x(t)) \;\text{ and }\; M\dot{x}_{x_0}(t) \overset{a.e.}{=} MAx_{x_0}(t) + Mf(x_{x_0}(t)),\; t \geq 0$$

as long as $0 < \|x_{x_0}(t)\|_{\mathbb{Y}} < +\infty$.

Sufficiency. Since $\frac{V(x_{x_0}(t+h)) - V(x_{x_0}(t))}{h} =$

$$\frac{V(x_{x_0}(t+h)) - V(x_{x_0}(t) + h(Ax_{x_0}(t) + f(x_{x_0}(t))))}{h} + \frac{V(x_{x_0}(t) + h(Ax_{x_0}(t) + f(x_{x_0}(t)))) - V(x_{x_0}(t))}{h}$$

and V is locally Lipschitz continuous in $\mathbb{Y}$ then, taking into account

$$\frac{|V\left(x_{x_0}(t+h)\right)-V(x_{x_0}(t)+h(Ax_{x_0}(t)+f(x_{x_0}(t))))|}{h}\leq$$

$$\frac{L\|x_{x_0}(t+h)-x_{x_0}(t)-h(Ax_{x_0}(t)+f(x_{x_0}(t)))\|_{\mathbb{Y}}}{h}=$$

$$L\left\|\frac{x_{x_0}(t+h)-x_{x_0}(t)}{h}-\dot{x}_{x_0}(t)\right\|_{\mathbb{Y}}\to 0 \text{ as } h\to 0 \text{ for almost all } t\geq 0,$$

we derive

$$\overline{D}^+V(x_{x_0}(t))\overset{a.e.}{=}\overline{D}^+V(x_{x_0}(t);\,Ax_{x_0}(t)+f(x_{x_0}(t)),\ \forall x_0\in M^{-1}\mathcal{D}(A)\backslash\{\mathbf{0}\}.$$

Finally, since the function $t\mapsto V(x_{x_0}(t))$ is locally Lipschitz continuous and $x_{x_0}(t)\overset{a.e.}{\in}M^{-1}\mathcal{D}(A)$ then (see Chap. 4 of Volume I) taking into account the inequality (6.24) we derive

$$\overline{D}^+V(x_{x_0}(t))\leq-\rho\|x_{x_0}(t)\|_{\mathbf{d},\mathbb{Y}}^{1+\mu}\tag{6.29}$$

as long as $0<\|x_{x_0}(t)\leq+\infty$.

Notice also that the existence of a **d**-homogeneous Lyapunov functional $V:\mathbb{Y}\mapsto\mathbb{R}$ of the degree $m>0$ satisfying (6.23), (6.24) is equivalent to the existence of a **d**-homogeneous Lyapunov functional $\tilde{V}:\mathbb{Y}\mapsto\mathbb{R}$ of the degree 1 satisfying

$$\underline{c}\|x\|_{\mathbf{d},\mathbb{Y}}\leq\tilde{V}(x)\leq\overline{c}\|x\|_{\mathbf{d},\mathbb{Y}},\quad\forall x\in\mathbb{Y},\quad\underline{c},\overline{c}>0\tag{6.30}$$

and

$$\overline{D}^+\tilde{V}(x_{x_0}(t))\leq-\|x_{x_0}(t)\|_{\mathbf{d},\mathbb{Y}}^{\mu+1},\quad t>0,\tag{6.31}$$

respectively. Indeed, selecting $\tilde{V}=m\underline{k}^{1-1/m}V^{1/m}$ for $m\in(0,1)$ (resp. $\tilde{V}=m\overline{k}^{1-1/m}V^{1/m}$ for $m>1$) we easily derive the mentioned inequalities with $\underline{c}=m\underline{k}$ and $\overline{c}=m\overline{k}^{1/m}\underline{k}^{1-1/m}$ (resp. $\underline{c}=m\overline{k}^{1-1/m}\underline{k}^{1/m}$ and $\overline{c}=m\overline{k}$). This means that without loss of generality we can fix $m=1$ for the proof.

If $\mu<0$ then Corollary 3.5 implies the uniform finite-time stability in $\mathbb{Y}$. For $\mu>0$ the same conclusion is guaranteed by Theorem 3.5. The case $\mu=0$ is treated by Theorem 3.2.

Necessity. We design the Lyapunov function in three steps: first, we construct a function $V_r:\mathbb{Y}\mapsto\mathbb{R}_+$ such that the function $t\mapsto V_r(x_0(t))$ is strictly decreasing as long as $x_{x_0}(t)\in K_{\mathbb{Y}}(r)$ for some $r>1$; next, using V_r we generate a **d**-homogeneous Lyapunov function $V:\mathbb{H}\mapsto\mathbb{R}_+$ by means of an integral transformation inspired by [14]. The obtained Lyapunov function satisfies the inequality (6.14) along trajectories of the system. Finally, using Radon–Nikodym property, which holds, in particular, for Hilbert spaces, we prove that the inequality (6.24) takes the place.

I. Let $s_0>0$,

$$r_0:=\sup_{s\in[-s_0,s_0]}\|\mathbf{d}(s)\|_{\mathbb{Y}}$$

and $r > r_0^2 > 1$. Let $\delta > r$ be defined by the formula (6.28). Then for any $x_0 \in \mathbb{Y}$ the system (6.25) has the unique mild solution $x_{x_0}^{\delta} : [0, +\infty) \mapsto \mathbb{Y}$ which, for $x_0 \in K_{\mathbb{Y}}(\delta)$, coincides with the unique solution x_{x_0} of (7.22) as long as $x_{x_0}^{\delta}(t) \in K_{\mathbb{Y}}(\delta)$. Moreover, as shown above, for any $T > 0$ there exists $L_{\delta,T} > 0$ such that

$$\|x_{x_1}^{\delta}(t) - x_{x_2}^{\delta}(t)\|_{\mathbb{Y}} \leq L_{\delta,T}\|x_1 - x_2\|_{\mathbb{Y}}$$

for all $t \in [0, T]$.

Since the origin of (7.22) is globally uniformly asymptotically stable then there exists $T_r > 0$ such that

$$\|x_0\|_{\mathbb{Y}} < r \quad \Rightarrow \quad \|x_{x_0}(t)\|_{\mathbb{Y}} \leq 1/r, \ \ \forall t \geq T_r.$$

We have $x_{x_0}^{\delta}(t) = x_{x_0}(t)$ as long as $\|x_{x_0}(t)\|_{\mathbb{Y}} \geq 1/\delta$ provided that $x_0 \in K_{\mathbb{Y}}(r)$. Indeed, the global uniform Lyapunov stability of (7.22) and the selection of the parameter $\delta > r$ imply that $\|x_{x_0}^{\delta}(t)\|_{\mathbb{Y}} \leq \delta$ for all $t \geq 0$ and all $x_0 \in K_{\mathbb{Y}}(r)$. Moreover, if there exists $t_1 > 0$ such that $\|x_{x_0}^{\delta}(t_1)\|_{\mathbb{Y}} \leq 1/\delta$ (resp., $\|x_{x_0}(t_1)\|_{\mathbb{Y}} \leq 1/\delta$) then $\|x_{x_0}^{\delta}(t)\|_{\mathbb{Y}} \leq 1/r$ (resp., $\|x_{x_0}(t)\|_{\mathbb{Y}} \leq 1/r$) for all $t \geq t_1$. In this case, for the functional $V_0 : \mathbb{Y} \mapsto \mathbb{R}_+$ given by

$$V_0(x_0) := \sup_{t\geq 0} \|x_{x_0}(t)\|_{\mathbb{Y}}$$

we have

$$V_0(x_0) = \sup_{t\in[0,T_r]} \|x_{x_0}^{\delta}(t)\|_{\mathbb{Y}}$$

for all $x_0 \in K(r)$. Hence, using the triangle inequality we derive

$$|V_0(x_1) - V_0(x_2)| \leq \sup_{t\in[0,T_r]} \|x_{x_1}^{\delta}(t) - x_{x_2}^{\delta}(t)\|_{\mathbb{Y}} \leq L_{\delta,T_r}\|x_1 - x_2\|_{\mathbb{Y}}, \ \forall x_1, x_2 \in K_{\mathbb{Y}}(r).$$

This means that the functional V_0 satisfy the Lipschitz condition on $K_{\mathbb{Y}}(r)$.

The global uniform Lyapunov stability of (7.22) implies that $\|x\|_{\mathbb{Y}} \leq V_0(x) \leq \varepsilon(\|x\|_{\mathbb{Y}})$, $\forall x \in \mathbb{Y}$, where $\varepsilon \in \mathcal{K}$ is given by (3.30). Moreover, the function $t \mapsto V_0(x_{x_0}(t))$ is decreasing as long as $x_{x_0}(t) \in K_{\mathbb{Y}}(r)$. Let us consider the functional

$$V_1(x_0) := \int_0^{+\infty} \tilde{a}(\|x_{x_0}(s)\|_{\mathbb{Y}})\|x_{x_0}(s)\|_{\mathbb{Y}} ds,$$

where $\tilde{a} \in C_c^{\infty}$ is such that $\tilde{a}(\rho) = 0$ for $s \notin (1/r, r)$, $0 < \tilde{a}(\rho) \leq 1$ for $\rho \in (1/r, r)$ and $\tilde{a}(\rho) = 1$ for $\rho \in [1/r_0, r_0]$.

Notice also $\tilde{a}(\|x_{x_0}(s)\|_{\mathbb{Y}})\|x_{x_0}(s)\|_{\mathbb{Y}} = 0$ if $x_{x_0}(s) \notin K_{\mathbb{Y}}(r)$. Repeating the above considerations we conclude that

$$V_1(x_0) = \int_0^{T_r} \tilde{a}(\|x_{x_0}(s)\|_{\mathbb{Y}})\|x_{x_0}(s)\|_{\mathbb{Y}} ds = \int_0^{T_r} \tilde{a}(\|x_{x_0}^{\delta}(s)\|_{\mathbb{Y}})\|x_{x_0}^{\delta}(s)\|_{\mathbb{Y}} ds$$

for all $x_0 \in K_{\mathbb{Y}}(r)$. Since $x \mapsto \tilde{a}(\|x\|_{\mathbb{Y}})\|x\|_{\mathbb{Y}}$ satisfy the Lipschitz condition on $\mathbb{Y}$ with some Lipschitz constant $\tilde{L} > 0$ then

$$|V_1(x_1) - V_1(x_2)| \le \left| \int_0^{T_r} \tilde{a}(\|x^\delta_{x_1}(s)\|_{\mathbb{Y}})\|x^\delta_{x_1}(s)\|_{\mathbb{Y}} - \tilde{a}(\|x^\delta_{x_2}(s)\|_{\mathbb{Y}})\|x^\delta_{x_2}(s)\|_{\mathbb{Y}} ds \right| \le$$

$$\int_0^{T_r} |\tilde{a}(\|x^\delta_{x_1}(s)\|_{\mathbb{Y}})\|x^\delta_{x_1}(s)\|_{\mathbb{Y}} - \tilde{a}(\|x^\delta_{x_2}(s)\|_{\mathbb{Y}})\|x^\delta_{x_2}(s)\|_{\mathbb{Y}}| ds \le$$

$$\tilde{L} \int_0^{T_r} \|x^\delta_{x_1}(s) - x^\delta_{x_2}(s)\|_{\mathbb{Y}} ds \le T_r L_{\delta,T_r} \tilde{L} \|x_1 - x_2\|_{\mathbb{Y}}, \quad \forall x_1, x_2 \in K_{\mathbb{Y}}(r),$$

i.e., V_1 also satisfies the Lipschitz condition on $K_{\mathbb{Y}}(r)$.

Moreover, if $h > 0$ and $t \ge 0$ are such that $x_{x_0}(t+\theta) \in K(r)$, $\forall \theta \in [0, h]$ then $x_{x_0}(t+\theta) = x^\delta_{x_{x_0}(t)}(\theta)$, $\forall \theta \in [0, h]$ and

$$V_1(x_{x_0}(t+h)) - V_1(x_{x_0}(t)) =$$

$$\int_0^{T_r} \tilde{a}(\|x^\delta_{x(t+h)}(s)\|_{\mathbb{Y}})\|x^\delta_{x(t+h)}(s)\|_{\mathbb{Y}} ds - \int_0^{T_r} \tilde{a}(\|x^\delta_{x(t)}(s)\|_{\mathbb{Y}})\|x^\delta_{x(t)}(s)\|_{\mathbb{Y}} ds =$$

$$\int_0^{T_r} \tilde{a}(\|x^\delta_{x(t+h)}(s)\|_{\mathbb{Y}})\|x^\delta_{x(t+h)}(s)\|_{\mathbb{Y}} ds - \int_0^{h} \tilde{a}(\|x^\delta_{x(t)}(s)\|_{\mathbb{Y}})\|x^\delta_{x(t)}(s)\|_{\mathbb{Y}} ds -$$

$$\int_h^{T_r+h} \tilde{a}(\|x^\delta_{x(t)}(s)\|_{\mathbb{Y}})\|x^\delta_{x(t)}(s)\|_{\mathbb{Y}} ds =$$

$$\int_0^{T_r} \tilde{a}(\|x^\delta_{x(t+h)}(s)\|_{\mathbb{Y}})\|x^\delta_{x(t+h)}(s)\|_{\mathbb{Y}} ds - \int_0^{h} \tilde{a}(\|x^\delta_{x(t)}(s)\|_{\mathbb{Y}})\|x^\delta_{x(t)}(s)\|_{\mathbb{Y}} ds -$$

$$\int_0^{T_r} \tilde{a}(\|x^\delta_{x(t)}(s+h)\|_{\mathbb{Y}})\|x^\delta_{x(t)}(s+h)\|_{\mathbb{Y}} ds =$$

$$-\int_0^{h} \tilde{a}(\|x^\delta_{x(t)}(s)\|_{\mathbb{Y}})\|x^\delta_{x(t)}(s)\|_{\mathbb{Y}} ds = -\int_0^{h} \tilde{a}(\|x_{x(t)}(s)\|_{\mathbb{Y}})\|x_{x(t)}(s)\|_{\mathbb{Y}} ds =$$

$$-\tilde{a}(\|x_{x_0}(t+\theta_h)\|_{\mathbb{Y}})\|x_{x_0}(t+\theta_h)\|_{\mathbb{Y}} h, \quad \theta_h \in [0, h],$$

where the mean value theorem for integrals is utilized in the last step. This means that

$$\limsup_{h \to 0^+} \tfrac{V_1(x_{x_0}(t+h)) - V_1(x_{x_0}(t))}{h} \le -\tilde{a}(\|x_{x_0}(t)\|_{\mathbb{Y}})\|x_{x_0}(t)\|_{\mathbb{Y}}$$

as long as $x_{x_0}(t) \in K_{\mathbb{Y}}(r)$. Notice that $0 \le V_1(x_0) \le T_r \varepsilon(\|x_0\|_{\mathbb{Y}})$.

In this case, the functional $V_r : \mathbb{B} \mapsto \mathbb{R}_+$ given by

$$V_r(x) := V_0(x) + V_1(x)$$

satisfies the Lipschitz condition on $K_{\mathbb{Y}}(r)$ and

$$\|x\|_{\mathbb{Y}} \leq V_r(x) \leq (1 + T_r)\varepsilon(\|x\|_{\mathbb{Y}})$$

for all $x \in \mathbb{Y}$. Moreover, for any $x_0 \in K_{\mathbb{Y}}(r)$ the function $t \mapsto V_r(x_{x_0}(t))$ is strictly decreasing,

$$\limsup_{h\to 0^+} \tfrac{V_r(x_{x_0}(t+h))-V_r(x_{x_0}(t))}{h} \leq -W_r(\|x_{x_0}(t)\|_{\mathbb{Y}})$$

as long as $x_{x_0}(t) \in K(r)$, where $W_r(\rho) := \tilde{a}(\rho)\rho$ with $\rho \in \mathbb{R}_+$. Due to homogeneity of the system (7.22) the solutions x_{x_0} are symmetric (see Theorem 6.1). Hence, if $\mathbf{d}(-s)x_{x_0}(t+\theta) \in K_{\mathbb{Y}}(r)$ for all $\theta \in [0, h]$ then using the identity $\mathbf{d}(-s)x_{x_0}(t) = x_{\mathbf{d}(-s)x_0}(e^{\mu s}t)$, $\forall s \in \mathbb{R}$ we derive

$$V_r(\mathbf{d}(-s)x_{x_0}(t+h)) - V_r(\mathbf{d}(-s)x_{x_0}(t)) =$$

$$V_r(x_{\mathbf{d}(-s)x_0}(e^{\mu s}(t+h))) - V_r(x_{\mathbf{d}(-s)x_0}(e^{\mu s}t)) \leq$$

$$-\tilde{a}(\|x_{\mathbf{d}(-s)x_0}(e^{\mu s}t+\theta_h)\|_{\mathbb{Y}})\|x_{\mathbf{d}(-s)x_0}(e^{\mu s}t+\theta_h)\|_{\mathbb{Y}}e^{\mu s}h,$$

where $\theta_h \in [0, e^{\mu s}h]$, and $\mu \in \mathbb{R}$ is the homogeneity degree of the system (7.22), i.e., the function $t \mapsto V_r(\mathbf{d}(-s)x_{x_0}(t))$ is strictly decreasing

$$\limsup_{h\to 0^+} \tfrac{V_r(\mathbf{d}(-s)x_{x_0}(t+h))-V_r(\mathbf{d}(-s)x_{x_0}(t))}{h} \leq$$

$$-\tilde{a}(\|\mathbf{d}(-s)x_{x_0}(t)\|_{\mathbb{Y}})\|\mathbf{d}(-s)x_{x_0}(t)\|_{\mathbb{Y}}e^{\mu s}$$

as long as $\mathbf{d}(-s)x_{x_0}(t) \in K_{\mathbb{Y}}(r)$.

II. Inspired by [14] let us consider the functional $V : \mathbb{Y} \mapsto \mathbb{R}_+$ given by

$$V(x) := \int_{-\infty}^{+\infty} e^{s-q}\hat{a}(V_r(\mathbf{d}(-s)x))ds$$

where $\hat{a} \in C^\infty$ is an increasing function such that

- $\hat{a}(\rho) = 1$ for $\rho \geq (1 + T_r)\varepsilon(r)$;
- $\hat{a}(\rho) = 0$ for $\rho \leq 1/r$;
- $\hat{a}'(\rho) > 0$ for $1/r < \rho < (1 + T_r)\varepsilon(r)$
- $\min\limits_{1/r_0 \leq \rho \leq (1+T_r)\varepsilon(r_0)} \hat{a}'(\rho) > 0$

and

$$q := \ln \frac{\min\{e^{(1+\mu)s_0}, e^{-(1+\mu)s_0}\} \min\limits_{1/r_0 \le \rho \le (1+T_r)\varepsilon(r_0)} \hat{a}'(\rho)}{r_0}.$$

For example, the function $\hat{a}$ can be selected as follows:

$$\hat{a}(\rho) = e^{-\frac{\rho_{\max}-\rho_{\min}}{\rho-\rho_{\min}}}\left(e^{-\frac{\rho_{\max}-\rho_{\min}}{\rho_{\max}-\rho}} + e\right)$$

for $\rho \in (\rho_{\min}, \rho_{\max})$, where $\rho_{\min} = 1/r$, $\rho_{\max} = (1+T_r)\varepsilon(r)$.

By construction, the functional $V : \mathbb{Y} \mapsto \mathbb{R}_+$ is $\mathbf{d}$-homogeneous of the degree 1. Moreover,

$$V(x) \le \int_{-\infty}^{+\infty} e^{s-q}\hat{a}((1+T_r)\varepsilon(\|\mathbf{d}(-s)x\|_{\mathbb{Y}}))ds \le$$

$$\int_{-\infty}^{\bar{s}(x)} e^{s-q}\hat{a}((1+T_r)\varepsilon(\|\mathbf{d}(-s)x\|_{\mathbb{Y}}))ds \le e^{-q}e^{\bar{s}(x)}$$

where $\bar{s}(x) := \ln\left(\frac{\|x\|_{\mathbf{d},\mathbb{Y}}}{\underline{\sigma}\left(\varepsilon^{-1}\left(\frac{1}{r(1+T_r)}\right)\right)}\right)$ and $\underline{\sigma} \in \mathcal{K}^{\infty}$ is defined in Lemma 4.5 Indeed, if $s \ge s(x)$ then we derive

$$\underline{\sigma}(\|\mathbf{d}(-s)x\|_{\mathbb{Y}}) \le \|\mathbf{d}(-s)x\|_{\mathbf{d},\mathbb{Y}} = e^{-s}\|x\|_{\mathbf{d},\mathbb{Y}} \le e^{-\bar{s}(x)}\|x\|_{\mathbf{d},\mathbb{Y}} =$$

$$\underline{\sigma}\left(\varepsilon^{-1}\left(\frac{1}{r(1+T_r)}\right)\right)$$

and taking into account $\underline{\sigma}, \varepsilon \in \mathcal{K}^{\infty}$ we conclude $\|\mathbf{d}(-s)x)\|_{\mathbb{Y}} \le \varepsilon^{-1}\left(\frac{1}{r(1+T_r)}\right)$ or, equivalently, $(1+T_r)\varepsilon(\|\mathbf{d}(-s)x)\|_{\mathbb{Y}}) \le 1/r$ for all $s \ge \bar{s}(x)$. This means $V_r(\mathbf{d}(-s)x) \le 1/r$ and $\hat{a}(V_r(\mathbf{d}(-s)x)) = 0$ for all $s \ge \bar{s}(x)$.

Similarly, we derive

$$V(x) \ge \int_{-\infty}^{+\infty} e^{s-q}\hat{a}(\|\mathbf{d}(-s)x)\|_{\mathbb{Y}})ds \ge \int_{-\infty}^{\underline{s}(x)} e^{s-q}ds = e^{-q}e^{\underline{s}(x)},$$

where $\underline{s}(x) := \ln\left(\frac{\|x\|_{\mathbf{d},\mathbb{Y}}}{\bar{\sigma}((1+T_r)\varepsilon(r))}\right)$. Indeed, if $s \le s(x)$ then we have

$$\bar{\sigma}(\|\mathbf{d}(-s)x\|_{\mathbb{Y}}) \ge \|\mathbf{d}(-s)x\|_{\mathbf{d},\mathbb{Y}} = e^{-s}\|x\|_{\mathbf{d}} \ge e^{-\underline{s}x}\|x\|_{\mathbf{d},\mathbb{Y}} = \bar{\sigma}((1+T_r)\varepsilon(r)).$$

Since $\bar{\sigma} \in \mathcal{K}^{\infty}$ then $\|\mathbf{d}(-s)x\|_{\mathbb{Y}} \ge (1+T_r)\varepsilon(r)$ and $\hat{a}(\|\mathbf{d}(-s)x\|_{\mathbb{Y}})) = 1$ for all $s \le \underline{s}(x)$. Therefore, we have proven that the inequality (6.23) holds for

$$\underline{k} := \frac{e^{-q}}{\bar{\sigma}((1+T_r)\varepsilon(r))}, \quad \bar{k} := \frac{e^{-q}}{\underline{\sigma}\left(\varepsilon^{-1}\left(\frac{1}{r(1+T_r)}\right)\right)}.$$

If $x(t+\theta) \in K_{\mathbb{Y}}(r_0)$ for all $\theta \in [0, h]$ then $\mathbf{d}(-s)x(t+\theta) \in K_{\mathbb{Y}}(r)$ for all $\theta \in [0, h]$ and for all $s \in [-s_0, s_0]$. Indeed,

$$\|\mathbf{d}(-s)x(t+\theta)\|_{\mathbb{Y}} \leq \|\mathbf{d}(-s)\|_{\mathbb{Y}} \cdot \|x(t+\theta)\|_{\mathbb{Y}} \leq r_0^2 = r$$

and

$$1/r_0 \leq \|x(t+\theta)\|_{\mathbb{Y}} = \|\mathbf{d}(s)\|_{\mathbb{Y}} \cdot \|\mathbf{d}(-s)x(t+\theta)\|_{\mathbb{Y}}.$$

The latter means that

$$1/r \leq V_r(\mathbf{d}(-s)x_{x_0}(t+\theta)) \leq (1+T_r)\varepsilon(r)$$

for all $s \in [-s_0, s_0]$ and all $\theta \in [0, h]$. Using

$$V_r(\mathbf{d}(-s)x_{x_0}(t+h)) - V_r(\mathbf{d}(-s)x_{x_0}(t)) \leq$$

$$-W_r(x_{\mathbf{d}(-s)x_0}(e^{\mu s}t+\theta_h))e^{\mu s}h =$$

$$-W_r(\mathbf{d}(-s)x_{x_0}(t+e^{-\mu s}\theta_h))e^{\mu s}h$$

we derive

$$V(x_{x_0}(t+h)) - V(x_{x_0}(t)) =$$

$$\int_{-\infty}^{+\infty} e^{s-q}[\hat{a}(V_r(\mathbf{d}(-s)x_{x_0}(t+h))) - \hat{a}(V_r(\mathbf{d}(-s)x_{x_0}(t)))]ds =$$

$$\int_{-s_0}^{s_0} e^{s-q}[\hat{a}(V_r(\mathbf{d}(-s)x_{x_0}(t+h))) - \hat{a}(V_r(\mathbf{d}(-s)x_{x_0}(t)))]ds =$$

$$\int_{-s_0}^{s_0} e^{s-q}\hat{a}'(\theta)[V_r(\mathbf{d}(-s)x_{x_0}(t+h)) - V_r(\mathbf{d}(-s)x_{x_0}(t))]ds \leq$$

$$-h\int_{-s_0}^{s_0} e^{s-q}\hat{a}'(\theta)W_r(\mathbf{d}(-s)x_{x_0}(t+e^{-\mu s}\theta_h))e^{\mu s}ds,$$

where the mean value theorem is utilized with

$$\theta \in [V_r(\mathbf{d}(-s)x_{x_0}(t+h)), V_r(\mathbf{d}(-s)x_{x_0}(t))].$$

Therefore, the function $t \to V(x_{x_0}(t))$ is strictly decreasing and

$$\limsup_{h\to 0^+} \frac{V(x_{x_0}(t+h))-V(x_{x_0}(t))}{h} \le$$

$$-\int_{-\infty}^{+\infty} e^{(1+\mu)s-q}\hat{a}'(V_r(\mathbf{d}(-s)x_{x_0}(t)))W_r(\|\mathbf{d}(-s)x_{x_0}(t)\|_{\mathbb{Y}})ds = -W(x_{x_0}(t)),$$

as long as $x_{x_0}(t) \in K_{\mathbb{Y}}(r_0)$, where

$$W(x) := \int_{-\infty}^{+\infty} e^{(1+\mu)s-q}\hat{a}'(V_r(\mathbf{d}(-s)x))W_r(\|\mathbf{d}(-s)x\|_{\mathbb{Y}})ds.$$

The functional W is, obviously, nonnegative and $\mathbf{d}$-homogeneous of the degree $\mu + 1$. If $\|x\|_{\mathbb{Y}} = 1$ then $\mathbf{d}(-s)x \in K_{\mathbb{Y}}(r_0)$ and $W_r(\|\mathbf{d}(-s)x\|_{\mathbb{Y}}) = \|\mathbf{d}(s)x\|_{\mathbb{Y}}$ for all $s \in [-s_0, s_0]$. Using the mean value theorem for integrals we derive

$$W(x) \ge \int_{-s_0}^{s_0} e^{(1+\mu)s-q}\hat{a}'(V_r(\mathbf{d}(-s)x))\|\mathbf{d}(-s)x\|_{\mathbb{Y}}ds \ge$$

$$\tfrac{1}{r_0}e^{(1+\mu)s^*}\hat{a}'(V_r(\mathbf{d}(-s^*)x)),$$

where $s^* \in [-s_0, s_0]$. Since $\mathbf{d}(-s^*)x \in K_{\mathbb{Y}}(r_0)$ then

$$\frac{1}{r} < \frac{1}{r_0} \le V_r(\mathbf{d}(-s^*)x) \le (1+T_r)\varepsilon(r_0) < (1+T_r)\varepsilon(r).$$

Hence, we derive

$$\inf_{\|x\|_{\mathbb{Y}}=1} W(x) \ge \frac{\min\{e^{(1+\mu)s_0}, e^{-(1+\mu)s_0}\} \min\limits_{1/r_0\le\rho\le(1+T_r)\varepsilon(r_0)} \hat{a}'(\rho)}{e^q r_0} = 1$$

and using $\mathbf{d}$-homogeneity of W we conclude

$$W(x) = W(\mathbf{d}(\ln\|x\|_{\mathbf{d},\mathbb{Y}})\mathbf{d}(-\ln\|x\|_{\mathbf{d},\mathbb{Y}})x) =$$

$$\|x\|_{\mathbf{d},\mathbb{Y}}^{1+\mu} W(\mathbf{d}(\ln\|x\|_{\mathbf{d},\mathbb{Y}})\mathbf{d}(-\ln\|x\|_{\mathbf{d},\mathbb{Y}})x) \le \|x\|_{\mathbf{d},\mathbb{Y}}^{1+\mu}.$$

Therefore, we have proven that

$$\limsup_{h\to 0^+} \frac{V(x_{x_0}(t+h))-V(x_{x_0}(t))}{h} \le -\|x_{x_0}(t)\|_{\mathbf{d},\mathbb{Y}}^{1+\mu}$$

as long as $x_{x_0}(t) \in K_{\mathbb{Y}}(r_0)$. If $x_{x_0}(t) \in \mathbf{d}(\tau)K_{\mathbb{Y}}(r_0)$, where $\tau \in \mathbb{R}$, then $\mathbf{d}(-\tau)x_{x_0}(t) \in K_{\mathbb{Y}}(r_0)$. Using homogeneity of V we derive

$$V(x_{x_0}(t)) = V(\mathbf{d}(-\tau)\mathbf{d}(\tau)x_{x_0}(t)) = e^{-\tau}V(\mathbf{d}(\tau)x_{x_0}(t)).$$

Since by Theorem 6.1 we have $x_{\mathbf{d}(\tau)x_0}(e^{-\mu\tau}t) = \mathbf{d}(\tau)x_{x_0}(t)$ then the function $t \to V(x_{x_0}(t))$ is strictly decreasing as well and

$$\limsup_{h\to 0^+} \frac{V(x_{x_0}(t+h))-V(x_{x_0}(t))}{h} = e^{-\tau s} \limsup_{h\to 0^+} \frac{V(x_{\mathbf{d}(\tau)x_0}(e^{-\mu\tau}(t+h)))-V(x_{\mathbf{d}(\tau)x_0}(e^{-\mu\tau}t))}{h}$$

$$\le -e^{-\tau(1+\mu)}\|x_{\mathbf{d}(\tau)x_0}(e^{-\mu\tau}t)\|_{\mathbf{d}}^{1+\mu} = -\|\mathbf{d}(-\tau)x_{\mathbf{d}(\tau)x_0}(e^{-\mu\tau}t)\|_{\mathbf{d}}^{1+\mu} = -\|x_{x_0}(t)\|_{\mathbf{d}}^{1+\mu}$$

as long as $x_{x_0}(t) \in \mathbf{d}(\tau)K_{\mathbb{Y}}(r_0)$, where $\tau \in \mathbb{R}$ is an arbitrary real number.

III. Let $t_j \to 0, t_j > 0$ as $j \to +\infty$ be an arbitrary a sequence of time instances such that $x(t_j) \in M^{-1}\mathcal{D}(A)$. Since any mild solution is continuous function of time in $\mathbb{Y}$ then for any $h > 0$ there exists $i : \|x(t_i) - x_0\|_{\mathbb{Y}} < h^2$ and $\|x(t_i + h) - x(h)\|_{\mathbb{Y}} < h^2$. Obviously, $i \to +\infty$ as $h \to 0$. Let us consider

$$\frac{V(x_0+h(Ax_0+f(x_0)))-V(x_0)}{h} = \frac{V(x_0+h(Ax_0+f(x_0)))-V(x_0+h(Ax(t_i)+f(x(t_i))))}{h} +$$

$$\frac{V(x_0+h(Ax(t_i)+f(x(t_i))))-V(x(t_i)+h(Ax(t_i)+f(x(t_i))))}{h} +$$

$$\frac{V(x(t_i)+h(Ax(t_i)+f(x(t_i))))-V(x(t_i+h))}{h} +$$

$$\frac{V(x(t_i+h))-V(x(h))}{h} + \frac{V(x(h))-V(x_0)}{h}.$$

The third term in the above sum tends to zero as $h \to 0^+$ (see above). Since the operators A and M are closed then $x(t_i) \to x_0, x(t_i) \in M^{-1}\mathcal{D}(A)$ implies $MAx(t_i) \to MAx_0$, $Ax(t_i) \to Ax_0$ as $t_j \to 0$ and and

$$\frac{|V(x_0+h(Ax_0+f(x_0))-V(x_0+h(Ax(t_i)+f(x(t_i))|}{h} \le$$

$$\|Ax_0 - Ax(t_i)\|_{\mathbb{Y}} + \|f(x_0) - f(x(t_i))\|_{\mathbb{Y}} \to 0 \text{ as } h \to 0.$$

Using the local Lipschitz continuity of V, we derive

$$\frac{|V(x_0+h(Ax(t_i)+f(x(t_i))))-V(x(t_i)+h(Ax(t_i)+f(x(t_i))))|}{h} \le$$

$$\frac{L\|x_0-x(t_i)\|_{\mathbb{Y}}}{h} \le Lh \to 0 \text{ as } h \to 0.$$

Similarly, we obtain

$$\frac{|V(x(t_i+h))-V(x(h))|}{h} \le \frac{L_{x_0}\|x_0-x(t_i)\|_{\mathbb{Y}}}{h} \le Lh \to 0 \text{ as } h \to 0.$$

Therefore, $\forall x_0 \in M^{-1}\mathcal{D}(A)$ we have

$$D^+(x_0; Ax_0 + f(x_0)) = \limsup_{h\to 0^+} \frac{V(x_0+h(Ax_0+f(x_0))-V(x_0)}{h} =$$

$$\limsup_{h\to 0^+} \frac{V(x(h))-V(x_0)}{h} \le -\|x_0\|_{\mathbf{d},\mathbb{Y}}^{\mu+1}.$$

The proof is complete. □

The following example demonstrates the possible design of homogeneous Lyapunov function for a Burgers equation with nonlinearity.

Example 6.7[4] Let us consider the nonlinear equation

$$\dot{x} = \frac{\partial^2 x}{\partial z^2} + x\frac{\partial x}{\partial z} - \frac{1}{3}\|x\|_{L^2}^4 x, \quad t > 0, \quad x(0) = x_0, \tag{6.32}$$

where $x(t) \in L^2(\mathbb{R};\mathbb{R})$. This equation admits the representation (6.3) with $A = \frac{\partial^2}{\partial z^2}$, $\mathbb{B} = L^2(\mathbb{R};\mathbb{R})$, $\mathcal{D}(A) = H^2(\mathbb{R};\mathbb{R})$, $f(x) = x\frac{\partial x}{\partial z} - \frac{1}{3}\|x\|_{L^2}^4 x$ and $\mathcal{D}(f) = \mathbb{Y} = H^1(\mathbb{R};\mathbb{R})$. Notice that f is unbounded but M-regular nonlinear operator with $M = \frac{\partial}{\partial z} + I$ and $(M^{-1}y)(z) = \int_{-\infty}^{z} e^{-(z-s)}y(s)ds$, $y \in L^2(\mathbb{R};\mathbb{R})$, $z \in \mathbb{R}$, respectively. The inequality (2.35) follows from $\|\frac{\partial}{\partial z}M^{-1}\| < \infty$ and

$$\|M^{-1}y\|_{L^\infty} \le \sup_{z\in\mathbb{R}} \int_{-\infty}^{z} e^{-(z-s)}|y(s)|ds \le \|y\|_{L^2} \sup_{z\in\mathbb{R}} \sqrt{\int_{-\infty}^{z} e^{-2(z-s)}ds} = \frac{\|y\|_{L^2}}{\sqrt{2}},$$

where $y \in L^2(\mathbb{R};\mathbb{R})$.

The considered system cannot be studied using Theorem 6.9, but Corollary 6.5 is applicable. Indeed, Eq. (6.32) satisfies the conditions of 6.5 with $(\Phi(t)y)(z) = \frac{1}{\sqrt{4\pi t}}\int_{\infty}^{+\infty} e^{-\frac{(z-s)^2}{4t}} y(s)ds$, so it has a unique mild solution for any $x_0 \in H^1(\mathbb{R};\mathbb{R})\backslash\{\mathbf{0}\}$ due to Theorem 2.15.

The system (6.32) is $\mathbf{d}$-homogeneous of the degree $\mu = 2$ with respect to the dilation $(\mathbf{d}(s)x)(z) = e^s x(e^s z)$, $z \in \mathbb{R}$, $x \in L^2(\mathbb{R};\mathbb{R})$. The linear dilation $\mathbf{d}$ is strictly monotone and strongly continuous in $L^2(\mathbb{R};\mathbb{R})$ and in $H^1(\mathbb{R};\mathbb{R})$.

Let us show that the system (6.32) is globally uniformly asymptotically stable and the canonical homogeneous norm $\|\cdot\|_{\mathbf{d},H^1}$ induced by the norm $\|\cdot\|_{H^1}$ is a $\mathbf{d}$-homogeneous Lyapunov function of the system. Indeed, for $v \in M^{-1}\mathcal{D}(A) = H^3(\mathbb{R};\mathbb{R})\backslash\{\mathbf{0}\}$ we have

[4] This example is reprinted from *Systems & Control Letters*, 148/2, A. Polyakov, Homogeneous Lyapunov functions for homogeneous infinite-dimensional systems with unbounded nonlinear operators, 104854, https://doi.org/10.1016/j.sysconle.2020.104854. ©2020 Elsevier, with permission from Elsevier.

$$\langle v, Av + f(v)\rangle_{H^1} =$$

$$\left\langle v, \tfrac{\partial^2 v}{\partial z^2} + v\tfrac{\partial v}{\partial z} - \tfrac{1}{3}\|v\|_{L^2}^4 v\right\rangle_{L^2} + \left\langle \tfrac{\partial v}{\partial z}, \tfrac{\partial}{\partial z}\left(\tfrac{\partial^2 v}{\partial z^2} + v\tfrac{\partial v}{\partial z} - \tfrac{1}{3}\|v\|_{L^2}^4 v\right)\right\rangle_{L^2} =$$

$$-\left\|\tfrac{\partial v}{\partial z}\right\|_{L^2}^2 - \tfrac{1}{3}\|v\|_{L^2}^6 - \left\|\tfrac{\partial^2 v}{\partial z^2}\right\|_{L^2}^2 - \left\langle \tfrac{\partial^2 v}{\partial z^2}, v\tfrac{\partial v}{\partial z}\right\rangle - \tfrac{1}{3}\|v\|_{L^2}^4\left\|\tfrac{\partial v}{\partial z}\right\|_{L^2}^2 =$$

$$-\left\|\tfrac{\partial v}{\partial z}\right\|_{L^2}^2 - \frac{\|v\|_{L^2}^6}{3} - \frac{\left\|\frac{\partial^2 v}{\partial z^2}\right\|_{L^2}^2}{2} - \frac{\left\|\frac{\partial^2 v}{\partial z^2} + v\frac{\partial v}{\partial z}\right\|_{L^2}^2}{2} + \frac{\left\|v\frac{\partial v}{\partial z}\right\|_{L^2}^2}{2} - \frac{\|v\|_{L^2}^4\left\|\frac{\partial v}{\partial z}\right\|_{L^2}^2}{3} =$$

$$-\left\|\tfrac{\partial v}{\partial z}\right\|_{L^2}^2 - \frac{\|v\|_{L^2}^6}{3} - \frac{\left\|\frac{\partial^2 v}{\partial z^2}\right\|_{L^2}^2}{2} - \frac{\left\|\frac{\partial^2 v}{\partial z^2} + v\frac{\partial v}{\partial z}\right\|_{L^2}^2}{2} - \frac{\left\langle v^3, \frac{\partial^2 v}{\partial z^2}\right\rangle}{6} - \frac{\|v\|_{L^2}^4\left\|\frac{\partial v}{\partial z}\right\|_{L^2}^2}{3} =$$

$$-\left\|\tfrac{\partial v}{\partial z}\right\|_{L^2}^2 - \tfrac{1}{3}\|v\|_{L^2}^6 - \tfrac{5}{12}\left\|\tfrac{\partial^2 v}{\partial z^2}\right\|_{L^2}^2 - \tfrac{1}{2}\left\|\tfrac{\partial^2 v}{\partial z^2} + v\tfrac{\partial v}{\partial z}\right\|_{L^2}^2 - \tfrac{1}{12}\left\|v^3 + \tfrac{\partial^2 v}{\partial z^2}\right\|_{L_2}^2 +$$

$$\tfrac{1}{12}\|v\|_{L_6}^6 - \tfrac{1}{3}\|v\|_{L^2}^4\left\|\tfrac{\partial v}{\partial z}\right\|_{L^2}^2.$$

Using the Gagliardo–Nirenberg–Sobolev inequality (see Appendix C or [17]) we derive $\|v\|_{L^6(\mathbb{R};\mathbb{R})}^6 \leq \sqrt{\frac{\pi}{2}}\|v\|_{L^2(\mathbb{R};\mathbb{R})}^4\left\|\frac{\partial v}{\partial z}\right\|_{L^2(\mathbb{R};\mathbb{R})}^2$ and conclude

$$\langle v, Av + f(v)\rangle_{H^1} \leq -\left\|\tfrac{\partial v}{\partial z}\right\|_{L^2}^2 - \tfrac{1}{3}\|v\|_{L^2}^6.$$

This means that $\|\cdot\|_{H^1}$ is a Lyapunov function of the system, but it is not **d**-homogeneous. Using **d**-homogeneity of operators A, f and the formula (4.27), for any $x \in M^{-1}\mathcal{D}(A)$ we obtain

$$\left(D\|x\|_{\mathbf{d},H^1}\right)(Ax + f(x)) = 2\frac{\left\langle \mathbf{d}(-\ln\|x\|_{\mathbf{d},H^1})(Ax+f(x)), v\right\rangle_{H^1}}{\|v\|_{L_2}^2 + \left\|\frac{\partial v}{\partial z}\right\|_{L_2}^2}\|x\|_{\mathbf{d},H^1} =$$

$$2\frac{\langle Av + f(v), v\rangle_{H^1}}{\|v\|_{L_2}^2 + \left\|\frac{\partial v}{\partial z}\right\|_{L_2}^2}\|u\|_{\mathbf{d},H^1}^3 \leq 2\frac{-\left\|\frac{\partial v}{\partial z}\right\|_{L^2}^2 - \frac{1}{3}\|v\|_{L^2}^6}{\|v\|_{L_2}^2 + \left\|\frac{\partial v}{\partial z}\right\|_{L_2}^2}\|x\|_{\mathbf{d},H^1}^3$$

where $v = \mathbf{d}(-\ln\|x\|_{\mathbf{d},H^1})x$. Taking into account $\|v\|_{H^1} = 1$ we conclude that

$$\left(D\|x\|_{\mathbf{d},H^1}\right)(Ax + f(x)) \leq -\frac{2}{3}\|x\|_{\mathbf{d},H^1}^{\mu+1},$$

so the functional $V : H^1(\mathbb{R};\mathbb{R}) \mapsto \mathbb{R}$ given by $V(x) = \frac{3\|x\|_{\mathbf{d},H^1}}{2}$ satisfies Corollary 6.5. Moreover, by Theorem 6.6, the system is nearly fixed-time stable in $H^1(\mathbb{R};\mathbb{R})$.

Below we show the existence of locally homogeneous Lyapunov function for locally homogeneous abstract evolution equation. The following theorem generalizes the result of [18] to locally homogeneous systems in Hilbert spaces.

Corollary 6.6 (Lyapunov functions for locally homogeneous system) *Let $\mathbf{d}_L$ be strongly continuous strictly monotone dilation in $\mathbb{H}$, where $L=0$ or $L=\infty$. Let a linear closed densely defined operator $A:\mathcal{D}(A)\subset\mathbb{H}\mapsto\mathbb{H}$ be a generator of a strongly continuous semigroup of linear bounded operators on $\mathbb{H}$. Let A be $\mathbf{d}_L$-homogeneous of degree μ_L.*

- *Let a locally Lipschitz continuous (on $\mathbb{H}\backslash\{\mathbf{0}\}$) operator $f\in C(\mathbb{H};\mathbb{H})$ be $\mathbf{d}_L$-homogeneous of degree $\mu_L\in\mathbb{R}$ at L-limit and a locally Lipshcitz continuous (on $\mathbb{H}\backslash\{\mathbf{0}\}$) operator $f_L:\mathbb{H}\mapsto\mathbb{H}$ be a $\mathbf{d}_L$-homogeneous approximation of f, where $L=0$ or $L=\infty$. If the $\mathbf{d}_L$-homogeneous system*

$$\dot{x}=Ax+f_L(x),\quad t>0$$

is asymptotically stable then for any $m_L>0$ there exist numbers $\rho_L>0, r_L>0$ and a locally Lipschitz continuous (on $\mathbb{H}\backslash\{\mathbf{0}\}$) positive define $\mathbf{d}_L$-homogeneous functional $V_L\in C(\mathbb{H};\mathbb{R})$ of degree m_L such that

$$D^+V_L(x;Ax+f(x))\leq-\rho_L V_L^{1+\frac{\mu_L}{m_L}}(x),\tag{6.33}$$

$\forall x\in\mathcal{D}(A)\backslash\{\mathbf{0}\}:\|x\|_{\mathbf{d}}\leq r_0$ *if* $L=0$ *or* $\forall x\in\mathcal{D}(A):\|x\|_{\mathbf{d}}\geq r_\infty$ *if* $L=\infty$.

- *Let a locally Lipschitz continuous (on $\mathbb{H}\backslash\{\mathbf{0}\}$) operator $f:\mathbb{H}\mapsto\mathbb{H}$ be $\mathbf{d}_0$-homogeneous of degree $\mu_0\in\mathbb{R}$ at 0-limit and $\mathbf{d}_\infty$-homogeneous of degree $\mu_\infty\in\mathbb{R}$ at ∞-limit. Let $f_0, f_\infty:\mathbb{H}\mapsto\mathbb{H}$ be the corresponding locally Lipschitz continuous (on $\mathbb{H}\backslash\{\mathbf{0}\}$) homogeneous approximations of f and $m_0>0$, $m_\infty>0$ be arbitrary numbers. If the systems*

$$\dot{x}=Ax+f(x),\quad \dot{x}=Ax+f_0(x),\quad \dot{x}=Ax+f_\infty(x),\quad t>0$$

are globally uniformly asymptotically stable, then there exist a globally proper locally Lipschitz continuous (on $\mathbb{H}\backslash\{\mathbf{0}\}$) positive definite function $V\in C(\mathbb{H};\mathbb{R})$ and a constant $\rho>0$ such that

$$D^+V(x;Ax+f(x))\leq-\rho\sigma(V(x)),\quad x\in\mathcal{D}(A)\backslash\{\mathbf{0}\},\tag{6.34}$$

where

$$\sigma(V)=\begin{cases}V^{1+\frac{\mu_0}{m_0}} & \text{if } V\in(0,1),\\ V^{1+\frac{\mu_\infty}{m_\infty}} & \text{if } V\geq 1.\end{cases}\tag{6.35}$$

Moreover, $V(x)=V_0(x)$ for all $x:\|x\|_{\mathbf{d}_0}\leq 0.5r_0$ and $V(x)=V_\infty(x)$ for all $x:\|x\|_{\mathbf{d}_\infty}\geq 2r_\infty$, where V_L is $\mathbf{d}_L$-homogeneous function of degree $m_L>0$ satisfying (6.33) and $L\in\{0,\infty\}$.

Proof (1) Let V_0 be the homogeneous Lyapunov function of degree $m_0 > \max\{-\mu_0, 0\}$ for $\dot{x} = Ax + f_0(x)$, which exists due to Theorem 6.9. Since $\mathbf{d}_0(-\ln\|x\|_{\mathbf{d}_0})x \in \mathcal{D}(A)$ for any $x \in \mathcal{D}(A)\backslash\{\mathbf{0}\}$ and $\|\mathbf{d}_0(-\ln\|x\|_{\mathbf{d}_0})x\| = 1$ then for any $x \in \mathcal{D}(A)\backslash\{\mathbf{0}\}$ there exists $h_1 = h_1(x) > 0$ such that

$$\tfrac{2}{3} \leq \left\|\mathbf{d}_0(-\ln\|x\|_{\mathbf{d}_0})x + h\|x\|_{\mathbf{d}_0}^{\mu_0} A\mathbf{d}_0(-\ln\|x\|_{\mathbf{d}_0})x\right\| \leq \tfrac{4}{3}, \quad \forall h \in (0, h_1].$$

Since f is locally Lipschitz continuous (and consequently locally bounded) on $\mathbb{H}\backslash\{\mathbf{0}\}$ then there exist $h_2 = h_2(x) \in (0, h_1]$ such that

$$\tfrac{1}{2} \leq \left\|\mathbf{d}_0(-\ln\|x\|_{\mathbf{d}_0})x + h\|x\|_{\mathbf{d}_0}^{\mu_0} A\mathbf{d}_0(-\ln\|x\|_{\mathbf{d}_0})x + h\mathbf{d}(-\ln\|x\|_{\mathbf{d}_0})f(x)\right\| \leq \tfrac{3}{2}$$

for all $h \in (0, h_2)$. Since f_0 is locally Lipschitz continuous (and consequently locally bounded) on $\mathbb{H}\backslash\{\mathbf{0}\}$ then there exist $h_3 = h_3(x) \in (0, h_2]$ such that

$$\tfrac{1}{3} \leq \|\mathbf{d}(-\ln\|x\|_{\mathbf{d}})x + h\|x\|_{\mathbf{d}}^{\mu_0} A\mathbf{d}(-\ln\|x\|_{\mathbf{d}})x + h\mathbf{d}(-\ln\|x\|_{\mathbf{d}})f_0(x)\| \leq \tfrac{5}{3}$$

for all $h \in (0, h_2)$. Let $L > 0$ be the Lipschitz constant for $V_0(y)$ with $1/3 \leq \|y\| \leq 5/3$. Since $V_0(x + h(Ax + f(x)) = \|x\|_{\mathbf{d}_0}^{m_0} V_0(\mathbf{d}_0(-\ln\|x\|_{\mathbf{d}_0})x + h\|x\|_{\mathbf{d}_0}^{\mu} A\mathbf{d}(-\ln\|x\|_{\mathbf{d}_0})x + h\mathbf{d}_0(-\ln\|x\|_{\mathbf{d}_0})f(x))$ and $V_0(x + h(Ax + f_0(x)) = \|x\|_{\mathbf{d}_0}^{m_0} V_0(\mathbf{d}_0(-\ln\|x\|_{\mathbf{d}_0})x + h\|x\|_{\mathbf{d}_0}^{\mu} A\mathbf{d}(-\ln\|x\|_{\mathbf{d}_0})x + h\mathbf{d}_0(-\ln\|x\|_{\mathbf{d}_0})f_0(x))$ then

$$\frac{V_0(x+h(Ax+f(x)))-V_0(x+h(Ax+f_0(x)))}{h} \leq$$

$$\|x\|_{\mathbf{d}_0}^{m_0} L\|\mathbf{d}_0(-\ln\|x\|_{\mathbf{d}_0})f(x) - \mathbf{d}_0(-\ln\|x\|_{\mathbf{d}_0})f_0(x)\| =$$

$$\|x\|_{\mathbf{d}_0}^{m_0+\mu_0} L\|\|x\|_{\mathbf{d}_0}^{-\mu}\mathbf{d}_0(-\ln\|x\|_{\mathbf{d}_0})f(x) - f_0(\mathbf{d}_0(-\ln\|x\|_{\mathbf{d}_0})x)\|$$

for any $h \in (0, h_3)$. Since f_0 is a $\mathbf{d}_0$-homogeneous approximation of f at 0-limit then the exists $\sigma \in \mathcal{K}$ such that

$$\|\|x\|_{\mathbf{d}_0}^{-\mu_0}\mathbf{d}_0(-\ln\|x\|_{\mathbf{d}_0})f(x) - f_0(\mathbf{d}_0(-\ln\|x\|_{\mathbf{d}_0})x)\| \leq \sigma_0(\|x\|_{\mathbf{d}_0}).$$

Hence, for any $x \in \mathcal{D}(A)$ and for any $h \in (0, h_3(x)]$ we derive

$$\frac{V_0(x+h(Ax+f(x)))-V_0(x)}{h} \leq \frac{V_0(x+h(Ax+f_0(x)))-V_0(x)}{h} + \|x\|_{\mathbf{d}_0}^{m_0+\mu_0} L\sigma_0(\|x\|_{\mathbf{d}_0}),$$

where Lipschitz constant L is independent of x and h. Taking the supremum limit in both sides of the above inequality for $h \to 0^+$, for any $x \in \mathcal{D}(A)$ we derive

$$D^+V_0(x; Ax + f(x)) \leq D^+V_0(x; Ax + f_0(x)) + \|x\|_{\mathbf{d}_0} L\sigma_0(\|x\|_{\mathbf{d}_0})) \leq$$

$$-\rho V_0^{1+\mu_0/m_0}(x) + \frac{L}{\underline{k}^{\frac{1}{m_0}}} V_0^{1+\mu_0/m_0}(x)\sigma_0(\|x\|_{\mathbf{d}_0}),$$

where the inequality (6.17) is utilized in the last step. Since $\sigma_0 \in \mathcal{K}$ then there exists $r_0 > 0$ such that

$$D^+V_0(x, Ax + f(x)) \le -\frac{\rho}{2} V_0^{1+\mu_0/m_0}(x)$$

for all $x \in \mathcal{D}(A)$ such that $\|x\|_{\mathbf{d}_0} \le r_0$.

Similarly, if V_∞ is the homogeneous Lyapunov function of degree $m_\infty > \max\{-\mu_\infty, 0\}$ for $\dot{x} = Ax + f_\infty(x)$, which exists due to Theorem 6.9, then, repeating the above considerations for V_∞, we derive

$$D^+V_\infty(x; Ax + f(x)) \le -\frac{\rho}{2} V_\infty^{1+\mu_\infty/m_\infty}(x)$$

for all $x \in \mathcal{D}(A)$ such that $\|x\|_{\mathbf{d}_\infty} \ge r_\infty$.

(2) If f is homogeneous in the bi-limit then, as shown above, there exist V_L, $\rho_L > 0$ and $r_L > 0$ such that (6.33) holds for both $L = 0$ and $L = \infty$.

Let $r > 0$ be such that $\|x\| \le 1/r \Rightarrow \|x\|_{\mathbf{d}_0} \le 0.5r_0$ and $\|x\| \ge r \Rightarrow \|x\|_{\mathbf{d}_\infty} \le 2r_\infty$.

By Corollary 3.3 there exist $\rho_r > 0$ and a globally proper locally Lipshcitz continuous (on $\mathbb{H}$) positive definite functional $V_r : \mathbb{H} \mapsto \mathbb{R}$ such that $\exists \epsilon \in \mathcal{K}_\infty : \|x\| \le V_r(x) \le \epsilon(\|x\|)$ for all $x \in \mathbb{H}$ and the inequality

$$\mathcal{D}^+V_r(x, Ax + f(x)) \le -1/r, \tag{6.36}$$

holds for all $x \in \mathcal{D}(A) \cap K(r)$, where $K(r) = \{x \in \mathbb{H} : r^{-1} \le \|x\| \le r\}$.

The relations (4.22) between the norms $\|\cdot\|$ and $\|\cdot\|_{\mathbf{d}_L}$ imply that there exists $\underline{k}, \overline{k} \in \mathcal{K}_\infty$ such that

$$\underline{k}(\|x\|_{\mathbf{d}_0}) \le V_r(x) \le \overline{k}(\|x\|_{\mathbf{d}_\infty}), \quad \forall x \in \mathbb{H}.$$

Without loss of generality, we assume that $r_0 > 0$ is selected small enough and $r_\infty > 0$ is selected large enough to guarantee $\underline{k}(r_0) < 1 < \overline{k}(r_\infty)$.

Inspired by [18], let us consider the function $V : \mathbb{H} \mapsto \mathbb{R}_+$ defined as

$$V = w_\infty \phi_\infty(V_r)V_\infty + (1 - \phi_\infty(V_r))\, \phi_0(V_r)V_r + w_0(1 - \phi_0(V_r))V_0,$$

where $\phi_0, \phi_\infty \in C^\infty$ are such that

$$\phi_0(\rho)=0 \text{ if } \rho\le\underline{k}\left(\tfrac{r_0}{2}\right),\ \phi_0'(\rho)>0 \text{ if } \rho\in\left(\underline{k}\left(\tfrac{r_0}{2}\right), \underline{k}(r_0)\right),\ \phi_0(\rho)=1 \text{ if } \rho\ge\underline{k}(r_0),$$

$$\phi_\infty(\rho)=0 \text{ if } \rho\le\overline{k}(r_\infty),\ \phi_\infty'(\rho)>0 \text{ if } \overline{k}(r_\infty)<\rho<\overline{k}(\tfrac{r_\infty}{0.5}),\ \phi_\infty(\rho)=1 \text{ if } \rho\ge\overline{k}(\tfrac{r_\infty}{0.5}),$$

$$w_0 = \inf_{\underline{k}(0.5r_0)\le V_r(x)\le\overline{k}(2r_\infty)} \frac{V_r(x)}{V_0(x)} \quad \text{and} \quad w_\infty = \sup_{\underline{k}(0.5r_0)\le V_r(x)\le\overline{k}(2r_\infty)} \frac{V_r(x)}{V_\infty(x)}.$$

The function V is positive definite, globally proper, and locally Lipschitz continuous on $\mathbb{H}\backslash\{\mathbf{0}\}$ by construction and

$$DV^{+}(x; Ax + f(x)) = (w_{\infty}\phi'_{\infty}V_{\infty} - \phi'_{\infty}\phi_0 V_r)DV_r^{+}(x; Ax + f(x))$$

$$+ w_{\infty}\phi_{\infty}DV_{\infty}^{+}(x; Ax + f(x)) + \left((1-\phi_{\infty})\,\phi'_0 V_r - w_0\phi'_0 V_0\right) DV_r^{+}(x; Ax + f(x))$$

$$+ (1-\phi_{\infty})\,\phi_0 DV_r^{+}(x; Ax + f(x)) + w_0(1-\phi_0)DV_0^{+}(x; Ax + f(x)).$$

Since $w_{\infty}\phi'_{\infty}(V_r)V_{\infty} \geq \phi'_{\infty}(V_r)\phi_0(V_r)V_r$ for $\underline{k}(0.5r_0) \leq V_r \leq \overline{k}(2r_{\infty})$, $(1-\phi_{\infty}(V_r))\,\phi'_0(V_r)V_r \geq w_0\phi'_0(V_r)V_0$ for $\underline{k}(0.5r_0) \leq V_r \leq \overline{k}(2r_{\infty})$ then

$$DV^{+}(x; Ax + f(x)) \leq w_0(1-\phi_0)DV_0^{+}(x; Ax + f(x))$$

$$+ \phi_0\,(1-\phi_{\infty})\,DV_r^{+}(x; Ax + f(x)) + w_{\infty}\phi_{\infty}DV_{\infty}^{+}(x; Ax + f(x)) \leq$$

$$-\rho_0 w_0(1-\phi_0)V_0^{1+\frac{\mu_0}{m_0}} - \rho_r\phi_0\,(1-\phi_{\infty})\,(V^{1+\frac{\mu_0}{m_0}} + V^{1+\frac{\mu_{\infty}}{m_{\infty}}}) - \rho_{\infty}w_{\infty}\phi_{\infty}V_{\infty}^{1+\frac{\mu_{\infty}}{m_{\infty}}},$$

where

$$\rho_r = \inf_{\underline{k}(0.5r_0)\leq V_r(x)\leq\overline{k}(2r_{\infty})} \frac{r^{-1}}{V^{1+\frac{\mu_0}{m_0}}(x)+V^{1+\frac{\mu_L}{m_l}}(x)} > 0.$$

We complete the proof taking into account the identities $V = V_0$ for $V_r(x) \leq \underline{k}(0.5r_0)$ and $V = V_{\infty}$ for $V_r(x) \geq \overline{k}(2r_{\infty})$, which hold by construction.

6.3 Robustness Issues

6.3.1 Homogeneous Domination Approach

Local Homogeneous Domination

In practice, homogeneous models may appear as local approximations of real physical systems. In this case, it is important to know if a local stability analysis of a system can be based on its homogeneous approximation. Let S, as before, the unit sphere $S = \{x \in \mathbb{H} : \|x\| = 1\}$.

Theorem 6.10 *Let a linear closed densely defined operator* $A : \mathcal{D}(A) \subset \mathbb{H} \mapsto \mathbb{H}$ *be a generator of a strongly continuous semigroup of linear bounded operators on* $\mathbb{H}$. *Let* $f : \mathbb{H} \mapsto \mathbb{H}$ *be locally Lipschitz continuous on* $\mathbb{H}\backslash\{\mathbf{0}\}$ *and* $f(\mathbf{0}) = \mathbf{0}$. *Let* f_0 *be a* $\mathbf{d}_0$*-homogeneous approximation of an operator* f *at* 0*-limit and* f_∞ *be a* $\mathbf{d}_\infty$*-homogeneous approximation of the operator* f *at* ∞*-limit. Let* $\mu_0 \in \mathbb{R}$ *and* $\mu_\infty \in \mathbb{R}$ *be homogeneity degrees of* f_0 *and* f_∞, *respectively. Let* f_0 *and* f_∞ *be locally Lipschitz continuous on* $\mathbb{H}\backslash\{\mathbf{0}\}$.

(1) If the operator A *is* $\mathbf{d}_0$*-homogeneous of the degree* $\mu_0 \in \mathbb{R}$ *with the strongly continuous strictly monotone linear dilation* $\mathbf{d}_0$ *in* $\mathbb{H}$ *and the system* $\dot{x} = Ax + f_0(x)$ *is uniformly asymptotically stable then the system* (6.3) *is*

- *locally uniformly finite-time stable for* $\mu_0 < 0$;
- *locally uniformly exponentially stable for* $\mu_0 = 0$;
- *locally uniformly asymptotically stable for* $\mu_0 > 0$.

(2) If the operator A *is* $\mathbf{d}_\infty$*-homogeneous of the degree* μ_∞ *with the strongly continuous strictly monotone linear dilation* $\mathbf{d}_\infty$ *in* $\mathbb{H}$ *and* $\dot{x} = Ax + f_\infty(x)$ *is uniformly asymptotically stable then there exists* $R>0$ *such that a* $\mathbf{d}_\infty$*-homogeneous ball* $B_{\mathbf{d}_\infty}(R) \subset \mathbb{H}$ *is*

- *a globally uniformly asymptotically attractive set for* $\mu_\infty < 0$;
- *a globally uniformly exponentially attractive set for* $\mu_\infty = 0$;
- *a globally uniformly fixed-time attractive set for* $\mu_\infty > 0$;

of the system (6.3).

(3) If the above conditions hold simultaneously for $\mu_0 < 0$ *and* $\mu_\infty > 0$ *then the globally uniformly asymptotically stable system* (6.3) *is globally uniformly fixed-time stable.*

Proof (1) By Corollary 6.6 the inequality (6.33) is fulfilled. Applying Corollary 3.5 for $\mu_0 < 0$ (resp., Corollary 3.2 for $\mu_0 = 0$ or Corollary 3.7 for $\mu_0 > 0$), we complete the proof of the case (1).

(2) The inequality (6.33) for $L = \infty$ implies that the conclusions of the case (2) hold for all strong solutions of the system (6.3). Applying Theorem 2.16 we extend the same conclusions to mild solutions as well.

(3) From (2) with $\mu_\infty > 0$ we derive the fixed-time convergence of all solutions of (6.3) to a ball $B(R)$, i.e., $x_{x_0}(t) \in B(R), \forall t \geq T_\infty, \forall x_0 \in \mathbb{B}$. From the case 1) with $\mu_0 < 0$ we conclude that there exists $r > 0$ such that all solutions with $\|x(t_0)\| \leq r$ converge to the origin in a uniformly bounded finite time, i.e., $x_{x_0}(t) = \mathbf{0}, \forall t \geq T_0$ provided that $\|x_0\| < r$. Finally, $\exists T' > 0 : x_0 \in B(R) \Rightarrow x_{x_0}(t) \in B(r), \forall r \geq T'$ due to the uniform asymptotic stability of (6.3), i.e., $x_{x_0}(t) = \mathbf{0}, \forall t \geq T_\infty + T' + T_0, \forall x_0 \in \mathbb{B}$. The proof is complete. □

Global Homogeneous Domination

As shown above, homogeneous approximations (if they exist) can be utilized for a local stability analysis (around the origin or "close to" the infinity point) of non homogeneous systems. Homogeneous extensions allow the similar conclusions to be made globally.

Corollary 6.7 *Let $F : \mathcal{D}(F) \subset \mathbb{B} \mapsto 2^{\mathbb{B}}$ be a (possibly set-valued) densely defined operator in $\mathbb{B}$ such that the evolution inclusion*

$$\dot{x} \in F(x), \tag{6.37}$$

satisfies Assumption 3.1. Let $\mathbf{d}$ *be strongly continuous strictly monotone linear dilation in $\mathbb{B}$. Let a* $\mathbf{d}$*-homogeneous operator $\tilde{F} : \mathcal{D}(\tilde{F}) \subset \mathbb{B} \mapsto 2^{\mathbb{B}}$ of degree $\mu \in \mathbb{R}$ be a* $\mathbf{d}$*-homogeneous extension of F. If the evolution inclusion*

$$\dot{x} \in \tilde{F}(x), \tag{6.38}$$

satisfying Assumption 3.1 is globally asymptotically stable then the evolution inclusion (6.37) *is*

- *globally uniformly finite-time stable for $\mu < 0$;*
- *globally uniformly exponentially stable for $\mu = 0$;*
- *globally uniformly practically fixed-time stable for $\mu > 0$.*

Proof The result immediately follows from the fact that the solution set of (6.37) is embedded into solution set of (6.38).

Corollary 6.8 *Let a linear closed densely defined operator $A : \mathcal{D}(A) \subset \mathbb{H} \mapsto \mathbb{H}$ be a generator of a strongly continuous semigroup Φ of linear bounded operators on $\mathbb{H}$. Let $f : \mathbb{H} \mapsto \mathbb{H}$ be locally Lipschitz continuous on $\mathbb{H}\backslash\{\mathbf{0}\}$ and $f(\mathbf{0}) = \mathbf{0}$. Let* $\mathbf{d}$ *be a strongly continuous strictly monotone linear dilation in $\mathbb{H}$. Let f and A be* $\mathbf{d}$*-homogeneous of degree $\mu \in \mathbb{R}$ such that*

$$\mu + \inf_{x \in S \cap \mathcal{D}(G_{\mathbf{d}})} \langle G_{\mathbf{d}} x, x \rangle > 0.$$

Let us consider the system

$$\dot{x} = Ax + f(x) + \|x\|_{\mathbf{d}}^{\mu} \mathbf{d}(\ln \|x\|_{\mathbf{d}}) q, \quad x(0) = x_0 \in \mathbb{H}. \tag{6.39}$$

Then

- *for any $q \in \mathcal{D}(G_{\mathbf{d}})$ and any $x_0 \in \mathcal{D}(A)$ the system* (6.39) *has a unique locally Lipschitz continuous strong solution;*
- *the system* (6.39) *has a mild solution $\forall q \in L^{\infty}(\mathbb{R}; \mathbb{H})$ and $\forall x_0 \in \mathbb{H}$;*

- *the uniform asymptotic stability of the system* (6.39) *with* $q = \mathbf{0}$ *implies its uniform asymptotic stability for small enough* $\|q\|_{L^\infty}$, *moreover, there exists* $L > 0$ *and* $\rho > 0$ *such that the inequality*

$$D^+V(x; Ax + f(x) + \|x\|_{\mathbf{d}}\mathbf{d}(\ln \|x\|_{\mathbf{d}})q) \overset{a.e.}{\leq} (L\|q\|_{L^\infty} - \rho)V^{1+\mu}(x), \tag{6.40}$$

holds for any $x \in \mathcal{D}(A)$ *and any* $q \in L^\infty(\mathbb{R}; \mathbb{H})$, *where* V *is a* **d**-*homogeneous Lyapunov functional of degree* 1 *satisfying all conditions of Theorem 6.9 for* $q = \mathbf{0}$.

Proof (1) Let $\beta = \inf_{x \in S \cap \mathcal{D}(G_{\mathbf{d}})} \langle G_{\mathbf{d}}x, x\rangle$ and $S = \{x \in \mathbb{B} : \|x\| = 1\}$. The system is **d**-homogeneous since all three homogeneous operators in the right-hand side have the same degree. The canonical homogeneous norm is continuous on $\mathbb{H}$. The dilation **d** is strongly continuous group of linear bounded operators on $\mathbb{H}$. Using Propositions 5.3 and 4.4 we conclude that the operators f and

$$x \mapsto f_q(x) := \|x\|^{\mu}_{\mathbf{d}}\mathbf{d}(\ln \|x\|_{\mathbf{d}})q$$

with $q \in \mathbb{H}$ are continuous on $\mathbb{H}$ provided that $f_q(\mathbf{0}) = \mathbf{0}$. Moreover, for any $x \in \mathbb{H}$: $\|x\| \leq 1$ we have

$$\|f(x)\| \leq C_f\sigma(\|x\|_{\mathbf{d}}) \quad \text{and} \quad \|f_q(x)\| = \left\| \|x\|^{\mu}_{\mathbf{d}}\mathbf{d}(\ln \|x\|_{\mathbf{d}})q \right\| \leq \|q\|\sigma(\|x\|_{\mathbf{d}}),$$

where $C_f = \sup_{\|x\|=1} \|f(x)\|$ and

$$\sigma(\|x\|_{\mathbf{d}}) = \max\{\|x\|^{\mu+\beta}_{\mathbf{d}}, M\|x\|^{\mu+\alpha}_{\mathbf{d}}\}$$

with $M \geq 1$ and $\alpha \geq \beta$ defined in Proposition 4.3. Moreover, for $q \in \mathcal{D}(G_{\mathbf{d}})$, by Corollary 2.6, we have

$$\|\mathbf{d}(\ln \|x_2\|_{\mathbf{d}})q - \mathbf{d}(\ln \|x_1\|_{\mathbf{d}})q\| = \left\| \int_{\ln \|x_1\|_{\mathbf{d}}}^{\ln \|x_2\|_{\mathbf{d}}} \mathbf{d}(\tau)G_{\mathbf{d}}q \, d\tau \right\| \leq$$

$$\|G_{\mathbf{d}}q\| \int_{\ln \|x_1\|_{\mathbf{d}}}^{\ln \|x_2\|_{\mathbf{d}}} \|\mathbf{d}(\tau)\| \, d\tau,$$

for all $x_1, x_2 \neq \mathbf{0}$. Hence, we conclude that the operator $x \mapsto \|x\|_{\mathbf{d}}\mathbf{d}(\ln \|x\|_{\mathbf{d}})q$ is locally Lipschitz continuous on $\mathbb{H}\backslash\{\mathbf{0}\}$ provided that $q \in \mathcal{D}(G_{\mathbf{d}})$.

(2) Let $q \neq \mathbf{0}$. Since $q \in L^\infty(\mathbb{R}, \mathbb{H})$ and $\mathcal{D}(G_{\mathbf{d}})$ is dense in $\mathbb{H}$ then for any $T > 0$ there exists a uniformly bounded sequence of piece-wise constant functions $q_n : \mathbb{R} \mapsto \mathcal{D}(G_{\mathbf{d}})$ such that

$$\|q_n - q\|_{L^1((0,T);\mathbb{H})} \to 0 \text{ as } n \to \infty$$

and $\|q_n\|_{L^\infty((0,T);\mathbb{H})} \leq 2\|q\|_{L^\infty((0,T);\mathbb{H})}$. Let x_{x_0,q_n} be mild solutions of the system with $x_0 \in \mathbb{H}\backslash\{\mathbf{0}\}$ and $q_n(t) \in \mathcal{D}(G_{\mathbf{d}})$, $\forall t \geq 0$:

$$x_{x_0,q_n}(t) = \Phi(t)x_0 + \int_0^t \Phi(t-\tau)(f(x_{x_0,q_n}(\tau)) + f_{q_n}(x_{x_0,q_n}(\tau))\,d\tau,$$

which exist and unique as long as $0 < \|x_{x_0,q_n}(t)\| < +\infty$ (since q_n is a piece-wise constant function).

Recall that, by Theorem 2.1, there exist $M_0 \geq 1$ and $\omega \geq 0$ such that $\|\Phi(t)\| \leq M_0 e^{\omega t}$, $\forall t \geq 0$. Let us show that the sequence x_{x_0,q_n} is uniformly bounded on some interval $[0, T]$ with $T > 0$:

$$\exists C_1 > 0 \;:\; \|x_{x_0,q_n}(t)\| \leq C_1, \quad \forall t \in [0, T].$$

Indeed, since

$$\begin{aligned}\|x_{x_0,q_n}(t)\| \leq & M_0 e^{\omega t}\|x_0\| \\ & + M_0(C_f + \|q_n\|_{L^\infty((0,T);\mathbb{H})}) \int_0^t e^{\omega(t-\tau)}\sigma(\|x_{x_0,q_n}(\tau))\|_{\mathbf{d}})d\tau\end{aligned}$$

and $\mu + \beta > 0$ then using the inequality (4.22) we derive

$$\begin{aligned}\|x_{x_0,q_n}(t)\| \leq & M_0 e^{\omega T}\|x_0\| \\ & + M_0(C_f + 2\|q\|_{L^\infty((0,T);\mathbb{H})}) \int_0^t e^{\omega(t-\tau)}\sigma\left(\underline{\sigma}^{-1}(\|x_{x_0,q_n}(t))\|\right)d\tau.\end{aligned}$$

Hence, using the Bihari–LaSalle inequality (see Appendix E) we conclude the uniform boundedness of the sequence $x_{x_0,q_n}(t)$ on a time interval $[0, T]$ with $T > 0$ independent of $n \in \mathbb{N}$. Moreover, for any $t \geq [0, T)$ and any $s \in (0, T-t)$ we have

$$\|x_{x_0,q_n}(t+s) - x_{x_0,q_n}(t)\| \leq \|\Phi(t+\delta)x_0 - \Phi(t)x_0\| + M_0 C_2 \int_t^{t+s} e^{\omega(t-\tau)}d\tau,$$

where

$$C_2 = \sup_{\|x\| \leq C_1, \|q_n\| \leq 2q} \|f(x) + \|x\|_{\mathbf{d}}^{\mu}\mathbf{d}(\ln\|x\|_{\mathbf{d}})q_n\| < +\infty.$$

Taking into account that

$$|\|x_{x_0,q_n}(t+\delta)\| - \|x_{x_0,q_n}(t)\|| \leq \|x_{x_0,q_n}(t+\delta) - x_{x_0,q_n}(t)\|$$

we conclude that the sequence $\|x_{x_0,q_n}(t)\|$ is equicontinuous.

Then by Arzelà–Ascoli theorem, there exists a sub-sequence q_{n_k} and function $\gamma \in C([0, T], \mathbb{R}_+)$ such that

$$\|x_{x_0,q_{n_k}}(t)\| \to \gamma(t) \quad \text{as} \quad n_k \to \infty$$

uniformly on $[0, T]$. Since $\gamma(0) = \|x_0\|$ then there exists $T' \in (0, T]$ such that $\gamma(t) > 0$ for all $t \in [0, T']$.

Since the canonical homogeneous norm is locally Lipschitz continuous on $\mathbb{H}\backslash\{\mathbf{0}\}$ then using the inequality (4.22) we conclude that the sequence

$$g_n(t) := \|x_{x_0,q_n}(t)\|_{\mathbf{d}}$$

is uniformly bounded and equicontinuous on $[0, T']$. Then, by Arzelà–Ascoli theorem, there exists a sub-sequence g_{n_k} and a function $g_0 \in C([0, T'], \mathbb{R})$ such that

$$g_{n_k}(t) \to g_0(t) \quad \text{as} \quad n_k \to \infty$$

uniformly on $[0, T']$. Notice that, in the view of the inequality (4.22), the uniform boundedness of γ on $[0, T']$ and the inequality $\gamma(t) > 0$ for all $t \in [0, T']$ imply that there exists $\delta \in (0, 1)$ such that $\delta \leq g_0(t) \leq 1/\delta$ for all $t \in [0, T']$. Moreover, since $g_{n_k} \in C([0, T']; \mathbb{R})$ and $g_{n_k} \to g_0$ as $n_k \to +\infty$ uniformly on $[0, T']$ then there exists n_k^* such that

$$0.5\delta \leq g_{n_k}(t) \leq 2/\delta, \quad \forall n_k \geq n_k^*.$$

Since $\mathbf{d}$ is a strongly continuous group of linear bonded operator then the function

$$(t, q) \mapsto g_{n_k}^{\mu}(t)\mathbf{d}(\ln g_{n_k}(t))q$$

is continuous. This implies that for any $q \in L^\infty((0, T); \mathbb{H})$ the function

$$g_{n_k}^{\mu}(\cdot)\mathbf{d}(\ln g_{n_k}(\cdot))q(\cdot) \in L^\infty((0, T'); \mathbb{H})$$

is uniformly essentially bounded on $[0, T']$ for all $n_k \geq n_k^*$. Moreover, the point-wise convergence

$$g_{n_k}^{\mu}(t)\mathbf{d}(\ln g_{n_k}(t))q(t) \to g_0^{\mu}(t)\mathbf{d}(\ln g_0(t))q(t) \text{ as } n_k \to +\infty$$

takes the place almost everywhere on $[0, T']$. By Dominated Convergence theorem for Bochner integrals (see Appendix C) we derive

$$\int_0^t \Phi(t-s)g_{n_k}^{\mu}(s)\mathbf{d}(\ln g_{n_k}(s))q(s)ds \to \int_0^t \Phi(t-s)g_0^{\mu}(s)\mathbf{d}(\ln g_0(s))q(s)ds$$

as $n_k \to +\infty$. Notice that

$$x_{x_0,g_{n_k}}(t) = \Phi(t)x_0 + \int_0^t \Phi(t-s)f(x_{x_0,g_{n_k}}(s))ds$$

$$+ \int_0^t \Phi(t-s)g^{\mu}_{n_k}(s)\mathbf{d}(\ln g_{n_k}(s))q(s)ds$$

is a unique mild solution of the system

$$\dot{x} = Ax + f(x) + g^{\mu}_{n_k}(t)\mathbf{d}(\ln g_{n_k}(t))q(t), \quad t > 0, \quad x(0) = x_0 \in \mathbb{H},$$

which always exists by Corollary 2.13. On the one hand, applying Grönwal–Bellman inequality to

$$\|x_{x_0,g_{n_k}}(t) - x_{x_0,g_0}(t)\| \le M_0 \int_0^t e^{\omega(t-\tau)}\|f(x_{x_0,g_{n_k}}(\tau)) - f(x_{x_0,g_0}(\tau))\|d\tau$$

$$+ \underbrace{\left\| \int_0^t \Phi(t-s)g^{\mu}_{n_k}(\tau)\mathbf{d}(\ln g_{n_k}(\tau))q(\tau) - g^{\mu}_0(\tau)\mathbf{d}(\ln g^{\mu}_0(\tau))q(\tau)d\tau \right\|}_{\to 0 \text{ as } n_k \to +\infty}$$

we conclude

$$\|x_{x_0,g_{n_k}}(t) - x_{x_0,g_0}(t)\| \to 0 \text{ as } n_k \to +\infty.$$

On the other hand,

$$\|x_{x_0,g_{n_k}}(t) - x_{x_0,q_k}(t)\| \le M_0 \int_0^t e^{\omega(t-\tau)}\|f(x_{x_0,g_{n_k}}(\tau)) - f(x_{x_0,q_{nk}}(\tau))\|d\tau$$

$$+ M_0 \int_0^t e^{\omega(t-\tau)}\|g^{\mu}_{n_k}(\tau))\mathbf{d}(\ln g_{n_k}(\tau))(q(\tau) - q_{n_k}(\tau))\|d\tau$$

$$\le M_0 e^{\omega T'}\left(L_\delta \int_0^t \|x_{x_0,g_{n_k}}(\tau) - x_{x_0,q_{nk}}(\tau)\|d\tau + C_0 \int_0^t \|q(\tau) - q_{n_k}(\tau)\|d\tau\right),$$

where $C_0 = \sup_{\delta \le r \le 1/\delta} \|r^{\mu}_{n_k}\mathbf{d}(\ln r)\| < +\infty$. So, using Grönwal–Bellman inequality, we derive

$$\|x_{x_0,g_{n_k}}(t) - x_{x_0,q_{n_k}}(t)\| \le M_0 e^{\omega T'} e^{M_0 e^{\omega T'} L_\delta T'} C_0 \int_0^t \|q(\tau) - q_{n_k}(\tau)\|d\tau.$$

Since $\|q(\tau) - q_{n_k}(\tau)\|_{L^\infty((0,T);\mathbb{H})} \to 0$ as $n_k \to +\infty$ then $\|x_{x_0,g_{n_k}}(t) - x_{x_0,q_{n_k}}(t)\| \to 0$ as $n_k \to +\infty$. Therefore,

$$\|x_{x_0,q_{n_k}}(t) - x_{x_0,g_0}(t)\| \leq \|x_{x_0,q_{n_k}}(t) - x_{x_0,g_{n_k}}(t)\| + \|x_{x_0,g_{n_k}}(t) - x_{x_0,g_0}(t)\| \to 0$$

$$\|x_{x_0,q_{n_k}}(t)\|_{\mathbf{d}} \to \|x_{x_0,q_0}(t)\|_{\mathbf{d}}$$

as $n_k \to +\infty$. Taking into account that $g_{n_k}(t) = \|x_{x_0,q_{n_k}}(t)\|_{\mathbf{d}} \to g_0(t)$. we conclude that $g_0(t) = \|x_{x_0,q_0}(t)\|_{\mathbf{d}}$ almost everywhere on $[0, T']$ and x_{x_0,g_0} is a mild solution of the system solution of the system (6.39) with $q \in L^\infty(\mathbb{R}; \mathbb{H})$ defined as long as $\delta \leq \|x_{x_0,q_0}(t)\|_{\mathbf{d}} \leq 1/\delta$ with $\delta \in (0, \delta)$. Tending $\delta \to 0$ we conclude that, for any $x_0 \in \mathbb{H}\backslash\{\mathbf{0}\}$ and any $q \in L^\infty(\mathbb{R}; \mathbb{H})$, the system (6.39) has a mild solution as long as $0 < \|x(t)\| < +\infty$. Since for $f(\mathbf{0}) = \mathbf{0}$ and $f_q(\mathbf{0}) = \mathbf{0}$ then the zero mild solution of the system with $x_0 = \mathbf{0}$ always exists. This means that any mild solution, which reaches the origin in a finite time, can always be prolonged.

(3) Let V be $\mathbf{d}$-homogeneous Lyapunov function of the system with $q = \mathbf{0}$, so, by Remark 6.1, we have

$$\sup_{y \in \mathcal{D}(A) \cap S} D^+V(y; Ay + f(y)) \leq -\rho < 0,$$

where $S = \{x \in \mathbb{H} : \|x\| = 1\}$. In this case, for any $y \in \mathcal{D}(A) \cap S$ we have $Ay + f(y) + f_q(x) = Ay + f(y) + q$ and

$$\frac{V(y+h(Ay+f(y)+q))-V(y)}{h} =$$

$$\frac{V(y+h(Ay+f(y)+q))-V(y+h(Ay+f(y)))}{h} + \frac{V(y+h(Ay+f(y)))-V(y)}{h}.$$

Since V is locally Lipschitz continuous on $\mathbb{H}\backslash\{\mathbf{0}\}$ then there exists a Lipschitz constant $L > 0$ for the set $\{x \in \mathbb{H} : 1/2 \leq \|x\| \leq 2\}$. So, for $y \in \mathcal{D}(A) \cap S$ and for a sufficiently small $h > 0$ we have

$$\frac{V(y + h(Ay + f(y) + q)) - V(y + h(Ay + f(y)))}{h} \leq L\|q\|$$

and

$$D^+V(y; Ay + f(y) + q) \leq L\|q\| + D^+V(y; Ay + f(y)) \leq L\|q\| - \rho. \quad (6.41)$$

Therefore, the inequality (6.40) holds and the system is globally uniformly asymptotically stable for any $q \in L^\infty(\mathbb{R}; \mathbb{H}) : \|q\|_{L^\infty} \leq 0.5\rho/L$. The proof is complete. □

6.3.2 Input-to-State Stability

Let us consider a system modeled by the evolution equation

$$\dot{x}(t) = Ax(t) + f(x(t), q(t)), \quad t > t_0, \tag{6.42}$$

where $x(t) \in \mathbb{H}$ is the system state, $q \in L^{\infty}((0, +\infty); \mathbb{V})$ describes a perturbation of the system, $A : \mathcal{D}(A) \subset \mathbb{H} \mapsto \mathbb{H}$ is a linear closed densely defined generator of a strongly continuous semigroup of bounded operators on $\mathbb{H}$ and the nonlinear mapping $f : \mathbb{H} \times \mathbb{V} \mapsto \mathbb{H}$ is locally Lipschitz continuous in the first argument and locally γ-Hölder continuous in the second one, i.e., for any $r > 0$ there exists $L_r > 0$ such that

$$\|f(x_1, q_1) - f(x_2, q_2)\| \leq L_r \left(\|x_1 - x_2\| + \|q_1 - q_2\|^{\gamma}\right), \tag{6.43}$$

for all $x_1, x_2 \in K(r)$ and $q_1, q_2 \in B(r)$, where $\gamma \in (0, 1]$, $K(r) = \{x \in \mathbb{H} : r^{-1} \leq \|x\| \leq r\}$ and $B(r) = \{q \in \mathbb{V} : \|q\| \leq r\}$ is a ball in $\mathbb{V}$ of the radius $r > 0$. By Theorem 2.14 the system (6.42) has a mild solution for any $x_0 \in \mathbb{H}$.

Let $x_{\mathbf{d}(s)x_0,\tilde{q}} : [t_0, t_0 + e^{-\mu s}T] \mapsto \mathbb{H}$ denote a mild solution of the system (6.42) for $x_0 \in \mathbb{H}$ replaced with $\mathbf{d}(s)x_0 \in \mathbb{B}$ and $q \in L^{\infty}((t_0, t_0 + T); \mathbb{V})$ replaced with $\tilde{q} \in L^{\infty}((t_0, t_0 + e^{-\mu s}T); \mathbb{V})$ given by

$$\tilde{q}(t) = \tilde{\mathbf{d}}(s)q(t_0 + e^{\mu s}(t - t_0)), \quad \forall t \in \left[t_0, t_0 + \tfrac{T}{e^{\mu s}}\right],$$

where $\tilde{\mathbf{d}}$ is a group of linear bounded operators on $\mathbb{V}$.

Theorem 6.11 *Let $\mathbf{d}$ be a group of linear bounded invertible operators in $\mathbb{H}$, the operator $A : \mathcal{D}(A) \subset \mathbb{H} \mapsto \mathbb{H}$ be $\mathbf{d}$-homogeneous of degree μ closed densely defined generator of a strongly continuous semigroup Φ of linear bounded operators on $\mathbb{H}$, the function $f : \mathbb{H} \times \mathbb{V} \mapsto \mathbb{H}$ satisfy* (6.43) *and*

$$f(\mathbf{d}(s)x, \tilde{\mathbf{d}}(s)q) = e^{\mu s}\mathbf{d}(s) f(x, q), \quad \forall x \in \mathbb{H}, \quad \forall q \in \mathbb{V}, \quad \forall s \in \mathbb{R}.$$

If $x_{x_0,q} : [t_0, t_0 + T] \mapsto \mathbb{H}$ is a mild solution of the system (6.42) *then for any $s \in \mathbb{R}$ one has*

$$x_{\mathbf{d}(s)x_0,\tilde{q}}(t) = \mathbf{d}(s)x_{x_0,q}(t_0 + e^{\mu s}(t - t_0)), \tag{6.44}$$

for all $t \in \left[t_0, t_0 + \frac{T}{e^{\mu s}}\right]$.

Proof Since the $\mathbf{d}$-homogeneous operator A generates a strongly continuous semigroup Φ, then according Lemma 6.1 we have

$$\Phi(\tau)\mathbf{d}(s) = \mathbf{d}(s)\Phi(e^{\mu s}\tau), \quad \forall \tau \geq 0, \quad \forall s \in \mathbb{R}.$$

Recall (see Appendix C) that $K\int_a^b \xi(s)ds = \int_a^b K\xi(s)ds$ for any bounded linear operator $K:\mathbb{H}\mapsto\mathbb{H}$ and any Bochner integrable function $\xi\in L^1((a,b);\mathbb{H})$.

Let $x_{x_0,q}$ be a solution of (6.42). Since $f(x_{x_0,q}(\cdot),q(\cdot))\in L^1((t_0,t_0+T);\mathbb{H})$ then

$$\begin{aligned}
&\mathbf{d}(s)x_{x_0,q}(t_0+e^{\mu s}(t-t_0))\\
&=\mathbf{d}(s)\Phi(e^{\mu s}(t-t_0))x_0+\int_{t_0}^{t_0+e^{\mu s}(t-t_0)}\mathbf{d}(s)\Phi(t_0+e^{\mu s}(t-t_0)-\tau)\circ\\
&\quad f(x_{x_0,q}(\tau),q(\tau))\,d\tau\\
&=\Phi(t-t_0)\mathbf{d}(s)x_0+e^{\mu s}\int_{t_0}^{t}\mathbf{d}(s)\Phi(e^{\mu s}(t-\sigma))\circ\\
&\quad f(x_{x_0,q}(t_0+e^{\mu s}(\sigma-t_0)),q(t_0+e^{\mu s}(\sigma-t_0)))\,d\sigma\\
&=\Phi(t-t_0)\mathbf{d}(s)x_0+e^{\mu s}\int_{t_0}^{t}\Phi(t-\sigma)\circ\\
&\quad \mathbf{d}(s)f(x_{x_0,q}(t_0+e^{\mu s}(\sigma-t_0)),q(t_0+e^{\mu s}(\sigma-t_0)))\,d\sigma\\
&=\Phi(t)\mathbf{d}(s)x_0+\int_{t_0}^{t}\Phi(t-\sigma)\circ\\
&\quad f(\mathbf{d}(s)x_{x_0,q}(t_0+e^{\mu s}(\sigma-t_0)),\tilde{\mathbf{d}}(s)q(t_0+e^{\mu s}(\sigma-t_0)))d\sigma,
\end{aligned}$$

where the linearity of the operator $\Phi(t-\sigma)$ and the **d**-homogeneity of Φ, f are utilized on the last two steps. The proof is complete. □

A bounded additive perturbation frequently destroys an asymptotic stability implying only a practical stability of the system. In this case, it is important to know if the size of an attractive set depends continuously on a magnitude of perturbations. *Input-to-State Stability (ISS)* is a well-established technique for the study of such a robustness of nonlinear ODEs (see [19] and references therein), PDEs (see [12] and references therein) as well as abstract evolution systems [20].

Definition 6.1 *(Input-to-State Stability)* The system (6.42) is said to be practically *input-to-state stable* (ISS), if there exist $c>0$ and two functions $\xi\in\mathcal{KL}$ and $\theta\in\mathcal{K}$ such that

$$\|x_{x_0,q}(t)\|\le c+\xi(\|x_0\|,t-t_0)+\theta(\|q\|_{L^\infty((t_0,t);\mathbb{V})}),\quad \forall t\ge t_0 \tag{6.45}$$

for any $t_0\in\mathbb{R}$, any solution $x_{x_0,q}$ of (6.42), any $q\in L^\infty(\mathbb{R};\mathbb{V})$ and any $x_0\in\mathbb{H}$. If the inequality holds for $c=0$ the system is ISS.

The local[5] ISS of the system (6.42) with f satisfying (6.43) is studied in [21]. The homogeneity allows a simple procedure for the ISS analysis to be developed [15].

[5] The system (6.42) is locally ISS if admissible x_0 and q in the above definition are restricted to some balls in $\mathbb{H}$ and $L^\infty(\mathbb{R};\mathbb{V})$, respectively.

The following theorem generalizes the results of [22, 23] to abstract differential equations.

Theorem 6.12 *Let* $\mathbf{d}$ *and* $\tilde{\mathbf{d}}$ *be strictly monotone strongly continuous dilations in* $\mathbb{H}$ *and* $\mathbb{V}$*, respectively, and let all conditions of Theorem 6.11 be fulfilled. If the origin of the system* (6.42) *with* $q = \mathbf{0}$ *is uniformly asymptotically stable then this system is ISS.*

Proof[6] **I.** Let $\bar{q} > 0$ be an arbitrary positive number. Let $r > 1$ and $\delta > r$ be defined as in the proof of Theorem 6.9. Since for $q = \mathbf{0}$ the system (6.42) is uniformly asymptotically stable then, due to homogeneity, it is globally uniformly asymptotically stable and exists $T_r > 0$ such that

$$x_{x_0,\mathbf{0}}(t_0) \in K(r) \quad \Rightarrow \quad \|x_{x_0,\mathbf{0}}(t)\| \leq 1/r, \quad \forall t > t_0 + T_r$$

and

$$x_{x_0,\mathbf{0}}(t_0) \in K(r) \quad \Rightarrow \quad \|x_{x_0,\mathbf{0}}(t)\| \leq \delta, \quad \forall t \geq t_0.$$

Moreover, $x_{x_0,\mathbf{0}}(t) = x^{\delta}_{x_0,\mathbf{0}}(t)$, as long as $x_{x_0,\mathbf{0}}(t) \in K(\delta)$, where $x^{\delta}_{x_0,\mathbf{0}}$ is the mild solution of

$$\dot{x} = Ax + f_{\delta}(x, q), \quad t > 0, \quad x(0) = x_0 \in \mathbb{H}, \quad f_{\delta}(x) = a_{\delta}(\|x\|) f(x, q), \tag{6.46}$$

which always exist due to Theorem 2.14. For any $x_0 \in \mathbb{H}$ we have

$$x^{\delta}_{x_0,q}(t) = \Phi(t)x_0 + \int_{t_0}^{t} \Phi(t-s) f_{\delta}(x^{\delta}_{x_0,q}(\tau), q(\tau))d\tau =$$

$$x^{\delta}_{x_0,\mathbf{0}}(t) + \int_{t_0}^{t} \Phi(t-s)(f_{\delta}(x^{\delta}_{x_0,q}(\tau), q(\tau)) - f_{\delta}(x^{\delta}_{x_0,\mathbf{0}}(\tau), \mathbf{0}))d\tau.$$

Let $q \in L^{\infty}((0, T_r); \mathbb{V}) : \|q\|_{\infty} \leq \bar{q}$. Let $M = \sup_{s \in [0,T_r]} \|\Phi(s)\|$ and $L_{\delta,T_r,q} = \tilde{L} > 0$ be defined as in Corollary 2.14. Then for any $q \in L^{\infty}((0, T_r); \mathbb{V}) : \|q\|_{\infty} \leq \bar{q}$ we have

$$\|x^{\delta}_{x_0,q}(t) - x^{\delta}_{x_0,\mathbf{0}}(t)\| \leq M L_{\delta,T_r,\bar{q}} \int_{t_0}^{t} \|x^{\delta}_{x_0,q}(\tau) - x^{\delta}_{x_0,\mathbf{0}}(\tau)\| +$$

$$M \int_{t_0}^{t} a_{\delta}(\|x^{\delta}_{x_0,\mathbf{0}}(\tau)\|) \|f(x^{\delta}_{x_0,\mathbf{0}}(\tau), q(\tau)) - f(x^{\delta}_{x_0,\mathbf{0}}(\tau), \mathbf{0})\| \, d\tau$$

[6] The proof is reprinted from *Automatica*, 129/7, A. Polyakov, Input-to-State Stability of homogeneous infinite-dimensional systems with locally Lipschitz nonlinearities, 109615, https://doi.org/10.1016/j.automatica.2021.109615. ©2021 Elsevier, with permission from Elsevier.

for all $t \in [t_0, t_0 + T_r]$ such that $\|x_{x_0,q}(t)\| < +\infty$. By assumption, there exist $\hat{L}_{\delta,\bar{q}} > 0$ such that

$$\|f(x_1, q_1) - f(x_2, q_2)\| \leq \hat{L}_{\delta,\bar{q}} \left(\|x_1 - x_2\| + \|q_1 - q_2\|^{\gamma}\right)$$

for all $x_1, x_2 \in K(2\delta)$ and for all $q_1, q_2 \in \mathbb{B}(\bar{q})$. Since $a_\delta(\|x\|) = 0$ for $x \notin K(2\delta)$ then

$$\|x^{\delta}_{x_0,q}(t) - x^{\delta}_{x_0,\mathbf{0}}(t)\| \leq M L_{\delta,T_r,\bar{q}} \int_{t_0}^{t} \|x^{\delta}_{x_0,q}(\tau) - x^{\delta}_{x_0,\mathbf{0}}(\tau)\| + M\hat{L}_{\delta,\bar{q}} \int_{t_0}^{t} \|q(\tau)\|^{\gamma}\, d\tau$$

and, using Grönwall–Bellman inequality, we derive

$$\|x^{\delta}_{x_0,q}(t) - x^{\delta}_{x_0,\mathbf{0}}(t)\| \leq \hat{L}_{\delta,\bar{q}} M e^{L_{\delta,T_r,\bar{q}} M t} \int_{t_0}^{t} \|q(\tau)\|^{\gamma} d\tau,$$

for all $t \in [0, T_r]$ and any $q \in L^{\infty}((0, T_r); \mathbb{V}) : \|q\|_{\infty} \leq \bar{q}$.

Using Theorem 6.9 (see also its proof) we conclude that there exists a **d**-homogeneous Lyapunov functional $V : \mathbb{H} \mapsto \mathbb{R}_+$ of degree 1 such that for any $x_0 \in \mathbb{H}\backslash\{\mathbf{0}\}$ we have

$$V(x_{x_0,\mathbf{0}}(h)) - V(x_{x_0,\mathbf{0}}(0)) \leq -\rho \int_0^h \|x_{x_0,\mathbf{0}}(\tau)\|_{\mathbf{d}}^{\mu+1} d\tau$$

as long as $x_{x_0,\mathbf{0}}(h) \neq \mathbf{0}$, where $\rho > 0$ is some positive number.

For any $h > 0 : x_{x_0,\mathbf{0}}(\theta), x_{x_0,q}(\theta) \in K(\delta), \forall \theta \in [0, h]$ we have

$$x_{x_0,\mathbf{0}}(\theta) = x^{\delta}_{x_0,\mathbf{0}}(\theta), \quad x_{x_0,q}(\theta) = x^{\delta}_{x_0,q}(\theta), \quad \theta \in [0, h].$$

Then taking into account the local Lipschitz continuity of V and $x_{x_0,q}(0) = x_{x_0,\mathbf{0}}(0) = x_0$ we derive

$$\begin{aligned}
&V(x_{x_0,q}(h)) - V(x_{x_0,q}(0)) \\
&= V(x_{x_0,\mathbf{0}}(h)) - V(x_{x_0,\mathbf{0}}(0)) + V(x_{x_0,q}(h)) - V(x_{x_0,\mathbf{0}}(h)) \\
&\leq -\rho \int_0^h \|x_{x_0,\mathbf{0}}(\tau)\|_{\mathbf{d}}^{\mu+1} d\tau + L_{V,\delta}\|x_{x_0,q}(h) - x_{x_0,\mathbf{0}}(h)\| \\
&\leq -\rho \int_0^h \|x_{x_0,\mathbf{0}}(\tau)\|_{\mathbf{d}} d\tau + L_{V,\delta}\hat{L}_{\delta,\bar{q}} M e^{L_{\delta,T_r,\bar{q}} M h} \int_0^h \|q(\tau)\|^{\gamma} d\tau,
\end{aligned}$$

for any $q \in L^{\infty}((0, h); \mathbb{V}) : \|q\|_{\infty} \leq \bar{q}$. Therefore, one has

$$\limsup_{h\to 0^+} \frac{V(x_{x_0,q}(h))-V(x_{x_0,q}(0))}{h} \leq -\rho\|x_0\|_{\mathbf{d}}^{1+\mu}+ \limsup_{h\to 0^+} \frac{C(\bar{q})\int_0^h \|q(\tau)\|^{\gamma}\, d\tau}{h},$$

where $C(\bar{q}) := L_{V,\delta}\hat{L}_{\delta,\bar{q}} M e^{L_{\delta,T_r,\bar{q}} M T_r}$. The function $\bar{q} \mapsto C(\bar{q})$ is uniformly bounded on any compact from $\mathbb{R}_+$.

Since $x_{x_0,q}(t+h) = x_{x^*,q^*}(h)$ with $x^* = x_{x_0,q}(t)$ and $q^*(h) := q(t+h)$ then the function $t \mapsto V(x_{x_0,q}(t))$ is strictly decreasing and

$$D^+V(x_{x_0,q}(t)) \leq -\rho\|x_{x_0,q}(t)\|_{\mathbf{d}}^{\mu+1}+ C(\bar{q}) \limsup_{h\to 0^+} \frac{1}{h}\int_0^h \|q(t+\tau)\|^{\gamma} d\tau$$

as long as $x_{x_0,q}(t) \in K(\delta)$ and

$$\rho\|x_{x_0,q}(t)\|_{\mathbf{d}}^{\mu+1} > C(\bar{q}) \limsup_{h\to 0^+} \frac{1}{h}\int_0^h \|q(t+\tau)\|^{\gamma} d\tau.$$

This conclusion holds for any $q \in L^{\infty}(\mathbb{R}; \mathbb{V})$ such that $\|q\|_{L^{\infty}((t_0,t),\mathbb{V})} \leq \bar{q}$.

II. Let $\mathbf{d}(s)x_{x_0,q}(t) \in K(r)$ or, equivalently, $x_{x_0,q}(t) \in \mathbf{d}(-s)K(r)$ for all $t \in [t_0, t_0+T]$ and some $T > 0$. From Theorem 6.11 we derive

$$x_{\mathbf{d}(s)x_0,\tilde{q}}(\tilde{t}) = \mathbf{d}(s)x_{x_0,q}(t_0+e^{\mu s}(\tilde{t}-t_0)), \quad \forall \tilde{t} \in \left[t_0, t_0+\tfrac{T}{e^{\mu s}}\right],$$

where $\tilde{t} := t_0 + e^{-\mu s}(t-t_0)$, $x_{\mathbf{d}(s)x_0,\tilde{q}} : [t_0, t_0+e^{-\mu s}T] \mapsto \mathbb{H}$ is a solution of the system (6.42) for x_0 replaced with $\mathbf{d}(s)x_0$ and $q \in L^{\infty}((t_0, t_0+T); \mathbb{V})$ replaced with $\tilde{q} \in L^{\infty}((t_0, t_0+e^{-\mu s}T); \mathbb{V})$ given by

$$\tilde{q}(\tilde{t}) = \tilde{\mathbf{d}}(s)q(t_0+e^{\mu s}(\tilde{t}-t_0)), \quad \forall \tilde{t} \in \left[t_0, t_0+\tfrac{T}{e^{\mu s}}\right].$$

Since $x_{\mathbf{d}(s)x_0,\tilde{q}}(\tilde{t}) = \mathbf{d}(s)x_{x_0,q}(t_0+e^{\mu s}(\tilde{t}-t_0)) \in K(r)$ then

$$D^+V(x_{\mathbf{d}(s)x_0,q}(\bar{t})) \leq -\|x_{\mathbf{d}(s)x_0,\tilde{q}}(\tilde{t})\|_{\mathbf{d}}^{1+\mu} + C(\bar{q}) \limsup_{h\to 0^+} \frac{\int_0^h \|\tilde{q}(\tilde{t}+\tau)\|^{\gamma} d\tau}{h}$$

for all $\tilde{t} \in [t_0, t_0+e^{-\mu s}T]$.

Taking into account homogeneity of V and $\|\cdot\|_{\mathbf{d}}$ we derive

$$\begin{aligned} V(x_{\mathbf{d}(s)x_0,\tilde{q}}(\tilde{t})) &= V(\mathbf{d}(s)x_{x_0,q}(t_0+e^{\mu s}(\tilde{t}-t_0))) \\ &= e^{s} V(x_{x_0,q}(t_0+e^{\mu s}(\tilde{t}-t_0))) \end{aligned}$$

and

$$\|x_{\mathbf{d}(s)x_0,\tilde{q}}(\tilde{t})\|_{\mathbf{d}}^{\mu+1} = e^{(\mu+1)s}\|x_{x_0,q}(t_0+e^{\mu s}(\tilde{t}-t_0))\|_{\mathbf{d}}.$$

This means that for any $q \in L^{\infty}(\mathbb{R};\mathbb{V}) : \|q\|_{L^{\infty}((t_0,t);\mathbb{V})} \le \bar{q}$ and any $x_0 \in \mathbb{H}\backslash\{\mathbf{0}\}$we have

$$D^+V(x_{x_0,q}(t)) \le -\rho\|x_{x_0,q}(t)\|_{\mathbf{d}}^{1+\mu}+ \limsup_{h\to 0^+}\frac{e^{-(\mu+1)s}C(\bar{q})\int_0^h\|\tilde{\mathbf{d}}(s)q(t+e^{\mu s}\tau)\|^{\gamma}d\tau}{h}$$

as long as $x_{x_0,q}(t) \in \mathbf{d}(-s)K(r)$.

Since the dilation $\tilde{\mathbf{d}}$ is strictly monotone then there exists $\tilde{\beta} > 0$ such that

$$\|\tilde{\mathbf{d}}(s)\| \le e^{\tilde{\beta}s}, \quad \forall s \le 0.$$

By Lemma 4.5 there exists $\sigma \in \mathcal{K}_{\infty}$ such that

$$\|q\|_{\tilde{\mathbf{d}}} \le \sigma(\|q\|) \le \sigma(\bar{q})$$

for all $q \in \mathbb{V} : \|q\| \le \bar{q}$. Hence, using the group property of $\tilde{\mathbf{d}}$, we derive

$$\tilde{\mathbf{d}}(s)q = \tilde{\mathbf{d}}(s+\ln\sigma(\bar{q}))\tilde{\mathbf{d}}(\ln\|q\|_{\tilde{\mathbf{d}}} - \ln\sigma(\bar{q}))\tilde{\mathbf{d}}(-\ln\|q\|_{\tilde{\mathbf{d}}})q$$

but the strict monotonicity of $\tilde{\mathbf{d}}$ yields

$$\|\tilde{\mathbf{d}}(s)q\| \le \|\tilde{\mathbf{d}}(s+\ln\sigma(\bar{q}))\| \le e^{\tilde{\beta}(s+\ln\sigma(\bar{q}))}, \quad \forall s \le -\ln\sigma(\bar{q}).$$

This means that

$$D^+V(x_{x_0,q}(t)) \le -\|x_{x_0,q}(t)\|_{\mathbf{d}}^{1+\mu} + e^{(\tilde{\beta}\gamma-(\mu+1))s}C(\bar{q})\sigma(\bar{q})$$

as long as $x_{x_0,q}(t) \in \mathbf{d}(-s)K(r)$ and $s \le -\ln\sigma(\bar{q})$. Selecting $s = -\ln\|x_{x_0,q}(t)\|_{\mathbf{d}}$ we derive

$$\|\mathbf{d}(-\ln\|x_{x_0,q}(t)\|_{\mathbf{d}})x_{x_0,q}(t)\| = 1$$

and $x_{x_0,q}(t) \in \mathbf{d}(-s)K(r)$. For any $q \in L^{\infty}(\mathbb{R};\mathbb{V}) : \|q\|_{L^{\infty}((t_0,t);\mathbb{V})} \le \bar{q}$ and any $x_0 \in \mathbb{H}\backslash\{\mathbf{0}\}$ we have

$$D^+V(x_{x_0,q}(t)) \le -\|x_{x_0,q}(t)\|_{\mathbf{d}}^{1+\mu}\left(\rho - \frac{C(\bar{q})\sigma^{\gamma}(\bar{q})}{\|x_{x_0,q}(t)\|_{\mathbf{d}}^{\tilde{\beta}\gamma}}\right)$$

as long as $\|x_{x_0,q}(t)\|_{\mathbf{d}} \geq \sigma(\bar{q})$. Since $C : \mathbb{R}_+ \mapsto \mathbb{R}_+$ is uniformly bounded on compacts from $\mathbb{R}_+$ and $\sigma \in \mathcal{K}_\infty$ then then there exists $\xi \in \mathcal{K}_\infty$ such that

$$\xi(\bar{q}) \geq \max\left\{\sigma(\bar{q}), (2\rho^{-1}C(\bar{q})\sigma(\bar{q}))^{1/(\tilde{\beta}\gamma)}\right\}.$$

Therefore, we have

$$D^+V(x_{x_0,q}(t)) \leq -0.5\rho\|x_{x_0,q}(t)\|_{\mathbf{d}}^{1+\mu}$$

as long as

$$\|x_{x_0,q}(t)\|_{\mathbf{d}} \geq \xi(\bar{q}),$$

where $x_{x_0,q}$ is any solution of (6.42) with an arbitrary $q \in L^\infty(\mathbb{R}; \mathbb{V})$: $\|q\|_{L^\infty((t_0,t);\mathbb{V})} \leq \bar{q}$. This means that V is the ISS Lyapunov function of the system. Repeating the conventional arguments (see, e.g., [11] for more details) we complete the proof. □

For ODE models the proof of the above theorem is much more simple due to differentiability of the homogeneous Lyapunov functions (see Chap. 7 of Volume I). However, the Lipschitz continuity of the Lyapunov function is known as the conventional requirement for the ISS analysis (see, e.g., [20]).

Example 6.8 Let $\mathbb{H} = L^2(\mathbb{R}; \mathbb{R})$ and $\mathbb{V} = \mathbb{R}$ and the system have the form

$$\dot{x}(t) = \Delta x(t) + (q(t) - \kappa\|x(t)\|)x, \quad t > 0,$$

where $\Delta : \mathcal{D}(\Delta) \subset \mathbb{H} \mapsto \mathbb{H}$ is the Laplace operator $\mathcal{D}(\Delta) = H^2(\mathbb{R}^2; \mathbb{R})$ and $\kappa \geq 0$ is a nonnegative constant. In this case we have

$$f(x,q) = -\kappa\|x\|x + qx \quad \text{and} \quad f(x,\mathbf{0}) = -\kappa\|x\|x.$$

In Example 5.2, it is shown that the operator Δ is $\mathbf{d}_x$-homogeneous of degree 1 with respect to the dilation $\mathbf{d}_x$ given by the formula (4.9) with $\Gamma = 1.25$ and $\Theta = 0.5$. It can be shown that $\tilde{f} = \binom{f}{0}$ is $\mathbf{d}$-homogeneous of degree 1 with respect the dilation $\mathbf{d}(s) = \mathrm{diag}\{\mathbf{d}_x(s), \mathbf{d}_q(s)\}$, where $\mathbf{d}_q(s) = e^s$.

For $\kappa > 0$ the unperturbed system is globally uniformly asymptotically stable since $\|\cdot\|^2$ is a strict Lapunov function of the system:

$$\langle \Delta x + f(x,\mathbf{0})x, x\rangle \leq -k\|x\|^3 < 0, \quad \forall x \neq \mathbf{0},$$

where the weak dissipativity of the Laplace operator $\langle \Delta x, x\rangle \leq 0$ is utilized in the last step.

According Theorem 6.12, the considered system is ISS for any $\kappa > 0$. If $\kappa = 0$ then the uniform asymptotic stability of the unperturbed system (and, consequently, the ISS of the perturbed system) cannot be proven.

Corollary 6.9 *Let $\mathbf{d}_x^L$ and $\mathbf{d}_q^L$ be linear strongly continuous strictly monotone dilations in a Hilbert space $\mathbb{H}$ and in a Banach space $\mathbb{V}$, respectively, where $L = 0$ or $L = \infty$. Let the closed densely defined operator $A : \mathcal{D}(A) \subset \mathbb{H} \mapsto \mathbb{H}$ be a generator of a strongly continuous semigroup Φ of linear bounded operators on $\mathbb{H}$. Let A be $\mathbf{d}_L$-homogeneous of degree $\mu_L \in \mathbb{R}$. Let the operator*

$$\begin{pmatrix} x \\ q \end{pmatrix} \in \mathbb{H} \times \mathbb{V} \mapsto \tilde{f}(x, q) := \begin{pmatrix} f(x, q) \\ \mathbf{0} \end{pmatrix} \in \mathbb{H} \times \mathbb{V}, \quad x \in \mathbb{H}, \quad q \in \mathbb{V}$$

be locally $\mathbf{d}_L$-homogeneous of degree μ_L in L-limit, where

$$\mathbf{d}_L = \begin{pmatrix} \mathbf{d}_x^L & \mathbf{0} \\ \mathbf{0} & \mathbf{d}_q^L \end{pmatrix},$$

and let $\tilde{f}_L = \left(\begin{smallmatrix} f_L \\ \mathbf{0} \end{smallmatrix}\right) : \mathbb{H} \times \mathbb{V} \mapsto \mathbb{H} \times \mathbb{V}$ be the corresponding homogeneous approximation of f. Let $f_L : \mathbb{H} \times \mathbb{V} \mapsto \mathbb{H}$ satisfy the condition (6.43) *with $\gamma = 1$. If $\mathbf{d}_L^*(s) = e^{\mu_L s}\mathbf{d}_L(s)$ with $s \in \mathbb{R}$ is a strictly monotone linear dilation in $\mathbb{H}$ and the system*

$$\dot{x} = Ax + f_L(x, \mathbf{0}), \quad t > 0 \tag{6.47}$$

is uniformly asymptotically stable then

- *the system* (6.42) *is locally ISS for $L = 0$;*
- *the system* (6.42) *is practically ISS for $L = \infty$.*

Proof Notice that, due to homogeneity of f_L, denoting

$$q^* = \|x\|_{\mathbf{d}_x^L}^{-\mu} \mathbf{d}_x^L(-\ln \|x\|_{\mathbf{d}_x^L}) f(x, q) - f_L(\mathbf{d}_x^L(-\ln \|x\|_{\mathbf{d}_x^L}), \mathbf{0}) \in \mathbb{H}$$

the system (6.42) can be rewritten as follows:

$$\dot{x} = Ax + \underbrace{f_L(x, \mathbf{0}) + \|x\|_{\mathbf{d}_x^L}^{\mu} \mathbf{d}_x^L(\ln \|x\|_{\mathbf{d}_x^L}) q^*}_{= f(x,q)}.$$

By Corollary 2.13, the considered system (6.42) has a mild solution x for any $q \in L^\infty(\mathbb{R}; \mathbb{H})$. So, $q^* \in L^\infty((0, T); \mathbb{H})$, at least, as long as $0 < \|x(t)\| < +\infty$, where $T > 0$ is any instant of time such that

$$0 < \inf_{t \in [0,T]} \|x(t)\| \le \sup_{t \in [0,T]} \|x(t)\| < +\infty.$$

By Corollary 6.8, for all $x \in \mathcal{D}(A)$ we have

$$D^+V(x; Ax + f_L(x, \mathbf{0}) + \|x\|^{\mu}_{\mathbf{d}^L_x} \mathbf{d}^L_x(\ln \|x\|_{\mathbf{d}^L_x})q^*) \le (L_V \|q^*\|_{L^\infty} - \rho)V^{1+\mu}(x),$$

provided that $q^* \in \mathbb{H}$, where $L_V > 0$ a Lipschitz constant of V on $\{x :\in \mathbb{H} : 0.5 \le \|x\|_{\mathbf{d}^L_x} \le 2\}$ and V is a $\mathbf{d}$-homogeneous Lyapunov function of degree 1 for the system (6.47), which always exists and satisfies the conditions of Theorem 6.9.

Case $L = 0$. Since f_0 is $\mathbf{d}$-homogeneous approximation of f at 0-limit then for sufficiently small $\delta_x > 0$ and $\delta_q > 0$ one holds

$$\| \|x\|^{-\mu}_{\mathbf{d}^0_x} \mathbf{d}^0_x(-\ln \|x\|_{\mathbf{d}^0_x}) f(x, q) - f_0(\mathbf{d}^0_x(-\ln \|x\|_{\mathbf{d}^0_x}), \mathbf{d}^0_q(-\ln \|x\|_{\mathbf{d}^0_x})q))\| \le \frac{\rho}{4L_V}$$

for all $x \in \mathbb{H} : \|x\| \le \delta_x$ and $q \in \mathbb{V} : \|q\| \le \delta_q$. On the other hand, since f_0 satisfy (6.43) and $\|\mathbf{d}^0_x(-\ln \|x\|_{\mathbf{d}^0_x})x\| = 1$ then

$$\| f_0(\mathbf{d}^0_x(-\ln \|x\|_{\mathbf{d}^0_x})x, \mathbf{d}^0_q(-\ln \|x\|_{\mathbf{d}^0_x})q)) - f_0(\mathbf{d}^0_x(-\ln \|x\|_{\mathbf{d}^0_x}), \mathbf{0})\| \le$$

$$L_1 \|\mathbf{d}^0_q(-\ln \|x\|_{\mathbf{d}^0_x})q\|$$

provided that $\|\mathbf{d}^0_q(-\ln \|x\|_{\mathbf{d}^0_x})q\| \le 1$, where $L_1 > 0$ is Lipschitz constant defined in the condition (6.43). Notice that, in the view of Lemma 4.5, one holds

$$L_1 \|\mathbf{d}^0_q(-\ln \|x\|_{\mathbf{d}^0_x})q\| \le \frac{\rho}{4L_V} \quad \Leftrightarrow \quad \left\| \tfrac{4L_V L_1}{\rho} q \right\|_{\mathbf{d}^0_q} \le \|x\|_{\mathbf{d}^0_x}.$$

Then for all $x \in \mathcal{D}(A) : \|x\| \le \delta_x$ and for all $q \in \mathbb{V} : \|q\| \le \delta_q$ we have

$$\left\| \tfrac{4L_V L_1}{\rho} q \right\|_{\mathbf{d}^0_q} \le \|x\|_{\mathbf{d}^0_x} \quad \Rightarrow \quad D^+V(x; Ax + f(x, q)) \le -0.5\rho V^{1+\mu}(x).$$

This means that V is a local ISS Lyapunov function of the system.

Case $L = \infty$. Similarly for all $x \in \mathcal{D}(A) : \|x\| \ge \delta_x$ and $q \in \mathbb{V} : \|q\| \ge \delta_q$ we derive

$$\left\| \tfrac{4LL_1}{\rho} q \right\|_{\mathbf{d}^0_q} \ge \|x\|_{\mathbf{d}^0_x} \quad \Rightarrow \quad D^+V(x; Ax + f(x, q)) \le -0.5\rho V^{1+\mu}(x).$$

This means that V is a practical ISS Lyapunov function of the system. The proof is complete. □

Example 6.9 Let $\mathbb{H} = L^2(\mathbb{R}; \mathbb{R})$ and $\mathbb{V} = \mathbb{R}$ and the system have the form

$$\dot{x}(t) = \Delta x(t) + (q(t) - \kappa\|x(t)\|)x(t) + q(t)\frac{x(t)}{\sqrt{\|x(t)\|}}, \quad t > 0,$$

where $\Delta : \mathcal{D}(\Delta) \subset \mathbb{H} \mapsto \mathbb{H}$ is the Laplace operator $\mathcal{D}(\Delta) = H^2(\mathbb{R}^2; \mathbb{R})$ and $\kappa \geq 0$ is a nonnegative constant. In this case we have

$$f(x, q) = -\kappa \|x\| x + qx + q\|x\|^{-1/2} x.$$

Let the dilation $\mathbf{d}_x$ be defined as in Example 6.8.

The nonlinear operator $\tilde{f} = \left(\begin{smallmatrix} f \\ 0 \end{smallmatrix}\right)$ is not homogeneous, but it admits a homogeneous approximation at zero for $\mathbf{d}_q = e^{1.5s} I$, $s \in \mathbb{R}$ and

$$\tilde{f}_0(x, q) = \left(\begin{smallmatrix} -\kappa\|x\|x + q\|x\|^{-1/2}x \\ 0 \end{smallmatrix}\right).$$

Since the unperturbed ($q = \mathbf{0}$) system is uniformly asymptotically stable then, by Corollary 6.9, the system is locally ISS.

The homogeneity in the bi-limit can be utilized for the ISS analysis too. The following theorem generalizes the result [18] to Hilbert spaces.

Corollary 6.10 *Let $\mathbf{d}_x^L$ and $\mathbf{d}_q^L$ be linear strongly continuous strictly monotone dilations in a Hilbert space $\mathbb{H}$ and in Banach space $\mathbb{V}$, respectively, where $L = 0$ or $L = \infty$. Let a closed densely defined operator $A : \mathcal{D}(A) \subset \mathbb{H} \mapsto \mathbb{H}$ be a generator of a strongly continuous semigroup Φ of linear bounded operators on $\mathbb{H}$. Let A be $\mathbf{d}_0$-homogeneous of degree $\mu_0 \in \mathbb{R}$ and $\mathbf{d}_\infty$-homogeneous of degree $\mu_\infty \in \mathbb{R}$. Let $f : \mathbb{H} \times \mathbb{V} \mapsto \mathbb{H}$ satisfy* (6.43) *with $\gamma = 1$. Let the operator*

$$\begin{pmatrix} x \\ q \end{pmatrix} \in \mathbb{H} \times \mathbb{V} \mapsto \tilde{f}(x, q) := \begin{pmatrix} f(x, q) \\ \mathbf{0} \end{pmatrix} \in \mathbb{H} \times \mathbb{V}, \quad x \in \mathbb{H}, \quad q \in \mathbb{V}$$

be locally $\tilde{\mathbf{d}}_0$-homogeneous of degree μ_0 in 0-limit and $\tilde{\mathbf{d}}_\infty$-homogeneous of degree μ_∞ at ∞-limit, where

$$\tilde{\mathbf{d}}_0 = \begin{pmatrix} \mathbf{d}_x^0 & \mathbf{0} \\ \mathbf{0} & \mathbf{d}_q^0 \end{pmatrix}, \quad \tilde{\mathbf{d}}_\infty = \begin{pmatrix} \mathbf{d}_x^\infty & \mathbf{0} \\ \mathbf{0} & \mathbf{d}_q^\infty \end{pmatrix},$$

and let $\tilde{f}_0 = \left(\begin{smallmatrix} f_0 \\ \mathbf{0} \end{smallmatrix}\right) : \mathbb{H} \times \mathbb{V} \mapsto \mathbb{H} \times \mathbb{V}$ and $\tilde{f}_\infty = \left(\begin{smallmatrix} f_\infty \\ \mathbf{0} \end{smallmatrix}\right) : \mathbb{H} \times \mathbb{V} \mapsto \mathbb{H} \times \mathbb{V}$ be the corresponding homogeneous approximations. Let f_0 and f_∞ satisfy (6.43). *If the systems*

$$\dot{x} = f_0(x, \mathbf{0}), \quad \dot{x} = f(x, \mathbf{0}), \quad \dot{x} = f_\infty(x, \mathbf{0}), \quad t > 0$$

are globally asymptotically stable then the system (6.42) *is ISS.*

Proof By Corollary 6.6, there exists a locally Lipschitz continuous (on $\mathbb{H}\backslash\{\mathbf{0}\}$) Lyapunov function V such that

$$D^+V(x; Ax + f(x, \mathbf{0})) \leq -\rho\sigma(V(x)), \quad \forall x \in \mathcal{D}(A)\backslash\{\mathbf{0}\},$$

where $\rho > 0$ and $\sigma \in \mathcal{K}_\infty$ has the form (6.35) with $m_0 > \max\{0, -\mu_0\}$ and $m_\infty > \max\{0, -\mu_\infty\}$.

Let $q : \mathbb{R} \mapsto \mathbb{V}$ be locally Lipschitz continuous and $\|q\|_{L^\infty(\mathbb{R},\mathbb{V})} \leq \bar{q}$ then, by Theorem 2.16, any solution of (6.42) with $x_0 \in \mathcal{D}(A)$ is strong and $x_{x_0,q}(t) \overset{a.e.}{\in} \mathcal{D}(A)$.

If $x_0 \in \mathcal{D}(A) \cap K$ with $\tilde{K}(r_0, r_\infty) := \{x \in \mathbb{H} : 0.5r_0 \leq \|x\|_{\mathbf{d}_0}, \|x\|_{\mathbf{d}_L} \leq 2r_\infty\}$ then for $x_{x_0,q}(t) \in \mathcal{D}(A) \cap \tilde{K}(r_0, r_\infty)$ there exists $\bar{h} > 0$ such that $x_{x_0,q}(t) + h(Ax_{x_0,q}(t) + f(x_{x_0,q}(t), q(t))) \in \tilde{K}(0.5r_0, 2r_\infty)$. Let $L > 0$ be a Lischitz constant for V on set $\tilde{K}(0.5r_0, 2r_\infty)$. Let $\tilde{L} > 0$ be a Lipschitz constant of f on $\tilde{K}(0.5r_0, 2r_\infty) \times B(\bar{q})$. Then for $h \in (0, \bar{h})$ we have

$$\frac{V(x_{x_0,q}(t)+h(Ax_{x_0,q}(t)+f(x_{x_0,q}(t),q(t))))-V(x_{x_0,q}(t)+h(Ax_{x_0,q}(t)+f(x_{x_0,q}(t),\mathbf{0})))}{h}$$

$$\leq L\|f(x_{x_0,q}(t), q(t)) - f(x_{x_0,q}(t), \mathbf{0})\| \leq L\tilde{L}\|q(t)\|.$$

Since

$$\frac{V(x_{x_0,q}(t) + h(Ax_{x_0,q}(t) + f(x_{x_0,q}(t), q(t))) - V(x_{x_0,q}(t)))}{h} =$$

$$\frac{V(x_{x_0,q}(t)+h(Ax_{x_0,q}(t)+f(x_{x_0,q}(t),q(t))))-V(x_{x_0,q}(t)+h(Ax_{x_0,q}(t)+f(x_{x_0,q}(t),\mathbf{0})))}{h}$$

$$+ \frac{V(x_{x_0,q}(t)+h(Ax_{x_0,q}(t)+f(x_{x_0,q}(t),\mathbf{0})))-V(x_{x_0,q}(t)+h(Ax_{x_0,q}(t)+f(x_{x_0,q}(t),\mathbf{0})))}{h}$$

then

$$D^+V(x_{x_0,q}(t); Ax_{x_0,q}(t) + f(x_{x_0,q}(t), q(t)))$$

$$\leq D^+V(x_{x_0,q}(t); Ax_{x_0,q}(t) + f(x_{x_0,q}(t), \mathbf{0})) + L\tilde{L}\|q(t)\|$$

$$\leq -\sigma(V(x_{x_0,q}(t))) + L\tilde{L}\|q(t)\|$$

taking $t = 0$ we derive

$$D^+V(x; Ax + f(x, q)) \leq -\sigma(V(x)) + L\tilde{L}\|q\|$$

for all $x \in \mathcal{D}(A) \cap \tilde{K}(r_0, r_\infty)$ and all $q \in B(\bar{q}) \subset \mathbb{V}$. Repeating the proof of Corollary 6.9 for both cases $L = 0$ and $L = \infty$, we conclude that V is an ISS Lyapunov function of the system. The proof is complete. □

Example 6.10 Considering the locally homogeneous system from Example 6.9 we conclude that this system also has a homogeneous approximation at infinity

$$\tilde{f}_\infty(x, q) = \begin{pmatrix} -\kappa\|x\|x+qx \\ 0 \end{pmatrix}.$$

The local ISS of this system is studied in Example 6.8. Applying Corollary 6.10 we conclude that the system is globally ISS.

Concluding Remarks

In this chapter, it is shown that the stability and robustness of infinite-dimensional homogeneous systems admits an analysis similar to the finite-dimensional case studied in Volume I. In particular, Zubov–Rosier theorem about existence of homogeneous Lyapunov functions for asymptotically stable homogeneous systems is extended to the infinite-dimensional setting. Forward completeness of the system can be analyzed using the homogeneity degree. The robustness (Input-to-State Stability) of a homogeneous system is proven to be equivalent to the uniform asymptotic stability of the unperturbed system. The stability and robustness of a non homogeneous system can be analyzed using local homogeneous approximations. The key feature of the infinite-dimensional analysis is the use of locally Lipschitz (possibly non-differentiable) Lyapunov functions.

References

1. Polyakov A, Efimov D, Fridman E, Perruquetti W (2016) On homogeneous distributed parameters equations. IEEE Trans Autom Control 61(11):3657–3662
2. Leray J (1938) Sur le mouvement d'un liquide visqueux que limitent des parois. Acta Math 63:193–248
3. Ozanski W, Pooley B (2018) Leray's fundamental work on the Navier–Stokes equations: a modern review of "Sur le mouvement d'un liquide visqueux emplissant l'espace", pp 11–203. London Mathematical Society Lecture Note Series: Partial Differential Equations in Fluid Mechanics, Series Number 452. Cambridge University Press
4. Zubov VI (1964) Methods of A.M. Lyapunov and Their Applications. Noordhoff, Leiden, (Translated from V.I. Zubov, Metody A.M. Lyapunova i ih primenenie, Izdatelstvo Leningradskogo universiteta, 1957)
5. Nakamura H, Yamashita Y, Nishitani H (2002) Smooth Lyapunov functions for homogeneous differential inclusions. In: Proceedings of the 41st SICE annual conference, pp 1974–1979
6. Pisano A, Orlov Y, Usai E (2011) Tracking control of the uncertain heat and wave equation via power-fractional and sliding-mode techniques. SIAM J Control Optim 49(2):363–382
7. Orlov Y (2020) Nonsmooth Lyapunov analysis in finite and infinite dimensions. Springer, Berlin
8. La Salle J, Lefschetz S (1961) Stability by Liapunov's direct method with applications. Elsevier Science
9. Zuyev A (2003) Partial asymptotic stability and stabilization of nonlinear abstract differential equations. In: IEEE conference on decision and control
10. Zubov VI (1958) On systems of ordinary differential equations with generalized homogeneous right-hand sides. Izvestia vuzov. Mathematica (available on-line). http://mi.mathnet.ru/ivm2874 1:80–88
11. Sontag ED, Wang Y (1996) On characterizations of the input-to-state stability property. Syst Control Lett 24(5):351–359

12. Karafyllis I, Krstic M (2018) Input-to-state stability for PDEs. Springer, Berlin
13. Mironchenko A (2013) Input-to-state stability: theory and applications. Springer, Berlin
14. Rosier L (1992) Homogeneous Lyapunov function for homogeneous continuous vector field. Syst Control Lett 19:467–473
15. Polyakov A (2021) Input-to-state stability of homogeneous infinite dimensional systems with locally lipschitz nonlinearities. Automatica 129:109615
16. Polyakov A (2021) Homogeneous Lyapunov functions for homogeneous infinite dimensional systems with unbounded nonlinear operators. Syst Control Lett 148(2):104854
17. Dolbeault J, Esteban MJ, Laptev A, Loss M (2014) One-dimensional Gagliardo–Nirenberg–Sobolev inequalities: remarks on duality and flows. J Lond Math Soc 90(10):525–550
18. Andrieu V, Praly L, Astolfi A (2008) Homogeneous approximation, recursive observer design, and output feedback. SIAM J Control Optim 47(4):1814–1850
19. Sontag ED (2007) Input to state stability: basic concepts and results. In: Nistri PP, Stefani G (eds) Nonlinear optimal control theory. Springer, Berlin, pp 163–220
20. Mironchenko A, Prieur C (2020) Input-to-state stability of infinite dimensional systems: Recent results and open questions. SIAM Rev 62(3):529–614
21. Dashkovskiy S, Mironchenko A (2013) Input-to-state stability of infinite-dimensional control systems. Math Control Signals Syst 25(1):1–35
22. Ryan EP (1995) Universal stabilization of a class of nonlinear systems with homogeneous vector fields. Syst Control Lett 26:177–184
23. Hong Y (2001) H_∞ control, stabilization, and input-output stability of nonlinear systems with homogeneous properties. Automatica 37(7):819–829

Chapter 7
Discretization of Infinite-Dimensional Homogeneous Systems

An approximation of an infinite-dimensional system by a finite-dimensional analog is required for numerical simulation and/or for a practical implementation of control/estimation algorithms in a digital device. This chapter develops a method of consistent discretization for homogeneous evolution equations in separable Hilbert spaces and proposes a special modification of the Galerkin method allowing the dilation symmetry of an infinite-dimensional system to be preserved in its finite-dimensional counterpart.

7.1 On Finite-Dimensional Approximation of Infinite-Dimensional System

7.1.1 Basics of Galerkin Method

Let us consider the weak formulation of the equation

$$Ax = f$$

in a Hilbert space $\mathbb{H}$, namely

$$\text{find } x \in \mathbb{H} \text{ such that } \langle Ax, v\rangle = \langle f, v\rangle \text{ for all } v \in \mathbb{H},$$

where $A : \mathcal{D}(A) \subset \mathbb{H} \mapsto \mathbb{H}$ is a linear operator and $f \in \mathbb{H}$. The Galerkin method (see, for example, [1]) deals with the weak formulation on a finite-dimensional linear subspace $V \subset \mathbb{H}$:

A. Polyakov, *Generalized Homogeneity in Systems and Control Volume II*, Communications and Control Engineering,
https://doi.org/10.1007/978-3-031-77496-6_7

$$\text{find } \tilde{x} \in V \subset \mathbb{H} \text{ such that for all } v \in V \subset \mathbb{H},\ \langle A\tilde{x}, v\rangle = \langle f, v\rangle.$$

If $\{h_i\}$ is an orthonormal basis in V, then taking into account

$$\tilde{x} = \sum_{j=1}^{n} \langle \tilde{x}, h_j\rangle h_j = \sum_{i=1}^{n} \alpha_j h_j, \quad v = \sum_{i=1}^{n} \langle v, h_i\rangle h_i = \sum_{i=1}^{n} \beta_i h_i,$$

$$\alpha_j := \langle \tilde{x}, h_j\rangle, \quad \beta_i := \langle v, h_i\rangle,$$

the latter problem can be reformulated in $\mathbb{R}^n$ as follows:

$$\text{find } \alpha \in \mathbb{R}^n \text{ such that } \sum_{j=1}^{n}\sum_{i=1}^{n} \langle Ah_j, h_i\rangle \alpha_j \beta_i = \sum_{i=1}^{n} \langle f, h_i\rangle \beta_i, \quad \forall \beta \in \mathbb{R}^n,$$

or, equivalently,

$$\text{find } \alpha \in \mathbb{R}^n \text{ such that } \tilde{A}_n \alpha = \tilde{f}_n,$$

where $\tilde{A}_n = \{a_{ij}\} \in \mathbb{R}^{n\times n}$ is a matrix with the elements $a_{ij} = \langle Ah_j, h_i\rangle$, and the components of the vector $\tilde{f}_n = (f_1, ..., f_n)^\top \in \mathbb{R}^n$ are $f_i = \langle f, h_i\rangle$, $i = 1, 2, ..., n$.

The obtained linear equation in $\mathbb{R}^n$ defines the Galerkin projection of the equation $Ax = f$ onto the finite-dimensional linear subspace V being isomorphic to $\mathbb{R}^n$, and the matrix $\tilde{A}_n$ is the Galerkin projection of the linear operator A onto V.

In the case of evolution equation, we have $f = \dot{x}$, so the Galerkin method gives the ODE

$$\dot{\alpha} = \tilde{A}_n \alpha,$$

which defines a projection of an infinite-dimensional dynamics $\dot{x} = Ax$ onto the finite-dimensional linear subspace V. The presented algorithm is referred to below as the *classical Galerkin method*.

The classical Galerkin method is well-developed for systems with linear operators. Under some mild restrictions on A, one can be shown that the approximate solution

$$\tilde{x} = \sum_{i=1}^{n} \alpha_i h_i \in V$$

converges to the exact solution $x \in \mathbb{H}$ as $n \to +\infty$. The classical Galerkin method is utilized for nonlinear systems [1] as well.

7.1.2 *Inconsistency of the Classical Galerkin Method for Homogeneous Systems*

As a motivating example, let us consider the viscous Burgers equation

$$\tfrac{\partial x}{\partial t}=\nu\tfrac{\partial^2 x}{\partial z^2}-x\tfrac{\partial x}{\partial z},\quad t>0,\quad x(0,z)=x_0(z),\quad z\in\mathbb{R},\quad x_0\in L^2(\mathbb{R};\mathbb{R}),\tag{7.1}$$

where $\nu>0$ is the viscosity coefficient. It is easy to see that this equation is **d**-homogeneous of the degree 2 with respect to the dilation (4.9) with $\Gamma=\Theta=1$. Indeed, the identity

$$\nu\tfrac{\partial^2\mathbf{d}(s)x}{\partial z^2}-(\mathbf{d}(s)x)\tfrac{\partial\mathbf{d}(s)\circ x}{\partial z}=e^{2s}\mathbf{d}(s)\left(\nu\tfrac{\partial^2 x}{\partial z^2}-x\tfrac{\partial x}{\partial z}\right)$$

holds $\forall s\in\mathbb{R}$, $\forall x\in H^2(\mathbb{R};\mathbb{R})$, $\forall z\in\mathbb{R}$. Therefore, solutions of the Burgers equation are symmetric with respect to the scaling $x_0\mapsto\mathbf{d}(s)x_0$ of the initial state (see Theorem 6.1).

Let $\langle\cdot,\cdot\rangle$ be an inner product in $L^2(\mathbb{R};\mathbb{R})$. The classical Galerkin projection of the Burgers equation is given by the following ODE (see [2] for more details):

$$\frac{d}{dt}\tilde{x}(t)=\nu A_n\tilde{x}(t)-B(\tilde{x}(t))\tilde{x}(t),\quad t>0,\tag{7.2}$$

where $\tilde{x}=(\tilde{x}_1,\dots,\tilde{x}_n)^\top\in\mathbb{R}^n$ approximates the solution of the original system $x(t)\in L^2(\mathbb{R};\mathbb{R})$ as follows:

$$x(t)\approx\sum_{i=1}^n\tilde{x}_i(t)h_i,\quad t\geq 0$$

with $h_i\in L^2, i=1,2,\dots,n$ being a (sufficiently smooth) orthonormal family in L^2, $A_n=\{\langle h_i,\tfrac{\partial^2}{\partial z^2}h_j\rangle\}_{i,j=1}^n\in\mathbb{R}^{n\times n}$, and $\tilde{x}\mapsto B(\tilde{x})\in\mathbb{R}^{n\times n}$ is a linear mapping satisfying

$$\tilde{y}^\top B(\tilde{x})\tilde{x}=\left\langle\sum_{j=1}^n\tilde{y}_jh_j,\left(\sum_{k=1}^n\tilde{x}_kh_k\right)\frac{\partial}{\partial z}\sum_{i=1}^n\tilde{x}_ih_i(z)\right\rangle,\quad\forall\tilde{x},\tilde{y}\in\mathbb{R}^n.$$

However, the dilation symmetry of the Burgers equation is destroyed by the classical Galerkin projection. Indeed, the first term on the right-hand side of (7.2) is standard homogeneous of the degree 0:

$$A_n\mathbf{d}(s)\tilde{x}=e^{0s}\mathbf{d}(s)A_n\tilde{x},\quad\forall s\in\mathbb{R},\quad\forall\tilde{x}\in\mathbb{R}^n$$

provided that $\mathbf{d}(s)=e^sI_n$. Notice that the matrix A_n is not nilpotent in the general case, so, in the view of Lemma 6.1 of Volume I, it cannot be **d**-homogeneous of nonzero degree. The second (nonlinear) term of the system is standard homogeneous of degree 1:

$$B(\mathbf{d}(s)\tilde{x})\mathbf{d}(s)\tilde{x} = e^{1s}\mathbf{d}(s)B(\tilde{x})\tilde{x}, \quad \forall s \in \mathbb{R}, \quad \forall \tilde{x} \in \mathbb{R}^n.$$

Therefore, the right-hand side of the system (7.2) is not $\mathbf{d}$-homogeneous if $\nu \neq 0$. Solutions of (7.2) are not symmetric with respect to homogeneous scaling of the initial condition. Therefore, the classical Galerkin method does not preserve the dilation symmetry of the viscous Burgers equation. The same conclusions can be made for the Galerkin projection of Navier–Stokes equations [3]. Therefore, *an approximation of the homogeneous infinite-dimensional system of nonzero degree by means of the classical Galerkin method is inconsistent in the sense that the corresponding finite-dimensional system is nonhomogeneous.*

There are two possible ways to preserve the homogeneity of an infinite-dimensional system in its finite-dimensional approximation. The first one is to use a Lie-symmetry-preserving numerical integration scheme based on finite-difference approximations [4] which would immediately provide some nonlinear discrete-time model of the system. The second way is to develop a Galerkin-like method which would preserve a dilation symmetry of the considered Burgers equation in its finite-dimensional reduced-order model (an ODE). The second option could be more useful for the theory of homogeneous control systems, since all design algorithms developed in Volume I are based on homogeneous ODEs.

In this chapter, we develop the so-called *homogeneous Galerkin method* which allows a homogeneity of an infinite-dimensional evolution equation to be preserved in its finite-dimensional counterpart (projection). First, we study the problem of finite-dimensional approximation of a linear dilation in a separable Hilbert space $\mathbb{H}$. Next, we construct homogeneous Galerkin method based on polar homogeneous coordinates introduced in $\mathbb{H}$. Finally, we illustrate the efficiency of the developed method on the Burgers equation considered above.

7.2 Approximation of Linear Dilation

7.2.1 Linear Dilation in a Separable Hilbert Space

First, let us design an approximation of a linear dilation group in $\mathbb{H}$. If the dilation is strictly monotone then its generator is a linear positive (but possibly non-symmetric) operator (see Proposition 4.7). The conventional Galerkin projection can be utilized for an approximation of the generator and the dilation group itself.

Let $h_i \in \mathbb{H}, i = 1, 2, \ldots$ be an orthonormal basis in $\mathbb{H}$. Such a countable basis exists if $\mathbb{H}$ is assumed to be separable (see Appendix A). Let the operator $\Pi_n : \mathbb{H} \mapsto \mathbb{R}^n$ be defined as follows:

$$\Pi_n x = \begin{pmatrix} \langle x, h_1 \rangle \\ \langle x, h_2 \rangle \\ \dots \\ \langle x, h_n \rangle \end{pmatrix}. \tag{7.3}$$

In this case, the orthogonal projector $P_n : \mathbb{H} \mapsto \text{span}\{h_1, ..., h_n\}$ to the linear subspace $\text{span}\{h_1, ..., h_n\}$ is given by

$$P_n x = \sum_{i=1}^{n} (\Pi_n x)_i h_i, \quad x \in \mathbb{H}.$$

Obviously that $P_n x \in \text{span}\{h_1, ..., h_n\}$ for any $x \in \mathbb{H}$ and $P_n^2 = P_n$.

Proposition 7.1 *Let* **d** *be a strictly monotone strongly continuous linear dilation in a separable Hilbert space* $\mathbb{H}$. *If* $h_i \in \mathcal{D}(G_{\mathbf{d}})$, $i = 1, 2, ...$ *being an orthonormal basis in* $\mathbb{H}$ *then*

- *for any* $x \in \mathcal{D}(G_{\mathbf{d}})$ *we have*

$$G_{\mathbf{d}} x = \sum_{i=1}^{+\infty} \sum_{j=1}^{+\infty} \langle G_{\mathbf{d}} h_i, h_j \rangle \langle x, h_i \rangle h_j; \tag{7.4}$$

- *for any* $x \in \mathbb{H}$ *we have*

$$\mathbf{d}(s)x = \sum_{i=1}^{+\infty} \sum_{j=1}^{+\infty} q_{ij}(s) \langle x, h_i \rangle h_j,$$

with $q_{ij}(s) = \langle \mathbf{d}(s) h_i, h_j \rangle$ *satisfying*

$$\frac{dq_{ij}(s)}{ds} = \sum_{k=1}^{\infty} \langle G_{\mathbf{d}} h_i, h_k \rangle q_{kj}(s), \quad q_{ij}(0) = \delta_{ij}, \quad i, j = 1, 2, ..., \tag{7.5}$$

where $\delta_{ij} = 0$ *if* $i \neq j$ *and* $\delta_{ii} = 1$;

- *the group*

$$\mathbf{d}_n(s) = e^{sG_{\mathbf{d}_n}}, \quad G_{\mathbf{d}_n} = \begin{pmatrix} \langle G_{\mathbf{d}} h_1, h_1 \rangle & \langle G_{\mathbf{d}} h_2, h_1 \rangle & \dots & \langle G_{\mathbf{d}} h_n, h_1 \rangle \\ \langle G_{\mathbf{d}} h_1, h_2 \rangle & \langle G_{\mathbf{d}} h_2, h_2 \rangle & \dots & \langle G_{\mathbf{d}} h_n, h_2 \rangle \\ \dots & \dots & \dots & \dots \\ \langle G_{\mathbf{d}} h_1, h_n \rangle & \langle G_{\mathbf{d}} h_2, h_n \rangle & \dots & \langle G_{\mathbf{d}} h_n, h_n \rangle \end{pmatrix} \in \mathbb{R}^{n \times n} \tag{7.6}$$

is a strictly monotone continuous linear dilation in $\mathbb{R}^n$, *where* $s \in \mathbb{R}$ *is a group parameter and*

$$G_{\mathbf{d}_n} + G_{\mathbf{d}_n}^\top \succeq 2\beta I_n \succ 0, \tag{7.7}$$

where the parameter $\beta > 0 : \|\mathbf{d}(s)\| \leq e^{\beta s}, \forall s \leq 0$ *always exists due to strict monotonicity of* $\mathbf{d}$*;*

- *for any* $x \in \mathbb{H}$ *we have*

$$\Pi_n G_{\mathbf{d}} P_n x = G_{\mathbf{d}_n} \Pi_n x, \tag{7.8}$$

$$(\Pi_n \mathbf{d}(s) P_n x)_k = (q_{1k}(s), q_{2k}(s), ..., q_{nk}(s)) \Pi_n x, \quad k = 1, ..., n. \tag{7.9}$$

Proof (1) If $x \in \mathcal{D}(G_{\mathbf{d}})$ then taking into account $h_i \in \mathcal{D}(G_{\mathbf{d}})$ we derive

$$\underbrace{\sum_{i=1}^{n} \langle x, h_i \rangle h_i}_{\in D(G_{\mathbf{d}})} \to x \in D(G_{\mathbf{d}}).$$

Since the operator $G_{\mathbf{d}}$ is closed then

$$\sum_{i=1}^{N} \langle x, h_i \rangle G_{\mathbf{d}} h_i = G_{\mathbf{d}} \sum_{i=1}^{N} \langle x, h_i \rangle h_i \to G_{\mathbf{d}} x \text{ as } N \to +\infty.$$

On the other hand, since $G_{\mathbf{d}} h_i \in \mathbb{H}$ then

$$G_{\mathbf{d}} h_i = \sum_{j=1}^{+\infty} \langle G_{\mathbf{d}} h_i, h_j \rangle h_j.$$

(2) For any $x \in \mathbb{H}$ we have $\mathbf{d}(s)x \in \mathbb{H}$ and

$$\mathbf{d}(s)x = \sum_{i=1}^{+\infty} \langle x, h_i \rangle \mathbf{d}(s) h_i = \sum_{i=1}^{+\infty} \sum_{j=1}^{+\infty} \langle \mathbf{d}(s) h_i, h_j \rangle \langle x, h_i \rangle h_j.$$

Show that $q_{ij}(s) = \langle \mathbf{d}(s) h_i, h_j \rangle$ satisfies (7.5). Indeed, for $q_{i,j}(0) = \langle \mathbf{d}(0) h_i, h_j \rangle = \delta_{ij}$, and since $h_i \in \mathcal{D}(G_{\mathbf{d}})$ then

$$\frac{dq_{ij}(s)}{ds} = \langle \mathbf{d}(s) G_{\mathbf{d}} h_i, h_j \rangle = \left\langle \mathbf{d}(s) \sum_{k=1}^{\infty} \langle G_{\mathbf{d}} h_i, h_k \rangle h_k, h_j \right\rangle =$$

$$\sum_{k=1}^{\infty} \langle G_{\mathbf{d}} h_i, h_k \rangle q_{k,j}(s).$$

(3) Let us show than the matrix $G_{\mathbf{d}_n} + G_{\mathbf{d}_n}^{\top}$ is positive definite. Since $\langle G_{\mathbf{d}} x, x \rangle \geq \beta \|x\|^2 > 0$ for any $x \in \mathcal{D}(G_{\mathbf{d}})$ then for any $\alpha \in \mathbb{R}^n$ we have

$$\left\langle G_{\mathbf{d}}\sum_{i=1}^{n}\alpha_i h_i, \sum_{j=1}^{n}\alpha_j h_j\right\rangle \geq \beta\left\langle\sum_{i=1}^{n}\alpha_i h_i, \sum_{j=1}^{n}\alpha_j h_j\right\rangle$$

or, equivalently,

$$\sum_{i=1}^{n}\sum_{j=1}^{n}\alpha_i\alpha_j\langle G_{\mathbf{d}}h_i, h_j\rangle \geq \beta\sum_{i=1}^{n}\alpha_i^2,$$

where the identities $\langle h_i, h_i\rangle = 0$ for $i = 1, 2, .., n$ and $\langle h_i, h_j\rangle = 0$ for $i \neq j$ are utilized in the last step. The obtained inequality immediately implies (7.7).

(4) For any $x \in \mathbb{H}$ we have $P_n x \in \mathcal{D}(G_{\mathbf{d}})$ and

$$G_{\mathbf{d}}P_n x = \sum_{i=1}^{n}(\Pi_n x)_i h_i = \sum_{i=1}^{n}(\Pi_n x)_i G_{\mathbf{d}}h_i = \sum_{i=1}^{n}\sum_{j=1}^{+\infty}(\Pi_n x)_i\langle G_{\mathbf{d}}h_i, h_j\rangle h_j.$$

Since $\{h_j\}$ is an orthonormal basis then

$$(\Pi_n G_{\mathbf{d}}P_n x)_k = \sum_{i=1}^{n}\sum_{j=1}^{+\infty}(\Pi_n x)_i\langle G_{\mathbf{d}}h_i, h_j\rangle\langle h_j, h_k\rangle = \sum_{i=1}^{n}(\Pi_n x)_i\langle G_{\mathbf{d}}h_i, h_k\rangle,$$

i.e., the identity (7.8) holds.

Similarly, for any $x \in \mathbb{H}$ we have

$$(\Pi_n\mathbf{d}(s)P_n x)_i = \langle\mathbf{d}(s)P_n x, h_i\rangle = \sum_{j=1}^{n}(\Pi_n x)_j\langle\mathbf{d}(s)h_j, h_i\rangle,$$

i.e., the identity (7.9) holds. The proof is complete. □

The identity (7.8) means that the matrix $G_{\mathbf{d}_n} \in \mathbb{R}^{n\times n}$ is a conventional Galerkin projection of the linear operator $G_{\mathbf{d}}$ on a finite-dimensional linear subspace $V = \text{span}\{h_1, ..., h_n\}$ of a Hilbert space $\mathbb{H}$.

Recall that $y(s) = \mathbf{d}(s)x$ is a solution of the linear evolution equation $\dot{y} = G_{\mathbf{d}}y$. The formula (7.9) suggests a way to project the dilation $\mathbf{d}$ on $\text{span}\{h_1, ..., h_n\}$. Indeed, since

$$\frac{d}{ds}(\mathbf{d}_n(s))_{ij} = \sum_{k=1}^{n}\langle G_{\mathbf{d}}h_i, h_k\rangle(\mathbf{d}_n(s))_{ij}, \quad (\mathbf{d}_n(0))_{ij} = \delta_{ij}, \quad i, j = 1, 2, ..., n$$

then $(\mathbf{d}_n(s))_{ij}$ can be considered as approximations of $q_{ij}(s)$, and, consequently, $\mathbf{d}_n(s)\Pi_n x$ can be treated as an approximation of $\Pi_n\mathbf{d}(s)P_n x$. In particular, if h_i are orthogonal eigenvectors of $G_{\mathbf{d}}$ then $\langle G_{\mathbf{d}}h_i, h_j\rangle = 0$ for $i \neq j$, i.e., the matrix $G_{\mathbf{d}_n}$ is diagonal for any $n \geq 1$, and we derive

$$\Pi_n \mathbf{d}(s) P_n x = \mathbf{d}_n(s) \Pi_n x.$$

This yields the conventional Galerkin projection, which guarantees that the projection error

$$e = \mathbf{d}(s)x - \sum_{i=1}^{n} (\mathbf{d}_n(s)\Pi_n x)_i h_i$$

is orthogonal to span$\{h_1, ..., h_n\}$:

$$\left\langle \mathbf{d}(s)x - \sum_{i=1}^{n} (\mathbf{d}_n(s)\Pi_n x)_i h_i, P_n y \right\rangle =$$

$$\left\langle \sum_{i=1}^{+\infty} \langle \mathbf{d}(s)h_i, h_i \rangle \langle x, h_i \rangle h_j - \sum_{i=1}^{n} \langle \mathbf{d}(s)h_i, h_i \rangle (\Pi_n x)_i h_i, P_n y \right\rangle = 0, \quad \forall y \in \mathbb{H}.$$

If $\Pi_n \mathbf{d}(s) P_n x \neq \mathbf{d}_n(s)\Pi_n x$ then the difference $\Pi_n \mathbf{d}(s) P_n x - \mathbf{d}_n(s)\Pi_n x$ characterizes a defect of the projection. For example, if for any n the matrix $G_{\mathbf{d}_n}$ has only m nonzero diagonals over the main diagonal

$$G_{\mathbf{d}_n} = \begin{pmatrix} * & * & \dots & * & 0 & 0 & \dots & 0 & 0 & 0 & \dots \\ * & * & \dots & * & * & 0 & \dots & 0 & 0 & 0 & \dots \\ \dots & \dots & \dots & \dots & \dots & \dots & \dots & \dots & \dots & \dots & \dots \\ * & * & \dots & * & \dots & \dots & \dots & * & 0 & 0 & \dots \\ * & * & \dots & * & \dots & \dots & \dots & * & * & 0 & \dots \\ \dots & \dots & \dots & \dots & \dots & \dots & \dots & \dots & \dots & \dots & \dots \end{pmatrix}$$

then, obviously, the identity

$$(\Pi_n \mathbf{d}(s) P_n x)_k = (\mathbf{d}_n(s)\Pi_n x)_k$$

holds for $k = 1, 2, n - m$, i.e., the only last m elements of the vector $(\Pi_n \mathbf{d}(s) P_n x \in \mathbb{R}^n$ are projected with defects. In this case, the projection error is orthogonal to the linear subspace $\mathbb{H} \setminus$ span$\{h_1, ..., h_{n-m}\}$.

7.2.2 Approximation of Linear Dilation in $L^2(\mathbb{R}; \mathbb{R})$

Let us consider the dilation $\mathbf{d}$ in $L^2(\mathbb{R}; \mathbb{R})$ given by (4.9) recalled here as

$$(\mathbf{d}(s)x)(z) = e^{\alpha s} x(e^{\beta s} z), \quad z, s \in \mathbb{R}, \quad x \in L^2(\mathbb{R}; \mathbb{R}),$$

where $\alpha > \beta/2$. The generator of the dilation $\mathbf{d}$ is given by

$$G_{\mathbf{d}}x = \alpha x + \beta z\frac{\partial x}{\partial z}, \quad x \in \mathcal{D}(G_{\mathbf{d}}) \subset L^2(\mathbb{R}; \mathbb{R}), \quad z \in \mathbb{R},$$

where $\mathcal{D}(G_{\mathbf{d}})$ is a closure of $C_c^\infty(\mathbb{R}; \mathbb{R})$ with respect to the norm $\|x\| = \|x\|_{L^2} + \|G_{\mathbf{d}}x\|_{L^2}$.

Let us consider the Hermite functions

$$h_i(z) = \frac{(-1)^{i-1}}{\sqrt{2^{i-1}(i-1)!\sqrt{\pi}}} e^{\frac{z^2}{2}} \frac{d^{i-1}}{dz^{i-1}} e^{-z^2}, \quad z \in \mathbb{R}, \quad i = 1, 2, \ldots, \tag{7.10}$$

which are known to be a smooth orthonormal basis in $L^2(\mathbb{R}; \mathbb{R})$, i.e., $\langle h_i, h_j \rangle = \delta_{ij}$ where $\delta_{ij} = 0$ for $i \neq j$ and $\delta_{ij} = 1$ for $i = j$. It is easy to see that $h_i \in \mathcal{D}(G_{\mathbf{d}})$.

The finite-dimensional projection of the dilation **d** can be given by (7.6), where $G_{\mathbf{d}_n} \in \mathbb{R}^{n\times n}$ is the generator of the dilation $\mathbf{d}_n$ in $\mathbb{R}^n$ with the following elements:

$$\langle G_{\mathbf{d}}h_i, h_j \rangle_{L^2} = \int_{\mathbb{R}} (G_{\mathbf{d}}h_i)(z)h_j(z)dz = \int_{\mathbb{R}} \left(\alpha h_i(z) + \beta z\frac{\partial h_i(z)}{\partial z}\right) h_j(z)dz =$$

$$\alpha\delta_{ij} + \int_{\mathbb{R}} \beta z\frac{\partial h_i(z)}{\partial z} h_j(z)dz.$$

Taking into account (see Appendix E)

$$\frac{\partial h_i(z)}{\partial z} = \sqrt{\frac{i-1}{2}} h_{i-1}(z) - \sqrt{\frac{i}{2}} h_{i+1}(z), \tag{7.11}$$

we obtain

$$\langle G_{\mathbf{d}}h_i, h_j \rangle_{L^2} = \alpha\delta_{ij} + \beta\sqrt{\frac{i-1}{2}}\langle h_{i-1}, yh_j\rangle - \beta\sqrt{\frac{i}{2}}\langle h_{i+1}, yh_j\rangle.$$

Finally, using the identity (see Appendix E)

$$yh_j(y) = \sqrt{\frac{j-1}{2}} h_{j-1}(y) + \sqrt{\frac{j}{2}} h_{j+1}(y),$$

we derive

$$\langle G_{\mathbf{d}}h_i, h_j \rangle_{L^2} = \tfrac{2\alpha-\beta}{2}\delta_{ij} + \tfrac{\beta\sqrt{(i-1)j}}{2}\delta_{i-1\, j+1} - \tfrac{\beta\sqrt{i(j-1)}}{2}\delta_{i+1\, j-1}.$$

Therefore, the matrix $G_{\mathbf{d}_n}$ has the form

$$G_{\mathbf{d}_n} = \begin{pmatrix} \frac{2\alpha-\beta}{2} & 0 & \beta\sqrt{\frac{1}{2}} & 0 & 0 & 0 & \dots \\ 0 & \frac{2\alpha-\beta}{2} & 0 & \beta\sqrt{\frac{3}{2}} & 0 & 0 & \dots \\ -\beta\sqrt{\frac{1}{2}} & 0 & \frac{2\alpha-\beta}{2} & 0 & \beta\sqrt{3} & 0 & \dots \\ 0 & -\beta\sqrt{\frac{3}{2}} & 0 & \frac{2\alpha-\beta}{2} & 0 & \beta\sqrt{5} & \dots \\ 0 & 0 & -\beta\sqrt{3} & 0 & \frac{2\alpha-\beta}{2} & 0 & \dots \\ 0 & 0 & 0 & -\beta\sqrt{5} & 0 & \frac{2\alpha-\beta}{2} & \dots \\ \dots & \dots & \dots & \dots & \dots & \dots & \dots \end{pmatrix} \tag{7.12}$$

and the finite-dimensional projection of the dilation $\mathbf{d}$ is given by

$$\mathbf{d}_n(s) = e^{sG_{\mathbf{d}_n}} = e^{(\alpha-0.5\beta)s}e^{s\Xi}, \quad s \in \mathbb{R}, \quad \Xi = -\Xi^\top = G_{\mathbf{d}_n} - (\alpha - 0.5\beta)I_n.$$

Notice that $e^{s\Xi}$ is the orthogonal (rotation) matrix: $e^{s\Xi}e^{s\Xi^\top} = e^{s\Xi^\top}e^{s\Xi} = I_n$. This means that $\|\tilde{x}\|_{\mathbf{d}_n} = |\tilde{x}|^{\frac{1}{\alpha-0.5\beta}}$ for any $\tilde{x} \in \mathbb{R}^n$.

7.3 Homogeneous Galerkin Method

7.3.1 Homogeneous Projection of Linear Evolution Equation

A behavior of any homogeneous evolution system is completely defined by its operator on a unit sphere. A homogeneity degree specifies a time scaling and growing/decay rate of the system (see Chap. 6 for more details). For construction of a homogeneous Galerkin projection, we consider a representation of the original evolution system in polar homogeneous coordinates. Such coordinates for ODE models were studied in [5] and in Chap. 8 of Volume I of this monograph.

Proposition 7.2 (Linear equation in polar homogeneous coordinates) *Let* $\mathbf{d}$ *be strongly continuous strictly monotone dilation in a real Hilbert space* $\mathbb{H}$. *Let* $A : \mathcal{D}(A) \subset \mathbb{H} \mapsto \mathbb{H}$ *be a densely defined* $\mathbf{d}$*-homogeneous operator of degree* $\nu \in \mathbb{R}$. *If* $x \in C([0,T);\mathbb{H})$ *is a classical (or a strong) solution of the system*

$$\dot{x}(t) = Ax(t), \quad t > 0 \tag{7.13}$$

such that $x(t) \in \mathcal{D}(A) \cap \mathcal{D}(G_{\mathbf{d}})$ *for (almost) all* $t \in [0,T)$ *and* $x(t) \neq \mathbf{0}$ *for all* $t \in [0,T)$ *then the continuous functions*

$$t \mapsto z(t) := \mathbf{d}(-\ln\|x(t)\|_{\mathbf{d}})x(t) \quad \text{and} \quad t \mapsto r(t) := \|x(t)\|_{\mathbf{d}} \tag{7.14}$$

satisfy the following system of equations:

$$\dot{z} = r^\nu Az - r^\nu \frac{\langle Az, z\rangle}{\langle G_{\mathbf{d}}z, z\rangle} G_{\mathbf{d}}z, \tag{7.15}$$

$$\dot{r} = r^{\nu+1}\frac{\langle Az, z\rangle}{\langle G_{\mathbf{d}}z, z\rangle} \tag{7.16}$$

for (almost) all $t \in (0, T)$, and vice versa; a classical (strong) solution of the system (7.15)–(7.16) such that $\|z(0)\| = 1$ and $r(t) \neq 0, \forall t \in [0, T)$ defines a classical (strong) solution of (7.13) by the formula

$$x(t) = \mathbf{d}(\ln r(t))z(t), \quad t \in [0, T) \tag{7.17}$$

such that $x(t) \in \mathcal{D}(A) \cap \mathcal{D}(G_{\mathbf{d}})$ for (almost) all $t \in [0, T)$. Moreover,

$$\frac{d}{dt}\|z(t)\|^2 = 0, \quad \forall t \in (0, T),$$

and the vector $z(t)$ belongs to the unit sphere S in $\mathbb{H}$ for all $t \in [0, T)$.

Proof (1) Let as show that the pair (r, z) given by (7.14) is a classical (strong) solution of (7.15)–(7.16). The continuity of r and z follows from the continuity of x, $\|\cdot\|_{\mathbf{d}}$ and the strong continuity of $\mathbf{d}(s)$.

Since $x(t) \in \mathcal{D}(G_{\mathbf{d}})$ for (almost) all $t \in [0, T)$ and $\mathcal{D}(G_{\mathbf{d}})$ is a $\mathbf{d}$-homogeneous cone then $z(t) \in \mathcal{D}(G_{\mathbf{d}})$ for (almost) all $t \in [0, T)$. By Proposition 4.4, we have $\langle G_{\mathbf{d}}z(t), z(t)\rangle$ as long as $z(t) \neq \mathbf{0}$.

Since x is a classical (strong solution) then from (4.25), we conclude that the identity

$$\dot{r} = \frac{d}{dt}\|x\|_{\mathbf{d}} = \frac{\langle \mathbf{d}(-\ln\|x\|_{\mathbf{d}})\dot{x}, \mathbf{d}(-\ln\|x\|_{\mathbf{d}})x\rangle}{\langle G_{\mathbf{d}}\mathbf{d}(-\ln\|x\|_{\mathbf{d}})x, \mathbf{d}(-\ln\|x\|_{\mathbf{d}})x\rangle}\|x\|_{\mathbf{d}} = \frac{\langle \mathbf{d}(-\ln\|x\|_{\mathbf{d}})Ax, z\rangle}{\langle G_{\mathbf{d}}z, z\rangle}r$$

is fulfilled for (almost) all $t \in [0, T)$. Using the $\mathbf{d}$-homogeneity of A we obtain

$$\mathbf{d}(-\ln\|x\|_{\mathbf{d}})Ax = \|x\|_{\mathbf{d}}^{\nu}A\mathbf{d}(-\ln\|x\|_{\mathbf{d}})x = r^{\nu}Az.$$

Since $\frac{d}{ds}\mathbf{d}(s)u = G_{\mathbf{d}}\mathbf{d}(s)u$ for all $s \in \mathbb{R}$ and all $u \in \mathcal{D}(G_{\mathbf{d}})$, then the identity

$$\dot{z} = \frac{d}{dt}(\mathbf{d}(-\ln\|x\|_{\mathbf{d}})x) = \mathbf{d}(-\ln\|x\|_{\mathbf{d}})\dot{x} - \frac{\frac{d}{dt}\|x\|_{\mathbf{d}}}{\|x\|_{\mathbf{d}}}G_{\mathbf{d}}\mathbf{d}(-\ln\|x\|_{\mathbf{d}})x$$

holds.

(2) Let us show that x given by the formula (7.17) is a solution of (7.13). Since (r, z) is a classical (strong) solution, then the identity (7.15) holds for (almost) all $t \in [0, T)]$ and $z(t) \in \mathcal{D}(A) \cap \mathcal{D}(G_{\mathbf{d}})$ for (almost) all $t \in [0, T)]$. Since A is a homogeneous operator then $\mathcal{D}(A)$ is a homogeneous cone. Hence, $\mathcal{D}(A) \cap \mathcal{D}(G_{\mathbf{d}})$ is a $\mathbf{d}$-homogeneous cone too, and $x(t)$ given by the formula (7.17) satisfies $x(t) \in \mathcal{D}(A) \cap \mathcal{D}(G_{\mathbf{d}})$ for (almost) all $t \in [0, T)$. Hence, using $\mathbf{d}$-homogeneity of A we derive

$$\dot{x}(t) = \mathbf{d}(\ln r(t))\dot{z}(t) + \frac{\dot{r}(t)}{r(t)}G_{\mathbf{d}}\mathbf{d}(\ln r(t))z(t) = r^{\nu}(t)\mathbf{d}(\ln r(t))Az(t) = Ax(t)$$

for (almost) all $t \in (0, T)$.

(3) Finally, taking into account

$$\frac{d\|z\|^2}{dt}=2\langle z,\dot z\rangle=\left\langle z, r^{\nu}Az-r^{\nu}\frac{\langle Az,z\rangle}{\langle G_{\mathbf{d}}z,z\rangle}G_{\mathbf{d}}z\right\rangle=r^{\nu}\langle z,Az\rangle-r^{\nu}\frac{\langle Az,z\rangle\langle z,G_{\mathbf{d}}z\rangle}{\langle G_{\mathbf{d}}z,z\rangle}=0$$

we complete the proof. □

To preserve the homogeneity, an approximation of the homogeneous system (7.13) can be based on its equivalent representation (7.15)–(7.16). Approximations of the linear operators A and $G_{\mathbf{d}}$ on the unit sphere are only required in order to design a reduced-order model.

Theorem 7.1 *Let a* $\mathbf{d}$*-homogeneous operator* $A : \mathcal{D}(A) \subset \mathbb{H} \mapsto \mathbb{H}$ *be an infinitesimal generator of a strongly continuous semigroup* Φ *of linear bounded operators on* $\mathbb{H}$*. Let* $\mathbf{d}$ *be a strongly continuous strictly monotone dilation in a separable Hilbert space* $\mathbb{H}$ *and* $h_i \in \mathcal{D}(A)\cap\mathcal{D}(G_{\mathbf{d}}), i = 1, 2, \ldots$ *being an orthonormal basis in* $\mathbb{H}$*.*

If $x_0 \in \mathbb{H}\backslash\{\mathbf{0}\}$ *is such that* $\Phi(t)x_0 \in \mathcal{D}(A)\cap\mathcal{D}(G_{\mathbf{d}})$ *then the following system of differential equations:*

$$\frac{d}{dt}\tilde z_i=r^{\nu}\sum_j\left(a_{ij}-\frac{\sum\limits_k\sum\limits_m a_{km}\tilde z_k\tilde z_m}{\sum\limits_k\sum\limits_m g_{km}\tilde z_k\tilde z_m}g_{ij}\right)\tilde z_j,\quad \tilde z_i(0)=\langle\mathbf{d}(-\ln\|x_0\|_{\mathbf{d}})x_0,h_i\rangle, \tag{7.18}$$

$$\dot r=r^{\nu+1}\frac{\sum\limits_k\sum\limits_m a_{km}\tilde z_k\tilde z_m}{\sum\limits_k\sum\limits_m g_{km}\tilde z_k\tilde z_m},\quad r(0)=\|x_0\|_{\mathbf{d}}, \tag{7.19}$$

$$a_{ij}=\langle Ah_i,h_j\rangle,\quad \text{and}\quad g_{ij}=\langle G_{\mathbf{d}}h_i,h_j\rangle,\quad i,j,k=1,2,\ldots \tag{7.20}$$

has a solution $\tilde z(t)=(\tilde z_1(t),\tilde z_2(t),\ldots)^{\top}\in\ell^2$ *defined at least on* $[0,T)$*, and satisfying*

$$\Phi(t)x_0=\sum_i\mathbf{d}(\ln r(t))\tilde z_i(t)h_i,\quad t\in[0,T)$$

where $T=T(x_0)$ *is the first instant of time when* $\Phi(T)x_0=\mathbf{0}$ *or* $T=\infty$ *if* $\Phi(t)x_0\neq\mathbf{0}$ *for all* $t>0$*. Moreover, if* $\{\tilde z_i(t)\}_{i=1}^{\infty}$ *with* $t\in[0,T)$ *is a solution to* (7.18)–(7.20) *then*

$$\frac{d}{dt}\sum_i\tilde z_i^2(t)=0\quad\text{and}\quad\sum_i\tilde z_i^2(t)=1.$$

Proof According to Proposition 7.2 $x(t)=\Phi(t)x_0$ satisfy (7.15) and (7.16) with $z(t)=\mathbf{d}(-\ln\|x(t)\|_{\mathbf{d}})x(t)$ and $r(t)=\|x(t)\|_{\mathbf{d}}$.

Since $\{h_i\}$ is an orthonormal basis in $\mathbb{H}$ then

$$z(t)=\sum_i\tilde z_i(t)h_i,$$

where $\tilde{z}_i(t) = \langle z(t), h_i\rangle$, $i = 1, 2, \ldots$. Hence, taking into account that A and $G_{\mathbf{d}}$ are closed operators and $h_i \in \mathcal{D}(A) \cap \mathcal{D}(G_{\mathbf{d}})$ we derive

$$\langle Az(t), z(t)\rangle = \left\langle A\sum_k \tilde{z}_k(t)h_k, \sum_j \tilde{z}_m(t)h_m\right\rangle = \sum_k\sum_m a_{km}\tilde{z}_k(t)\tilde{z}_m(t)$$

and

$$\langle G_{\mathbf{d}}z(t), z(t)\rangle = \left\langle G_{\mathbf{d}}\sum_k \tilde{z}_k(t)h_k, \sum_m \tilde{z}_m(t)h_m\right\rangle = \sum_k\sum_m g_{km}\tilde{z}_k(t)\tilde{z}_m(t).$$

Since $z(t)$ satisfies (7.15) then

$$\sum_j \frac{d}{dt}\tilde{z}_j(t)h_j = r^{\nu}(t)A\sum_j \tilde{z}_j(t)h_j - r^{\nu}(t)\frac{\langle Az, z\rangle}{\langle G_{\mathbf{d}}z, z\rangle}\sum_j \tilde{z}_j(t)G_{\mathbf{d}}h_j$$

and

$$\left\langle h_i, \sum_j \tfrac{d}{dt}\tilde{z}_j(t)h_j\right\rangle = \tfrac{d\tilde{z}_i(t)}{dt} =$$

$$r^{\nu}(t)\left\langle h_i, A\sum_j \tilde{z}_j(t)h_j\right\rangle - r^{\nu}(t)\tfrac{\langle Az,z\rangle}{\langle G_{\mathbf{d}}z,z\rangle}\left\langle h_i, \sum_j \tilde{z}_j(t)G_{\mathbf{d}}h_j\right\rangle.$$

Therefore, we conclude that $\tilde{z}_i$ and r satisfy the system (7.18)–(7.20).

Finally, since $\|\mathbf{d}(-\ln\|x_0\|_{\mathbf{d}})x_0\| = 1$ then

$$1 = \langle\|\mathbf{d}(-\ln\|x_0\|_{\mathbf{d}})x_0, \|\mathbf{d}(-\ln\|x_0\|_{\mathbf{d}})x_0\rangle\left\langle\sum_i \tilde{z}_i(0)h_i, \sum_j \tilde{z}_j(0)h_j\right\rangle = \sum_i \tilde{z}_i^2(0),$$

but

$$\frac{d}{dt}\sum_i \tilde{z}_i^2 = \sum_i \tilde{z}_i\frac{d}{dt}\tilde{z}_i =$$

$$r^{\nu}\sum_i\sum_j \tilde{z}_i a_{ij}\tilde{z}_j - r^{\nu}\frac{\sum_k\sum_j a_{ij}\tilde{z}_k\tilde{z}_j}{\sum_k\sum_j g_{ij}\tilde{z}_k\tilde{z}_j}\sum_i\sum_j g_{ij}\tilde{z}_i\tilde{z}_j = 0$$

provided that $\tilde{z}_i$ satisfies the system (7.18)–(7.20). The proof is complete. □

The above theorem motivates us to consider the system (7.18)–(7.20) with $i, j, k = 1, \ldots, n \in \mathbb{N}$ as a possible finite-dimensional approximation of the $\mathbf{d}$-homogeneous linear evolution equation (7.13). Indeed, if we denote

$$\tilde{x} = \mathbf{d}_n(r)\tilde{z}, \quad \tilde{z} = (\tilde{z}_1, \tilde{z}_2, ..., \tilde{z}_n)^\top \in \mathbb{R}^n, \quad \tilde{A}_n = \{a_{i,j}\}_{i,j=1}^n \in \mathbb{R}^{n\times n}$$

then $\|\tilde{x}\|_{\mathbf{d}} = r$ and simple computations show

$$\frac{d}{dt}\tilde{x} = \dot{r}\mathbf{d}_n(r)G_{\mathbf{d}_n}\tilde{z} + \mathbf{d}_n(r)\frac{d}{dt}\tilde{z} =$$

$$= \|\tilde{x}\|_{\mathbf{d}}^{\nu}\mathbf{d}_n(\ln\|\tilde{x}\|)\tilde{A}_n\mathbf{d}_n(-\ln\|\tilde{x}\|_{\mathbf{d}})\tilde{x}, \quad t > 0$$

where the identity $\|\tilde{x}\|_{\mathbf{d}} = \tilde{r}$ is utilized in the last step. According to Proposition 7.1, the dilation $\mathbf{d}_n$ in $\mathbb{R}^n$ can be treated as an approximation of the dilation $\mathbf{d}$ in $\mathbb{H}$. Taking into account $\mathbf{d}$-homogeneity of the operator A ($e^{\nu s}\mathbf{d}(s)A\mathbf{d}(-s)x = Ax$) we conclude that the system

$$\frac{d}{dt}\tilde{x} = \|\tilde{x}\|_{\mathbf{d}_n}^{\nu}\mathbf{d}_n(\ln\|\tilde{x}\|)\tilde{A}_n\mathbf{d}_n(-\ln\|\tilde{x}\|_{\mathbf{d}_n})\tilde{x}, \quad t > 0 \tag{7.21}$$

is a finite-dimensional approximation of the $\mathbf{d}$-homogeneous evolution system

$$\dot{x} = Ax = \|x\|_{\mathbf{d}}^{\nu}\mathbf{d}(\ln\|x\|_{\mathbf{d}})A\mathbf{d}(-\ln\|x\|)x,$$

obtained by means of sequential approximations of the dilation group $\mathbf{d}$, the canonical homogeneous norm $\|x\|_{\mathbf{d}}$, and the operator A.

Corollary 7.1 *Let all conditions of Theorem 7.1 be fulfilled and for some* $\gamma > 0$ *we have*

$$\langle Ax + \gamma G_{\mathbf{d}}x, x\rangle \le 0, \quad \forall x \in \mathcal{D}(A) \cap \mathcal{D}(G).$$

Then

$$\tilde{x}^\top(\tilde{A} + \gamma G_{\mathbf{d}_n})\tilde{x} \le 0, \quad \forall \tilde{x} \in \mathbb{R}^n$$

and the origin of the system (7.21) *is globally uniformly asymptotically stable.*

Proof Since $\mathcal{D}(A) \cap \mathcal{D}(G)$ is a linear subspace of $\mathbb{H}$ then under conditions of Theorem 7.1, we have $\text{span}\{h_1, ..., h_n\} \subset \mathcal{D}(A) \cap \mathcal{D}(G)$. In this case, for any $\tilde{x} \in \mathbb{R}^n$ we have

$$x = \sum_{i=1}^n \tilde{x}_i h_i \in \mathcal{D}(A) \cap \mathcal{D}(G)$$

and

$$\langle Ax + \gamma G_{\mathbf{d}}x, x\rangle = \tilde{x}^\top(\tilde{A} + \gamma G_{\mathbf{d}_n})\tilde{x}.$$

Hence, $\|\cdot\|_{\mathbf{d}_n}$ is a Lyapunov function of the system (7.21). The proof is complete. □

The above corollary shows that the homogeneous Galerkin projection preserves the stability of the original system.

Notice that the obtained system (7.21) is $\mathbf{d}_n$-homogeneous of the same degree ν. Therefore, *the proposed homogeneous Galerkin projection allows us to preserve the homogeneity of the original system and its homogeneity degree as well.* However, this scheme approximates a linear infinite-dimensional system by a nonlinear n-dimensional one. So, it is expectable that the linearity should be somehow recovered again as $n \to +\infty$.

Example 7.1 *(On a homogeneous projection of a heat equation in* $\mathbb{R}$*)* Let us consider the heat system on $\mathbb{R}$

$$\frac{\partial x}{\partial t} = \frac{\partial^2 x}{\partial y^2}, \quad y \in \mathbb{R}.$$

In Example 5.2 the Laplace operator is $\mathbf{d}$-homogeneous of the degree 2β with respect to the dilation $\mathbf{d}$ given by (4.9).

Let us take the Hermite functions

$$h_i(y) = \frac{(-1)^{i-1}}{\sqrt{2^{i-1}(i-1)!\sqrt{\pi}}} e^{\frac{y^2}{2}} \frac{d^{i-1}}{dy^{i-1}} e^{-y^2}, \quad y \in \mathbb{R}, i = 1, 2, ...$$

which are known to be the orthonormal basis in $L^2(\mathbb{R}, \mathbb{R})$.

We construct an approximation of the heat system in the form (7.18)–(7.20) with $i, k, j = 1, 2, ..., n$, where $n < +\infty$. Next, using the transformation $\tilde{x} = \mathbf{d}_n(r)\tilde{z}$ we derive an approximation in the form (7.21). The generator of the dilation $\mathbf{d}_n$ is given by (7.12) and the matrix $\tilde{A}$ has the elements

$$a_{ij} = \langle Ah_i, h_j \rangle = \left\langle \frac{d^2}{dy^2} h_i, h_j \right\rangle.$$

Taking into account

$$\frac{d^2}{dy^2} h_i = -(2i-1)h_i + y^2 h_i$$

we derive

$$a_{ij} = -(2i-1)\delta_{ij} + \langle yh_i, yh_j \rangle,$$

where $\delta_{ij} = 0$ for $i \neq j$ and $\delta_{ij} = 0$ for $i = j$. Finally, using

$$yh_j(y) = \sqrt{\frac{j-1}{2}} h_{j-1}(y) + \sqrt{\frac{j}{2}} h_{j+1}(y)$$

we derive

$$a_{ij} = -(i - 1/2)\delta_{ij} + \sqrt{\frac{i-1}{2}}\sqrt{\frac{j}{2}}\delta_{i-1\,j+1} + \sqrt{\frac{i}{2}}\sqrt{\frac{j-1}{2}}\delta_{i+1\,j-1}.$$

Recall that the matrix $\tilde{A}_n = \{a_{ij}\} \in \mathbb{R}^{n\times n}$ is $\mathbf{d}_n$-homogeneous of the degree 2β if and only if $\tilde{A}_n G_{\mathbf{d}_n} = (G_{\mathbf{d}_n} + 2\beta I_n)\tilde{A}_n$. Calculations show that for any natural $n > 3$ we have

$$\tilde{A}_n G_{\mathbf{d}} - (G_{\mathbf{d}_n} + 2\beta I_n)\tilde{A}_n = \begin{pmatrix} \mathbf{0} & \mathbf{0} \\ \mathbf{0} & \Lambda_n \end{pmatrix},$$

where $\Lambda_n \in \mathbb{R}^{2\times 2}$ is a diagonal matrix. This means that the matrix $\tilde{A}_n$ is "close" to a homogeneous one.

Moreover, $\tilde{A}_n$ becomes a *$\mathbf{d}_\infty$-homogeneous linear operator* $\tilde{A}_\infty : \mathcal{D}(\tilde{A}_\infty) \subset \ell^2 \mapsto \ell^2$ as $n \to \infty$, where $\mathbf{d}_\infty$ is a dilation in ℓ^2 obtained as a limit of $\mathbf{d}_n$ as $n \to +\infty$ and $\tilde{A}_\infty$ is a counterpart (in ℓ^2) of the operator A. Notice that, due to a specific structure of the generator $G_{\mathbf{d}_n}$ (see the formula (7.12)), the generator $G_{\mathbf{d}_\infty}$ has the same structure

$$G_{\mathbf{d}_\infty} = \frac{2\alpha - \beta}{2} I_\infty + \Xi_\infty,$$

where I_∞ is the identity operator in ℓ^2 and $\Xi_\infty : \mathcal{D}(\Xi_\infty) \subset \ell^2 \mapsto \ell^2$ is such that

$$\langle \Xi_\infty x_\infty, x_\infty \rangle = 0, \quad \forall x_\infty \in \mathcal{D}(\Xi_\infty) \subset \ell^2.$$

7.3.2 Homogeneous Projection of Nonlinear Evolution Equation

Let us consider the nonlinear system

$$\dot{x} = Ax + f(x), \quad t > 0, \quad x(0) = x_0, \tag{7.22}$$

where $x(t) \in \mathbb{H}$ is a system state at the time instant t, $\mathbb{H}$ is a real Hilbert space, $x_0 \in \mathbb{H}$ is an initial state, $A : \mathcal{D}(A) \subset \mathbb{H} \mapsto \mathbb{H}$ is a linear (possibly unbounded) closed densely defined $\mathbf{d}$-homogeneous operator of degree $\nu \in \mathbb{R}$, which generates a strongly continuous semigroup Φ of linear bounded operators on $\mathbb{H}$, $f : \mathcal{D}(f) \subset \mathbb{H} \mapsto \mathbb{H}$ is a nonlinear (possibly unbounded) closed densely defined $\mathbf{d}$-homogeneous operator of degree ν such that $f(\mathbf{0}) = \mathbf{0}$, and $\mathcal{D}(f) \subset \mathbb{H}$ is a linear subspace dense in $\mathbb{H}$. The assumption $f(\mathbf{0}) = \mathbf{0}$ implies the existence of the zero solution of (7.22) with $x_0 = \mathbf{0}$.

Assuming that the linear dilation $\mathbf{d}$ in $\mathbb{H}$ is strongly continuous and strictly monotone, let us introduce the *polar homogeneous coordinates* in $\mathbb{H}$:

$$\phi = \mathbf{d}(-\ln \|x\|_{\mathbf{d}})x, \quad r = \|x\|_{\mathbf{d}},$$

where $x \in \mathbb{H}$. Recall that $\pi_{\mathbf{d}}(x) = \mathbf{d}(-\ln \|x\|_{\mathbf{d}})x$ is a $\mathbf{d}$-homogeneous projector on the unit sphere (see Chap. 4). Since $\|\cdot\|_{\mathbf{d}}$ is continuous, then $x \mapsto (\phi, r)$ is a homeomorphism $\mathbb{H}\backslash\{\mathbf{0}\} \mapsto S_{\mathbb{H}} \times (0, +\infty)$, where $S_{\mathbb{H}} = \{x \in \mathbb{H} : \|x\| = 1\}$ is the unit sphere in $\mathbb{H}$. The inverse operator is given by $x = \mathbf{d}(\ln r)\phi$.

A behavior of a **d**-homogeneous system (7.22) is completely defined by the properties of the operator $A + f$ on a unit sphere [6] and the homogeneity degree of the system. Indeed, if the operators A, f are **d**-homogeneous of degree $\nu \in \mathbb{R}$ and $x : [0, T) \mapsto \mathbb{H}$ is a strong solution of (7.22) then simple calculations show that the polar coordinates of $x(t)$ satisfy

$$\frac{\mathbf{d}(-\ln r(t))\frac{d(\mathbf{d}(\ln r(t))\phi(t))}{dt}}{r^{\nu}(t)} \overset{a.e.}{=} A\phi(t)+f(\phi(t)), \quad t\in(0,T), \quad \phi(0)=\mathbf{d}(-\ln\|x_0\|_{\mathbf{d}})x_0. \tag{7.23}$$

$$\dot{r}(t) \overset{a.e.}{=} r^{\nu+1}(t)\frac{\langle A\phi(t)+f(\phi(t)),\phi(t)\rangle}{\langle G_{\mathbf{d}}\phi(t),\phi(t)\rangle}, \quad t \in (0, T), \quad r(0) = \|x_0\|_{\mathbf{d}}, \tag{7.24}$$

where the formula (4.25) is utilized for the derivation of the latter identity. The system (7.23), (7.24) is an equivalent representation of the evolution equation (7.22) in the polar homogeneous coordinates [5, 6]. Since $\phi(t)$ belongs to the unit sphere and $r(t)$ is nonnegative, then the weak formulation of (7.23)–(7.24) is given as follows:

$$\begin{array}{c}\textbf{find } \phi_v \in C([0,T),S_V) \textbf{ and } \tilde{r} \in C([0,T),\mathbb{R}_+) \textbf{ such that}\\ \left\langle\frac{\mathbf{d}(-\ln\tilde{r}(t))\frac{d}{dt}(\mathbf{d}(\ln\tilde{r}(t))\phi_v(t))}{\tilde{r}^{\nu}(t)} - A\phi_v(t) - f(\phi_v(t)), v\right\rangle \overset{a.e.}{=} 0, \quad \forall v\in V, \forall t\in(0,T),\\ \frac{d}{dt}\tilde{r}(t) \overset{a.e.}{=} \tilde{r}^{\nu+1}(t)\frac{\langle A\phi_v(t)+f(\phi_v(t)),\phi_v(t)\rangle}{\langle G_{\mathbf{d}}\phi_v(t),\phi_v(t)\rangle}, \quad \forall t \in (0,T)\\ \langle\phi_v(0), v\rangle = \langle\mathbf{d}(-\ln\|x_0\|_{\mathbf{d}})x_0, v\rangle, \forall v \in V, \quad \text{and} \quad \tilde{r}(0) = \|x_0\|_{\mathbf{d}},\end{array} \tag{7.25}$$

where $V \subset \mathbb{H}$ is a linear subspace of $\mathbb{H}$ and $S_V = \{z \in V : \|z\|_{\mathbb{H}} = 1\}$ is the unit sphere in V.

The pair $\phi_v(t), \tilde{r}(t)$ is a Galerkin-like projection of the pair $(\phi(t), r(t)) \in S_{\mathbb{H}} \times \mathbb{R}_+$ on $S_V \times \mathbb{R}_+$. In this case, the quantity

$$x_v(t) = \mathbf{d}(\ln\tilde{r}(t))\phi_v(t), \quad t \in [0, T) \tag{7.26}$$

can be interpreted as a Galerkin-like projection of the strong solution $x : [0, T) \mapsto \mathbb{H}$ of (7.22) on the **d**-homogeneous cone

$$\mathcal{D}_V := \bigcup_{s\in\mathbb{R}} \mathbf{d}(s)S_V, \tag{7.27}$$

since $x_v \in C([0, T); \mathcal{D}_V)$ and $\|x_v(t)\|_{\mathbf{d}} = \tilde{r}(t), t \geq 0$. Notice that $\mathcal{D}_V$ is not a linear subspace of $\mathbb{H}$ in the general case. If S_V is dense in $S_{\mathbb{H}}$ then $\mathcal{D}_V$ is dense in $\mathbb{H}$ due to [6, Lemma 6].

Theorem 7.2 (On existence of a homogeneous Galerkin projection) *Let* **d** *be a strongly continuous strictly monotone dilation in* $\mathbb{H}$*, the operators* A, f *be* **d***-homogeneous of the degree* $\nu \in \mathbb{R}$*,* $\mathcal{D}(A) \subset \mathcal{D}(f)$*, the operator* f *be* M*-regular,* $h_i \in \mathcal{D}(A) \cap \mathcal{D}(G_{\mathbf{d}})$*,* $i = 1, 2, ..., n$ *being an orthonormal family in*

$\mathbb{H}$, *and* $V = \text{span}\{h_1, ..., h_n\}$. *Then for any* $x_0 \in \mathcal{D}_V$ *there exists a pair* $\phi_v, \tilde{r}$ *satisfying* (7.25) *such that*

$$\phi_v(t) = \sum_{i=1}^{n} \tilde{\phi}_i(t)h_i, \quad t \in [0, T),$$

where the pair $\tilde{\phi}(t) = (\tilde{\phi}_1(t), ..., \tilde{\phi}_n(t))^\top \in \mathbb{R}^n$, $\tilde{r}(t) \in \mathbb{R}_+$ *is the unique classical solution of the following ODE:*

$$\begin{cases} \frac{d}{dt}\tilde{\phi}(t) = \tilde{r}^\nu(t)\left(\tilde{A}_n\tilde{\phi}(t) + \tilde{f}(\tilde{\phi}(t))\right) - \tilde{r}^\nu(t)\frac{\tilde{\phi}^\top(t)A_n\tilde{\phi}(t) + \tilde{\phi}^\top(t)\tilde{f}(\tilde{\phi}(t))}{\tilde{\phi}^\top(t)G_{\mathbf{d}_n}\tilde{\phi}(t)}G_{\mathbf{d}_n}\tilde{\phi}(t), \\ \frac{d}{dt}\tilde{r}(t) = \tilde{r}^{\nu+1}(t)\frac{\tilde{\phi}^\top(t)\tilde{A}_n\tilde{\phi}(t) + \tilde{\phi}^\top\tilde{f}(\tilde{\phi}(t))}{\tilde{\phi}^\top(t)G_{\mathbf{d}_n}\tilde{\phi}(t)}, \quad t \in (0, T), \\ \tilde{\phi}(0) = \Pi_n\mathbf{d}(-\ln\|x_0\|_{\mathbf{d}})x_0, \quad \tilde{r}(0) = \|x_0\|_{\mathbf{d}}, \end{cases} \tag{7.28}$$

$G_{\mathbf{d}_n}$ *given by* (7.6) *is a generator of a linear dilation* $\mathbb{R}^n$,

$$\tilde{A}_n = \begin{pmatrix} \langle Ah_1, h_1\rangle & \langle Ah_2, h_1\rangle & \dots & \langle Ah_n, h_1\rangle \\ \langle Ah_1, h_2\rangle & \langle Ah_2, h_2\rangle & \dots & \langle Ah_n, h_2\rangle \\ \dots & \dots & \dots & \dots \\ \langle Ah_1, h_n\rangle & \langle Ah_2, h_n\rangle & \dots & \langle Ah_n, h_n\rangle \end{pmatrix} \in \mathbb{R}^{n\times n}, \tag{7.29}$$

$$\tilde{f}(\tilde{\phi}) = \begin{pmatrix} \left\langle f\left(\sum_{k=1}^{n}\tilde{\phi}_k h_k\right), h_1\right\rangle \\ \left\langle f\left(\sum_{k=1}^{n}\tilde{\phi}_k h_k\right), h_2\right\rangle \\ \dots \\ \left\langle f\left(\sum_{k=1}^{n}\tilde{\phi}_k h_k\right), h_n\right\rangle \end{pmatrix}, \quad \tilde{\phi} = (\tilde{\phi}_1, ..., \tilde{\phi}_n)^\top \in \mathbb{R}^n, \tag{7.30}$$

and $T \le +\infty$ *such that*

- $T = +\infty$ *if* $\nu \ge 0, \tilde{\phi}^\top(A_n\tilde{\phi} + \tilde{f}(\tilde{\phi})) \le 0, \forall\tilde{\phi} \in S_{\mathbb{R}^n}$ *or* $\nu \le 0, \tilde{\phi}^\top(A_n\tilde{\phi} + \tilde{f}(\tilde{\phi})) \ge 0, \forall\tilde{\phi} \in S_{\mathbb{R}^n}$;
- $T < +\infty : \lim_{t\to T}\tilde{r}(t) = 0$ *if* $\nu < 0, \tilde{\phi}^\top(A_n\tilde{\phi} + \tilde{f}(\tilde{\phi})) < 0, \forall\tilde{\phi} \in S_{\mathbb{R}^n}$;
- $T < +\infty : \lim_{t\to T}\tilde{r}(t) = +\infty$ *if* $\nu > 0, \tilde{\phi}^\top(A_n\tilde{\phi} + \tilde{f}(\tilde{\phi})) > 0, \forall\tilde{\phi} \in S_{\mathbb{R}^n}$.

Proof Since h_i is an orthonormal basis in V then any $v \in V$ can be written as $v = \sum_{i=1}^{n}\tilde{v}_i h_i$ with some $\tilde{v} = (\tilde{v}_1, ..., \tilde{v}_n)^\top \in \mathbb{R}^n$ and any $\phi_v : [0, T) \mapsto S_V$ can be rewritten as $\phi_v(t) = \sum_{i=1}^{n}\tilde{\phi}_i(t)h_i$ with some $\tilde{\phi} : [0, T) \mapsto S_{\mathbb{R}^n}$, $\tilde{\phi} = (\phi_1, ..., \phi_n)^\top$. Taking into account

$$\left\langle \tilde{r}^{-\nu}(t)\mathbf{d}(-\ln\tilde{r}(t))\frac{d}{dt}(\mathbf{d}(\ln\tilde{r}(t))\phi_v(t)) - A\phi_v(t) - f(\phi_v(t)), v\right\rangle =$$

$$\left\langle \frac{\mathbf{d}(-\ln\tilde{r}(t))\frac{d}{dt}\left(\mathbf{d}(\ln\tilde{r}(t))\sum_{k=1}^{n}\tilde{\phi}_k(t)h_k\right)}{\tilde{r}^\nu(t)} - A\sum_{k=1}^{n}\tilde{\phi}_k(t)h_k - f\left(\sum_{k=1}^{n}\tilde{\phi}_k(t)h_k\right), v\right\rangle =$$

$$\sum_{i=1}^{n}\tilde{v}_i\left\langle\frac{\mathbf{d}(-\ln\tilde{r}(t))\frac{d}{dt}\left(\mathbf{d}(\ln\tilde{r}(t))\sum_{k=1}^{n}\tilde{\phi}_k(t)h_k\right)}{\tilde{r}^{\nu}(t)}-A\sum_{i=1}^{n}\tilde{\phi}_k(t)h_k-f\left(\sum_{k=1}^{n}\tilde{\phi}_k(t)h_k\right),h_i\right\rangle$$

we conclude that the problem (7.25) can be rewritten as follows:

$$\begin{cases}\textbf{find } \tilde{\phi}\in C([0,T),S_{\mathbb{R}^n}) \textbf{ and } \tilde{r}\in C([0,T),\mathbb{R}_+) \textbf{ such that}\\ \sum\limits_{i=1}^{n}\tilde{v}_i\left\langle\frac{\mathbf{d}(-\ln\tilde{r}(t))}{\tilde{r}^{\nu}(t)}\frac{d\left(\mathbf{d}(\ln\tilde{r}(t))\sum\limits_{k=1}^{n}\tilde{\phi}_k(t)h_k\right)}{dt}-A\sum\limits_{k=1}^{n}\tilde{\phi}_k(t)h_k-f\left(\sum\limits_{k=1}^{n}\tilde{\phi}_k(t)h_k\right),h_i\right\rangle=0,\\ \frac{d}{dt}\tilde{r}(t)\overset{a.e.}{=}\tilde{r}^{\nu+1}(t)\frac{\left\langle A\sum\limits_{k=1}^{n}\tilde{\phi}_k(t)h_k+f(\sum\limits_{k=1}^{n}\tilde{\phi}_k(t)h_k),\sum\limits_{k=1}^{n}\tilde{\phi}_k(t)h_k\right\rangle}{\left\langle G_{\mathbf{d}}\sum\limits_{k=1}^{n}\tilde{\phi}_k(t)h_k,\sum\limits_{k=1}^{n}\tilde{\phi}_k(t)h_k\right\rangle},\quad \forall t\in(0,T),\quad\forall\tilde{v}\in\mathbb{R}^n,\\ \sum\limits_{i=1}^{n}\tilde{v}_i\left\langle\sum\limits_{k=1}^{n}\tilde{\phi}_k(0)h_k-\mathbf{d}(-\ln\|x_0\|_{\mathbf{d}})x_0,h_i\right\rangle=0,\quad\tilde{r}(0)=\|x_0\|_{\mathbf{d}},\end{cases}$$

or, equivalently,

$$\begin{cases}\textbf{find } \tilde{\phi}:[0,T)\to S_{\mathbb{R}^n}\subset\mathbb{R}^n \textbf{ and } \tilde{r}:[0,T)\to\mathbb{R} \textbf{ such that}\\ \left\langle\frac{\mathbf{d}(-\ln\tilde{r}(t))}{\tilde{r}^{\nu}(t)}\frac{d\left(\mathbf{d}(\ln\tilde{r}(t))\sum\limits_{k=1}^{n}\tilde{\phi}_k(t)h_k\right)}{dt}-A\sum\limits_{k=1}^{n}\tilde{\phi}_k(t)h_k-f\left(\sum\limits_{k=1}^{n}\tilde{\phi}_k(t)h_k\right),h_i\right\rangle=0,\\ \frac{d}{dt}\tilde{r}(t)\overset{a.e.}{=}\tilde{r}^{\nu+1}(t)\frac{\left\langle A\sum\limits_{k=1}^{n}\tilde{\phi}_k(t)h_k+f(\sum\limits_{k=1}^{n}\tilde{\phi}_k(t)h_k),\sum\limits_{k=1}^{n}\tilde{\phi}_k(t)h_k\right\rangle}{\left\langle G_{\mathbf{d}}\sum\limits_{k=1}^{n}\tilde{\phi}_k(t)h_k,\sum\limits_{k=1}^{n}\tilde{\phi}_k(t)h_k\right\rangle},\quad\forall t\in(0,T),\\ \left\langle\sum\limits_{k=1}^{n}\tilde{\phi}_k(0)h_k-\mathbf{d}(-\ln\|x_0\|_{\mathbf{d}})x_0,h_i\right\rangle=0,\quad i=1,2,\ldots,n,\quad\tilde{r}(0)=\|x_0\|_{\mathbf{d}}.\end{cases}$$

Since $h_i\in\mathcal{D}(G_{\mathbf{d}})$ implies $\frac{d}{ds}\mathbf{d}(s)h_i=\mathbf{d}(s)G_{\mathbf{d}}h_i$ then

$$\left\langle\frac{\mathbf{d}(-\ln\tilde{r}(t))}{\tilde{r}^{\nu}(t)}\frac{d\left(\mathbf{d}(\ln\tilde{r}(t))\sum\limits_{k=1}^{n}\tilde{\phi}_k(t)h_k\right)}{dt}-A\sum_{k=1}^{n}\tilde{\phi}_k(t)h_k-f\left(\sum_{k=1}^{n}\tilde{\phi}_k(t)h_k\right),h_i\right\rangle=$$

$$\left\langle\sum_{k-1}^{n}\frac{1}{\tilde{r}^{\nu}(t)}\frac{d\tilde{\phi}_k(t)}{dt}h_k+\frac{\frac{d\tilde{r}(t)}{dt}}{\tilde{r}^{\nu+1}(t)}\tilde{\phi}_k(t)G_{\mathbf{d}}h_k-\tilde{\phi}_k(t)Ah_k-f\left(\sum_{k-1}^{n}\tilde{\phi}_k(t)h_k\right),h_i\right\rangle.$$

Hence, we derive that the problem (7.25) can be rewritten as follows:

find $\tilde{\phi}\in C([0,T);S_{\mathbb{R}^n})$ **and** $\tilde{r}\in C([0,T);\mathbb{R}_+)$ satisfying (7.28).

Since f is M-regular and $h_i\in\mathcal{D}(A)\subset\mathcal{D}(f)$ then $\tilde{f}$ is locally Lipschitz continuous on $\mathbb{R}^n\backslash\{\mathbf{0}\}$. Due to Proposition 4.4 we have $\langle G_{\mathbf{d}}\phi_v(t),\phi_v(t)\rangle=\tilde{\phi}^{\top}(t)G_{\mathbf{d}_n}\tilde{\phi}(t)\geq\gamma\langle\phi_v(t),\phi_v(t)\rangle=\gamma\tilde{\phi}^{\top}(t)\tilde{\phi}(t)$ and, consequently, then the right-hand side of (7.28) locally Lipschitz continuous on $(\mathbb{R}^n\backslash\{\mathbf{0}\})\times(\mathbb{R}_+\backslash\{0\})$. The latter means that for any

$x_0 \in \mathbb{H} : \Pi_n \mathbf{d}(-\ln \|x_0\|_{\mathbf{d}})x_0 \neq \mathbf{0}$ the system has a unique classical solution defined as long as $0 < \tilde{r}(t) < +\infty$. Moreover, if $x_0 \in \bigcup_{s\in\mathbb{R}} \mathbf{d}(s)S_V$ then $\|\tilde{\phi}(0)\|_{\mathbb{R}^n} = 1$ and $\|\tilde{\phi}(t)\|_{\mathbb{R}_n} = 1$ for all $t \in (0, T)$ due to

$$\frac{d}{dt}\left(\tilde{\phi}^\top(t)\tilde{\phi}(t)\right) = 2\left(\tilde{\phi}^\top(t)\frac{d}{dt}\tilde{\phi}(t)\right) = 0.$$

Since $h_i \in \mathcal{D}(A) \subset D(f)$ and $\|\tilde{\phi}(t)\|_{\mathbb{R}^n} = 1$ then

$$\frac{|\tilde{\phi}^\top(t)(\tilde{A}_n\tilde{\phi}(t) + \tilde{f}(\tilde{\phi}(t)))|}{|\tilde{\phi}^\top(t)G_{\mathbf{d}_n}\tilde{\phi}(t)|} \leq C_n/\gamma$$

for some $C_n > 0$ and $r(t)$ converges to zero in a finite time $T < +\infty$ if $\nu < 0$, $\tilde{\phi}^\top(t)(A_n\tilde{\phi}(t) + \tilde{f}(\tilde{\phi}(t))) < 0$ or $r(t)$ blows up in a finite time $T < +\infty$ if $\nu > 0$, $\tilde{\phi}^\top(t)(A_n\tilde{\phi}(t) + \tilde{f}(\tilde{\phi}(t))) > 0$. For $\nu \geq 0$, $\tilde{\phi}^\top(t)(A_n\tilde{\phi}(t) + \tilde{f}(\tilde{\phi}(t))) \leq 0$ or $\nu \leq 0$, $\tilde{\phi}^\top(t)(A_n\tilde{\phi}(t) + \tilde{f}(\tilde{\phi}(t))) \geq 0$ we have $T = +\infty$. □

Introducing the notation

$$\tilde{x}(t) = \mathbf{d}_n(\ln \tilde{r}(t))\tilde{\phi}(t)$$

we derive that the system (7.28) is homeomorphic to

$$\begin{cases} \frac{d\tilde{x}}{dt} = \|\tilde{x}\|_{\mathbf{d}_n}^{\nu}\mathbf{d}_n(\ln \|\tilde{x}\|_{\mathbf{d}})\left(\tilde{A}_n\mathbf{d}_n(-\ln \|\tilde{x}\|_{\mathbf{d}_n})\tilde{x} + \tilde{f}\left(\mathbf{d}_n(-\ln \|\tilde{x}\|_{\mathbf{d}_n})\tilde{x}\right)\right), \\ \tilde{x}(0) = \mathbf{d}_n(\ln \|x_0\|_{\mathbf{d}})\Pi_n\mathbf{d}(-\ln \|x_0\|_{\mathbf{d}})x_0, \quad t \in (0, T), \end{cases} \tag{7.31}$$

where $\|\cdot\|_{\mathbf{d}_n}$ is the canonical $\mathbf{d}_n$-homogeneous norm induced by the Euclidean norm $\|\cdot\|_{\mathbb{R}^n}$ and Π_n is given by (7.3). Taking into account $\mathbf{d}$-homogeneity of the operator $A + f$ we derive

$$Ax + f(x) = \|x\|_{\mathbf{d}}^{\nu}\mathbf{d}(\ln \|x\|_{\mathbf{d}})\left(A\mathbf{d}(-\ln \|x\|_{\mathbf{d}})x + f(\mathbf{d}(-\ln \|x\|_{\mathbf{d}})x)\right);$$

therefore, the system (7.31) can be treated as a finite-dimensional projection of the $\mathbf{d}$-homogeneous evolution equation (7.22) obtained by means of sequential approximations of the dilation group $\mathbf{d}$, the canonical homogeneous norm $\|x\|_{\mathbf{d}}$, and the operators A, f. The formula (7.26) can be rewritten as

$$x_v(t) = \mathbf{d}(\ln \|\tilde{x}(t)\|_{\mathbf{d}_n})\sum_{i=1}^{n}[\mathbf{d}_n(-\ln \|\tilde{x}(t)\|_{\mathbf{d}_n})\tilde{x}(t)]_i h_i, \quad t \geq 0,$$

where $[v]_i$ denotes the ith element of the vector $v \in \mathbb{R}^n$. Since

$$|\mathbf{d}_n(-\ln \|\tilde{x}(t)\|_{\mathbf{d}_n})\tilde{x}(t)| = 1$$

then $\left\|\sum_{i=1}^{n}[\mathbf{d}_n(-\ln\|\tilde{x}(t)\|_{\mathbf{d}_n})\tilde{x}(t)]_i h_i\right\|_{\mathbb{H}} = 1$ and

$$\|x_\upsilon(t)\|_{\mathbf{d}} = \|\tilde{x}(t)\|_{\mathbf{d}_n}, \quad t \geq 0.$$

The system (7.31) is $\mathbf{d}_n$-homogeneous of the degree ν. Therefore, *the homogeneous Galerkin projection preserves the homogeneity of the original system as well as its homogeneity degree.*

Corollary 7.2 *Let all conditions of Theorem 7.2 be fulfilled and $x_\upsilon \in C([0,T);\mathcal{D}_V)$ be a solution of (7.25) for $x_0 \in \mathcal{D}_V$. Then for any $s \in \mathbb{R}$ the function $x_\upsilon^s \in C([0, e^{-\nu s}T);\mathcal{D}_V)$ given by*

$$x_\upsilon^s(t) = \mathbf{d}(s)x_\upsilon(e^{\nu s}t), \quad t \geq 0$$

is a solution of (7.25) with the scaled initial condition $x(0) = \mathbf{d}(s)x_0$.

Proof Let $\tilde{x}$ be the solution of (7.31) for some $x_0 \in \mathcal{D}_V$. Denote $x_0^s = \mathbf{d}(s)x_0$. Since $\mathcal{D}_V$ is a $\mathbf{d}$-homogeneous cone then $x_0^s \in \mathcal{D}_V$. Since

$$\mathbf{d}(-\ln\|x_0^s\|_{\mathbf{d}})\mathbf{d}(s)x_0^s = \mathbf{d}(-\ln\|x_0\|_{\mathbf{d}})x_0$$

and $\|x_0^s\|_{\mathbf{d}} = e^s\|x_0\|_{\mathbf{d}}$ then

$$\begin{aligned}\mathbf{d}_n(\ln\|x_0^s\|_{\mathbf{d}})\Pi_n\mathbf{d}(-\ln\|x_0^s\|_{\mathbf{d}})x_0^s &= \mathbf{d}_n(s)\mathbf{d}_n(\ln\|x_0\|_{\mathbf{d}})\Pi_n\mathbf{d}(-\ln\|x_0\|_{\mathbf{d}})x_0\\ &= \mathbf{d}(s)\tilde{x}(0).\end{aligned}$$

Due to Theorem 6.1, the solution of (7.31) for the case when x_0 is replaced with x_0^s is given by $\tilde{x}^s(t) = \mathbf{d}(s)\tilde{x}(e^{\nu s}t)$. Hence, the solution of (7.25) with the scaled initial condition $x(0) = \mathbf{d}(s)x_0$ has the form

$$x_\upsilon^s(t) = \mathbf{d}(\ln\|\tilde{x}^s(t)\|_{\mathbf{d}_n})\sum_{i=1}^{n}[\mathbf{d}_n(-\ln\|\tilde{x}^s(t)\|_{\mathbf{d}_n})\tilde{x}^s(t)]_i h_i =$$

$$\mathbf{d}(s)\mathbf{d}(\ln\|\tilde{x}(e^{\nu s}t)\|_{\mathbf{d}_n})\sum_{i=1}^{n}[\mathbf{d}_n(-\ln\|\tilde{x}(e^{\nu s}t)\|_{\mathbf{d}_n})\tilde{x}(e^{\nu s}t)]_i h_i = \mathbf{d}(s)x_\upsilon(e^{\nu s}t).$$

□

The obtained finite-dimensional projection of the nonlinear evolution equation (7.22) preserves some stability properties of the original system.

Corollary 7.3 *If*

$$\exists \rho \geq 0: \quad \langle A\phi + f(\phi), \phi \rangle \leq -\rho \langle G_{\mathbf{d}}\phi, \phi \rangle, \quad \forall \phi \in \mathcal{D}(A) \cap \mathcal{D}(G_{\mathbf{d}}) : \|\phi\| = 1$$

then both the systems (7.22) *and* (7.31) *are*

- *finite-time stable if* $\nu < 0$*;*
- *exponentially stable if* $\nu = 0$*;*
- *nearly fixed-time stable if* $\nu > 0$.

Proof Indeed, for all $\tilde{\phi} \in S_{\mathbb{R}^n}$ we have

$$\langle A\phi_v + f(\phi_v), \phi_v \rangle = \tilde{\phi}^\top A_n \tilde{\phi} + \tilde{\phi}^\top f(\tilde{\phi}) \leq -\rho \langle G_{\mathbf{d}}\phi_v, \phi_v \rangle = -\rho \tilde{\phi}^\top G_{d_n} \tilde{\phi},$$

where $\phi_v = \sum_{i=1}^n \tilde{\phi}_i h_i \in S_V$. This means

$$\frac{d}{dt}\|\tilde{x}(t)\|_{\mathbf{d}_n} \leq -\rho \|\tilde{x}(t)\|_{\mathbf{d}_n}^{\nu+1},$$

i.e., the canonical homogeneous norm is the Lyapunov function for the system (7.31). Due to the formula (4.25) the inequality $\langle A\phi + f(\phi), \phi \rangle \leq -\rho, \forall \phi \in \mathcal{D}(A) \cap \mathcal{D}(G_{\mathbf{d}}) : \|\phi\| = 1$ implies

$$\frac{d}{dt}\|x(t)\|_{\mathbf{d}} \overset{a.e.}{\leq} -\rho \|x(t)\|_{\mathbf{d}}^{\nu+1}$$

for any strong solution x of (7.22), i.e., the canonical homogeneous norm $\|\cdot\|_{\mathbf{d}}$ is the Lyapunov function of the original system (7.22). Moreover, the convergence rates of the original and the projected systems are identical, namely both systems are finite-time stable if $\nu < 0$; both systems are exponentially stable if $\nu = 0$; both systems are nearly fixed-time stable if $\nu > 0$. The proof is complete. □

Notice that if the system (7.22) is standard homogeneous (i.e., $\mathbf{d}(s) = e^s I$) of degree $\nu = 0$, then the formula (7.31) defines the classical Galerkin projection of (7.22):

$$\frac{d\tilde{x}(t)}{dt} = \tilde{A}_n \tilde{x}(t) + \tilde{f}(\tilde{x}(t)), \quad \tilde{x}(0) = \Pi_n x_0.$$

7.3.3 *Classical versus Homogeneous Galerkin Projection*

For comparison reasons, let us consider the Burgers equation (7.1). This equation can be represented in the form (7.22) with $A = \frac{\partial^2}{\partial z^2}$, $x_0 \in \mathbb{H} := L^2(\mathbb{R}, \mathbb{R})$, $x(t) \in L^2(\mathbb{R}, \mathbb{R})$, $\mathcal{D}(A) = H^2(\mathbb{R}, \mathbb{R})$, $f(x) = -x\frac{\partial x}{\partial z}$, and $\mathcal{D}(f) = \mathbb{Y} := H^1(\mathbb{R}, \mathbb{R})$. Notice (see Example 6.7 for more details) that f is M-regular with $M = \frac{\partial}{\partial z} + I$ and

$(M^{-1}y)(z) = \int_{-\infty}^{z} e^{-(z-s)}y(s)ds$, $y \in L^2(\mathbb{R}, \mathbb{R}), z \in \mathbb{R}$, respectively. The considered system satisfies Theorem 2.15 and the semigroup generated by A is $(\Phi(t)y)(z) = \frac{1}{\sqrt{4\pi t}} \int_{\infty}^{+\infty} e^{-\frac{(z-s)^2}{4t}} y(s)ds$, $y \in \mathbb{H}$. Moreover, the Burgers equation has the explicit solution

$$[x(t)](z) = -2\frac{\partial}{\partial z} \ln \left\{ \frac{1}{\sqrt{4\pi t}} \int_{-\infty}^{+\infty} e^{-\frac{(z-\sigma)^2}{4t} - \frac{1}{2}\int_0^{\sigma} x_0(s)ds} d\sigma \right\} \tag{7.32}$$

due to the Hopf–Cole transformation. This exact solution is utilized below in order to compare the approximation precision of the classical and homogeneous Galerkin method.

The system (7.1) is $\mathbf{d}$-homogeneous of the degree $\nu = 2$ with respect to the dilation $(\mathbf{d}(s)x)(z) = e^s x(e^s z), z \in \mathbb{R}, x \in L^2(\mathbb{R}, \mathbb{R})$. The dilation $\mathbf{d}$ is strictly monotone and strongly continuous in $L^2(\mathbb{R}, \mathbb{R})$ and $H^1(\mathbb{R}, \mathbb{R})$.

Taking the Hermite functions (7.10) as an orthonormal basis $h_1, h_2, ..., h_n$ in $L^2(\mathbb{R}, \mathbb{R})$, we derive the following homogeneous Galerkin projection of the Burgers equation:

$$\begin{cases} \frac{d\tilde{x}}{dt} = \|\tilde{x}\|_{\mathbf{d}_n}^{\nu} \mathbf{d}_n(\ln \|\tilde{x}\|_{\mathbf{d}_n}) \left(\tilde{A}_n + \tilde{B}_n \left(\mathbf{d}_n(-\ln \|\tilde{x}\|_{\mathbf{d}_n})\tilde{x}\right)\right) \mathbf{d}_n(-\ln \|\tilde{x}\|_{\mathbf{d}_n})\tilde{x}, \\ \tilde{x}(0) = \mathbf{d}_n(\ln \|x_0\|_{\mathbf{d}}) \Pi_n \mathbf{d}(-\ln \|x_0\|_{\mathbf{d}}) x_0, \end{cases} \tag{7.33}$$

where G_n, $\mathbf{d}_n$ and $\|\cdot\|_{\mathbf{d}_n}$ are defined in Sect. 7.2.2, $A_n \in \mathbb{R}^{n\times n}$ has the elements

$$(\tilde{A}_n)_{ij} = \langle Ah_j, h_i\rangle = \left\langle \frac{d^2}{dz^2} h_j, h_i \right\rangle,$$

and the linear matrix-valued function $\tilde{x} \mapsto B_n(\tilde{x}) \in \mathbb{R}^{n\times n}$ is defined as follows:

$$[\tilde{B}_n(\tilde{\phi})]_{i,j} = \frac{1}{2}\sum_{k=1}^{n} \tilde{\phi}_k \left\langle h_k h_j, \frac{\partial h_i}{\partial y}\right\rangle, \quad \tilde{\phi} = (\tilde{\phi}_1, \tilde{\phi}_2, ..., \tilde{\phi}_k)^{\top} \in \mathbb{R}^n, \quad i, j = 1, ..., n.$$

The matrix-valued function $\tilde{x} \to B(\tilde{x}), \tilde{x} \in \mathbb{R}^n$ is obtained from (7.30) taking into account the identity

$$-\left\langle \left(\sum_{k=1}^{n} \tilde{x}_k h_k\right) \frac{\partial}{\partial z} \left(\sum_{k=1}^{n} \tilde{\phi}_k h_k\right), h_i \right\rangle = \frac{1}{2}\left\langle \left(\sum_{k=1}^{n} \tilde{x}_k h_k\right)^2, \frac{\partial}{\partial z} h_i \right\rangle,$$

which comes from the integration by parts. Using the identity

$$\frac{d^2}{dz^2} h_j(z) = -(2i-1)h_j(z) + y^2 h_j(z), \quad z \in \mathbb{R},$$

we derive

$$(\tilde{A}_n)_{ij} = -(2i-1)\delta_{ij} + \langle yh_i, yh_j \rangle,$$

where $\delta_{ij} = 0$ for $i \neq j$ and $\delta_{ij} = 1$ for $i = j$. From the identity

$$zh_j(z) = \sqrt{\tfrac{j-1}{2}}h_{j-1}(z) + \sqrt{\tfrac{j}{2}}h_{j+1}(z), \quad z \in \mathbb{R},$$

we conclude

$$(\tilde{A}_n)_{ij} = -(i-1/2)\delta_{ij} + \sqrt{\tfrac{i-1}{2}}\sqrt{\tfrac{j}{2}}\delta_{i-1\,j+1} + \sqrt{\tfrac{i}{2}}\sqrt{\tfrac{j-1}{2}}\delta_{i+1\,j-1}.$$

Notice that the classical Galerkin projection gives the following ODE:

$$\begin{cases} \frac{d\tilde{x}^{cl}(t)}{dt} = \left(\tilde{A}_n + \tilde{B}_n\left(\tilde{x}^{cl}(t)\right)\right)\tilde{x}^{cl}(t), \quad t > 0, \\ \tilde{x}^{cl}(0) = \Pi_n x_0, \end{cases} \tag{7.34}$$

which approximates the solution for the Burgers equation as

$$x_{cl}(t) = \sum_{i=1}^{n} \tilde{x}_i^{cl}(t)h_i, t \geq 0.$$

Figures 7.1 and 7.2 presents the comparison results of the exact solution x_{exact} of the Burgers equation (given by (7.32)) with the approximate solutions x_{cl} and

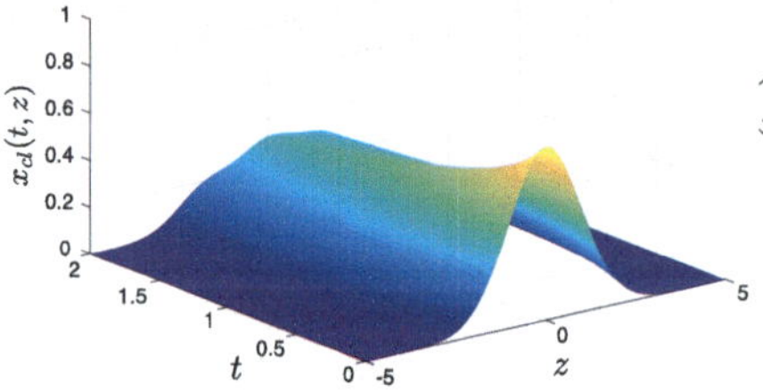

(a) Classical Galerkin

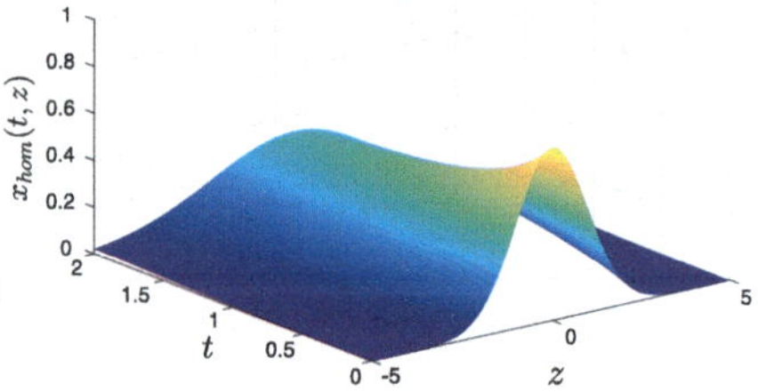

(b) Homogeneous Galerkin

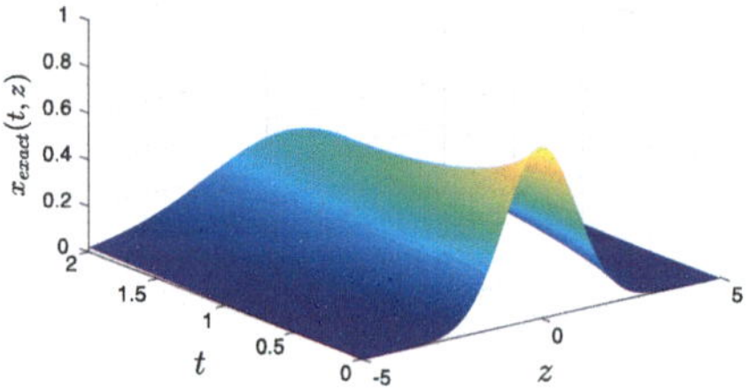

(c) Exact solution

Fig. 7.1 Approximate and exact solutions of viscous Burgers equation with $x_0 = h_1$

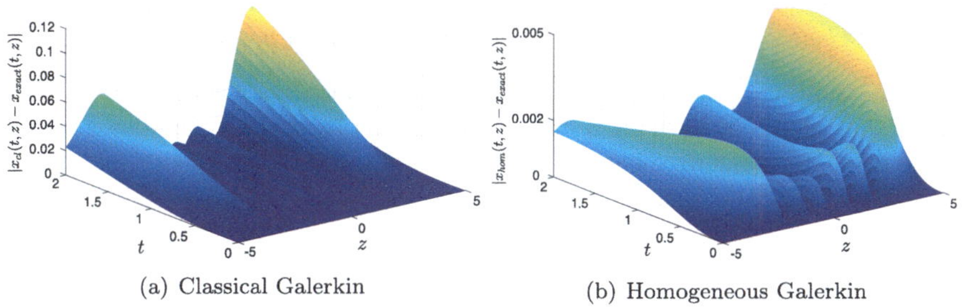

Fig. 7.2 Approximation errors of viscous Burgers equation with $x_0 = h_1$

$$x_{hom}(t) = \mathbf{d}(\ln \|\tilde{x}(t)\|_{\mathbf{d}_n}) \sum_{i=1}^{n} [\mathbf{d}_n(-\ln \|\tilde{x}(t)\|_{\mathbf{d}_n})\tilde{x}(t)]_i h_i,$$

obtained by means of the classical and the homogeneous Galerkin methods, respectively. The initial condition was selected as follows $x_0 = h_1 \in V$. The number of basis functions was identical for both classical and homogeneous Galerkin projections: $n = 5$.

In both cases, the numerical simulations were made using semi-implicit Euler scheme with the time step $\tau = 0.01$:

$$\frac{\tilde{x}_{cl}^{[p+1]} - \tilde{x}_{cl}^{[p]}}{\tau} = \left(\tilde{A}_n + B\left(\tilde{x}_{cl}^{[p]}\right)\right)\tilde{x}_{cl}^{[p+1]}, \quad \tilde{x}_{cl}^{[0]} = \Pi_n x_0, \quad p = 0, 1, \ldots$$

In the view of the results [7], the semi-implicit discretization is more appropriate for homogeneous systems with positive degrees. The equivalent representation (7.28) of the system (7.33) is utilized for numerical simulations, since it is computationally less consuming:

$$\begin{cases} \frac{\tilde{\phi}^{[p+1]} - \tilde{\phi}^{[p]}}{\tau} = \left(\tilde{r}^{[p]}\right)^{\nu}\left(\tilde{A}_n + B\left(\tilde{\phi}^{[p]}\right) - \frac{(\tilde{\phi}^{[p]})^{\top}\left(\tilde{A}_n + B\left(\tilde{\phi}^{[p]}\right)\right)\tilde{\phi}^{[p]}}{(\tilde{\phi}^{[p]})^{\top} G_{\mathbf{d}_n}\tilde{\phi}^{[p]}} G_{\mathbf{d}_n}\right)\tilde{\phi}^{[p+1]}, \\ \frac{\tilde{r}^{[p+1]} - \tilde{r}^{[p]}}{\tau} = \left(\tilde{r}^{[p]}\right)^{\nu} \frac{(\tilde{\phi}^{[p]})^{\top}\left(\tilde{A}_n + B\left(\tilde{\phi}^{[p]}\right)\right)\tilde{\phi}^{[p]}}{(\tilde{\phi}^{[p]})^{\top} G_{\mathbf{d}_n}\tilde{\phi}^{[p]}}\tilde{r}^{[p+1]}, \quad p = 0, 1, \ldots, \\ \tilde{\phi}^{[0]} = \Pi_n \mathbf{d}(-\ln \|x_0\|_{\mathbf{d}})x_0, \quad \tilde{r}^{[0]} = \|x_0\|_{\mathbf{d}}. \end{cases}$$

Since $\tilde{\phi}(t) \in S_{\mathbb{R}^n}, \forall t \geq 0$, then $\tilde{\phi}^{[p+1]}$ must belong to the unit sphere as well. To preserve this property the discrete-time approximation of $\tilde{\phi}$ is combined with the projection of $\tilde{\phi}^{[p]}$ on the unit sphere $S_{\mathbb{R}^n}$ after each time step. Therefore, we have

$$\tilde{x}^{[p]} = \mathbf{d}(\ln \tilde{r}^{[p]})\tilde{\phi}^{[p]},$$

where $\tilde{x}^{[p]}$ is the discrete-time approximation of $\tilde{x}(t)$.

> *The simulations for $x_0 = h_1$ show that the homogeneous Galerkin method has 20 times better precision in L^∞-norm and 10 times better precision in L^2-norm than the classical Galerkin method.*

Notice that such a good approximate solution was obtained by the homogeneous Galerkin method with just 5 basis functions. To obtain a similar precision for $(t, z) \in [0, 2] \times [-5, 5]$, the classical Galerkin method needs at least 20 basis functions.

Figures 7.3 and 7.4 present comparison results for the initial condition $x_0 = \xi$:

$$\xi(z) = \begin{cases} 1 & \text{if } |z| \le 1, \\ 0 & \text{if } |z| > 1, \end{cases} \qquad z \in \mathbb{R}.$$

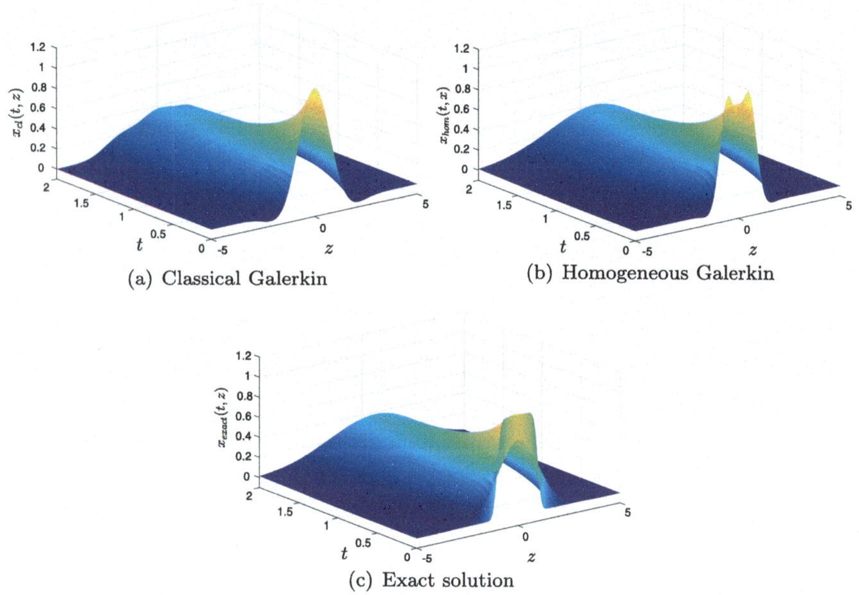

Fig. 7.3 Approximate and exact solutions of viscous Burgers equation with $x_0 = \xi$

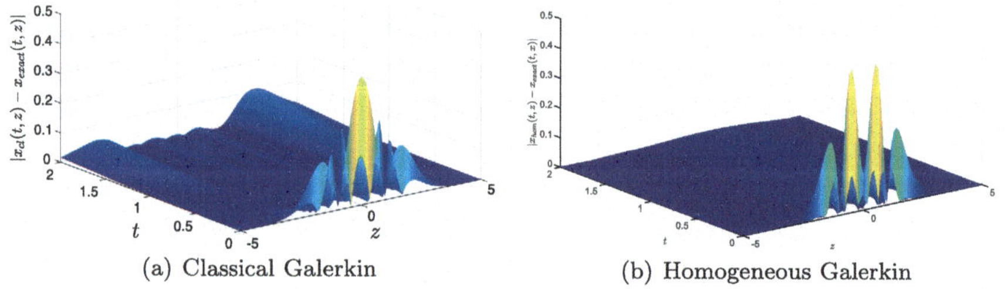

Fig. 7.4 Approximation errors for viscous Burgers equation with $x_0 = \xi$

The approximation error of the initial state is very quickly filtered by the homogeneous ODE (see Fig. 7.4), so the homogeneous Galerkin projection method again shows much better approximation precision than the classical one.

Concluding Remarks

A consistent approximation of infinite-dimensional homogeneous system is presented. A special version of Galerkin method is developed for this purpose. The proposed method preserves a dilation symmetry of infinite-dimensional homogeneous system (evolution equation) in its finite-dimensional counterpart (ODE). Such an approximation allows the convergence rate (like finite-time or nearly fixed-time stability) to be preserved as well. Moreover, in an example, it is demonstrated that the consistency of the homogeneity properties of the finite-dimensional (approximate) and infinite-dimensional (original) systems essentially improves the approximation precision (in 10 times for the Burgers equation).

References

1. Dolejsi V, Feistauer M (2015) Discontinuous Galerkin method. Springer
2. Davies AM (1978) Application of the Galerkin method to the solution of Burgers' equation. Comput Methods Appl Mech Eng 14(3):305–321
3. Brooks A, Hughes T (1982) Streamline upwind/Petrov-Galerkin formulations for convection dominated flows with particular emphasis on the incompressible Navier–Stokes equations. Comput Methods Appl Mech Eng 32:199–259
4. Bihlo A, Valiquette F (2017) Symmetry-preserving numerical schemes. In: Symmetries and integrability of difference equations. Springer, pp 261–324
5. Praly L (1997) Generalized weighted homogeneity and state dependent time scale for linear controllable systems. In: Proceedings of the IEEE conference on decision and control, pp 4342–4347, San Diego, USA
6. Polyakov A (2020) Generalized homogeneity in systems and control. Springer
7. Polyakov A, Efimov D, Brogliato B (2019) Consistent discretization of finite-time and fixed-time stable systems. SIAM J Control Optim 57(1):78–103

Chapter 8
Homogeneous Stabilization of Infinite-Dimensional Systems

In this chapter, the problem of homogeneous stabilization of an infinite-dimensional control system is studied.

Definition 8.1 Let $\mathbb{H}$ be a Hilbert space and $\mathbb{U}$ be a Banach space. An evolution system

$$\dot{x} = f(x, u), \quad t > 0, \quad f : \mathcal{D}(f) \subset (\mathbb{H} \times \mathbb{U}) \mapsto \mathbb{B}$$

is said to be **d**-homogeneously stabilizable with degree $\mu \in \mathbb{R}$ if there exists a (locally or globally bounded) feedback law $u : \mathbb{H} \mapsto \mathbb{U}$ such that the closed-loop system is globally uniformly asymptotically stable and **d**-homogeneous of the degree μ, where **d** is a dilation in $\mathbb{B}$.

An algorithm of homogeneous stabilization of a generalized linear plant in a Hilbert space is developed. Issues of homogenization of linear plants are discussed. Algorithms for homogeneous stabilization of nonlinear systems with locally Lipschitz continuous nonlinearities are developed. Partial homogeneous finite-time stabilization of systems with modal operators is considered. Theoretical results are illustrated by various examples of control design for PDE models.

A. Polyakov, *Generalized Homogeneity in Systems and Control Volume II*, Communications and Control Engineering,
https://doi.org/10.1007/978-3-031-77496-6_8

8.1 Stabilization of Linear Plant

8.1.1 Implicit Homogeneous Stabilizer for Linear Plant

Let us consider the following abstract control system

$$\dot{x}(t) = Ax(t) + Bu(x(t)), \quad t > 0, \tag{8.1}$$

$$x(0) = x_0, \tag{8.2}$$

where $A : \mathcal{D}(A) \subset \mathbb{H} \mapsto \mathbb{H}$ is a (possibly unbounded) closed linear operator with the domain $\mathcal{D}(A)$ dense in $\mathbb{H}$, $B : \mathbb{U} \mapsto \mathbb{H}$ is a linear bounded operator, $\mathbb{H}$ is a real Hilbert space and $\mathbb{U}$ is a real Banach space, $x(t) \in \mathbb{H}$ is the system state, $u : \mathbb{H} \mapsto \mathbb{U}$ is a (*locally or globally*) bounded feedback control.

Algorithms of linear stabilization $u = Kx$ with $K \in \mathcal{L}(\mathbb{H}; \mathbb{U})$ for such a system are developed for both finite-dimensional (see, e.g., [1–3]) and infinite-dimensional (see, e.g., [4, 5] and references wherein) system (8.1). In this case, the linear closed-loop system is exponentially stable and $\mathbf{d}$-homogeneous of the degree $\mu = 0$ with respect to the standard dilation $\mathbf{d}(s) = e^s I$. The problem of homogeneous stabilization with a non zero degree $\mu \neq 0$ is studied below. In the view of Theorem 6.6, this implies finite-time (or nearly fixed-time) stability of the closed-loop system for $\mu < 0$ (resp., $\mu > 0$). The finite-time stabilization of PDEs is also studied in the literature (see, e.g., [6–11]). This problem is liked with a controllability analysis of an evolution system [4] as well as with sliding-mode control method [12, 13].

In Volume I, it is shown that any controllable linear system can be homogeneously stabilized using a feedback $u = K(\|x\|_{\mathbf{d}})x$ with the gain K dependent of a canonical homogeneous norm $\|\cdot\|_{\mathbf{d}}$ in $\mathbb{H}$. In this chapter, we design similar controllers for infinite-dimensional systems. The following theorem refines the results of the paper [11].

Theorem 8.1 *Let $A : \mathcal{D}(A) \subset \mathbb{H} \mapsto \mathbb{H}$ be a generator of a strongly continuous semigroup Φ of linear bounded operators on $\mathbb{H}$, $B : \mathbb{U} \mapsto \mathbb{H}$ be a linear bounded operator and*

(A) $\mathbf{d}_1$ *be a uniformly continuous group of linear bounded operators in $\mathbb{H}$ with the generator $G_{\mathbf{d}_1} \in \mathcal{L}(\mathbb{H}; \mathbb{H})$;*

(B) $\mathbf{d}$ *be a strongly continuous strictly monotone dilation in $\mathbb{H}$, where $G_{\mathbf{d}} : \mathcal{D}(G_{\mathbf{d}}) \subset \mathbb{H} \mapsto \mathbb{H}$ is the generator of the dilation group* $\mathbf{d}$ *such that $\langle G_{\mathbf{d}}x, x\rangle \geq \beta\|x\|^2$ for all $x \in \mathcal{D}(G_{\mathbf{d}})$ and some $\beta > 0$;*

(C) there exist a linear bounded operator $K : \mathbb{H} \mapsto \mathbb{U}$ such that

$$\exists \rho > 0 : \langle (A + BK + \rho G_{\mathbf{d}})x, x\rangle \leq 0, \quad \forall x \in \mathcal{D}, \tag{8.3}$$

where a set $\mathcal{D} \subset (\mathcal{D}(A) \cap \mathcal{D}(G_{\mathbf{d}}))$ is a $\mathbf{d}$*-homogeneous cone dense in $\mathbb{H}$.*

(D) the operator A be **d***-homogeneous of degree* $\mu \in \mathbb{R}$ *with* $\mu > -\beta$ *and*

$$\mathbf{d}(s)BK\mathbf{d}_1(s) = BK\mathbf{d}(s), \quad \forall s \in \mathbb{R}.$$

Then the following holds.

- *The feedback control* $u : \mathbb{H} \mapsto \mathbb{U}$ *defined as*

$$u(x) = \begin{cases} \|x\|_{\mathbf{d}}^{\mu} K\mathbf{d}_1(-\ln(\|x\|_{\mathbf{d}}))x & \text{if } x \neq \mathbf{0}, \\ \mathbf{0} & \text{if } x = \mathbf{0} \end{cases} \tag{8.4}$$

is locally Lipschitz continuous on $\mathbb{H}\backslash\{\mathbf{0}\}$ *and Fréchet differentiable on* $\mathcal{D}(G_{\mathbf{d}})\backslash\{\mathbf{0}\}$. *Moreover,*

  - *if* $\beta + \mu > \gamma_{\max}$ *then u is continuous at* $\mathbf{0} \in \mathbb{H}$*;*
  - *if* $\beta + \mu \geq \gamma_{\max}$ *then*

$$\sup_{\|x\| \leq 1} \|u(x)\| \leq \|K\|;$$

  - *if* $\exists M \geq 1 : \|\mathbf{d}(s)\| \leq Me^{(-\mu+\gamma_{\min})s}$, $\forall s > 0$ *then*

$$\sup_{\|x\| \geq 1} \|u(x)\| \leq M\|K\|,$$

where $\gamma_{\min} := \inf_{z \in S} \langle G_{\mathbf{d}_1} z, z \rangle$ *and* $\gamma_{\max} := \sup_{z \in S} \langle G_{\mathbf{d}_1} z, z \rangle$.

- *The closed-loop system* (8.1)–(8.4) *has*
  - *a unique mild solution defined on* $[0, +\infty)$ *for any* $x_0 \in \mathbb{H}$*;*
  - *a unique locally Lipschitz continuous strong solution defined on* $[0, +\infty)$ *for any* $x_0 \in \mathcal{D}(A)$*;*
  - *a unique classical solution defined on* $[0, +\infty)$ *for any* $x_0 \in \mathcal{D}(A)$ *provided that* $BK\mathcal{D}(A) \subset \mathcal{D}(A)$.
- *The closed-loop system* (8.1)–(8.4) *is*
  - *globally uniformly finite-time stable for* $\mu < 0$ *and the settling-time functional* $T : \mathbb{H} \mapsto [0, +\infty)$ *admits the estimate*

$$T(x_0) \leq \frac{\|x_0\|_{\mathbf{d}}^{-\mu}}{-\rho\mu}; \tag{8.5}$$

  - *globally uniformly exponentially stable for* $\mu = 0$ *and*

$$\|x(t)\|_{\mathbf{d}} \leq \|x_0\|_{\mathbf{d}} e^{-\rho t}; \tag{8.6}$$

  - *globally uniformly nearly fixed-time stable for* $\mu > 0$ *and*

$$\forall r > 0 \quad : \quad \|x(t)\|_{\mathbf{d}} \leq r, \quad \forall t \geq \frac{1}{\rho\mu r^{\mu}} \tag{8.7}$$

independently of initial condition $x((0) = x_0$.

- *The canonical homogeneous norm* $\|\cdot\|_{\mathbf{d}}$ *is a Lyapunov function of the closed-loop system* (8.1)–(8.4) *such that*

$$\frac{d}{dt}\|x(t)\|_{\mathbf{d}} \overset{a.e.}{\leq} -\rho\|x(t)\|_{\mathbf{d}}^{1+\mu}, t > 0 \tag{8.8}$$

for any strong solution of the closed-loop system (8.1)–(8.4).

Proof **I.** Since $\|\mathbf{d}(-\ln\|x\|_{\mathbf{d}})x\| = 1$ and $\mathbf{d}(\ln\|x\|_{\mathbf{d}})\mathbf{d}(-\ln\|x\|_{\mathbf{d}}) = I$ then

$$\|u(x)\| \leq \|K\|\|x\|_{\mathbf{d}}^{\mu}\|\mathbf{d}_1(-\ln\|x\|_{\mathbf{d}})\|\|\mathbf{d}(\ln\|x\|_{\mathbf{d}})\|, \quad \forall x \neq \mathbf{0}.$$

Since $\mathbf{d}_1$ is a uniformly continuous group of linear bonded operators on $\mathbb{H}$ then from Corollary 4.1, we conclude $\|\mathbf{d}_1(s)\| \leq e^{\gamma_{\max}s}$ for $s \geq 0$ and $\|\mathbf{d}_1(s)\| \leq e^{\gamma_{\min}s}$ for $s \leq 0$.

Since the dilation $\mathbf{d}$ is strictly monotone then $\|\mathbf{d}(s)\| \leq e^{\beta s}$ for $s \leq 0$ (see Proposition 4.4) and for $\|x\| \leq 1$, we derive

$$\|u(x)\|_{\mathbb{U}} \leq \|K\|\|x\|_{\mathbf{d}}^{\mu}\|x\|_{\mathbf{d}}^{-\gamma_{\max}}\|x\|_{\mathbf{d}}^{\beta}.$$

Therefore, if $\beta + \mu - \gamma_{\max} > 0$ then $\|u(x)\|_{\mathbb{U}} \to 0$ as $x \to \mathbf{0}$ and $\|u(x)\|_{\mathbb{U}} \leq \|K\|$ for $\|x\| \leq 1$. For $\beta + \mu = \gamma_{\max}$ the control u remains bounded in the unit ball by the same constant $\|K\|$, but it becomes possibly discontinuous at $\mathbf{0}$.

If $\exists M \geq 1 : \|\mathbf{d}(s)\| \leq Me^{(-\mu+\gamma_{\min})s}$ then for $\|x\| \geq 1$ we have

$$\|u(x)\|_{\mathbb{U}} \leq M\|K\|\|x\|_{\mathbf{d}}^{\mu}\|x\|_{\mathbf{d}}^{-\gamma_{\min}}\|x\|_{\mathbf{d}}^{-\mu+\gamma_{\min}} = M\|K\|.$$

Let us show that the control law u is locally Lipschitz continuous on $\mathbb{H}\backslash\{\mathbf{0}\}$. Indeed, since K is a linear bounded operator then

$$\|u(x_1) - u(x_2)\| \leq$$

$$\|K\| \cdot \left\| \|x_1\|_{\mathbf{d}}^{\mu}\mathbf{d}_1(-\ln\|x_1\|_{\mathbf{d}})x_1 - \|x_2\|_{\mathbf{d}}^{\mu}\mathbf{d}_1(-\ln\|x_2\|_{\mathbf{d}})x_2 \right\| \leq$$

$$\|K\| \cdot \|\mathbf{d}_1(-\ln\|x_1\|_{\mathbf{d}})x_1\| \cdot \left|\|x_1\|_{\mathbf{d}}^{\mu} - \|x_2\|_{\mathbf{d}}^{-\mu}\right| +$$

$$\|x_2\|_{\mathbf{d}}^{\mu}\|\mathbf{d}_1(-\ln\|x_1\|_{\mathbf{d}})x_1 - \mathbf{d}_1(-\ln\|x_2\|_{\mathbf{d}})x_1\| +$$

$$\|x_2\|_{\mathbf{d}}^{\mu}\|\mathbf{d}_1(-\ln\|x_2\|_{\mathbf{d}})\| \cdot \|x_1 - x_2\|,$$

for all $x_1, x_2 \in \mathbb{H}\backslash\{\mathbf{0}\}$. Without of loss of generality, we can assume that $\|x_1\|_{\mathbf{d}} \leq \|x_2\|_{\mathbf{d}}$. Since the group $\mathbf{d}_1$ is uniformly continuous then

$$\mathbf{d}_1(-\ln \|x_1\|_{\mathbf{d}}) - \mathbf{d}_1(-\ln \|x_2\|_{\mathbf{d}}) = \mathbf{d}_1(-\ln \|x_2\|_{\mathbf{d}})(\mathbf{d}_1(\ln \|x_2\|_{\mathbf{d}} - \ln \|x_1\|_{\mathbf{d}}) - I) =$$

$$\mathbf{d}_1(-\ln \|x_2\|_{\mathbf{d}}) \int_0^{\ln \|x_2\|_{\mathbf{d}} - \ln \|x_1\|_{\mathbf{d}}} G_{\mathbf{d}_1} \mathbf{d}_1(s) ds$$

and $\|\mathbf{d}_1(s)\| \le e^{\|G_{\mathbf{d}_1}\| s}$ for all $s \ge 0$. Hence, we derive

$$\|\mathbf{d}_1(-\ln \|x_1\|_{\mathbf{d}})x_1 - \mathbf{d}_1(-\ln \|x_2\|_{\mathbf{d}})x_1\| \le$$

$$\|\mathbf{d}_1(-\ln \|x_2\|_{\mathbf{d}})\| \cdot \|G_{\mathbf{d}_1}\| \int_0^{\ln \|x_2\|_{\mathbf{d}} - \ln \|x_1\|_{\mathbf{d}}} e^{\|G_{\mathbf{d}}\| s} ds \le$$

$$\frac{\|x_1\| \cdot \|\mathbf{d}_1(-\ln \|x_2\|_{\mathbf{d}})\|}{\|x_1\|_{\mathbf{d}}^{\|G_{\mathbf{d}_1}\|}} \cdot \left| \|x_2\|_{\mathbf{d}}^{\|G_{\mathbf{d}_1}\|} - \|x_1\|_{\mathbf{d}}^{\|G_{\mathbf{d}_1}\|} \right|.$$

Taking into account local Lipschitz continuity of the canonical homogeneous norm (see Lemma 4.5) on $\mathbb{H}\backslash\{\mathbf{0}\}$, we derive local Lipschitz continuity of the control u on $\mathbb{H}\backslash\{\mathbf{0}\}$. The Fréchet differentiability of u on $\mathcal{D}(G_{\mathbf{d}})\backslash\{\mathbf{0}\}$ follows from Lemma 4.6.

If $BK\mathcal{D}(A) \subset \mathcal{D}(A)$ then the inclusion $Bu(\mathcal{D}(A)) \subset \mathcal{D}(A)$ follows from Conditions (C) and (D). Indeed, $\mathbf{d}(s)\mathcal{D}(A) \subset \mathcal{D}(A)$ and $BK\mathcal{D}(A) \subset \mathcal{D}(A)$ imply $\mathbf{d}(-s)BK\mathbf{d}(s)\mathcal{D}(A) \subset \mathcal{D}(A)$, $\forall s \in \mathbb{R}$. From Condition D) we derive $BK\mathbf{d}_1(s) = \mathbf{d}(-s)BK\mathbf{d}(s)$.

II. From Corollary 2.13 we conclude that the closed-loop nonlinear system has mild solutions, which exist on $\mathbb{H}\backslash\{\mathbf{0}\}$. Condition (D) yields

$$BK\mathbf{d}_1(-\ln \|x\|_{\mathbf{d}}) = \mathbf{d}(\ln \|x\|_{\mathbf{d}})BK\mathbf{d}(-\ln \|x\|_{\mathbf{d}}).$$

Since $\beta + \mu > 0$ and $\|\mathbf{d}(\ln \|x\|_{\mathbf{d}})\| \le \|x\|_{\mathbf{d}}^{\beta}$ for $\|x\| \le 1$ then taking into account $\|\mathbf{d}(-\ln \|x\|_{\mathbf{d}})x\| = 1$ we conclude $\|Bu(x)\| \to \mathbf{0}$ as $\|x\|_{\mathbf{d}} \to \mathbf{0}$. Therefore, if there exists an instant of time $T(x_0)$ such that the mild solution x vanishes to $\mathbf{0}$ as $t \to T(x_0)$ then this mild solution, obviously, can be prolonged as $x(t) = \mathbf{0}$ for $t \ge T(x_0)$. This prolongation is unique if the origin of the closed-loop system is Lyapunov stable (see, Proposition 3.1).

From Theorem 2.16, we conclude for any $x_0 \in \mathcal{D}(A)\backslash\{\mathbf{0}\}$ the mild solution is a locally Lipschitz continuous strong solution of the closed-loop system.

Moreover, if $BK\mathcal{D}(A) \subset \mathcal{D}(A)$ then for any $x_0 \in \mathcal{D}(A)\backslash\{\mathbf{0}\}$ the mild solution is a classical solution. Indeed, since the group $\mathbf{d}_1$ is uniformly continuous then for any continuous function $g \in C([0, T'], (0, +\infty))$ the family of linear bounded operators

$$t \mapsto g^{\mu}(t)BK\mathbf{d}_1(-\ln g(t))$$

is uniformly continuous too.

Let x be a mild solution of the closed-loop system on $[0, T']$ with $x_0 \in \mathcal{D}(A)\backslash\{\mathbf{0}\}$, where $T' > 0$ is such that $\|x(t)\|_{\mathbf{d}} > 0$ for all $t \in [0, T']$. Then x is also a solution of the system

$$\dot{x} = (A + g^{\mu}(t)BK\mathbf{d}_1(-\ln g(t)))x, \quad t > 0 \tag{8.9}$$

with the initial condition $x(0) = x_0$, where $g(t) = \|x(t)\|_{\mathbf{d}}$.

According to Theorem 2.12, the latter system has a unique classical solution on $[0, T']$. Tending T' to $T(x_0)$ we conclude that the mild solution of the closed-loop system with $x_0 \in \mathcal{D}(A)$ is, indeed, a classical solution provided that $BK\mathcal{D}(A) \subset \mathcal{D}(A)$.

III. Conditions (B) and (D) imply that the nonlinear operator $f : \mathcal{D}(A) \subset \mathbb{H} \mapsto \mathbb{H}$ given by $f(x) = Ax + Bu(x)$ is $\mathbf{d}$-homogeneous of the degree μ. Indeed, from Condition (C) we have that $\mathcal{D}(A)$ is a $\mathbf{d}$-homogeneous cone.

From Condition (D) we derive

$$f(\mathbf{d}(s)x) = A\mathbf{d}(s)x + \|\mathbf{d}(s)x\|_{\mathbf{d}}^{\mu}\mathbf{d}(\ln\|\mathbf{d}(s)x\|_{\mathbf{d}})BK\mathbf{d}(-\ln\|\mathbf{d}(s)x\|_{\mathbf{d}})\mathbf{d}(s)x =$$

$$e^{\mu s}\mathbf{d}(s)f(x), \quad \forall x \in \mathcal{D}(A).$$

On the other hand, Condition (C) and the formula (4.25) yield

$$\begin{aligned} D\|x\|_{\mathbf{d}}(Ax + Bu(x)) &= \frac{\|x\|_{\mathbf{d}}\langle \mathbf{d}(-\ln\|x\|_{\mathbf{d}})(Ax+Bu(x)), \mathbf{d}(-\ln\|x\|_{\mathbf{d}})x\rangle}{\langle G_{\mathbf{d}}\mathbf{d}(-\ln\|x\|_{\mathbf{d}})x, \mathbf{d}(-\ln\|x\|_{\mathbf{d}})x\rangle} \\ &= \frac{\|x\|_{\mathbf{d}}^{1+\mu}\langle (A+BK)\mathbf{d}(-\ln\|x\|_{\mathbf{d}})x, \mathbf{d}(-\ln\|x\|_{\mathbf{d}})x\rangle}{\langle G_{\mathbf{d}}\mathbf{d}(-\ln\|x\|_{\mathbf{d}})x, \mathbf{d}(-\ln\|x\|_{\mathbf{d}})x\rangle} \\ &= -\rho\|x\|_{\mathbf{d}}^{1+\mu} \\ &\quad + \frac{\|x\|_{\mathbf{d}}^{1+\mu}\langle (A+BK+\rho G_{\mathbf{d}})\mathbf{d}(-\ln\|x\|_{\mathbf{d}})x, \mathbf{d}(-\ln\|x\|_{\mathbf{d}})x\rangle}{\langle G_{\mathbf{d}}\mathbf{d}(-\ln\|x\|_{\mathbf{d}})x, \mathbf{d}(-\ln\|x\|_{\mathbf{d}})x\rangle} \end{aligned}$$

for $\mathbf{0} \neq x \in \mathcal{D} \subset (\mathcal{D}(A) \cap \mathcal{D}(G_{\mathbf{d}}))$, where $D\|\cdot\|_{\mathbf{d}} \in \mathcal{L}(\mathbb{H}, \mathbb{R})$ is a Fréchet derivative of $\|\cdot\|_{\mathbf{d}}$. Since the set $\mathcal{D}$ is dense in $\mathcal{D}(A)$ then

$$\sup_{z \in \mathcal{D}(A)\setminus\{\mathbf{0}\}} \left\langle \frac{(A+BK+\rho G_{\mathbf{d}})z}{\|z\|}, \frac{z}{\|z\|} \right\rangle = \sup_{z \in \mathcal{D}\setminus\{\mathbf{0}\}} \left\langle \frac{(A+BK+\rho G_{\mathbf{d}})z}{\|z\|}, \frac{z}{\|z\|} \right\rangle \leq 0$$

and

$$\sup_{z \in \mathcal{D}(A)\setminus\{\mathbf{0}\}} \left\langle \frac{G_{\mathbf{d}}z}{\|z\|}, \frac{z}{\|z\|} \right\rangle = \sup_{z \in \mathcal{D}\setminus\{\mathbf{0}\}} \left\langle \frac{G_{\mathbf{d}})z}{\|z\|}, \frac{z}{\|z\|} \right\rangle \geq \beta > 0.$$

Therefore, we derive

$$D\|x\|_{\mathbf{d}}(Ax + Bu(x)) \leq -\rho\|x\|_{\mathbf{d}}^{1+\mu} \quad \text{for all} \quad x \neq \mathbf{0}.$$

Taking into account $\|\mathbf{d}(-\ln\|x\|_{\mathbf{d}})x\| = 1$ we derive

$$\frac{d}{dt}\|x(t)\|_{\mathbf{d}} \stackrel{a.e.}{=} D\|x\|_{\mathbf{d}}(Ax + Bu(x))|_{x=x(t)} \stackrel{a.e.}{\leq} -\rho\|x(t)\|_{\mathbf{d}}^{1+\mu} \quad \text{if} \quad x(t) \neq \mathbf{0}$$

for any strong solution of the closed-loop system. Since the strong solution and the canonical homogeneous norm are locally Lipschiz continuous then the function $t \mapsto \|x(t)\|_{\mathbf{d}}$ is locally Lipschitz continuous too and the obtained differential inequality immediately implies that $\|x(t)\|_{\mathbf{d}}$ decreases monotonically to zero along any strong solution of the closed-loop system. The same property holds for all mild solutions since $\mathcal{D}(A)$ is dense in $\mathbb{H}$ and all solutions of the system depend continuously on initial conditions (see Corollary 2.14).

Finally, from Theorem 6.9 we conclude that $\|\cdot\|_{\mathbf{d}}$ is a Lyapunov function of the closed-loop system and the estimates (8.5)–(8.7) take the place. The proof is complete. □

In the general case, the canonical homogeneous norm is defined implicitly (see the formula (4.21)). A practical implementation of the homogeneous control (8.4) requires special computational algorithms to be applied.

Corollary 8.1 *Theorem 8.1 remains true for $\mu = -\beta$ provided that mild Filippov solutions are considered for the closed-loop system.*

Proof Indeed, the existence, uniqueness, and convergence to $\mathbf{0}$ of mild solutions on $\mathbb{H}\backslash\{\mathbf{0}\}$ does not require the restriction $-\beta < \mu$. For $\beta + \mu = 0$ the function $x \mapsto Bu(x)$ remains bounded, but possibly discontinuous at the origin:

$$\|Bu(x)\| = \|x\|_{\mathbf{d}}^{\mu}\|\mathbf{d}(\ln\|x\|_{\mathbf{d}})BK\mathbf{d}(-\ln\|x\|_{\mathbf{d}})x\| \le \|BK\|, \quad x \neq \mathbf{0}.$$

In this case, the mild Filippov solution (see Definition 2.13) of the evolution equation is obtained by means of the evolution inclusion

$$\dot{x} - Ax \in G(x),$$

where $G : \mathbb{H} \mapsto 2^{\mathbb{H}}$ is given by

$$G(x) = \bigcap_{\varepsilon>0} \overline{\text{co}} Bu(x \dot{+} B(\varepsilon)\backslash\{\mathbf{0}\}),$$

where $B(\varepsilon)$ denotes a ball of the radius $\varepsilon > 0$ centered at the origin.

Obviously, for $x \neq \mathbf{0}$ we have $G(x) = Bu(x)$ due to continuity of u. Taking into account, $\|\mathbf{d}(-\ln\|x\|_{\mathbf{d}})x\| = 1$ we conclude

$$G(\mathbf{0}) \subset BKB(1),$$

where $B(1)$ is a unit ball in $\mathbb{H}$. Notice that, $\mathbf{0} \in G(\mathbf{0})$ due to symmetry of the control $u(-x) = -u(x)$. By Theorem 8.1 all mild solutions of the obtained evolution inclusion converge to zero in a finite time. Since $\mathbf{0} \in G(\mathbf{0})$ then these solutions can be prolonged as $x(t) = \mathbf{0}$ for $t \ge T(x_0)$.

Since $\|\cdot\|_{\mathbf{d}}$ is a Lyapunov function of the system (see the proof of Theorem 8.1), then the origin of the system is Lyapunov stable and, in the view of Proposition 3.1, the evolution inclusion has the unique zero solution $x(t) = \mathbf{0}, t \geq t_0$ for the zero initial condition $x(t_0) = \mathbf{0}$. The proof is complete. □

In the general case, the operator A may not be homogeneous. In the next section, we show that the non homogeneous system sometimes may be homogenized by means of a linear static feedback. In some other cases, the finite-time and nearly fixed-time stability of the closed-loop system (8.1), (8.4) with $\mu \neq 0$ can be proven under the assumption that this system admits an appropriate homogeneous extension.

Corollary 8.2 *Theorem 8.1 remains true if Condition (D) is replaced with*

(D)* *there exists a set-valued operator* $F : \mathcal{D}(F) \subset \mathbb{H} \mapsto 2^{\mathbb{H}}$ *such that*

$$Ax + Bu(x) \in F(x) \text{ for } \forall x \in \mathcal{D}(A) \subset \mathcal{D}(F),$$

F *is* **d**-*homogeneous of a degree* $\mu \in \mathbb{R}$ *with* $\mu \leq \gamma_{\max} - \beta$ *and*

$$\rho \langle G_{\mathbf{d}} z, z \rangle + \sup_{y \in F(z)} \langle y, z \rangle \leq 0, \quad \forall z \in \mathcal{D} : \|z\| = 1 \tag{8.10}$$

for some $\rho > 0$, *where* $\mathcal{D} \subset (\mathcal{D}(F) \cap \mathcal{D}(G))$ *is a* **d**-*homogeneous cone dense in* $\mathcal{D}(A)$.

Proof Obviously, all properties of the control law (8.4) claimed in Theorem 8.1 hold. So the closed-loop system (8.1)–(8.4) has a unique mild solution for any $x_0 \in \mathbb{H}\backslash\{\mathbf{0}\}$. Since $\mu \leq \gamma_{\max} - \beta$ then $u(x)$ continuous at zero and any mild solution (if it converges to zero in a finite time) can be prolonged. According to Theorem 2.16 a mild solution is a locally Lipschitz continuous strong solution x provided that $x(0) \in \mathcal{D}(A)\backslash\{\mathbf{0}\}$. In this case, the function $t \mapsto \|x(t)\|_{\mathbf{d}}$ is locally Lipschitz continuous and

$$D\|x\|_{\mathbf{d}}(Ax + Bu(x)) = \frac{\|x\|_{\mathbf{d}} \langle \mathbf{d}(-\ln \|x\|_{\mathbf{d}})(Ax+Bu(x)), \mathbf{d}(-\ln \|x\|_{\mathbf{d}})x \rangle}{\langle G_{\mathbf{d}} \mathbf{d}(-\ln \|x\|_{\mathbf{d}})x, \mathbf{d}(-\ln \|x\|_{\mathbf{d}})x \rangle}$$
$$\leq \|x\|_{\mathbf{d}}^{1+\mu} \sup_{y \in F(x)} \frac{\|x\|^{-\mu} \langle \mathbf{d}(-\ln \|x\|_{\mathbf{d}})y, \mathbf{d}(-\ln \|x\|_{\mathbf{d}})x \rangle}{\langle G_{\mathbf{d}} \mathbf{d}(-\ln \|x\|_{\mathbf{d}})x, \mathbf{d}(-\ln \|x\|_{\mathbf{d}})x \rangle}$$

for any $x \in \mathcal{D}$. Since $\mathcal{D}$ is dense in $\mathcal{D}(A)$ then taking into account homogeneity of F and the identity $\|\mathbf{d}(-\ln \|x\|_{\mathbf{d}})x\| = 1$ we conclude $F(x) = \|x\|_{\mathbf{d}}^{\mu} \mathbf{d}(\ln \|x\|_{\mathbf{d}}) F(\mathbf{d}(-\ln \|x\|_{\mathbf{d}})x)$ and

$$\tfrac{d}{dt}\|x(t)\|_{\mathbf{d}} \overset{a.e.}{\leq} \|x(t)\|_{\mathbf{d}}^{1+\mu} \sup_{z \in \mathcal{D} \cap S, \tilde{y} \in F(z)} \frac{\langle \tilde{y}, z \rangle}{\langle G_{\mathbf{d}} z, z \rangle} \overset{a.e.}{\leq} -\rho \|x(t)\|_{\mathbf{d}}^{1+\mu} \text{ if } x(t) \neq \mathbf{0}$$

for any strong solutions of the system. The obtained estimate implies finite-time (resp., nearly fixed-time) stability of the system. The proof is complete. □

Below we present an example of the finite-time stabilization of a non homogeneous PDE using Corollary 8.2. The corresponding PDE model is defined on a bounded domain $\Omega \subset \mathbb{R}^n$. First, we design a homogeneous controller (e.g., using Theorem 8.1) for the system defined on whole $\mathbb{R}^n$, but next we construct the homogeneous extension (see Chap. 5), and, finally, we apply Corollary 8.2 in order to show that the designed control stabilizes the original system to zero as well. Due to Corollary 8.2, *the homogeneous stabilizers designed for systems in a space of functions with unbounded support may be applicable to systems in a space with compactly supported functions.*

Fixed-Time Stabilization

Combining the cases $\mu < 0$ and $\mu > 0$ in Theorem 8.1, a fixed-time stabilizer for the system (8.1) can be derived similarly to the finite-dimensional case (see Chap. 9 of Volume I or [14]).

Theorem 8.2 *Let all conditions of Theorem 8.1 hold with parameters* $(\mathbf{d}, \mathbf{d}_1, \mu, \rho)$ *and* $(\tilde{\mathbf{d}}, \tilde{\mathbf{d}}_1, \tilde{\mu}, \tilde{\rho})$. *If* $\mu < 0 < \tilde{\mu}$ *then the following holds.*

- *The control law*

$$u(x) = \begin{cases} \|x\|_{\tilde{\mathbf{d}}}^{\tilde{\mu}} K\tilde{\mathbf{d}}_1(-\ln(\|x\|_{\tilde{\mathbf{d}}}))x & \text{if} \quad \|x\| \geq 1, \\ \|x\|_{\mathbf{d}}^{\mu} K\mathbf{d}_1(-\ln(\|x\|_{\mathbf{d}}))x & \text{if} \ 0 < \|x\| \leq 1, \\ \mathbf{0} & \text{if} \quad x = \mathbf{0} \end{cases} \tag{8.11}$$

  is locally Lipschitz continuous on $\mathbb{H}\backslash\{\mathbf{0}\}$, *Fréchet differentiable on* $\{x \in \mathcal{D}(A) : \|x\| \in (0, 1) \cap (1, +\infty)\}$ *and continuous at the origin if* $\beta > \mu + \gamma_{\max}$ *(see Theorem 8.1 for the definition of parameters* β *and* $\gamma_{\max}$*).*
- *The closed-loop system* (8.1), (8.11) *has a unique mild (strong) solution defined on* $[0, +\infty)$ *for any* $x_0 \in \mathbb{H}$ *(resp.,* $x_0 \in \mathcal{D}(A)$*).*
- *The origin of the closed-loop system* (8.1), (8.11) *is fixed-time stable and the settling-time functional* $T : \mathbb{H} \mapsto [0, +\infty)$ *admits the estimate*

$$T(x_0) \leq -\frac{1}{\rho\mu} + \frac{1}{\tilde{\rho}\tilde{\mu}}. \tag{8.12}$$

- *The closed-loop system* (8.1), (8.11) *is* $\mathbf{d}$*-homogeneous at* 0*-limit with the negative degree* $\mu < 0$ *and* $\tilde{\mathbf{d}}$*-homogeneous at* ∞*-limit with the positive degree* $\tilde{\mu} > 0$.

Proof Local homogeneity at 0-limit and ∞-limit is obtained by construction. Even more, the closed-loop system coincide with a $\mathbf{d}$-homogeneous system of degree $\mu < 0$ for $\|x\| \leq 1$ and with a $\tilde{\mathbf{d}}$-homogeneous system of degree $\tilde{\mu} > 0$ for $\|x\| \geq 1$.

Since $\mathbf{d}_1$ and $\tilde{\mathbf{d}}_1$ are uniformly continuous groups then the mappings $q \mapsto q^{\tilde{\mu}} K\tilde{\mathbf{d}}(-\ln q)$ and $q \mapsto q^{\mu} K\mathbf{d}(-\ln q)$ are uniformly continuous too. For $\|x\| = 1$ we have $\|x\|_{\mathbf{d}} = \|x\|_{\tilde{\mathbf{d}}} = 1$ and

$$\|x\|_{\tilde{\mathbf{d}}}^{\tilde{\mu}} K\tilde{\mathbf{d}}_1(-\ln(\|x\|_{\tilde{\mathbf{d}}}))x = \|x\|_{\mathbf{d}}^{\mu} K\mathbf{d}_1(-\ln(\|x\|_{\mathbf{d}}))x = Kx.$$

The canonical homogeneous norms $\|\cdot\|_{\mathbf{d}}$ and $\|\cdot\|_{\tilde{\mathbf{d}}}$ are locally Lipschitz continuous on $\mathbb{H}\backslash\{\mathbf{0}\}$. Then repeating considerations of Theorem 8.1 we conclude that u is locally Lipschitz continuous as well. Local differentiability of u can be derived from Lemma 4.6.

The existence of mild and classical solutions as well as local finite-time stability (for $\|x_0\| \leq 1$) can be proven similarly to Theorem 8.1. Combining the convergence results obtained in the mentioned theorem for the cases of positive and negative homogeneity degrees we complete the proof. □

Corollaries 8.2 and 8.1 can be extended to the case of the fixed-time stability and proven similarly to Theorem 8.2.

On Norm-Quantized Realization of the Implicit Homogeneous Control

For a practical implementation of the control laws (8.4) and (8.11) in a digital device it is important to know if the feedback can be applied in a sampled way assuming that $\|x(t)\|_{\mathbf{d}}$ is replaced with $\|x(t_j)\|_{\mathbf{d}}$ for $t \in [t_j, t_{j+1})$. The value $\|x(t_j)\|_{\mathbf{d}}$ can be computed or estimated by means of an appropriate numerical algorithm. The control (8.4) (resp., (8.11)) becomes linear on each interval of time $[t_j, t_{j+1})$.

Let us consider the following switched linear feedback

$$u_{\sigma(t)}(x(t)) = r_{\sigma(t)}^{\mu} K\mathbf{d}_1(-\ln r_{\sigma(t)})x(t), \tag{8.13}$$

with the state-dependent switching function σ defined as follows: $\sigma(0) \in \mathbb{Z}$,

$$\sigma(t) = S(\sigma(t+0^-), x(t+0^-)), \quad S : \mathbb{Z} \times \mathbb{H} \mapsto \mathbb{Z},$$

$$S(\sigma, x) := \begin{cases} i-1 & \text{if } \sigma=i \text{ and } \|x\|_{\mathbf{d}} \geq r_{i-1}, \\ i & \text{if } \sigma=i \text{ and } r_{i+1} \leq \|x\|_{\mathbf{d}} \leq r_{i-1}, \\ i+1 & \text{if } \sigma=i \text{ and } r_{i+1} \leq \|x\|_{\mathbf{d}}, \end{cases} \tag{8.14}$$

where $r_i \in \mathbb{R}_+ : r_{i+1} < r_i$, $i \in \mathbb{Z}$ and $\sigma(t+0^-) := \lim_{\tau\to 0^-} \sigma(t+\tau)$ (resp., $x(t+0^-) := \lim_{\tau\to 0^-} x(t+\tau)$). The switching variable σ has its own evolution equation given by (8.14) so the pair $(x(t), \sigma(t))$ can be treated as an augmented state of the closed-loop switched systems. We refer the reader to [15] for more details about switched systems.

Corollary 8.3 (On switched linear finite-time feedback law) *If all conditions of Theorem 8.1 hold for* $\mu < 0$, *the parameters* $M \geq 1$ *and* $\alpha > 0$ *are defined as in Lemma 4.5 and the control law is given by* (8.13), (8.14) *with*

$$r_0 = 1, \quad r_{i+1} = M^{1/\alpha} e^{-q_i} r_i, \quad i \in \mathbb{Z}$$

and $q_i > \frac{1}{\alpha} \ln M$ *such that*

$$\sum_{i=0}^{+\infty} q_i = +\infty \quad \text{and} \quad \lim_{i \to +\infty} q_{i+1} e^{\mu q_i} / q_i < 1,$$

then

- *for any* $x_0 \in \mathbb{H} \backslash \{\mathbf{0}\}$ *(resp.* $x_0 \in \mathcal{D}(A) \backslash \{\mathbf{0}\}$*) the closed-loop system* (8.1), (8.2), (8.13), (8.14) *has a unique mild (resp. strong) solution and*

$$\|x(t)\| \to 0 \text{ as } t \to T(x_0),$$

where the settling-time functional $T : \mathbb{H} \mapsto \mathbb{R}$ *admits the estimate*

$$T(x_0) \leq \frac{\alpha}{\beta \rho} \sum_{j=i_0}^{\infty} r_j^{-\mu} q_j < +\infty$$

provided that

$$\|x_0\|_{\mathbf{d}} \in (r_{i_0+1}, r_{i_0}];$$

- *the time instants* $t_j : \|u(t_j)\|_{\mathbf{d}} = r_j$ *are isolated and the set* $\{t_j\}_{j=i_0+1}^{+\infty}$ *has the unique accumulation point* $T(x_0)$*;*
- *the control is a linear static feedback on each interval* (t_j, t_{j+1}) *and* $\|u_{\sigma(t)}(t)\| \leq M \|K\|$ *provided that* $\|\mathbf{d}(s)\| \leq M e^{(\mu + \gamma_{\min})s}$ *for* $s > 0$.

Proof Let us denote $\mathcal{K}(a, b) = \{u \in \mathcal{D}(A) : a < \|u\|_{\mathbf{d}} \leq b\}$. Since $BK_i \in \mathcal{L}(\mathbb{H}, \mathbb{H})$ for any $i \in \mathbb{Z}$, where $K_i = r_i^{\mu} K \mathbf{d}_1(-\ln r_i)$, then for any initial condition $x(0) = x^* \in \mathcal{D}(A)$ the evolution equation

$$\dot{x}_i(t) = (A + BK_i) x_i(t) \tag{8.15}$$

has a unique classical solution $x_{i,x_0}(t) \in \mathcal{D}(A)$ $x_{i,x^*}(t) \in \mathcal{D}(A)$ defined for all $t \geq 0$. We have $r_{i+1} < r_i$, $i \in \mathbb{Z}$ by construction. Since $\|x_{i,x_0}(\cdot)\|_{\mathbf{d}} : \mathbb{R} \to \mathbb{R}_+$ is continuous then for $x(t_i) = x_0 \in \mathcal{K}(r_{i+1}, r_i)$, $\sigma(t_i) \equiv i$ the closed-loop system (8.1), (8.2), (8.13), (8.14) has a locally Lipschitz continuous strong solution x defined at least on $[t_i, t_{i+1})$ and $x(t) = x_{i,x_0}(t - t_i)$ for $t \in [t_i, t_{i+1})$, where $t_{i+1} = +\infty$ or

$\lim_{t\to t_{i+1}} \|x(t)\|_{\mathbf{d}} = r_{i-1}$ or $\lim_{t\to t_{i+1}} \|x(t)\|_{\mathbf{d}} = r_{i+1}$. Let us show that the latter case is only possible and $t_{i+1} < +\infty$. Since

$$\begin{gathered}\tfrac{d}{dt}\|\mathbf{d}(-\ln r_i)x(t)\|^2 \overset{a.e.}{=} 2\langle \mathbf{d}(-\ln r_i)\dot{x}(t), \mathbf{d}(-\ln r_i)x(t)\rangle \overset{a.e.}{=} \\ 2e^{\mu \ln r_i}\langle (A+BK)\mathbf{d}(-\ln r_i)x(t), \mathbf{d}(-\ln r_i)x(t)\rangle \overset{a.e.}{\leq} \\ -\tfrac{2\rho\langle G_{\mathbf{d}}\mathbf{d}(-\ln r_i)x(t), \mathbf{d}(-\ln r_i)x(t)\rangle}{r_i^{-\mu}} \overset{a.e.}{\leq} -\tfrac{2\rho\beta\|\mathbf{d}(-\ln r_i)x(t)\|^2}{r_i^{-\mu}}\end{gathered}$$

then the locally Lipschitz continuous function $t \mapsto \|\mathbf{d}(-\ln r_i)x(t)\|$ decays exponentially. Taking into account $\|\mathbf{d}(-\ln r_i)x\|^2 = 1$ if $\|x\|_{\mathbf{d}} = r_i$ and $\|\mathbf{d}(-\ln r_i)x\|^2 < 1$ if $\|x\|_{\mathbf{d}} < r_i$, we derive $\|x(t)\|_{\mathbf{d}} < r_i$ for $t > t_i$ and $\lim_{t\to t_{i+1}} \|x(t)\|_{\mathbf{d}} = r_{i+1}$. Moreover,

$$t_{i+1} \leq t_i + r_i^{-\mu}\beta^{-1}\rho^{-1}\ln\left(\frac{\|\mathbf{d}(-\ln r_i)x(t_i)\|}{\varepsilon_i}\right) \leq t_i + r_i^{-\mu}\beta^{-1}\rho^{-1}\ln\varepsilon_i^{-1},$$

where $\varepsilon_i > 0$ is such that

$$\|\mathbf{d}(-\ln r_i)x\| \leq \varepsilon_i \quad \Rightarrow \quad \|x\|_{\mathbf{d}} \leq r_{i+1}.$$

Therefore, $x(t_{i+1}) = x_{i,x_0}(t_{i+1} - t_i) \in \mathcal{D}(A)$. Repeating the same considerations for $[t_{i+1}, t_{i+2}]$, for $[t_{i+2}, t_{i+3}]$, …and taking into account uniqueness and continuity of solutions of the (8.15) we derive that for any initial condition $x_0 \in \mathcal{D}(A)\backslash\{0\}$, $\sigma(0) = i_0$ such that $r_{i_0+1} < \|x_0\|_{\mathbf{d}} < r_{i_0-1}$ the closed-loop system (8.1), (8.2), (8.13), (8.14) has a unique strong solution defined on $[0, T(x_0))$, which may not have a time derivative at time instances $t_i : \|x(t_i)\|_{\mathbf{d}} = r_i$, where

$$T(x_0) = \sum_{j=i_0}^{+\infty} t_{i+1} - t_i \leq \sum_{j=i_0}^{+\infty} r_i^{-\mu}\beta^{-1}\rho^{-1}\ln\varepsilon_i^{-1}$$

is the unique accumulation point of the set $\{t_i\}$. In addition, $\sigma(t) = const$ for $t \in (t_i, t_{i+1}]$. Since $\sum_{i=0}^{+\infty} q_i = +\infty$ then $r_i = \frac{1}{e^{\sum_{j=0}^{i-1} q_j}} \to 0$ as $i \to +\infty$ and $\|x(t)\| \to 0$ as $t \to T(x_0)$.

Let us show that

$$\|\mathbf{d}(-\ln r_i)x\| \leq e^{-\alpha q_i} \Rightarrow \|\mathbf{d}(-\ln r_{i+1})x\|_{\mathbf{d}} \leq 1.$$

Indeed, using the $\mathbf{d}$-homogeneity of $\|\cdot\|_{\mathbf{d}}$ and $\|\mathbf{d}(-\ln r_i)x\| \leq 1$ form Lemma 4.5 we derive

$$\|\mathbf{d}(-\ln r_{i+1})x\|_{\mathbf{d}}^{\alpha} = \|\mathbf{d}(q_i - \ln M^{1/\alpha})\mathbf{d}(-\ln r_i)x\|_{\mathbf{d}}^{\alpha} =$$

$$\left(e^{q_i - \ln M^{1/\alpha}}\|\mathbf{d}(-\ln r_i)x\|_{\mathbf{d}}\right)^{\alpha} = e^{\alpha q_i}M^{-1}\|\mathbf{d}(-\ln r_i)x\|_{\mathbf{d}}^{\alpha} \leq e^{\alpha q_i}\|\mathbf{d}(-\ln r_i)x\|.$$

Therefore, for $\varepsilon_i = e^{-\alpha q_i}$ we derive

$$t_{i+1} \le t_i + \alpha\rho^{-1}\beta^{-1}r_i^{-\mu}q_i$$

and the settling time $T(x_0)$ is finite since the ratio test

$$\lim_{i\to+\infty}\frac{r_{i+1}^{-\mu}q_{i+1}}{r_i^{-\mu}q_i} = \lim_{i\to+\infty} q_{i+1}e^{\mu q_i}/q_i < 1$$

implies that the series, which estimates $T(x_0)$, is convergent.

Finally, a boundedness of the control and convergence of mild solutions to zero can be proven similarly. The proof is complete. □

Corollary 8.3 proposes a scheme for homogeneity-based finite-time control design in the form of switched linear feedback. A similar algorithm of the fixed-time stabilization can be designed by means of a modification of the control law (8.11).

8.1.2 Generalized Homogenization of Linear Plant

Recall that the condition of the **d**-homogeneity of the operator A is given by (5.8). Adding a linear feedback term to a control law

$$u(x) = K_0 x + u_{\mathrm{hom}}(x), \quad K_0 \in \mathcal{L}(\mathbb{H};\mathbb{U})$$

may allow the operator $A_0 = A + BK_0$ to become **d**-homogeneous with a nonzero degree, so that all conditions of Theorem 8.1 hold for A_0 and a control u_{hom} of the form (8.4) guarantees a homogeneous stabilization of the closed-loop system. In the finite-dimensional case (see Volume I), the gain K_0 is defined together with the generator $G_{\mathbf{d}}$ of the dilation by solving a special linear algebraic equation, which is shown to be feasible for any controllable pair $\{A, B\}$. The existence of solution of the corresponding linear equation is given by the assumption.

Assumption 8.1 There exist a linear closed densely defined operator $G_0 : \mathcal{D}(G_0) \subset \mathbb{H} \mapsto \mathbb{H}$, a linear closed densely defined operator $\tilde{G} : \mathcal{D}(\tilde{G}) \subset \mathbb{U} \mapsto \mathbb{U}$ and a linear bounded operator $K_0 \in \mathcal{L}(\mathbb{H};\mathbb{U})$ such that

- G_0 is an infinitesimal generator of a strongly continuous group $\mathbf{d}_0$ of linear bounded operators on $\mathbb{H}$ such that $\mathbf{d}_0(s)\mathcal{D}(A) \subset \mathcal{D}(A)$, $\forall s \in \mathbb{R}$;
- $\tilde{G}$ is an infinitesimal generator of a strongly continuous group $\tilde{\mathbf{d}}$ of linear bounded operators on $\mathbb{U}$;

- there exist a set $\mathcal{D}_0 \subset \mathcal{D}_{\exp}(G_0^\infty)$ dense in $\mathbb{H}$ such that $G_0^n \mathcal{D}_0 \subset \mathcal{D}(A), \forall n = 0, 1, 2, \ldots$ and a set $\tilde{\mathcal{D}} \subset \mathcal{D}(\tilde{G}^\infty)$ dense in $\mathbb{U}$ such that $B\tilde{\mathcal{D}} \subset \mathcal{D}_{\exp}(G_0^\infty)$,

$$A_0 G_0 \sigma - G_0 A_0 \sigma = A_0 \sigma, \quad \forall \sigma \in \mathcal{D}_0, \tag{8.16}$$

$$G_0 B u = B \tilde{G} u, \quad \forall u \in \tilde{\mathcal{D}}, \tag{8.17}$$

where

$$A_0 := A + BK_0. \tag{8.18}$$

In the view of Corollary 2.8, the set $\mathcal{D}_{\exp}(G_0^\infty)$ of entire vector generator of the strongly continuous group is always dense in $\mathbb{H}$, so the above assumption is feasible in some cases. Below we introduce a class of differential operators satisfying the above assumption.

The system of operator equations (8.16), (8.17) is similar to the algebraic system (see Chaps. 9 and 13 of Volume I) discovered for the finite-dimensional homogeneous control design. Below we show that Assumption 8.1 is sufficient to guarantee that the operator $A + BK_0$ is homogeneous of degree μ with respect to the dilation $\mathbf{d}$ in $\mathbb{H}$ generated by the operator $G_{\mathbf{d}} = I + \mu G_0$ provided that $\mu \in \mathbb{R}$ satisfies the inequality (4.4). In the view of Lemma 5.1, the existence of solution of the operator equation (8.16) is necessary for homogenization of the operator A. The solution of the operator equation (8.17) is required for existence of the homogeneous feedback u_{hom}. In the finite-dimensional case, the system of operator equations (8.16), (8.17) is feasible if the pair $\{A, B\}$ is controllable. In the infinite-dimensional case, there is not clear way to link a null controllability[1] of the plant with feasibility of the system of operator equations (8.16), (8.17).

Let us define the feedback control in the form

$$u(x) = K_0 x + \|x\|_{\mathbf{d}}^{1+\mu} \tilde{\mathbf{d}}(\mu \ln \|x\|_{\mathbf{d}}) K \mathbf{d}(-\ln \|x\|_{\mathbf{d}}) x, \tag{8.19}$$

where $\|\cdot\|_{\mathbf{d}}$ is a canonical homogeneous norm induced by an appropriate norm in $\mathbb{H}$, $K_0 \in \mathcal{L}(\mathbb{H}; \mathbb{U})$ and the strongly continuous group $\tilde{\mathbf{d}}$ is defined in Assumption 8.1 whereas $\mu \in \mathbb{R}$ and $\mathbf{d}$ are defined by Lemma 4.1. The controller (8.19) differs from its finite-dimensional counterpart. For $\mathbb{H} = \mathbb{R}^n$, the term $\tilde{\mathbf{d}}$ is omitted, i.e., $\tilde{\mathbf{d}}(s) = I, \forall s \in \mathbb{R}$ (see, [16]).

[1] The issues of controllability of infinite-dimensional systems are studied, for example, in [4].

Lemma 8.1 *If Assumption 8.1 is fulfilled and $\mu \in \mathbb{R}$ satisfies (4.4) then the operator $g : \mathbb{H} \subset \mathcal{D}(A) \mapsto \mathbb{H}$ given by $g(\mathbf{0}) = \mathbf{0}$,*

$$g(x) = (A + BK_0)x + \|x\|_{\mathbf{d}}^{1+\mu} B\tilde{\mathbf{d}}(\mu \ln \|x\|_{\mathbf{d}}) K \mathbf{d}(-\ln \|x\|_{\mathbf{d}})x,$$

is $\mathbf{d}$-homogeneous of degree $\mu \in \mathbb{R}$ and

$$\mathbf{d}(s)B = e^s B\tilde{\mathbf{d}}(\mu s), \quad \forall s \in \mathbb{R}. \tag{8.20}$$

Proof Since the linear operator $A + BK_0$ is $\mathbf{d}$-homogeneous (see, Lemma 5.1) then we just need to show that the nonlinear operator $g_0 : \mathbb{H} \mapsto \mathbb{H}$, $g_0(\mathbf{0}) = \mathbf{0}$ and

$$g_0(x) = \|x\|_{\mathbf{d}}^{\mu} B\tilde{\mathbf{d}}(\mu \ln \|x\|_{\mathbf{d}}) K \mathbf{d}(-\ln \|x\|_{\mathbf{d}})x, \quad x \neq \mathbf{0}$$

is $\mathbf{d}$-homogeneous of degree μ. Indeed, using $\|\mathbf{d}(s)x\|_{\mathbf{d}} = e^s \|x\|_{\mathbf{d}}$ we derive

$$\|\mathbf{d}(s)x\|_{\mathbf{d}}^{1+\mu} B\tilde{\mathbf{d}}(\mu \ln \|\mathbf{d}(s)x\|_{\mathbf{d}}) K \mathbf{d}(-\ln \|\mathbf{d}(s)x\|_{\mathbf{d}})\mathbf{d}(s)x$$

$$= e^{(\mu+1)s} \|x\|_{\mathbf{d}}^{1+\mu} B\tilde{\mathbf{d}}(\mu s)\tilde{\mathbf{d}}(\mu \ln \|x\|_{\mathbf{d}}) K \mathbf{d}(-\ln \|x\|_{\mathbf{d}})x.$$

Using (8.17) we derive $G_{\mathbf{d}} Bu = B(I + \mu\tilde{G})u$ for any $u \in \tilde{\mathcal{D}} \subset \mathcal{D}(\tilde{G}^{\infty})$. Since, by Assumption 8.1, $B\tilde{\mathcal{D}} \subset \mathcal{D}(G_0^{\infty})$ then

$$\mathbf{d}(s)Bu = \sum_{i=0}^{\infty} \tfrac{s^n G_{\mathbf{d}}^n x}{n!} Bu = B \sum_{i=0}^{\infty} \tfrac{s^n (I + \mu\tilde{G})^n u}{n!} = e^s B\tilde{\mathbf{d}}(\mu s).$$

Taking into account the identity (8.20) we obtain $g_0(\mathbf{d}(s)x) = e^{\mu s}\mathbf{d}(s)g_0(x)$. The proof is complete. □

The following assumption is involved to guarantee the well-posedness of the closed-loop system (8.1), (8.19).

Assumption 8.2 For any $R > 0$ there exist $M \geq 1$ and $\omega \in \mathbb{R}$ such that the family of linear bounded operators

$$s \mapsto \mathbf{d}(s)BK\mathbf{d}(-s) \in \mathcal{L}(\mathbb{H}, \mathbb{U}), \quad s \in \mathbb{R}$$

satisfies the inequality

$$\|\mathbf{d}(s_1)BK\mathbf{d}(-s_1) - \mathbf{d}(s_2)BK\mathbf{d}(-s_2)\|_{\mathcal{L}(\mathbb{H},\mathbb{U})} \leq Me^{\omega|s_1 - s_2|},$$

for all $s_i \in \mathbb{R} : |s_i| < R, i = 1, 2$, where the dilation $\mathbf{d}$ is generated by $G_{\mathbf{d}} = I_n + \mu G_0$ with G_0 satisfying Assumption 8.1.

The above assumption is fulfilled, for example, if $\mathbf{d}$ is a uniformly continuous dilation in $\mathbb{H}$ or if the generator $G_{\mathbf{d}}$ of the dilation $\mathbf{d}$ commutes with BK.

Lemma 8.2 *Let a linear closed densely defined operator $A : \mathcal{D}(A) \subset \mathbb{H} \mapsto \mathbb{H}$ be an infinitesimal generator of a strongly continuous semigroup Φ of linear bounded operators on $\mathbb{H}$ and $B : \mathbb{U} \mapsto \mathbb{H}$ be a linear bounded operator. Under Assumptions 8.1–8.2, for any $x_0 \in \mathbb{H}\backslash\{\mathbf{0}\}$, (resp., $x_0 \in \mathcal{D}(A)\backslash\{\mathbf{0}\}$), the closed-loop system* (8.1), (8.19) *has a unique mild (resp., strong) solution defined as long as $0 < \|x(t)\| < +\infty$. For $x_0 = \mathbf{0}$, the system has (possibly non unique) zero solution.*

Proof Since the canonical homogeneous norm $\|\cdot\|_{\mathbf{d}}$ is locally Lipschitz on $\mathbb{H}\backslash\{\mathbf{0}\}$ then, due to Assumption 8.2, the mapping $x \mapsto Bu(x)$ is locally Lipschitz on $\mathbb{H}\backslash\{\mathbf{0}\}$. In the view of Corollary 2.13, the closed-loop system has a unique mild solution for any $x_0 \in \mathbb{H}\backslash\{\mathbf{0}\}$. For $x_0 \in \mathcal{D}(A)\backslash\{\mathbf{0}\}$), the corresponding solution turns out to be strong due to Theorem 2.16. If $A\,\mathbf{0}+Bu(\mathbf{0}) + f(t,\mathbf{0}) = \mathbf{0}$ then, for $x_0 = \mathbf{0}$, the system always possesses the zero solution, which may be non unique. □

The parameters μ and group $\tilde{\mathbf{d}}$ specify the properties of the feedback law at zero. Indeed, since $\pi(x) = \mathbf{d}(-\ln\|x\|_{\mathbf{d}})x$ is uniformly bounded by virtue of $\|\pi(x)\| = 1$ then the condition

$$e^{(1+\mu)s}\|B\tilde{\mathbf{d}}(\mu s)K\|_{\mathcal{L}(\mathbb{H};\mathbb{U})} \to \mathbf{0} \text{ as } s \to -\infty \tag{8.21}$$

guarantees $Bu(x) \to 0$ as $x \to \mathbf{0}$. In the finite-dimensional case (see Volume I), the group $\tilde{\mathbf{d}}$ may be ignored by taking $\tilde{\mathbf{d}}(s) = I, \forall s \in \mathbb{R}$, so the continuity of the control may depend only on the homogeneity degree μ.

Theorem 8.3 *Let a linear closed densely defined operator $A : \mathcal{D}(A) \subset \mathbb{H} \mapsto \mathbb{H}$ be an infinitesimal generator of a strongly continuous semigroup Φ of linear bounded operators on $\mathbb{H}$ and $B : \mathbb{U} \mapsto \mathbb{H}$ be a linear bounded operator. Let Assumptions 8.1–8.2 be fulfilled and the parameter $\mu \in \mathbb{R}$ satisfy (4.4). If there exist numbers $\beta, \rho \in \mathbb{R}_+$, an operator $K \in \mathcal{L}(\mathbb{H}, \mathbb{U})$ and a self-adjoint positive operator $P \in \mathcal{L}(\mathbb{H}, \mathbb{H})$ such that*

$$\langle P(A_0 + BK + \rho G_{\mathbf{d}})x, x\rangle \le 0, \quad \forall x \in \mathcal{D}_0, \tag{8.22}$$

$$\langle PG_{\mathbf{d}}x, x\rangle \ge \beta\langle x, x\rangle, \quad \forall x \in \mathcal{D}_0 \tag{8.23}$$

then the closed-loop system (8.1), (8.19) *is*

- *finite-time stable for $\mu<0$: $x(t)=\mathbf{0}$, $\forall t \ge t_0 + \frac{\|x_0\|_{\mathbf{d}}^{-\mu}}{(-\mu)(\rho-\kappa)}$,*
- *exponentially stable for $\mu = 0$:*

$$\|x(t)\|_{\mathbf{d}} \le e^{-(\rho-\kappa)(t-t_0)}\|x_0\|_{\mathbf{d}}, \quad \forall t \ge t_0,$$

- *nearly fixed-time stable for* $\mu > 0$:

$$\|x(t)\|_{\mathbf{d}} \leq r, \quad \forall t \geq \tfrac{1}{(\rho-k)r^{\mu}}, \quad \forall r > 0, \quad \forall x_0 \in \mathbb{H},$$

where the canonical homogeneous norm $\|x\|_{\mathbf{d}}$ *is induced by the norm* $\|x\| = \sqrt{\langle Px, x\rangle}$ *in* $\mathbb{H}$.

Proof First, let us denote $g(x) = Ax + Bu(x)$. On one hand, by taking into account $\mathcal{D}_0 \subset \mathcal{D}(G_{\mathbf{d}}) \cap \mathcal{D}(A)$ and using (4.25) we derive

$$D\|x\|_{\mathbf{d}}(g(x)) = \|x\|_{\mathbf{d}} \tfrac{\langle P\mathbf{d}(-\ln \|x\|_{\mathbf{d}})g(x), \pi(x)\rangle}{\langle G_{\mathbf{d}}\pi(x), \pi(x)\rangle}, \quad x \in \mathcal{D}_0 \backslash \{\mathbf{0}\},$$

where $\pi(x) = \mathbf{d}(-\ln \|x\|_{\mathbf{d}})x$ is the homogeneous projector on the unit sphere (i.e., $\|\pi(x)\| = 1$) and $D\|\cdot\|_{\mathbf{d}} \in \mathcal{L}(\mathbb{H}; \mathbb{R})$ is a Fréchet derivative of $\|\cdot\|_{\mathbf{d}}$. On the other hand, using Lemma 8.1 we conclude that g is $\mathbf{d}$-homogeneous operator of degree μ, thereby arriving at $g(\pi(x)) = \|x\|_{\mathbf{d}}^{-\mu}\mathbf{d}(-\ln \|x\|_{\mathbf{d}})g(x)$ and

$$D\|x\|_{\mathbf{d}}(g(x)) = \|x\|_{\mathbf{d}}^{1+\mu} \tfrac{\langle P(A_0+BK)\pi(x), \pi(x)\rangle}{\langle G_{\mathbf{d}}\pi(x), \pi(x)\rangle} \leq -\rho\|x\|_{\mathbf{d}}^{1+\mu},$$

where inequalities (8.22), (8.23) are utilized in the last step. By Lemma 8.2, the closed-loop system has a strong solution $x_{x_0}(t)$ for any $x_0 \in \mathcal{D}(A)\backslash\{\mathbf{0}\}$, which is locally Lipschitz continuous function of time and satisfies (8.1) almost everywhere (as long as $x_{x_0}(t) \neq \mathbf{0}$). Since $x \mapsto \|x\|_{\mathbf{d}}$ is locally Lipschitz continuous on $\mathbb{H}\backslash\{\mathbf{0}\}$ then the function $t \mapsto \|x_{x_0}(t)\|_{\mathbf{d}}$ is absolutely continuous. Taking into account that $\mathcal{D}_0$ is dense in $\mathbb{H}$ we derive

$$\tfrac{d\|x_{x_0}(t)\|_{\mathbf{d}}}{dt} \overset{a.e.}{=} D\|y\|_{\mathbf{d}}(g(y))|_{y=x_{x_0}(t)} \leq -\rho\|x_{x_0}(t)\|_{\mathbf{d}}^{1+\mu}$$

as long as $x_{x_0}(t) \neq \mathbf{0}$. Hence, dependently of μ, we derive the finite-time, exponential, or nearly fixed-time convergence of the strong solutions to zero. The same conclusions hold for the mild solutions in the view of Theorem 2.10. So, $\|\cdot\|_{\mathbf{d}}$ is a Lyapunov function of the closed-loop system. The proof is complete. □

Homogeneous Control for a Class of PDE Models

As an example, let us consider the differential operator $A : \mathcal{D}(A) \subset L^2(\mathbb{R}; \mathbb{R}^k) \mapsto L^2(\mathbb{R}; \mathbb{R}^k)$ (see Example 5.4) given by

$$(Ax)(z) = \sum_{|\alpha| \leq q} \Xi_{\alpha}(z) \tfrac{\partial^{\alpha_1} \ldots \partial^{\alpha_n} x(z)}{\partial z_1^{\alpha_1} \ldots z_n^{\alpha_n}}, \tag{8.24}$$

where $q \in \mathbb{N}$, $\alpha = (\alpha_1, \ldots, \alpha_n)^{\top} \in \mathbb{R}^n$, $z = (z_1, \ldots, z_n)^{\top} \in \mathbb{R}^n$, and $\Xi_{\alpha} \in C^{\infty}(\mathbb{R}^n; \mathbb{R}^{k \times k})$.

Let $B \in \mathcal{L}(\mathbb{U}; \mathbb{H})$ be given by

$$(Bu)(z) = \Omega(z)u(z), \quad z \in \mathbb{R}^n, \quad \Omega \in C^\infty(\mathbb{R}^n; \mathbb{R}^{k\times m}), \tag{8.25}$$

where $\mathbb{U}$ is a Banach space of functions $\mathbb{R}^n \mapsto \mathbb{R}^m$ such that $C^\infty(\mathbb{R}^n; \mathbb{R}^k)$ is dense in $\mathbb{U}$. Let the operator $K_0 \in \mathcal{L}(\mathbb{H}; \mathbb{U})$ be defined in the form:

$$(K_0 x)(z) = \Upsilon_0(z)x(z), \; z \in \mathbb{R}^n, \; \Upsilon_0 \in C^\infty(\mathbb{R}^n; \mathbb{R}^{m\times k}). \tag{8.26}$$

In this case, the operator $A_0 = A + BK$ has the same structure as the operator A, but the term Ξ_0 in (8.24) is replaced with

$$\tilde{\Xi}_0(z) = \Xi_0(z) + \Omega(z)\Upsilon_0(z), \quad z \in \mathbb{R}^n. \tag{8.27}$$

Selecting the generator of the group G_0 in the form

$$G_0 x = \Gamma_0 x + ((\Theta z) \cdot \nabla) x, \quad z \in \mathbb{R}^n, \tag{8.28}$$

where $\Xi_0 \in \mathbb{R}^{n\times n}$ and $\Gamma_0 \in \mathbb{R}^{k\times k}$, the condition (8.16) becomes the algebraic system

$$\mathbf{0} = \left(\tilde{\Xi}_0(z)\Gamma_0 - (\Gamma_0 + I_k)\tilde{\Xi}_0(z)\right) - \sum_{i=1}^{n} (\Theta z)_i \tfrac{\partial}{\partial z_i} \tilde{\Xi}_0(z), \tag{8.29}$$

$$\begin{gathered}\mathbf{0} = \sum_{1 \le |\alpha| \le q} (\Xi_\alpha(z)\Gamma_0 - (\Gamma_0 + I_k)\Xi_\alpha(z)) \frac{\partial^{\alpha_1} \ldots \partial^{\alpha_n}}{\partial z_1^{\alpha_1} \ldots z_n^{\alpha_n}} + \\ \Xi_\alpha(z) \frac{\partial^{\alpha_1} \ldots \partial^{\alpha_n}}{\partial z_1^{\alpha_1} \ldots z_n^{\alpha_n}} ((\Theta z) \cdot \nabla) - ((\Theta z) \cdot \nabla) \left(\Xi_\alpha(z) \frac{\partial^{\alpha_1} \ldots \partial^{\alpha_n}}{\partial z_1^{\alpha_1} \ldots z_n^{\alpha_n}}\right).\end{gathered} \tag{8.30}$$

The condition (8.17) is algebraic too:

$$\Gamma_0 \Omega(z) + \sum_{i=1}^{n} (\Theta z)_i \tfrac{\partial}{\partial z_i} \Omega(z) = \Omega(z)\tilde{\Gamma}, \quad \tilde{\Gamma} \in \mathbb{R}^{m\times m} \tag{8.31}$$

provided that

$$\tilde{G}u = \tilde{\Gamma}u + ((\Theta z) \cdot \nabla)u, \quad \tilde{\Gamma} \in \mathbb{R}^{m\times m}. \tag{8.32}$$

Therefore, to define a homogeneous control (8.4) for the system (8.1) with the differential operator A given by (8.24) and the linear bounded operator B given by (8.25), we need to solve the system of algebraic equations (8.29), (8.30), (8.31). Moreover, the operator inequalities (8.3) and $\langle G_{\mathbf{d}} x, x\rangle \ge \beta\|x\|$ may also be reduced to linear matrix inequalities (LMIs) in various particular cases.

Remark 8.1 If $\Xi_0(z) = \Xi_0 \in \mathbb{R}^{k\times k}$ and $\Omega(z) = \Omega \in \mathbb{R}^{k\times m}$ are constant matrices then conditions (8.29) and (8.31) become

$$(\Xi_0 + \Omega\Upsilon_0)\Gamma_0 = (\Gamma_0 + I_k)(\Xi_0 + \Omega\Upsilon_0), \quad \Gamma_0\Omega = \Omega\tilde{\Gamma} \tag{8.33}$$

with $\Upsilon_0 \in \mathbb{R}^{m\times k}$, $\Gamma_0 \in \mathbb{R}^{k\times k}$, $\tilde{\Gamma} \in \mathbb{R}^{m\times m}$.

Remark 8.2 If $\Omega(z) = \Omega \in \mathbb{R}^{k\times m}$ is a constant matrix, the conditions (8.29), (8.30), (8.31) are fulfilled and the operator $K \in \mathcal{L}(\mathbb{H}; \mathbb{U})$ is defined as

$$Kx = \Upsilon x, \quad \Upsilon \in \mathbb{R}^{k\times m}$$

then Assumption 8.2 holds and the control (8.19) has the form

$$u = \Upsilon_0 x + \|x\|_{\mathbf{d}}^{1+\mu}\Upsilon e^{-\Gamma \ln \|x\|_{\mathbf{d}}} x. \tag{8.34}$$

Indeed, since, in the considered case, we have

$$(\mathbf{d}(s)x)(z) = e^{s\Gamma} x(e^{s\mu\Theta} z), \quad x \in \mathbb{H},$$

and

$$(\tilde{\mathbf{d}}(s)u)(z) = e^{s\tilde{\Gamma}} u(e^{s\Theta} z), \quad u \in \mathbb{U}, \quad z \in \mathbb{R}^n,$$

then

$$\tilde{\mathbf{d}}(\mu \ln \|x\|_{\mathbf{d}})\Upsilon \mathbf{d}(-\ln \|x\|_{\mathbf{d}})x = \Upsilon e^{-\Gamma \ln \|x\|_{\mathbf{d}}} x$$

and the formula (8.34) holds. Moreover, since $\mathbf{d}(s)BK\mathbf{d}(-s) = e^{s\Gamma}\Omega\Upsilon e^{-s\Gamma}$ then the mapping $s \mapsto \mathbf{d}(s)BK\mathbf{d}(-s)$ is uniformly continuous and Assumption 8.2 is fulfilled.

Remark 8.3 If $n = 1$ and the linear matrix inequality (LMI)

$$(\Gamma - \tfrac{\mu\Theta}{2} I_k)X + X(\Gamma - \tfrac{\mu\Theta}{2} I_k)^\top \succ 0, \quad X \succ 0 \tag{8.35}$$

is fulfilled then the operator inequality (8.23) holds for $Px = X^{-1}x$.

Indeed, using the integration by parts we derive

$$\langle zX^{-1}\tfrac{\partial}{\partial z}x, x\rangle = -\tfrac{1}{2}\langle X^{-1}x, x\rangle$$

for all $x \in C_c^\infty$. Hence, we have

$$\langle X^{-1}G_{\mathbf{d}}x, x\rangle = \langle X^{-1}(\Gamma - \tfrac{\mu\Theta}{2} I_k)x, x\rangle,$$

so that the LMI (8.35) is equivalent to (8.23) in the considered case.

Remark 8.4 Let conditions of Remarks 8.1, 8.2, 8.3 be fulfilled and $q = 2$, $\tilde{\Xi}_0 = \Xi_0 + \Omega\Upsilon_0$, $\Xi_1(z) = \gamma z I_k$ with $\gamma \geq 0$, $\Xi_2(z) = \tilde{\Xi}_2 \in \mathbb{R}^{k\times k}$, $z \in \mathbb{R}$. If the system of LMIs (8.35) and

$$\tilde{\Xi}_0 X + \tilde{\Xi}_0^\top X + \Omega Y + Y^\top \Omega^\top - \gamma X \prec 0, \quad \tilde{\Xi}_2 X + X\tilde{\Xi}_2^\top \succ 0 \tag{8.36}$$

is feasible with respect to $X \in \mathbb{R}^{k\times k}$ and $Y \in \mathbb{R}^{m\times k}$ then the inequality (8.22) is fulfilled for $Px = X^{-1}x$ and $Kx = YX^{-1}x$ for some $\rho > 0$.

Indeed, using integration by parts we derive

$$\langle z\gamma X^{-1}\tfrac{\partial x}{\partial z}, x\rangle = \tfrac{-\gamma\langle X^{-1}x, x\rangle}{2}, \quad \langle X^{-1}\Xi_2 \tfrac{\partial^2 x}{\partial z^2}, x\rangle = -\langle Qx, x\rangle,$$

where $2Q = X^{-1}\tilde{\Xi}_2 + \tilde{\Xi}_2^\top X^{-1}$. Since, by assumption, $A_0 x = \tilde{\Theta}_0 x + z\gamma\frac{\partial x}{\partial z} + \tilde{\Xi}_2\frac{\partial^2 x}{\partial z^2}$ then $2\langle PA_0x, x\rangle = \langle (X^{-1}\tilde{\Xi}_0 + \tilde{\Xi}_0^\top X^{-1} + 2Q - \gamma X^{-1})x, x\rangle$. Since $\langle X^{-1}G_{\mathbf{d}}x, x\rangle = \langle X^{-1}(\Gamma - \frac{\mu\Theta}{2}I_k)x, x\rangle$ then the operator inequality (8.22) becomes

$$\langle (X^{-1}\tilde{\Xi}_0 + \tilde{\Xi}_0^\top X^{-1} + X^{-1}\tilde{\Xi}_0 + \Xi_0^\top X^{-1} - 2Q - \gamma X^{-1})x, x\rangle \leq$$

$$-2\rho\langle X^{-1}(\Gamma - \tfrac{\mu\Theta}{2}I_k)x, x\rangle$$

and it holds for a sufficiently small $\rho > 0$ provided that (8.36) is fulfilled.

If the pair $\{\Xi, \Omega\}$ is controllable and $\Xi_2 = \nu I_n$ with some $\nu \geq 0$ then the LMIs (8.35), (8.36) are always feasible (see Chap. 9 of Volume I) and the inequalities (8.22), (8.23) are feasible for any $\rho > 0$ and $\mu \in \mathbb{R}$. By Theorem 8.3, the latter implies null controllability of the system (8.1) in the considered case.

8.1.3 *Examples of Homogeneous Stabilization*

Homogeneous Stabilization of Heat Equation on $\mathbb{R}^n$

Let $\mathbb{H} = \mathbb{U} = L^2(\mathbb{R}^n; \mathbb{R})$ and $A = \Delta : \mathcal{D}(\Delta) \subset \mathbb{H} \mapsto \mathbb{H}$ be the Laplace operator and $B = I$ be an identity operator, where $\mathcal{D}(\Delta) = H^2(\mathbb{R}^n; \mathbb{R})$ is the domain of the Laplace operator Δ.

Let $\mathbf{d}$ be selected in the form (4.9) with $\Gamma = \gamma I_n$ with $\gamma > -\frac{n}{4}$ and $\Theta = -\frac{1}{2}$. In this case, from (4.21) we derive $\|x\|_{\mathbf{d}} = \|x\|^{\frac{4}{4\alpha+n}}$.

In Example 5.2, we have shown that the Laplace operator is $\mathbf{d}$-homogeneous of the degree $2\Theta = -1$. The Condition (B) of Theorem 8.1 holds.

Let us select $K \in \mathcal{L}(\mathbb{H}; \mathbb{U})$ as follows:

$$Kx = -\left(\alpha + \tfrac{n}{4}\right)x, \quad x \in \mathbb{H}.$$

The generator of the dilation $\mathbf{d}$ is given by (see Lemma 4.3)

$$(G_{\mathbf{d}}x)(z) = \gamma x(z) - \frac{1}{2}z \cdot \nabla x(z)$$

where $z = (z_1, \dots, z_n)^\top \in \mathbb{R}^n$ and $x \in \mathcal{D}(G_{\mathbf{d}})$.

In this case, for any $x \in C_c^\infty(\mathbb{R}^n; \mathbb{R})$ and any open bounded connected set $\Omega \subset \mathbb{R}^n$ with a smooth boundary, which contains a support of x, we have

$$\langle x, (A + BK + G_{\mathbf{d}})x\rangle = \langle x, \Delta x\rangle + \langle x, BKx\rangle + \alpha\langle x, x\rangle - \tfrac{1}{2}\int_\Omega \sum_{i=1}^n z_i x(z)\frac{\partial x(z)}{\partial z_i}dz =$$

$$\langle x, \Delta x\rangle - \tfrac{n}{4}\langle x, x\rangle - \tfrac{1}{4}\int_\Omega \sum_{i=1}^n z_i \frac{\partial x^2(z)}{\partial z_i}dz =$$

$$-\int_\Omega |\nabla x(z)|^2 dz + \frac{n}{4}\langle x, x\rangle - \frac{n}{4}\int_\Omega x^2(z)dz \le 0$$

where the integration by parts (for the first and the third term) has been utilized in the last step. Since, by Lemma 4.2, $C_c^\infty(\mathbb{R}^n; \mathbb{R})$ is $\mathbf{d}$-homogeneous cone dense in $L^2(\mathbb{R}^n; \mathbb{R})$ then Condition (C) of Theorem 8.1 is fulfilled.

Finally, selecting the group $\mathbf{d}_1(s) = I$ (with the generator $G_{\mathbf{d}_1} = \mathbf{0}$) we conclude that all conditions of Theorem 8.1 are satisfied. This means that the control

$$u(x) = -\frac{4\alpha + n}{4}\frac{x}{\|x\|^{\frac{4}{4\alpha+n}}}$$

is continuous for $\alpha > 1 - \frac{n}{4}$, globally bounded for $\alpha = 1 - \frac{n}{4}$ and stabilizes the heat system

$$\dot{x} = \Delta x + u(x)$$

in a finite time $T(x_0) \le \frac{4\|x_0\|^{\frac{4}{4\alpha+n}}}{4\alpha+n}$. For $\alpha = 1 - \frac{n}{4}$ the control u is discontinuous at the origin, so mild Filippov solutions have to be considered for $t \ge T(x_0)$ in this case.

Homogeneous Control for a Wave Equation on a Line

Let us consider a Hilbert space

$$\mathbb{H} = H^1(\mathbb{R}; \mathbb{R}) \times L^2(\mathbb{R}; \mathbb{R})$$

with an inner product to be defined below.

Let us consider a wave equation represented (see Example 2.11 for more details) in the form (8.1) using the notation:

$$A = \begin{pmatrix} O & I \\ \frac{\partial^2}{\partial z^2} & O \end{pmatrix} : \mathcal{D}(A) \subset \mathbb{H} \mapsto \mathbb{H},$$

where

$$\mathcal{D}(A) = H^2(\mathbb{R};\mathbb{R}) \times H^1(\mathbb{R};\mathbb{R}),$$

$$O \in \mathcal{L}(L^2(\mathbb{R};\mathbb{R}); L^2(\mathbb{R};\mathbb{R})) \quad \text{and} \quad I \in \mathcal{L}(L^2(\mathbb{R};\mathbb{R}); L^2(\mathbb{R};\mathbb{R}))$$

are zero and identity operators, respectively, and

$$B = \begin{pmatrix} O \\ I \end{pmatrix} : L^2(\mathbb{R};\mathbb{R}) \mapsto \mathbb{H}.$$

Let the inner product in $\mathbb{H}$ be defined as follows:

$$\langle x, \tilde{x} \rangle = \int_{\mathbb{R}} x^\top(z) P \tilde{x}(z) + p_{22} \frac{\partial x_1}{\partial z} \frac{\partial \tilde{x}_1}{\partial z} dz,$$

where

$$P := \begin{pmatrix} p_{11} & p_{12} \\ p_{12} & p_{22} \end{pmatrix} = Q^{-1}$$

with

$$Q = \begin{pmatrix} q_{11} & q_{12} \\ q_{12} & q_{22} \end{pmatrix} \in \mathbb{R}^{2\times 2}$$

being a positive definite solution of the system of linear matrix inequalities (LMIs) and equations:

$$\begin{pmatrix} \nu-1.5\mu & 1 \\ 0 & \nu-0.5\mu \end{pmatrix} Q + Q \begin{pmatrix} \nu-1.5\mu & 0 \\ 1 & \nu-0.5\mu \end{pmatrix} + y^\top b^\top + by = 0, \tag{8.37}$$

$$\begin{pmatrix} \nu-1.5\mu & 0 \\ 0 & \nu-0.5\mu \end{pmatrix} Q + Q \begin{pmatrix} \nu-1.5\mu & 0 \\ 0 & \nu-0.5\mu \end{pmatrix} \succ 0, \quad Q \succ 0, \tag{8.38}$$

where $\nu > \max\{0, 1.5\mu\}$, $b = (0 \;\; 1)^\top \in \mathbb{R}^2$, $y = (y_1 \;\; y_2) \in \mathbb{R}^{1\times 2}$.

The operator A is a generator of a strongly continuous semigroup of linear bounded operators on $\mathbb{H}$ (see, e.g., [17, Sect. 7.4] for more details).

This system of LMIs is always feasible with respect to Q and y (see Chap. 9 of Volume I). Since from (8.37) we conclude

$$(\nu - 1.5\mu) q_{11} + q_{12} = 0$$

then taking into account $P = Q^{-1} \succ 0$ it is easy to see that

$$0 < (\nu - 1.5\mu)p_{22} = p_{12}.$$

Let the operator $K : \mathbb{H} \mapsto L^1(\mathbb{R}; \mathbb{R})$ be defined as

$$Kx := yQ^{-1}x, \quad x \in \mathbb{H},$$

where the pair (Q, y) is a solution of the LMIs (8.37), (8.38).

(A) Let the uniformly continuous semigroup $\mathbf{d}_l$ of linear bounded operators on $\mathbb{H}$ be defined as follows:

$$\mathbf{d}_l(s)x = \begin{pmatrix} e^{-\mu s} & 0 \\ 0 & 1 \end{pmatrix} x.$$

Its generator

$$G_{\mathbf{d}_l}x = \begin{pmatrix} -\mu & 0 \\ 0 & 0 \end{pmatrix} x,$$

obviously, satisfies the condition (A) of Theorem 8.1 with $\gamma_{\min} = \min\{0, -\mu\}$ and $\gamma_{\max} = \max\{0, -\mu\} \geq 0$.

(B) Let us introduce the dilation $\mathbf{d}$ on $\mathbb{H}$ as follows:

$$(\mathbf{d}(s)x)(z) = \left(\begin{smallmatrix} e^{s(\nu-\mu)} & 0 \\ 0 & e^{s\nu} \end{smallmatrix}\right) x(e^{\mu s}z), \quad x \in \mathbb{H}, \quad z \in \mathbb{R}.$$

The dilation $\mathbf{d}$ has the generator $G_{\mathbf{d}} : \mathcal{D}(G_{\mathbf{d}}) \subset \mathbb{H} \mapsto \mathbb{H}$ defined as

$$G_{\mathbf{d}}x = \left(\begin{smallmatrix} \nu-\mu & 0 \\ 0 & \nu \end{smallmatrix}\right) x + \mu z \frac{\partial x}{\partial z},$$

where $\mathcal{D}(G_{\mathbf{d}}) = \left\{x \in \mathbb{H} : z\frac{\partial x}{\partial z} \in \mathbb{H}\right\}$.

According to Proposition 4.4 the dilation $\mathbf{d}$ is strictly monotone on $\mathbb{H}$ if there exists $\beta > 0$ such that $\langle G_{\mathbf{d}}x, x\rangle \geq \beta\|x\|^2$ for any $x \in C_c^\infty(\mathbb{R}; \mathbb{R}) \times C_c^\infty(\mathbb{R}; \mathbb{R})$, where $\|x\|^2 = \sqrt{\langle x, x\rangle}$. Using integration by parts we derive

$$\langle G_{\mathbf{d}}x, x\rangle = \int_{\mathbb{R}} x^\top(z)P\left(\left(\begin{smallmatrix} \nu-\mu & 0 \\ 0 & \nu \end{smallmatrix}\right) x(z) + \mu z \tfrac{\partial x(s)}{\partial z}\right) dz +$$

$$p_{22}\int_{\mathbb{R}} (\nu - \mu)\left(\tfrac{\partial x_1(z)}{\partial z}\right)^2 + \mu \tfrac{\partial x_1(z)}{\partial z}\tfrac{\partial}{\partial z}\left(z\tfrac{\partial x_1(z)}{\partial z}\right) dz -$$

$$\int_{\mathbb{R}} x^\top(z)P\left(\begin{smallmatrix} \nu-\mu & 0 \\ 0 & \nu \end{smallmatrix}\right) x(z)dz + \tfrac{\mu}{2}\int_{\mathbb{R}} z\frac{\partial}{\partial z}\left(x^\top(z)Px(z)\right) dz +$$

$$p_{22}\nu\int_{\mathbb{R}}\left(\tfrac{\partial x_1(z)}{\partial z}\right)^2 dz + \tfrac{\mu}{2}\int_{\mathbb{R}} z\frac{\partial}{\partial z}\left(\tfrac{\partial x_1(z)}{\partial z}\right)^2 dz =$$

$$\int_{\mathbb{R}} x^\top(z)P\left(\begin{smallmatrix} \nu-1.5\mu & 0 \\ 0 & \nu-0.5\mu \end{smallmatrix}\right) x(z) + p_{22}(\nu - 0.5\mu)\left(\tfrac{\partial x_1(z)}{\partial z}\right)^2 dz \geq \beta\|x\|^2,$$

where

$$\beta \geq \min\left\{\lambda_{\min}\left(P^{\frac{1}{2}}\begin{pmatrix}\nu-\frac{3\mu}{2} & 0\\ 0 & \nu-\frac{\mu}{2}\end{pmatrix}P^{-\frac{1}{2}}+P^{-\frac{1}{2}}\begin{pmatrix}\nu-\frac{3\mu}{2} & 0\\ 0 & \nu-\frac{\mu}{2}\end{pmatrix}P^{\frac{1}{2}}\right), \nu-\frac{\mu}{2}\right\}.$$

The LMI (8.38) implies $\beta > 0$. Since $C_c^\infty(\mathbb{R};\mathbb{R}) \times C_c^\infty(\mathbb{R};\mathbb{R})$ is dense in $\mathbb{H}$ then Condition (B) of Theorem 8.1 is fulfilled.

(C) Given $x = (x_1, x_2)^\top \in C_c^\infty(\mathbb{R};\mathbb{R}) \times C_c^\infty(\mathbb{R};\mathbb{R})$ and $A_0 = \begin{pmatrix}0 & I\\ 0 & 0\end{pmatrix}$ we have

$$\langle (A+BK+G_{\mathbf{d}})x, x\rangle = \langle (A_0+BK+G_{\mathbf{d}})x, x\rangle + \left\langle \begin{pmatrix}\mathbf{0}\\ \frac{\partial^2 x_1}{\partial z^2}\end{pmatrix}, x\right\rangle.$$

Using (8.37), we derive

$$\langle x, (A_0+BK+G_{\mathbf{d}})x\rangle = p_{22}\int_{\mathbb{R}} \frac{\partial x_1(z)}{\partial z}\frac{\partial x_2(z)}{\partial z}dz + p_{22}(\nu-0.5\mu)\int_{\mathbb{R}}\left(\frac{\partial x_1(z)}{\partial z}\right)^2 dz.$$

Using integration by parts, we derive

$$\left\langle x, \begin{pmatrix}\mathbf{0}\\ \frac{\partial^2 x_1}{\partial z^2}\end{pmatrix}\right\rangle = \int_{\mathbb{R}} -p_{12}\left(\frac{\partial x_1(z)}{\partial z}\right)^2 - p_{22}\frac{\partial x_1(z)}{\partial z}\frac{\partial x_2(z)}{\partial z}dz.$$

Therefore, taking into account the identity $0 < (\nu - 1.5\mu)p_{22} = p_{12}$ we conclude

$$\langle x, (A+BK+G_{\mathbf{d}})x\rangle = \mu p_{22}\int_{\mathbb{R}}\left(\frac{\partial x_1(z)}{\partial z}\right)dz.$$

For $\mu \leq 0$ we, obviously, have $\langle x, (A+BK+G_{\mathbf{d}})x\rangle \leq 0$. On the other hand, if $\nu \geq 2.5\mu > 0$ then using the representation for $\langle x, G_{\mathbf{d}}x\rangle$ we derive

$$\langle x, (A+BK+0.5G_{\mathbf{d}})x\rangle \leq -\frac{p_{22}(\nu-2.5\mu)}{2}\int_{\mathbb{R}}\left(\frac{\partial x_1(z)}{\partial z}\right)dz \leq 0.$$

Therefore, the condition (C) of Theorem 8.1 holds for any $\mu \in \mathbb{R}$ and any $\nu \geq \max\{0, 2.5\mu\}$.

(D) Finally, $\mathbf{d}(s)BK\mathbf{d}_1(s) = BK\mathbf{d}(s)$ for any $s \in \mathbb{R}$ and the operator A is $\mathbf{d}$-homogeneous of the degree μ. Indeed,

$$(A\mathbf{d}(s)x)(z) = A\begin{pmatrix}e^{(\nu-\mu)s}x_1(e^{\mu s}z)\\ e^{\nu s}x_2(e^{\mu s}z)\end{pmatrix} = \begin{pmatrix}e^{\nu s}x_2(e^{\mu s}z)\\ e^{(\nu+\mu)s}\left.\frac{\partial^2 x_1(q)}{\partial q^2}\right|_{q=e^{\mu s}z}\end{pmatrix} = e^{\mu s}(\mathbf{d}(s)Ax)(z)$$

and the condition (D) of Theorem 8.1 is fulfilled.

Therefore, the homogeneous feedback control of the form (8.4) steers all trajectories of the wave system to

- the origin in a finite time if $\mu < 0$;
- a neighborhood of the origin in a fixed time independently of the initial state if $\mu > 0$.

In the general case, the homogeneous norm $\|\cdot\|_{\mathbf{d}}$ is defined implicitly, see (4.21). For $\mu = -1$, $\nu = \frac{1}{2}$ the canonical homogeneous norm can be found as a unique positive solution of the quartic equation

$$V^4 = aV^2 + bV + c$$

with $V = e^{s_x} = \|x\|_{\mathbf{d}}$,

$$a = p_{22}\int_{\mathbb{R}} x_2^2(z) + (x_1'(z))^2 dz, \quad b = 2p_{12}\int_{\mathbb{R}} x_1(z)x_2(z)dz, \quad c = p_{11}\int_{\mathbb{R}} x_1^2(z)dz.$$

In this case $\|x\|_{\mathbf{d}}$ can be computed using Ferrari formulas, i.e., the homogeneous finite-time stabilizing feedback (8.4) for the wave equation admits an explicit representation.

A Distributed Control for 1-D Heat System on a Segment [0, 1]

To demonstrate a possible way of application of Corollary 8.2, we design a distributed finite-time control for the following heat system on [0, 1]:

$$\frac{\partial x}{\partial t} = \frac{\partial^2 x}{\partial z^2} + \phi(z)u(t,z), \;\; x(t,0)=x(t,1)=0, \;\; x(0,z)=x_0(z),$$

where x is the system state, u is the distributed control, $\phi : [0,1] \mapsto \mathbb{R}_+$ is a continuous function such that

$$c\,z^2 \leq \phi(z) \quad \text{for} \quad z \in [0,1], \tag{8.39}$$

for some $c > 0$. The function ϕ represents possibly non uniform feedback gain on $[0,1]$. The design procedure is realized in several steps

- First, we extend the considered system virtually to $\mathbb{R}_+$ (with the zero operator on $(1,+\infty)$);
- Next, we design a feedback control using conditions (A)–(C) of Theorem 8.1 and Condition (D*) of Corollary 8.2;
- Finally, we reduce the extended closed-loop system to the original one assuming that the initial condition of the extended system has a support only in $(0,1)$.

Homogeneous finite-time control design on $\mathbb{R}_+$. Let $\mathbb{H} = L^2(\mathbb{R}_+;\mathbb{R})$ and $x_{(a,b)}$ denotes a restriction of a function $x : \mathbb{R}_+ \mapsto \mathbb{R}$ to the interval (a,b). For any $x \in \mathbb{H}$ we have $x_{(0,1)} \in L^2((0,1);\mathbb{R})$ and $x_{(1,+\infty)} \in L^2((1,+\infty);\mathbb{R})$.

Let the operator $A : \mathcal{D}(A) \subset \mathbb{H} \mapsto \mathbb{H}$ be defined as the connection of the Laplace operator with Dirichlet boundary conditions on [0, 1] and the zero operator on $[1,+\infty)$, i.e., $(Ax)(z) = \frac{\partial^2 x(z)}{\partial z^2}$ if $z \in (0,1)$ and $(Ax)(z) = 0$ if $z \geq 1$ for

$$x \in \mathcal{D}(A) := \left\{x \in \mathbb{H} : x_{(0,1)} \in H_0^1((0,1);\mathbb{R}) \cap H^2((0,1);\mathbb{R})\right\}.$$

The operator A is closed and densely defined. It generates the strongly continuous semigroup Φ of linear bounded operators on the Hilbert space $\mathbb{H}$:

$$(\Phi(t)u)(x) = \begin{cases} 2\sum_{n=1}^{\infty} e^{-n^2\pi^2 t} \sin(\pi n x)\int_0^1 \sin(\pi n y)u(y)dy & \text{if } x \in (0,1), \\ u(x) & \text{if } x \in [1,+\infty). \end{cases}$$

Let us consider the equation (8.1) with A given above and $B : \mathbb{H} \mapsto \mathbb{H}$ defined as follows: $(Bx)(z) = \phi(z)x(z)$ if $z \in (0,1)$ and $(Bx)(z) = \phi(1)x(z)$ if $z \geq 1$ for $x \in \mathbb{H}$.

Let us show that conditions (A)–(C) of Theorem 8.1 and the condition D*) of Corollary 8.2 hold for $K = -\tilde{k}I \in \mathcal{L}(\mathbb{H};\mathbb{H})$ with a sufficiently large scalar $\tilde{k} > 0$.

(A) Let a uniformly continuous group $\mathbf{d}_1$ be defined as follows:

$$(\mathbf{d}_1(s)x)(z) = e^{-\mu s}x(z) \text{ if } z \in (0,1) \quad \text{and} \quad (\mathbf{d}_1(s)x)(z) = x(z) \text{ if } z \geq 1,$$

where $\mu < 0$. Obviously, $\gamma_{\min} = 0$ and $\gamma_{\max} = -\mu$, where $\gamma_{\min}$ and $\gamma_{\max}$ are defined in Theorem 8.1.

(B) Let the dilation $\mathbf{d}(s)$ in $\mathbb{H}$ be defined as

$$(\mathbf{d}(s)x)(z) = e^s\, x(e^{0.5\mu s}z).$$

It has the generator $G_{\mathbf{d}}x = x + 0.5\mu z\frac{\partial x}{\partial z}$ with the domain $\mathcal{D}(G_{\mathbf{d}}) = \{x \in \mathbb{H} : x + 0.5\mu z\frac{\partial x}{\partial z} \in \mathbb{H}\}$. Obviously,

$$\|\mathbf{d}(s)x\| = \sqrt{\int_{\mathbb{R}_+} e^{2s}x^2(e^{0.5\mu s}z)dz} = e^{(1-0.25\mu)s}\|x\|$$

and $(1 - 0.25\mu)\|x\|^2 = \langle G_{\mathbf{d}}x, x\rangle$ if $x \in \mathcal{D}(G_{\mathbf{d}})$, where, as before, $\|\cdot\|$ denotes the norm in $\mathbb{H}$. Hence, we derive

$$\|x\|_{\mathbf{d}} = \|x\|^{\frac{1}{1-0.25\mu}}$$

and the formula (8.4) gives

$$u(x) = -\tilde{k}\|x\|_{\mathbf{d}}^{\mu}\mathbf{d}_1(-\ln\|x\|_{\mathbf{d}})x = -\tilde{k}\|x\|^{\frac{2\mu}{1-0.25\mu}}x.$$

Theorem 8.1, guarantees that this feedback control is locally Lipschitz continuous on $\mathbb{H}\backslash\{\mathbf{0}\}$. Moreover, it is continuous at $\mathbf{0}$ if $-4/7 < \mu < 0$ and globally bounded if $\mu = -4/7$.

(C)–(D*) Let us consider the homogeneous extension (5.21):

$$F(x) = \bigcup_{s\in\mathbb{R}:\mathbf{d}(s)x\in\mathcal{D}(A)} \{A_s x + q_s(x))\},$$

where $A_s = e^{-\mu s}\mathbf{d}(-s)A\mathbf{d}(s)$ and $g_s(x) = e^{-\mu s}\mathbf{d}(-s)Bu(\mathbf{d}(s)x)$. It is easy to see that $(A_s x)(z) = \frac{\partial^2 x(z)}{\partial z^2}$ if $z \in (0, e^{0.5\mu s})$ and $(A_s x)(z) = 0$ if $z \geq e^{0.5\mu s}$,

$$(g_s(x))(z) = -\tilde{k}\begin{cases} \phi(e^{-0.5\mu s}z)e^{\mu s}\frac{x(z)}{\|x\|_{\mathbf{d}}^{-2\mu}}, & z\in(0, e^{0.5\mu s}),\\ \phi(1)\frac{x(z)}{\|x\|_{\mathbf{d}}^{-\mu}}, & z\geq e^{0.5\mu s}.\end{cases}$$

The set-valued operator $F : \mathcal{D}(F) \subset \mathbb{H} \mapsto 2^{\mathbb{H}}$ *is* $\mathbf{d}$*-homogeneous of degree* $\mu < 0$ *and* $Ax + Bu(x) \in F(x)$ *for any* $x \in \mathcal{D}(A)$, *where* $\mathcal{D}(F) = \bigcup_{s\in\mathbb{R}}\mathbf{d}(s)\mathcal{D}(A)$.

Let us denote $q_s = e^{0.5\mu s}$ and consider the set

$$\tilde{\mathcal{D}}_s = \left\{ x \in \mathbb{H} : \begin{array}{c} x_{(0,q_s)} \in C_c^{\infty}((0, q_s); \mathbb{R}), \\ x_{(q_s,+\infty)} \in C_c^{\infty}((q_s, +\infty); \mathbb{R}) \end{array} \right\}.$$

Obviously, $\tilde{\mathcal{D}}_0 \subset \mathcal{D}(A)$ is dense in $\mathcal{D}(A)$ and $\mathbf{d}(s)x \in \tilde{\mathcal{D}}_0 \Rightarrow x \in \tilde{\mathcal{D}}_s$.

Let us denote $\langle x, y\rangle_{(a,b)} = \int_a^b x(z)y(z)dz$. Using integration by parts for any $s \in \mathbb{R}$ and for any $x \in \tilde{\mathcal{D}}_s$ we derive

$$\left\langle z\frac{\partial x}{\partial z}, x\right\rangle_{(0,q_s)} = \frac{-\langle x, x\rangle_{(0,q_s)}}{2}$$

and

$$\left\langle z\frac{\partial x}{\partial z}, x\right\rangle_{(q_s,+\infty)} = \frac{-\langle x, x\rangle_{(q_s,+\infty)}}{2}.$$

Due to (8.39) one has

$$\phi(e^{-0.5\mu s}z)e^{\mu s} \geq cz^2 \quad \text{if } z \in [0, q_s]$$

and

$$\langle \phi(e^{-0.5\mu s}z)e^{\mu s}x, x\rangle_{(0,e^{0.5\mu s})} \geq c\langle zx, zx\rangle_{(0,q_s)}.$$

Therefore, for any $x \in \tilde{\mathcal{D}}_s$, $\|x\| = 1$ we obtain

$$\begin{gathered}\langle (A_s x + g_s(x) + \rho G_{\mathbf{d}})x, x\rangle \leq \\ -\langle \tfrac{\partial x}{\partial z}, \tfrac{\partial x}{\partial z}\rangle_{(0,q_s)} - 2\rho(1 - 0.25\mu)\left\langle z\tfrac{\partial x}{\partial z}, x\right\rangle_{(0,q_s)} - \tilde{k}c\langle zx, zx\rangle_{(0,q_s)} \\ -(\tilde{k}\phi(1) - \rho(1 - 0.25\mu))\langle x, x\rangle_{(q_s,+\infty)} \leq 0\end{gathered}$$

provided that $\tilde{k} \geq \max\left\{\frac{\rho(1-0.25\mu)}{\phi(1)}, \frac{\rho^2(1-0.25\mu)^2)}{c}\right\}$.

The condition (D*) of Corollary 8.2 holds for

$$\mathcal{D} = \bigcup_{s\in\mathbb{R}} \tilde{\mathcal{D}}_s$$

since

- $\mathcal{D}$ is a **d**-homogeneous cone by construction;
- $\mathcal{D}$ contains the set $\tilde{\mathcal{D}}_0$ dense in $\mathcal{D}(A)$;
- $x \in \mathcal{D} : \mathbf{d}(s)x \in \mathcal{D}(A) \;\Rightarrow\; x \in \tilde{\mathcal{D}}_s$.

Therefore, the feedback law (8.4) steers any solution of the extended evolution system (8.1) to zero in a finite time $T \le (-\rho\mu)^{-1}\|x_0\|_{\mathbf{d}}^{-\mu}$, provided that $\mu \in [-4/7, 0)$.

Restriction of the finite-time control to the segment $[0, 1]$. If $\mathrm{supp}(x) \subset [0, 1]$ and $x \in \mathbb{H}\backslash\{\mathbf{0}\}$ then $\mathrm{supp}(\Phi(t)x) \subset [0, 1]$, $\mathrm{supp}(u(x)) \subset [0, 1]$ and $\|x\| = \sqrt{\int_0^1 x^2(z)dz}$, where $\mathrm{supp}(x)$ denotes the support of a function $x \in \mathcal{D}(A)$. Therefore, any solution of the closed-loop system (8.1), (8.4) with initial condition $x_0 \in \{x \in \mathcal{D}(A) : \mathrm{supp}(x) \subset [0, 1]\}$ coincides on $[0, 1]$ with the solution to the equation

$$\frac{\partial x}{\partial t} = \frac{\partial^2 x}{\partial z^2} + \phi(z)\frac{-\tilde{k}}{\|x\|^{\frac{-2\mu}{1-0.25\mu}}}x, \qquad \begin{array}{c} x(t,0)=x(t,1)=0, \\ x(0,z)=u_0(x),\;\; z\in[0,1]. \end{array}$$

Consequently, any solution of the latter heat system vanishes in a finite time

$$T(x_0) \le (-\rho\mu)^{-1}\left(\int_0^1 x_0^2(z)dz\right)^{-0.5\mu/(1-0.25\mu)}.$$

Piecewise linear finite-time control on $[0, 1]$. Since $\|x\|_{\mathbf{d}} = \|x\|^{\frac{1}{1-0.25\mu}}$ then the parameters of Lemma 4.5 are $M = 1, \alpha = \beta = 1 - 0.25\mu$. Applying Corollary 8.3 for $q_i = \ln 2$ we derive $u(t) = -\lambda(t)x(t)$, where the switched gain $\lambda(t) = \tilde{k}/r_{\sigma(t)}^{-2\mu/1-0.25\mu}$ with $r_i = 2^{-i}$ and $\sigma(t)$ defined by (8.14). Therefore, for any $x_0 \in H_0^1((0, 1); \mathbb{R}) \cap H^2((0, 1); \mathbb{R})$ the equation

$$\frac{\partial x}{\partial t} = \frac{\partial^2 x}{\partial z^2} - \lambda(t)\phi(z)x, \qquad \begin{array}{c} x(t,0)=x(t,1)=0, \\ x(0,z)=u_0(z),\;\; z\in[0,1]. \end{array} \tag{8.40}$$

has a unique strong solution which vanishes in a finite-time

$$T(x_0) \le \frac{\ln 2}{\rho}\sum_{j=i_0}^{+\infty} 2^{\mu j}$$

for $\|x_0\|_{\mathbf{d}} \in \left(r_{i_0+1}, r_{i_0}\right]$.

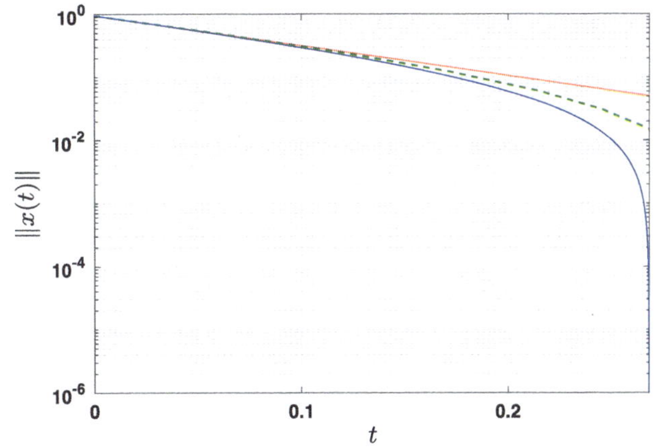

Fig. 8.1 Evolution of L^2-norm of the heat equation with a linear (red dotted line), switched linear (green dashed line), and a nonlinear (blue solid line) homogeneous control (plotted in the logarithmic scale)

Numerical simulations. The simulation results of the system (8.40) (for $\phi(z) = z^2$) with homogeneous and switched linear finite-time controls ($\mu = -0.5, \tilde{k} = 1$) as well as with the static linear feedback ($\lambda = \tilde{k}$) are depicted at Fig. 8.1. The initial condition is given by $x_0(z) = 5.25z(1 - z)$. For numerical simulations, the system has been discretized by means of divided differences on a uniform grid with the step $h = 0.005$ for the space variable. The discretization with respect to time has been done using implicit scheme with the step size $\tau = h^2$. The norm $\|x\|$ is calculated by Simpson rule. Numerical simulations show that the heat system with homogeneous control has a fast decay (typical for finite-time stable systems) with the settling time $T(x_0) \leq 0.3$.

Homogeneous Stabilization of Interconnected Heat Systems

Let us consider the system

$$\frac{\partial x(z,t)}{\partial t} = \nu \frac{\partial^2 x(z,t)}{\partial z^2} + \Xi_0 x(z,t) + \Omega u(z,t), \tag{8.41}$$

where $\nu > 0$ is a diffusion coefficient, $z \in \mathbb{R}$ is the space variable, $t \geq 0$ is the time variable, $x(\cdot, t) \in L^2(\mathbb{R}; \mathbb{R}^k)$ is the distributed state of the system, $u(\cdot, t) \in L^2(\mathbb{R}; \mathbb{R}^m)$ is a distributed control input be defined in the state feedback form, $\Xi_0 \in \mathbb{R}^{k\times k}$, $\Omega \in \mathbb{R}^{n\times m}$ are constant matrices. The system (8.41) written in a componentwise manner can be interpreted as a system of interconnected heat equations with the interconnection weight given by the matrix Ξ_0. Such a system may model thermogalvanic cells having some heat exchange with neighbors.

Let us first design a linear feedback which homogenize the system (8.41). In the view of Remark 8.1, we define the operator

$$K_0 \in \mathcal{L}(L^2(\mathbb{R};\mathbb{R}^m); L^2(\mathbb{R};\mathbb{R}^k))$$

by solving the system of equations

$$\Xi_0\Gamma_0 - \Gamma_0\Xi_0 + \Omega Y_0 = \Xi_0, \quad \Gamma_0\Omega = \mathbf{0}$$

with respect to $\Gamma_0 \in \mathbb{R}^{k\times k}$, $Y_0 \in \mathbb{R}^{m\times k}$. This system of equations is linear and it always has a solution if the pair $\{\Xi_0, \Omega\}$ is controllable (see, Chap. 9 of Volume I for more details). Moreover, the matrix $\Gamma_0 - I_k$ is invertible and K_0 is given by

$$K_0 x = Y_0(\Gamma_0 - I_n)^{-1}x.$$

Hence, to fulfill Assumption 8.1 we need to find $\Theta \in \mathbb{R}$ such that the algebraic condition (8.30) holds. In our case, the required condition is

$$0 = -\tfrac{\partial^2}{\partial z^2} + 2\Theta\tfrac{\partial^2}{\partial z^2}$$

i.e., $\Theta = 0.5$. Using Lemma 4.1 we define the dilation in $L^2(\mathbb{R};\mathbb{R}^k)$ as follows:

$$\mathbf{d}(s) = e^{s\Gamma}x(e^{0.5\mu s}z), \quad \Gamma = I_k + \mu\Gamma_0$$

with $\mu < 2/3$. The group $\tilde{\mathbf{d}}$ is defined as follows:

$$(\tilde{\mathbf{d}}(s)u)(z) = u(e^{0.5s}z), \quad z \in \mathbb{R}.$$

Let us denote $\tilde{\Xi}_0 = \Xi_0 + \Omega Y_0(\Gamma_0 - I_n)^{-1}$. In the view of Remark 8.4, to design the homogeneous stabilizer (8.34) it is sufficient to solve the system of LMIs (8.35), (8.36) with $\gamma = 0$, $\Xi_2 = \nu I_k$. The corresponding system of matrix inequalities is always feasible provided that the pair of matrices $\{\Xi_0, \Omega\}$ is controllable. So, the feedback gain in (8.34) is defined as $\Upsilon = YX^{-1}$ and the canonical homogeneous norm $\|x\|_{\mathbf{d}}$ is induced by the norm $\|x\| = \sqrt{\langle X^{-1}x, x\rangle}$. By Theorem 8.3, the closed-loop system is finite-time stable for $\mu < 0$, exponentially stable for $\mu = 0$, and nearly-fixed-time stable for $\mu > 0$. By Theorem 6.12, the closed-loop system is Input-to-State Stable with respect to measurement noises and additive perturbations.

8.1.4 Robustness Analysis

In Chap. 6, we have discovered that the homogeneous systems are robust with respect to a rather large class of nonlinear perturbations. Let us characterize a class of additive perturbations which can be rejected by means of the homogeneous control (8.4).

Let us consider the perturbed linear control system in $\mathbb{H}$

$$\dot{x} = Ax + Bu(x) + f(t, x), \quad t > t_0, \quad x(t_0) = x_0 \tag{8.42}$$

where the state $x(t) \in \mathbb{H}$, the control $u : \mathbb{H} \mapsto \mathbb{U}$ and the operators $A : \mathcal{D}(A) \subset \mathbb{H} \mapsto \mathbb{H}$, $B : \mathbb{U} \mapsto \mathbb{H}$ are as before and $f : \mathbb{R} \times \mathbb{H} \mapsto \mathbb{H}$.

Corollary 8.4 *Let all conditions of Theorem 8.1 (or Theorem 8.3) hold and there exists $\kappa \in (0, \rho)$ such that*

$$\frac{\|\mathbf{d}(-\ln\|x\|_{\mathbf{d}})f(t,x)\|}{\langle G_{\mathbf{d}}\mathbf{d}(-\ln\|x\|_{\mathbf{d}})x, \mathbf{d}(-\ln\|x\|_{\mathbf{d}})x\rangle} \le \kappa\|x\|_{\mathbf{d}}^{\mu}, \quad \forall x \in \mathbb{H}\backslash\{\mathbf{0}\}, \forall t \ge 0. \tag{8.43}$$

If $f \in C(\mathbb{R} \times (\mathbb{H}\backslash\{\mathbf{0}\}); \mathbb{H})$ is locally Lipschitz continuous then all conclusions (except the existence of classical solutions) of Theorem 8.1 (resp., Theorem 8.3) remain valid for the closed-loop system (8.42), (8.4).

Proof Indeed, since f is locally Lipschitz continuous on $\mathbb{R} \times (\mathbb{H}\backslash\{\mathbf{0}\})$ then the closed-loop system has a unique locally Lipschitz continuous strong solution x for any $x(0) \in \mathcal{D}(A)\backslash\{\mathbf{0}\}$ (see Theorem 2.16). In this case, we have

$$\begin{aligned}
\tfrac{d}{dt}\|x\|_{\mathbf{d}} :=& D\|x\|_{\mathbf{d}}(Ax + Bu(x) + f(t, x)) \\
=& \frac{\|x\|_{\mathbf{d}}\langle \mathbf{d}(-\ln\|x\|_{\mathbf{d}})(Ax+Bu(x)+f(t,x)), \mathbf{d}(-\ln\|x\|_{\mathbf{d}})x\rangle}{\langle G_{\mathbf{d}}\mathbf{d}(-\ln\|x\|_{\mathbf{d}})x, \mathbf{d}(-\ln\|x\|_{\mathbf{d}})x(t)\rangle} \\
=& \frac{\|x\|_{\mathbf{d}}^{1+\mu}\langle (A+BK)\mathbf{d}(-\ln\|x\|_{\mathbf{d}})x, \mathbf{d}(-\ln\|x\|_{\mathbf{d}})x\rangle}{\langle G_{\mathbf{d}}\mathbf{d}(-\ln\|x\|_{\mathbf{d}})x, \mathbf{d}(-\ln\|x\|_{\mathbf{d}})x\rangle} \\
&+ \frac{\|x\|_{\mathbf{d}}\langle \mathbf{d}(-\ln\|x\|_{\mathbf{d}})f(t,x), \mathbf{d}(-\ln\|x\|_{\mathbf{d}})x\rangle}{\langle G_{\mathbf{d}}\mathbf{d}(-\ln\|x\|_{\mathbf{d}})x, \mathbf{d}(-\ln\|x\|_{\mathbf{d}})x\rangle}
\end{aligned}$$

for $\mathbf{0} \ne x \in \mathcal{D} \subset (\mathcal{D}(A) \cap \mathcal{D}(G_{\mathbf{d}}))$. Using Cauchy–Schwarz inequality and the identity $\|\mathbf{d}(-\ln\|x\|_{\mathbf{d}})x\| = 1$ we derive

$$\langle \mathbf{d}(-\ln\|x\|_{\mathbf{d}})f(t, x), \mathbf{d}(-\ln\|x\|_{\mathbf{d}})x\rangle \le \|\mathbf{d}(-\ln\|x\|_{\mathbf{d}})f(t, x)\|.$$

Since $\mathcal{D}$ is dense in $\mathcal{D}(A)$ then, taking into account the inequality (8.43), we conclude

$$\tfrac{d}{dt}\|x(t)\|_{\mathbf{d}} \overset{a.e.}{\le} -(\rho - \kappa)\|x(t)\|_{\mathbf{d}}^{1+\mu} \text{ if } x(t) \ne \mathbf{0}$$

for any strong solution of the closed-loop system. Therefore, all considerations of the proof of Theorem 8.1 (resp., Theorem 8.3) can be repeated for the closed-loop system (8.42), (8.4) to complete the proof. □

Example 8.1 Let us consider the evolution control system

$$\dot{x}(t) = Ax(t) + Bu(t) + \underbrace{\begin{pmatrix} \delta \mathcal{S}_h x_2(t) \\ 0 \end{pmatrix}}_{f(t,x(t))}, \quad A = \begin{pmatrix} 0 & 1 \\ 0 & 0 \end{pmatrix}, B = \begin{pmatrix} 0 \\ 1 \end{pmatrix}, \quad t > 0$$

where $x(t) = (x_1(t), x_2(t))^\top \in L^2((-\infty, 0); \mathbb{R}^2)$, is the system state, $u(t) \in L^2((-\infty, 0); \mathbb{R})$, $\delta \in (0, 1)$ and $\mathcal{S}_h : L^2((-\infty, 0); \mathbb{R}) \mapsto L^2((-\infty, 0); \mathbb{R})$ is the linear bounded shift operator

$$(\mathcal{S}_h y)(\sigma) = y(\sigma - h), \quad y \in L^2((-\infty, 0); \mathbb{R}), \quad \sigma < 0, \quad h > 0.$$

Let $\mathbb{H} = L^2((-\infty, 0); \mathbb{R}^2)$ with the inner product

$$\langle u, v \rangle = \int_{-\infty}^{0} u^\top(\sigma) P v^\top(\sigma) d\sigma, \qquad u, v \in \mathbb{H},$$

where $P = P^\top \in \mathbb{R}^{2\times 2}$ satisfies the following linear matrix inequality

$$PG_{\mathbf{d}} + G_{\mathbf{d}}P \succ 0, \quad P \succ 0, \quad G_{\mathbf{d}} = \begin{pmatrix} 2 & 0 \\ 0 & 1 \end{pmatrix}.$$

In this case, the uniformly continuous dilation

$$\mathbf{d}(s) = \begin{pmatrix} e^{2s} & 0 \\ 0 & e^{s} \end{pmatrix},$$

is strictly monotone in $\mathbb{H}$ (see Proposition 4.4). According to Theorem 8.1 (with $\mu = -1$) the controller

$$u(x) = K\mathbf{d}(-\ln \|x(t)\|_{\mathbf{d}})x(t)$$

with $K \in \mathbb{R}^{1\times 2}$ satisfying

$$PA + A^\top P + PBK + K^\top B^\top P + (\rho + \kappa)G_{\mathbf{d}}P + PG_{\mathbf{d}} \preceq 0, \quad \kappa \geq 0, \rho > 0$$

stabilizes the nominal system (with $f = \mathbf{0}$) in a finite-time $T(x_0) \leq \|x_0\|_{\mathbf{d}}/\rho$. According to Corollary 8.4 the robust stability ($f \neq 0$) can be guaranteed if the following inequality

$$\frac{\delta \|\mathcal{S}_h x_2\|_{L^2((-\infty,0);\mathbb{R})}}{\|x\|_{\mathbf{d}}} \leq \kappa \langle G_{\mathbf{d}}\mathbf{d}(-\ln \|x\|_{\mathbf{d}})x, \mathbf{d}(-\ln \|x\|_{\mathbf{d}})x \rangle$$

is fulfilled for any $x = (x_1, x_2)^\top \in L^2((-\infty, 0); \mathbb{R}^2)$. Taking into account

$$\langle G_{\mathbf{d}}\mathbf{d}(-\ln \|x\|_{\mathbf{d}})x, \mathbf{d}(-\ln \|x\|_{\mathbf{d}})x \rangle \leq \frac{\lambda_{\min}(P^{1/2}G_{\mathbf{d}}P^{-1/2} + P^{1/2}G_{\mathbf{d}}P^{-1/2})}{2}$$

and $\|S_h\| = 1$ we conclude the robust stability of the closed-loop system for a sufficiently small δ.

The Input-to-State Stability (ISS) analysis of the closed-loop system follows from Theorem 6.12.

Corollary 8.5 *Let conditions of Theorem 8.1 (or Theorem 8.3) hold. Then the closed-loop system*

$$\dot{x}(t) = Ax(t) + Bu(x(t) + q_1(t)) + q_2(t), \quad t > 0, \tag{8.44}$$

where $q_1, q_2 \in L^\infty(\mathbb{R}; \mathbb{H})$, is

- *ISS with respect to $q = (q_1, q_2)$ provided that the group $\mathbf{d}_q(s) = e^{\mu s}\mathbf{d}(s)$ is a dilation in $\mathbb{H}$;*
- *ISS with respect to q_1 provided that $q_2 = \mathbf{0}$.*

Proof Since the unperturbed system ($q = \mathbf{0}$) is globally uniformly asymptotically stable and $\tilde{\mathbf{d}}(s) = \mathrm{diag}\{\mathbf{d}(s), \mathbf{d}_q(s)\}$ is a dilation in $\mathbb{H} \times \mathbb{H}$ then, by Theorem 6.12, the closed-loop system (8.44) is ISS with respect to q. The ISS with respect to q_1 for $q_2 = \mathbf{0}$ is studied similarly by taking $\tilde{\mathbf{d}}(s) = \mathbf{d}(s)$. The proof is complete. □

8.1.5 On Finite-Dimensional Homogeneous Stabilizer for Infinite-Dimensional System

Let us consider the linear system (8.1) admitting a modal decomposition [18].

Assumption 8.3 Let us assume that

- $A : \mathcal{D}(A) \subset \mathbb{H} \mapsto \mathbb{H}$ is a *modal operator*:

$$Ax = \sum_{i=1}^{\infty} \lambda_i \langle x, \phi_i \rangle \phi_i, \quad \forall x \in \mathcal{D}(A), \tag{8.45}$$

  where $\lambda_i \in \mathbb{R}$ are eigenvalues of A and $\phi_i \in \mathbb{H}$ are eigenvectors of A

$$A\phi_i = \lambda_i \phi_i, \quad i = 1, 2, \ldots \tag{8.46}$$

  such that

$$\lambda_1 \geq \lambda_2 \geq \lambda_3 \geq \cdots$$

  and $\{\phi_i\}_{i=1}^{\infty}$ is an orthonormal basis in $\mathbb{H}$:

$$\langle \phi_i, \phi_i \rangle = 1, \quad \langle \phi_i, \phi_j \rangle = 0, \quad \text{for all} \quad i \neq j;$$

- for any $\delta > 0$ there exists $n > 1$ such that

$$\lambda_i < -\delta, \quad \forall i \geq n; \tag{8.47}$$

- $\mathbb{U} = \mathbb{R}^m$;
- $\sum_{i=1}^{\infty} |b_i^\top|^2 < +\infty$, where

$$b_i = \left(\langle Be_1, \phi_i \rangle \ \langle Be_2, \phi_i \rangle \ \ldots \ \langle Be_m, \phi_i \rangle \right) \in \mathbb{R}^{1\times m} \tag{8.48}$$

and $e_k = (0, \ldots, 0, 1, 0, \ldots, 0)^\top \in \mathbb{R}^m$ is the unit vector in $\mathbb{R}^m$, $k = 1, 2, \ldots, m$.

Under this assumption, the system (8.1) can be decomposed into two subsystems (see, e.g., [18]). The first subsystem is finite-dimensional

$$\frac{d\bar{x}}{dt} = A_n\bar{x} + B_n u, \quad t > 0, \tag{8.49}$$

but possibly unstable, where

$$\bar{x}(t) = \begin{pmatrix} \langle x(t), \phi_1 \rangle \\ \langle x(t), \phi_2 \rangle \\ \ldots \\ \langle x(t), \phi_n \rangle \end{pmatrix} \in \mathbb{R}^n, \ A_n = \mathrm{diag}\{\lambda_1, \ldots, \lambda_n\} \in \mathbb{R}^{n\times n}, \ B_n = \begin{pmatrix} b_1 \\ b_2 \\ \ldots \\ b_n \end{pmatrix} \in \mathbb{R}^{n\times m}.$$

The second subsystem is infinite-dimensional but stable. It has the form

$$\dot{x}_i = \lambda_i x_i + b_i u, \quad t > 0, \quad i > n. \tag{8.50}$$

If $x \in \mathbb{H}$ then, by Parseval's identity (see Appendix A),

$$\|x\|^2 = \sum_{i=1}^{+\infty} x_i^2 < +\infty.$$

On the one hand, this means that the existence/uniqueness of the so-called week solution[2] (see, e.g., Chap. 7) of the system (8.1), (8.2)

$$x_{x_0}(t) = \sum_{i=1}^{+\infty} x_i(t)\phi_i \tag{8.51}$$

follows from the existence/uniqueness of solution $x_i(t), i = 1, 2, \ldots,$ of the system(8.49), (8.50) with

$$\{x_i(t)\}_{i=1}^{\infty} \in \ell^2, \quad \forall t \geq 0 \tag{8.52}$$

[2] Notice that any mild solution is a weak solution(see, e.g., [17, Chap. 2]).

with the initial condition

$$x_i(0) = \langle x_0, \phi_i \rangle, \quad i = 1, 2, \ldots. \tag{8.53}$$

On the other hand, since

$$x_i(t) = e^{\lambda_i t} x_i(0) + \int_0^t e^{\lambda_i (t-\tau)} b_i u(\tau) d\tau$$

then, under Assumption 8.3, for any $u \in L^\infty(\mathbb{R}; \mathbb{R}^m)$ we have

$$\|\{x_i(t)\}_{i=1}^\infty\|_{\ell^2} \leq \|\{|e^{\lambda_i t} x_i(0)|\}_{i=1}^\infty\|_{\ell^2} + \left\| \left\{ \left| \int_0^t e^{\lambda_i (t-\tau)} b_i u(\tau) d\tau \right| \right\}_{i=1}^\infty \right\|_{\ell^2}$$

$$\leq C(t) \left(\|\{|x_i(0)|\}_{i=1}^\infty\|_{\ell^2} + t \left\| \left\{ |b_i^\top| \right\}_{i=1}^\infty \right\|_{\ell^2} \|u\|_{L^\infty((0,t),\mathbb{R}^m)} \right) < +\infty,$$

where

$$C(t) = \max_{i>1, s\in[0,t]} e^{\lambda_i s}.$$

This means that the system (8.1) has a unique weak solution for any $u \in L^\infty(\mathbb{R}; \mathbb{R}^m)$.

Assumption 8.4 Let the pair $\{A_n, B_n\}$ be controllable.

All unstable modes of the system are contained in its finite-dimensional part (8.49), so, using the homogeneous controllers developed in Chap. 9 of Volume I, they can be homogeneously stabilized.

Theorem 8.4 *If Assumptions 8.3, 8.4 be fulfilled and $\delta > 0$, $n \in \mathbb{N}$ are defined as in Assumption 8.3, then*

(1) the linear algebraic equation

$$A_n G_0 - G_0 A_n + B_n Y_0 = A_n, \quad G_0 B_n = \mathbf{0} \tag{8.54}$$

has a solution $Y_0 \in \mathbb{R}^{m\times n}$, $G_0 \in \mathbb{R}^{n\times n}$ and for any solution of (8.54)

- *the matrix*

$$G_{\mathbf{d}} = I_n + \mu G_0 \tag{8.55}$$

  is anti-Hurwitz for $\mu < 1/(\tilde{n} - 1)$, where $\tilde{n} \in \mathbb{N}$ is a minimal number such that $\operatorname{rank}(B_n, A_n B_n, \ldots, A_n^{\tilde{n}-1} B_n) = n$;
- *the matrix $G_0 - I_n$ is invertible and the matrix*

$$A_0 = A_n + B_n Y_0 (G_0 - I_n)^{-1} \tag{8.56}$$

satisfies the identity

$$A_0 G_{\mathbf{d}} = (G_{\mathbf{d}} + \mu I_n) A_0 \quad \text{and} \quad G_{\mathbf{d}} B_n = B_n; \tag{8.57}$$

(2) *the linear algebraic system*

$$\begin{array}{c} A_0 X + X A_0^\top + B_n Y + Y^\top B_n^\top + \rho(G_{\mathbf{d}} X + X G_{\mathbf{d}}^\top) = \mathbf{0}, \\ G_{\mathbf{d}} X + X G_{\mathbf{d}}^\top \succ 0, \quad X = X^\top \succ 0 \end{array} \tag{8.58}$$

has a solution $X \in \mathbb{R}^{n\times n}$, $Y \in \mathbb{R}^{m\times n}$ *for any* $\rho > 0$;

(3) *the finite-dimensional feedback control for the system* (8.1) *can be defined as follows:*

$$u(\bar{x}) = K_0 \bar{x} + \|\bar{x}\|_{\mathbf{d}}^{1+\mu} K \mathbf{d}(-\ln \|\bar{x}\|_{\mathbf{d}}) \bar{x}, \quad \bar{x} = \begin{pmatrix} \langle x, \phi_1 \rangle \\ \langle x, \phi_2 \rangle \\ \dots \\ \langle x, \phi_n \rangle \end{pmatrix}, \tag{8.59}$$

where $\mathbf{d}(s) = e^{sG_{\mathbf{d}}}$, $s \in \mathbb{R}$ *is a linear dilation generated by* $G_{\mathbf{d}}$, *the canonical homogeneous norm* $\|\cdot\|_{\mathbf{d}}$ *is induced by* $\|\bar{x}\| = \sqrt{\bar{x}^\top X^{-1} \bar{x}}$ *and*

$$K_0 = Y_0 (G_0 - I_n)^{-1}, \qquad K = Y X^{-1}; \tag{8.60}$$

(4) *the function* $\bar{x} \mapsto u(\bar{x})$ *is analytic on* $\mathbb{R}^n \backslash \{\mathbf{0}\}$ *for* $\mu \in [-1, 1/(\tilde{n}-1))$ *and* $u \in C(\mathbb{R}^n; \mathbb{R}^m)$ *for* $\mu \in (-1, 1/(\tilde{n}-1))$,

$$|u(\bar{x})| \le |K_0 \bar{x}| + \|K\| \cdot \|\bar{x}\|_{\mathbf{d}}^{1+\mu}, \quad \forall \bar{x} \in \mathbb{R}^n \backslash \{\mathbf{0}\}, \tag{8.61}$$

where $\|K\| := \sqrt{\lambda_{\max}\left(X^{1/2} K^\top K X^{1/2}\right)}$;

(5) *the closed-loop system* (8.1), (8.59) *is globally uniformly asymptotically stable such that*

- *for* $\mu \in [-1, 0)$ *it holds*

$$\begin{pmatrix} \langle x(t), \phi_1 \rangle \\ \langle x(t), \phi_2 \rangle \\ \dots \\ \langle x(t), \phi_n \rangle \end{pmatrix} = \mathbf{0}, \quad \forall t \ge T := \frac{\|\bar{x}_0\|_{\mathbf{d}}^{-\mu}}{-\mu\rho}, \tag{8.62}$$

and there exists $\xi \in \mathcal{K}_\infty$ *such that*

$$\|\{\langle x(t), \phi_i \rangle\}_{i>n}\|_{\ell^2} \le \xi(\|\{\langle x(0), \phi_i \rangle\}_{i\ge 1}\|_{\ell^2}) e^{-\delta t}, \forall t \ge 0, \tag{8.63}$$

where

$$\bar{x}_0 = \begin{pmatrix} \langle x(0), \phi_1 \rangle \\ \langle x(0), \phi_2 \rangle \\ \dots \\ \langle x(0), \phi_n \rangle \end{pmatrix};$$

- *for $\mu = 0$ there exists $C > 0$ (possibly dependent on ρ and δ):*

$$\|\{\langle x(t), \phi_i\rangle\}_{i\geq 1}\|_{\ell^2} \leq C\|\{\langle x(0), \phi_i\rangle\}_{i\geq 1}\|_{\ell^2} e^{-\min\{\delta,\rho\}t}, \quad \forall t \geq 0 \qquad (8.64)$$

provided that $\rho \neq \delta$.

Proof The claimed properties of $\bar{x}$ and the finite-time (resp., exponential) stability for $\mu < 0$ (resp., for $\mu = 0$) of the finite-dimensional subsystem follows from Theorem 9.1 of Volume I of this monograph, where, in particular, it is proven that

$$\tfrac{d}{dt}\|\bar{x}\|_{\mathbf{d}} = \tfrac{\partial\|\bar{x}\|_{\mathbf{d}}}{\partial \bar{x}}(A_n\bar{x} + B_n u(\bar{x})) = -\rho\|\bar{x}\|_{\mathbf{d}}^{1+\mu}, \quad \forall x \neq \mathbf{0}; \qquad (8.65)$$

(1) For $\mu < 0$ the above inequality implies the finite-time stability of the finite-dimensional system (8.49). Moreover, we have

$$x_i(t) = 0, \quad \text{and} \quad u(t) = 0, \quad \forall t \geq T(\bar{x}_0), \quad i = 1, \ldots, n,$$

and

$$x_i(t) = e^{\lambda_i(t-T)}x(T), \quad \forall i > n, \quad \forall t \geq T(\bar{x}_0).$$

Since

$$\|\{x_i(t)\}_{i>n}\|_{\ell^2} \leq \|\{|e^{\lambda_i t}x_i(0)|\}_{i>n}\|_{\ell^2} + \left\|\left\{\left|\int_0^t e^{\lambda_i(t-\tau)}b_i u(\bar{x}(\tau))d\tau\right|\right\}_{i>n}^{\infty}\right\|_{\ell^2}$$

$$\leq e^{-\delta t}\|\{|x_i(0)|\}_{i>n}\|_{\ell^2} + \left\|\{b_i\}_{i>n}^{\infty}\right\|_{\ell^2}\int_0^t e^{-\delta(t-\tau)}\,|u(\bar{x}(\tau))|\,d\tau$$

$$= e^{-\delta t}\left(\|\{|x_i(0)|\}_{i>n}^{\infty}\|_{\ell^2} + \left\|\{b_i\}_{i>n}^{\infty}\right\|_{\ell^2}\int_0^{T(\bar{x}_0)} e^{\delta\tau}\,|u(\bar{x}(\tau))|\,d\tau\right).$$

Taking into account $T = \frac{\|\bar{x}_0\|_{\mathbf{d}}^{-\mu}}{-\mu\rho}$ and the estimate (8.61) we conclude that there exists $\xi \in \mathcal{K}_\infty$ such that

$$\int_0^{T(\bar{x}_0)} e^{\delta\tau}\,|u(\bar{x}(\tau))|\,d\tau \leq \xi(\|\bar{x}_0\|), \quad \forall x_0 \in \mathbb{R}^n.$$

(2) For $\mu = 0$ we have $\|\cdot\|_{\mathbf{d}} = \|\cdot\|$ and the feedback law is a linear stabilizer

$$u(\bar{x}) = (K_0 + K)\bar{x}.$$

In this case, the inequality (8.65) implies

$$\|\bar{x}(t)\| = \|\bar{x}_0\|e^{-\rho t}, \quad \forall t \geq 0.$$

Hence, using the estimate (8.61), we derive

$$\begin{aligned}\|\{x_i(t)\}_{i>n}\|_{\ell^2} \leq & e^{-\delta t}\|\{|x_i(0)|\}_{i>n}^{\infty}\|_{\ell^2} \\ & + \left\|\{b_i\}_{i>n}^{\infty}\right\|_{\ell^2}\int_0^t (\|K_0\|+\|K\|)\|\bar{x}_0\|e^{-\delta(t-\tau)}e^{-\rho\tau}d\tau \\ \leq & e^{-\delta t}\|\{|x_i(0)|\}_{i>n}^{\infty}\|_{\ell^2} + \frac{\|\{b_i\}_{i>n}^{\infty}\|_{\ell^2}(\|K_0\|+\|K\|)\|\bar{x}_0\|(e^{-\rho t}-e^{-\delta t})}{\delta-\rho}.\end{aligned}$$

Taking into account $\rho \neq \delta$ we complete the proof. □

Remark 8.5 The above theorem remains valid if the linear algebraic equation in the system (8.58) is replaced with the LMI:

$$\begin{aligned}A_0X+XA_0^{\top}+B_nY+Y^{\top}B_n^{\top}+\rho(G_{\mathbf{d}}X+XG_{\mathbf{d}}^{\top})\preceq\mathbf{0},\\ G_{\mathbf{d}}X+XG_{\mathbf{d}}^{\top}\succ 0, \quad X=X^{\top}\succ 0.\end{aligned} \tag{8.66}$$

Therefore, a controllable linear evolution equation with a modal operator can be exponentially stabilized with an arbitrary fast decay rate by means of homogeneous ($\mu < 0$) or linear ($\mu = 0$) finite-dimensional feedback laws. The advantages of the homogeneous feedback with negative degree have been studied in Volume I, where, in particular, an essential reduction of the peaking effect has been discovered. This advantage remains valid, at least, for some infinite-dimensional models as shown on the following example.

Example 8.2 *(Partial finite-time stabilization of 1-d reaction–diffusion equation on [0, 1] by homogeneous boundary feedback)* Let us consider the system

$$\frac{\partial x(t,z)}{\partial t} = \frac{\partial^2 x(t,z)}{\partial z^2} + qx(t,z), \tag{8.67}$$

$$x(t,0) = 0, \tag{8.68}$$

$$x(t,1) = U(t), \tag{8.69}$$

$$x(0,z) = x_0(z), \tag{8.70}$$

where $(t,z) \in \mathbb{R}_+ \times [0,1]$, $x(t,\cdot)$ is the state of the reaction–diffusion PDE at the time instant $t > 0$, the function $x_0 \in H_0^1((0,1);\mathbb{R})$ defines the initial state at the time instant $t = 0$, $q > 0$ is the reaction coefficient, $U(t) \in \mathbb{R}$ is the control input. We assume that U satisfies the compatibility condition, i.e.,

$$U(0) = 0. \tag{8.71}$$

Our aim is

- to design the homogeneous partial finite-time stabilizer for the system (8.67)–(8.70) using modal decomposition approach introduced above;
- to investigate if, similarly to finite-dimensional case, the use of the homogeneous feedback results in a smaller peaking comparing with the linear feedback. More precisely, considering the control problem: *given initial state* $x_0 \in L^2((0,1);\mathbb{R})$, *a stabilization precision* $\epsilon > 0$, *a prescribed time* $T > 0$ *and a maximal control amplitude* $\bar{U}$, *the closed-loop system has the desired stabilization precision:*

$$\|x(t,\cdot)\|_{L^2} \le \epsilon, \quad \forall t > T \tag{8.72}$$

with the restricted control magnitude

$$\sup_{t>0} |U(t)| \le \bar{U}, \tag{8.73}$$

we need to investigate if (for some given $x_0 \in L^2(0,1)$) the homogeneous feedback may solve this problem, while linear feedback always violates one of the above conditions.

We consider change of variable (see, e.g., [19, Definition 3.3.2]):

$$w(t,z) = x(t,z) - zU(t) \tag{8.74}$$

and obtain the following equivalent PDE:

$$\frac{\partial w(t,z)}{\partial t} = \frac{\partial^2 w(t,z)}{\partial z^2} + qw(t,z) - z\dot{U}(t) + qzU(t), \tag{8.75}$$

$$w(t,0) = 0, \tag{8.76}$$

$$w(t,1) = 0, \tag{8.77}$$

$$w(0,x) = x_0(z), \tag{8.78}$$

taking into the identity

$$w(0,z) = x_0(z) - zU(0) = x_0(z). \tag{8.79}$$

By introducing a new control input $u(t) - \dot{U}(t)$, one obtains a coupled ODE–PDE system:

$$\dot{U}(t) = u(t), \tag{8.80}$$

$$\frac{\partial w(t,z)}{\partial t} = \frac{\partial^2 w(t,z)}{\partial z^2} + qw(t,z) - zu(t) + qzU(t), \tag{8.81}$$

$$w(t,0)=0, \tag{8.82}$$
$$w(t,1)=0, \tag{8.83}$$
$$w(0,z)=x_0(z). \tag{8.84}$$

The solution to (8.82) can be represented as follows:

$$w(t,z)=\sum_{i=1}^{\infty} w_i(t)\phi_i(z), \quad w_i(t)=\langle w(t,\cdot),\phi_i\rangle \tag{8.85}$$

with $\phi_i(z)=\sqrt{2}\sin(i\pi z)$. Projecting onto the basis $\{\phi_i\}_{i=1}^{\infty}$, one has

$$\dot{U}(t)=u(t), \tag{8.86}$$
$$\dot{w}_i(t)=(q-i^2\pi^2)w_i(t)+b_iu(t)-qb_iU(t), \tag{8.87}$$
$$w_i(0)=\langle x_0,\phi_i\rangle, \tag{8.88}$$

where $-i^2\pi^2$ is the eigenvalue corresponding to the eigenvector ϕ_i of the the operator $\frac{\partial^2}{\partial z^2}$ with the domain $\mathcal{D}\left(\frac{\partial^2}{\partial z^2}\right)=H^2(0,1)\cap H_0^1(1,0)$,

$$\tilde{b}_i=-\int_0^1 z\phi_i(z)dz, \quad i\geq 1. \tag{8.89}$$

Since $\lambda_i\to-\infty$ when $i\to+\infty$, then for any $\delta>0$ there exists an integer $n\geq 1$ such that

$$\lambda_i:=q-i\pi^2<-\delta, \quad \forall i>n. \tag{8.90}$$

Denoting

$$x_i=w_i+\tfrac{q\tilde{b}_i}{\lambda_i}U, \quad \text{for} \quad i\geq 1, \tag{8.91}$$

we derive the system

$$\dot{x}_i(t)=\lambda_i x_i(t)+b_iu(t), \tag{8.92}$$

where

$$b_i=\tilde{b}_i+\tfrac{q\tilde{b}_i}{\lambda_i}. \tag{8.93}$$

Therefore, for a properly selected $n\geq 1$, the finite-dimensional subsystem

$$\frac{d\bar{x}(t)}{dt}=A_n\bar{x}(t)+B_nu(t), \tag{8.94}$$

with $\bar{x} = (U, x_1, \ldots, x_n)$,

$$A_n = \mathrm{diag}\{\lambda_0, \lambda_1, \ldots, \lambda_N\}, \quad \lambda_0 = 0, \tag{8.95}$$

$$B_n = (b_0, b_1, \ldots, b_N)^\top, \; b_0 = -1, \tag{8.96}$$

contains all unstable modes of the infinite-dimensional system (8.80)–(8.84).

The pair $\{A_n, B_n\}$ is controllable [20]. In the view of Theorem 8.4, we can homogeneously stabilize the subsystem (8.94) in a finite time such that the whole system is exponentially stable with the decay rate $\delta > 0$. Let us compare linear and homogeneous stabilizers.

Numerical simulations. Let $q = 1.1\pi^2$. In this case, it is sufficient to stabilize only one mode to ensure the exponential stabilization of the reaction–diffusion equation. Therefore, we can select $n = 1$. In this case, we have $A = \mathrm{diag}\{0, -8.8826\})$ and $B = (-1, 5.4019)^\top$.

The homogeneous stabilizer is designed for $\mu = -0.5$. The generator of the dilation in this case is

$$G_{\mathbf{d}} = \begin{pmatrix} 1.5 & 0.0926 \\ 0 & 1 \end{pmatrix}.$$

The gain of the homogenizing linear feedback is given by

$$K_0 = (0, -0.1827).$$

The gain of the homogeneous stabilizer is defined as follows:

$$K = (-29.6050, -7.0340).$$

For comparison reasons, we also consider a linear stabilizer

$$u_{\mathrm{lin}} = K_{\mathrm{lin}}\bar{x}, \quad \bar{x} = (U, x_1)^\top$$

with the gain

$$K_{\mathrm{lin}} = (-32.1603, -7.7406).$$

The modes with the numbers $i = 2, 3, \ldots$ are exponentially stable. We compute an approximate solution of the closed-loop system by numerical simulation of 10 modes:

$$x(t, z) \approx zU(t) + \sum_{i=1}^{10} \left(x_i - \frac{q\tilde{b}_i}{\lambda_i} \right) \sqrt{2} \sin(i\pi z), \quad t \geq 0, \quad z \in [0, 1],$$

where b_i is given by the formula (8.89) and $\lambda_i := q - i\pi^2$. Since $\phi_i(z) = \sqrt{2}\sin(i\pi z)$ is an orthonormal basis in $L^2((0, 1); \mathbb{R})$ then

$$\|x(t, \cdot)\|_{L^2} \approx \sqrt{\sum_{i=1}^{10} \left(x_i(t) - U(t)\frac{q\tilde{b}_i}{\lambda_i} - U(t)\tilde{b}_i \right)^2} = \sqrt{\sum_{i=1}^{10} (x_i(t) - U(t)b_i)^2},$$

where $\tilde{b}_i$ is given by the formula (8.93). The plots for the L^2-norm of the state are given in Fig. 8.2.

The gain of the linear stabilizer is tuned such that the homogeneous controller and linear controller have the same overshoot during the transient. The homogeneous controller stabilizes $(U(t), x_1(t))$ to zero in a finite time $T(\bar{x}_0) \approx 0.9$ (see Fig. 8.3). However, at the time instant $t = 1$, the homogeneous control has much better stabilization precision than the linear controller. To improve the precision, the gain of the linear controller has to be increased. We consider also the higher-gain linear controller (with $K_{\text{lin}} = (-76.2875, -16.6363)$), which has a similar stabilization precision with the homogeneous controller. However, the higher-gain linear stabilizer has a larger overshoot (see Fig. 8.4), so the physical restrictions maximal magnitude of U may not allow to implement such a controller in practice. Notice also the precision of the higher-gain linear controller is still worse than the precision of the homogeneous control. The further increase of the linear feedback gain increase the overshoot drastically.

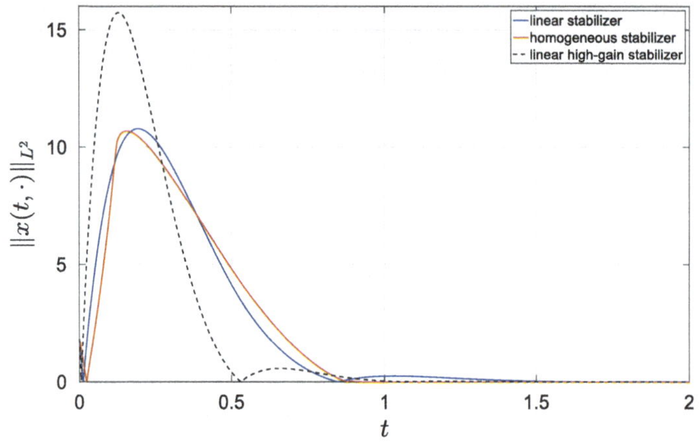

Fig. 8.2 The L^2-norms of solutions with homogeneous and linear stabilizers

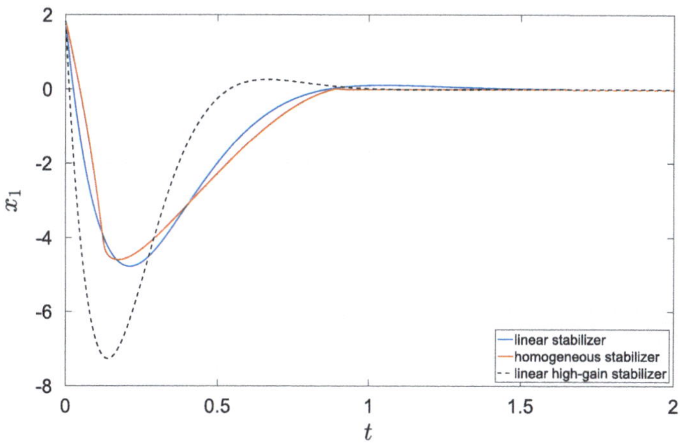

Fig. 8.3 The coordinate x_1 for the system with homogeneous and linear stabilizers

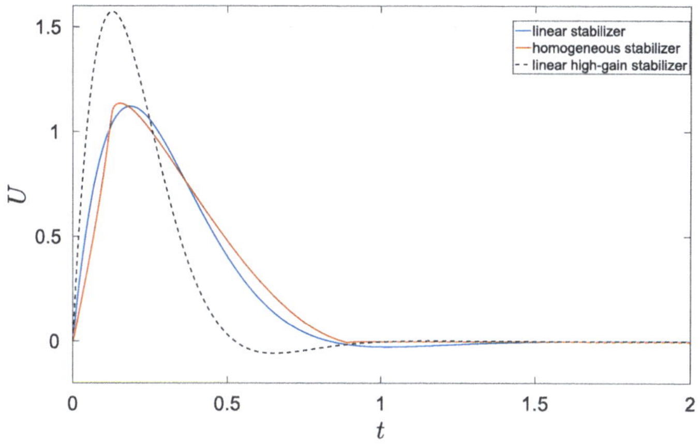

Fig. 8.4 The control signal U for the system with homogeneous and linear stabilizers

8.2 Stabilization of Nonlinear Plant

8.2.1 Static Homogeneous Feedback for a Homogeneous Nonlinear Plant

Let us consider the following nonlinear evolution control system

$$\dot{x} = Ax + f(x, u), \quad t > 0, \quad x(0) = x_0, \tag{8.97}$$

where, as before, x and u are the system state and control, respectively, $A : \mathcal{D}(A) \subset \mathbb{H} \mapsto \mathbb{H}$ is a (possibly) unbounded linear operator, $f : \mathbb{H} \times \mathbb{U} \mapsto \mathbb{H}$ is a nonlinear operator and $\mathbb{H}$ (resp., $\mathbb{U}$) is a real Hilbert (resp., Banach) space.

Theorem 8.5 *Let A be a generator of a strongly continuous semigroup of linear bounded operators on $\mathbb{H}$ and f be locally Lipschitz continuous on $\mathbb{H} \times \mathbb{U}$ and*

(a) $\mathbf{d}_1$ be a uniformly continuous group of linear bounded operators in $\mathbb{H}$ with a generator $G_{\mathbf{d}_1} \in \mathcal{L}(\mathbb{H}; \mathbb{H})$;
(b) $\mathbf{d}$ be a strongly continuous strictly monotone dilation in $\mathbb{H}$ such that $\mathcal{D}(A) \subset \mathcal{D}(G_{\mathbf{d}})$, where $G_{\mathbf{d}} : \mathcal{D}(G_{\mathbf{d}}) \subset \mathbb{H} \mapsto \mathbb{H}$ is a generator of the dilation group $\mathbf{d}$;
(c) there exist a linear bounded operator $K : \mathbb{H} \mapsto \mathbb{U}$ such that

$$\exists \rho > 0 : \langle (Az + f(z, Kz) + \gamma G_{\mathbf{d}})z, z\rangle \leq 0, \quad \forall z \in S \cap \mathcal{D}, \tag{8.98}$$

where $\mathcal{D} \subset \mathcal{D}(A) \cap \mathcal{D}(G_{\mathbf{d}})$ is a $\mathbf{d}$-homogeneous cone dense in $\mathbb{H}$;
(d) the operator A be $\mathbf{d}$-homogeneous of degree $\mu \in \mathbb{R}$ with $\mu > -\beta$ and

$$f(\mathbf{d}(s)x, K\mathbf{d}(s)x) = e^{\mu s}\mathbf{d}(s) f(x, e^{-\mu s} K\mathbf{d}_1(s)x), \quad \forall s \in \mathbb{R}, \forall x \in \mathcal{D}(A)\backslash\{\mathbf{0}\}.$$

Then all conclusions (except the existence of classical solutions) of Theorem 8.1 hold for the closed-loop system (8.97), (8.4).

Proof The properties of local Lipschitz continuityon $\mathbb{H}\backslash\{\mathbf{0}\}$ and differentiability on $\mathcal{D}(G_{\mathbf{d}})\backslash\{\mathbf{0}\}$ of the control law (8.4) are proven in Theorem 8.1. The local Lipschitz continuity of f implies local Lipschitz continuity of $f(\cdot, u(\cdot)) : \mathbb{H}\backslash\{\mathbf{0}\} \mapsto \mathbb{H}$. From Theorem 2.15 we conclude that the closed-loop nonlinear system has mild solutions, which exist on $\mathbb{H}\backslash\{\mathbf{0}\}$. The existence of locally Lipschitz continuous strong solution for $x_0 \in \mathcal{D}(A)$ follows from Theorem 2.16. In this case, taking into account Condition (d) we derive

$$\|x\|^{\mu} f(\mathbf{d}(-\ln \|x\|_{\mathbf{d}})x, K\mathbf{d}(-\ln \|x\|_{\mathbf{d}})x) = \|x\|^{\mu}(A(z) + f(z, Kz)),$$

where $z = \mathbf{d}(-\ln \|x\|_{\mathbf{d}})x \in S$. Hence, for $\mathbf{0} \neq x \in \mathcal{D}$ we have

$$\begin{aligned} D\|x\|_{\mathbf{d}}(Ax + f(x, u(x))) &= \frac{\|x\|_{\mathbf{d}}\langle \mathbf{d}(-\ln \|x\|_{\mathbf{d}})(Ax + f(x, u(x)), \mathbf{d}(-\ln \|x\|_{\mathbf{d}})x\rangle}{\langle G_{\mathbf{d}}\mathbf{d}(-\ln \|x\|_{\mathbf{d}})x, \mathbf{d}(-\ln \|x\|_{\mathbf{d}})x\rangle} \\ &= \frac{\|x\|_{\mathbf{d}}\langle \mathbf{d}(-\ln \|x\|_{\mathbf{d}})(Ax + f(x, u(x)), \mathbf{d}(-\ln \|x\|_{\mathbf{d}})x\rangle}{\langle G_{\mathbf{d}}\mathbf{d}(-\ln \|x\|_{\mathbf{d}})x, \mathbf{d}(-\ln \|x\|_{\mathbf{d}})x\rangle} \\ &= \frac{\|x\|_{\mathbf{d}}^{1+\mu}\langle (Az + f(z, Kz)), z\rangle}{\langle G_{\mathbf{d}}z, z\rangle} \leq -\rho\|x\|_{\mathbf{d}}^{1+\mu}, \end{aligned}$$

where Condition (c) is utilized in the last step. The further constructions repeat the proof of Theorem 8.1. The proof is complete. □

Notice that if f is independent of the first argument and linear with respect to the second one then Condition (d) of Theorem 8.5 becomes Condition (D) of Theorem 8.1. Therefore, the control design for the nonlinear (8.97) is essentially inspired by Theorem 8.1. For nonlinear operators f, which admit homogeneous convex embedding (see Chap. 6 of Volume I) the more constructive stabilization conditions can be obtained.

Corollary 8.6 *Let $A : \mathcal{D}(A) \subset \mathbb{H} \mapsto \mathbb{H}$ be a generator of a strongly continuous semigroup of linear bounded operators on $\mathbb{H}$ and $\mathbf{d}$ be a strongly continuous strictly monotone dilation in $\mathbb{H}$ such that $\mathcal{D}(A) \subset \mathcal{D}(G_{\mathbf{d}})$, where $G_{\mathbf{d}} : \mathcal{D}(G_{\mathbf{d}}) \subset \mathbb{H} \mapsto \mathbb{H}$ is a generator of the dilation group $\mathbf{d}$. Let $\mathbb{U} = \mathbb{U}_1 \times \mathbb{U}_2 \times \cdots \times \mathbb{U}_N$ be a Banach space. Let the operator f admits the following representation*

$$f(x,u)=\|x\|_{\mathbf{d}}^{\mu}\mathbf{d}(\ln\|x\|_{\mathbf{d}})\sum_{i=1}^{N}\alpha_i(x)(A_i\mathbf{d}(-\ln\|x\|_{\mathbf{d}})x+B_iu_i),\quad x\neq\mathbf{0},\ u\neq\mathbf{0} \tag{8.99}$$

where $\mu \in \mathbb{R}$, $N \in \mathbb{N}$, $u = (u_1, u_2, \ldots, u_n) \in \mathbb{U}$, $A_i \in \mathcal{L}(\mathbb{H};\mathbb{H})$, $B_i \in \mathcal{L}(\mathbb{U}_i;\mathbb{H})$ and $\alpha_i : \mathbb{H}\backslash\{\mathbf{0}\} \mapsto [0,+\infty)$ are nonlinear locally Lipschitz continuous functionals such that

$$\sum_{i=1}^{N}\alpha_i(x)=1,\quad \forall x\neq\mathbf{0}.$$

Let $f(\mathbf{0},\mathbf{0}) = \mathbf{0}$. If the operator A is $\mathbf{d}$-homogeneous of the degree μ and there exists $K_i \in \mathcal{L}(\mathbb{H};\mathbb{U}_i)$ such that

$$\exists\rho>0\ :\ \langle(A+A_i+B_iK_i+\rho G_{\mathbf{d}})x,x\rangle\leq 0,\quad \forall x\in\mathcal{D},\forall i=1,\ldots,N, \tag{8.100}$$

where $\mathcal{D} \subset (\mathcal{D}(A) \cap \mathcal{D}(G_{\mathbf{d}}))$ is a $\mathbf{d}$-homogeneous cone dense in $\mathbb{H}$, then the feedback control $u : \mathbb{H} \mapsto \mathbb{U}$ given by $u_i(\mathbf{0}) = \mathbf{0}$ and

$$u_i(x)=K_i\mathbf{d}(-\ln\|x\|_{\mathbf{d}})x,\quad i=1,2,\ldots,N \tag{8.101}$$

is globally bounded, locally Lipschitz continuous on $\mathbb{H}\backslash\{\mathbf{0}\}$ and Frechét differentiable on $\mathcal{D}(G_{\mathbf{d}})\backslash\{\mathbf{0}\}$, the closed-loop system (8.97), (8.101)

- *is $\mathbf{d}$-homogeneous of the degree μ;*
- *has a unique mild (strong) solution defined on $[0,+\infty)$ for any $x_0 \in \mathbb{H}$ (resp., $x_0 \in \mathcal{D}(A)$);*
- *is globally uniformly finite-time for $\mu < 0$;*

$$x(t)=\mathbf{0},\quad \forall t\geq\frac{\|x_0\|_{\mathbf{d}}^{-\mu}}{-\mu\rho},$$

- *is globally uniformly exponentially for* $\mu = 0$*:*

$$\|x(t)\|_{\mathbf{d}} \leq e^{-\rho t}\|x_0\|_{\mathbf{d}}, \quad \forall t \geq 0,$$

- *is globally uniformly nearly fixed-time stable for* $\mu > 0$*:*

$$\|x(t)\|_{\mathbf{d}} \leq r, \quad \forall t \geq \tfrac{1}{\rho r^{\mu}}, \quad \forall r > 0, \quad \forall x_0 \in \mathbb{H}.$$

Proof Since the function $x \mapsto f(x, u(x))$ is locally Lipschitz continuous then from Theorems 8.1 and 2.16 we conclude a local-in-time existence of mild and locally Lipschitz continuous strong solutions of the closed-loop system for any $x_0 \in \mathbb{H}\backslash\{\mathbf{0}\}$ and $x_0 \in \mathcal{D}(A)\backslash\{\mathbf{0}\}$, respectively. For $\mathbf{0} \neq x \in \mathcal{D}$ we derive

$$\begin{aligned} D\|x\|_{\mathbf{d}}(Ax+f(x,u(x))) &= \frac{\|x\|_{\mathbf{d}}\langle \mathbf{d}(-\ln\|x\|_{\mathbf{d}})(Ax+f(x,u(x))), \mathbf{d}(-\ln\|x\|_{\mathbf{d}})x\rangle}{\langle G_{\mathbf{d}}\mathbf{d}(-\ln\|x\|_{\mathbf{d}})x, \mathbf{d}(-\ln\|x\|_{\mathbf{d}})x\rangle} \\ &= \frac{\|x(t)\|_{\mathbf{d}}\langle \mathbf{d}(-\ln\|x\|_{\mathbf{d}})Ax, \mathbf{d}(-\ln\|x\|_{\mathbf{d}})x\rangle}{\langle G_{\mathbf{d}}\mathbf{d}(-\ln\|x\|_{\mathbf{d}})x, \mathbf{d}(-\ln\|x\|_{\mathbf{d}})x\rangle} \\ &+ \frac{\sum_{i=1}^{N}\alpha_i(x)\|x\|_{\mathbf{d}}^{1+\mu}\langle A_i\mathbf{d}(-\ln\|x\|_{\mathbf{d}})x+B_iu_i(x), \mathbf{d}(-\ln\|x\|_{\mathbf{d}})x\rangle}{\langle G_{\mathbf{d}}\mathbf{d}(-\ln\|x\|_{\mathbf{d}})x, \mathbf{d}(-\ln\|x\|_{\mathbf{d}})x\rangle} \end{aligned}$$

and, taking into account the identity $\sum_{i=1}^{N}\alpha_i(x) = 1$, we conclude

$$D\|x\|_{\mathbf{d}}(Ax + f(x,u(x))) = \|x\|_{\mathbf{d}}^{1+\nu}\frac{\sum_{i=1}^{N}\alpha_i(x)\langle (A+A_i+BK_i)\mathbf{d}(-\ln\|x\|_{\mathbf{d}})x, \mathbf{d}(-\ln\|x\|_{\mathbf{d}})x\rangle}{\langle G_{\mathbf{d}}\mathbf{d}(-\ln\|x\|_{\mathbf{d}})x, \mathbf{d}(-\ln\|x\|_{\mathbf{d}})x\rangle}$$

for $\mathbf{0} \neq x \in \mathcal{D}$. Since $\mathcal{D}$ is dense in $\mathcal{D}(A)$ then, using stability condition (8.100), for any strongly solution x we have

$$\tfrac{d\|x(t)\|_{\mathbf{d}}}{dt} \overset{a.e.}{\leq} -\rho\|x(t)\|_{\mathbf{d}}^{1+\mu} \quad \text{if} \quad x(t) \neq \mathbf{0}.$$

This means that the locally Lipschitz continuous function $t \mapsto \|x(t)\|_{\mathbf{d}}$ decays monotonically to zero as the time is growing. The same conclusion can be established for mild solutions, which are the uniform limits of strong solutions. This implies the global uniform asymptotic stability of the origin of the closed-loop system and the uniqueness of solutions prolonged zero in the case of a finite-time reaching of the origin. Since the closed-loop system is homogeneous by construction, then, by Theorem 6.6, the homogeneity degree specifies the convergence rate of the asymptotically stable system. The estimates of the settling times follow from the differential inequality $\frac{d\|x(t)\|_{\mathbf{d}}}{dt} \overset{a.e.}{\leq} -\rho\|x(t)\|_{\mathbf{d}}^{1+\mu}$. The proof is complete. □

The stabilization condition (8.100) is a linear operator inequality, which may be represented by LMIs in the case of PDE models (see Sect. 8.1.2).

8.2.2 Homogeneous Affine-in-Control System

Let us consider the affine control system

$$\dot{x} = Ax + f_0(x) + \sum_{i=1}^{m} u_i f_i(x), \quad t > 0, \tag{8.102}$$

where $x(t) \in \mathbb{H}$ and $u(t) = (u_1(t), \ldots, u_m(t) \in \mathbb{R}^m$ are the system state and the control, respectively, $A : \mathcal{D}(A) \subset \mathbb{H} \mapsto \mathbb{H}$ is a linear possibly unbounded operator, $f_i : \mathbb{H}\backslash\{\mathbf{0}\} \mapsto \mathbb{H}$ is a nonlinear locally bounded operator, $i = 1, 2, \ldots, m$ and $\mathbb{H}$ is a real Hilbert space.

The following theorem presents an analog of Sontag's universal controller [21] for the generalized homogeneous evolution equation (8.102).

Theorem 8.6 *Let the operators f_i be locally Lipschitz continuous on $\mathbb{H}\backslash\{\mathbf{0}\}$ such that $f_i(\mathbf{0}) = 0$ for all $i = 0, 1, \ldots, m$ and A be a generator of a strongly continuous semigroup Φ of linear bounded operators on $\mathbb{H}$. If*

(a) $\mathbf{d}$ is a strongly continuous strictly monotone dilation in $\mathbb{H}$ such that $\mathcal{D}(A) \subset \mathcal{D}(G_{\mathbf{d}})$, where $G_{\mathbf{d}} : \mathcal{D}(G_{\mathbf{d}}) \subset \mathbb{H} \mapsto \mathbb{H}$ is a generator of the dilation group $\mathbf{d}$;

(b) the operators f_i are $\mathbf{d}$-homogeneous of degrees $\mu_i, i = 0, 1, \ldots, m$, respectively, and the operator A be $\mathbf{d}$-homogeneous of the degree μ_0;

(c) there exist $\rho > 0$ and $\delta \in \mathbb{R}$ such that

$$\sup_{z\in\mathcal{D}\cap S} \langle (A + \rho G_{\mathbf{d}}), z\rangle \leq \delta, \tag{8.103}$$

and

$$z \in S, \langle f_i(z), z\rangle = 0, \forall i = 1, \ldots, m \quad \Rightarrow \quad \langle f_0(z), z\rangle < -\delta, \tag{8.104}$$

where $\mathcal{D} \subset (\mathcal{D}(A) \cap \mathcal{D}(G_{\mathbf{d}}))$ is a $\mathbf{d}$-homogeneous cone dense in $\mathbb{H}$ and $S \in \mathbb{H}$ is the unit sphere, then

- *the functionals $u_i : \mathbb{H}\backslash\{\mathbf{0}\} \mapsto \mathbb{R}$ given by $u_i(\mathbf{0}) = \mathbf{0}$ and*

$$u_i(x) = -\|x\|_{\mathbf{d}}^{\mu_0-\mu_i} \frac{a(z)+\delta+\sqrt{(a(z)+\delta)^2+\left(b^\top(z)b(z)\right)^2}}{b^\top(z)b(z)} b_i(z), \quad x \neq \mathbf{0} \tag{8.105}$$

are locally Lipschitz continuous on $\mathbb{H}\backslash\{\mathbf{0}\}$ and $\mathbf{d}$-homogeneous of the degree $\mu_0 - \mu_i$, where

$$z = \mathbf{d}(-\ln \|x\|_{\mathbf{d}})x, \quad b = (b_1, \ldots, b_m)^\top, \quad b_i(z) = \langle f_i(z), z\rangle, \quad a(z) = \langle f_0(z), z\rangle;$$

- *then the closed-loop system* (8.102)–(8.105)
  - *has a unique strong solution for any initial condition* $x(0)=x_0\in\mathcal{D}(A)$;
  - *has a unique mild solution for any initial condition* $x(0)=x_0\in\mathbb{H}$
  - *is* **d**-*homogeneous of the degree* μ_0;
  - *is globally uniformly finite-time stable for* $\mu_0<0$:

$$x(t)=\mathbf{0},\quad \forall t\geq \frac{\|x_0\|_{\mathbf{d}}^{-\mu_0}}{-\mu_0\rho},$$

  - *is globally uniformly exponentially stable for* $\mu_0=0$:

$$\|x(t)\|_{\mathbf{d}}\leq e^{-\rho t}\|x_0\|_{\mathbf{d}},\quad \forall t\geq 0,$$

  - *is globally uniformly nearly fixed-time stable for* $\mu>0$:

$$\|x(t)\|_{\mathbf{d}}\leq r,\quad \forall t\geq \frac{1}{\rho r^{\mu_0}},\quad \forall r>0,\quad \forall x_0\in\mathbb{H}.$$

Proof First, let us recall (see [21] for more details) that the function $(a,b)\mapsto \frac{a+\sqrt{a^2+b^4}}{b^2}$ is real analytic on the domain $Q=\{(a,b)\in\mathbb{R}^2: a<0 \text{ or } b\neq 0\}$. The condition (8.104) implies that the functions $z\mapsto (a(z)+\delta, |b(z)|)$ maps $S\subset\mathbb{H}$ into Q. Hence, the local Lipschitz continuity of f_i and $\|\cdot\|_{\mathbf{d}}$ on $\mathbb{H}\backslash\{\mathbf{0}\}$ (see, Lemma 4.5) yield the local Lipschitz continuity of the feedbacks u_i on $\mathbb{H}\backslash\{\mathbf{0}\}$. In this case, the existence of mild solutions of the closed-loop system follows from Corollary 2.13. The existence of locally Lipschitz continuous strong solutions of the closed-loop system (8.102), (8.105) with $x(0)\in\mathcal{D}(A)\backslash\{\mathbf{0}\}$ follows from Theorem 2.16. By construction, the closed-loop system is **d**-homogeneous of a degree ν_0. If $x\in\mathcal{D}$ then

$$D\|x\|_{\mathbf{d}}\left(Ax+f_0(x)+\sum_{i=1}^{m}u_i f_i(x)\right)$$

$$=-\rho|x\|_{\mathbf{d}}^{1+\mu_0}+\|x\|_{\mathbf{d}}^{1+\mu_0}\frac{+\langle Az+\gamma G_{\mathbf{d}}z,z\rangle+\langle f_0 z,z\rangle+\sum_{i=1}^{n}u_i(z)\langle f_i(z),z\rangle}{\langle G_{\mathbf{d}}z,z\rangle},$$

where $z=\mathbf{d}(-\ln\|x\|_{\mathbf{d}})x$ and $x\neq\mathbf{0}$. Using (8.105) we derive

$$\sum_{i=1}^{n}u_i(z)\langle f_i(z),z\rangle=-\frac{a(z)+\delta+\sqrt{(a(z)+\delta)^2+\left(b^\top(z)b(z)\right)^2}}{b^\top(z)b(z)}\sum_{i=1}^{n}b_i(z)\langle f_i(z),z\rangle=$$

$$-a(z)-\delta-\sqrt{(a(z)+\delta)^2+\left(b^\top(z)b(z)\right)^2}$$

and from (8.103) we conclude

$$D\|x\|_{\mathbf{d}}\left(Ax+f_0(x)+\sum_{i=1}^{m}u_i f_i(x)\right)\leq\|x\|_{\mathbf{d}}^{1+\mu_0}\frac{-\rho\langle G_{\mathbf{d}}z,z\rangle-\sqrt{(a(z)+\delta)^2+\left(b^{\top}(z)b(z)\right)^2}}{\langle G_{\mathbf{d}}z,z\rangle}$$
$$\leq-\rho\|x\|_{\mathbf{d}}^{1+\mu_0}$$

for $\mathbf{0}\neq x\in\mathcal{D}$. Since $\mathcal{D}$ is dense in $\mathcal{D}(A)$ then for any strong solution x we have

$$\frac{d}{dt}\|x(t)\|_{\mathbf{d}}\overset{a.e.}{\leq}-\rho\|x(t)\|_{\mathbf{d}}^{1+\mu_0}\quad\text{if}\quad x(t)\neq\mathbf{0}.$$

Hence, we conclude conclude that the locally Lipschitz continuous functions $t\mapsto\|x(t)\|_{\mathbf{d}}$ decrease monotonically to zero along any strong solution of the closed-loop system. Taking into account continuous dependence of solutions on initial conditions (see Corollary 2.14) we prove the same behavior for all mild solutions, i.e., the origin of the closed-loop system is globally uniformly asymptotically stable. The estimates of the settling times follow from the differential inequality $\frac{d\|x(t)\|_{\mathbf{d}}}{dt}\overset{a.e.}{\leq}-\rho\|x(t)\|_{\mathbf{d}}^{1+\mu_0}$. Using Theorem 6.6 we complete the proof. □

Notice that the above theorem does not guarantee the continuity and boundedness of the control functionals u_i at $\mathbf{0}$. The corresponding properties depend on μ_i and can be studied using Proposition 5.1. Namely, if $\mu_0>\mu_i$ then u_i is continuous at $x=\mathbf{0}$.

Remark 8.6 The presented result can be combined with the homogeneous domination approach in order to stabilize locally a nonlinear evolution system which admits a $\mathbf{d}$-homogeneous approximation at zero in the form (8.102).

Concluding Remarks

In this chapter, the homogeneous stabilizers for infinite-dimensional plants modeled by linear and nonlinear evolution equations are designed. The design procedures are similar to the finite-dimensional systems. In the linear case, the control law is obtained by homogeneous scaling of a linear stabilizer. The control parameters are chosen by solving linear operator inequalities, which are equivalent to LMIs discovered for the finite-dimensional stabilization. Moreover, these operator inequalities may reformulated as LMIs if the plant model is described by some "homogenizable" differential operator. Robustness (Input-to-State Stability) of the designed control with respect to additive perturbations and noisy measurements is proven. Possible extensions of the homogeneous control design methodology to nonlinear infinite-dimensional systems are discussed. Stability conditions in terms of linear operator inequalities are obtained for nonlinear plants admitting a homogeneous convex embedding. An analog of the Sontag's universal controller is designed for an infinite-dimensional homogeneous affine-in-control system.

References

1. Kalman R (1960) On the general theory of control systems. In: Proceedings of the first international congress of the IFAC, Moscow
2. Wonham WM (1985) Linear multivariable control: a geometric approach. Springer
3. Boyd S, Ghaoui E, Feron E, Balakrishnan V (1994) Linear matrix inequalities in system and control theory. SIAM, Philadelphia
4. Tucsnak M, Weiss G (2009) Observation and control for operator semigroups. Birkhauser
5. Cheremensky A, Fomin V (2013) Operator approach to linear control systems. Springer, B.V
6. Korobov VI (2007) Method of controlability functions. NIC, Moscow (in Russian)
7. Perrollaz V, Rosier L (2014) Finite-time stabilization of 2x2 hyperbolic systems on tree-shaped networks. SIAM J Control Optim 52(1):143–163
8. Coron J-M, Hu L, Olive G (2016) Stabilization and controllability of first-order integro-differential hyperbolic equations. J Funct Anal 271(12):3554–3587
9. Coron J-M, Nguyen H-M (2017) Null controllability and finite time stabilization for the heat equations with variable coefficients in space in one dimension via backstepping approach. Archive for rational mechanics and analysis. https://doi.org/10.1007/s00205-017-1119-y
10. Hu L, Di Meglio F (2015) Finite-time backstepping boundary stabilization of 3x3 hyperbolic systems. In: European control conference (ECC), pp 67–73
11. Polyakov A, Coron J-M, Rosier L (2018) On homogeneous finite-time control for linear evolution equation in hilbert space. IEEE Trans Autom Control 63(9):3143–3150
12. Orlov Y (1983) Application of Lyapunov method in distributed systems. Autom Remote Control 44:426–430
13. Pisano A, Orlov Y, Usai E (2011) Tracking control of the uncertain heat and wave equation via power-fractional and sliding-mode techniques. SIAM J Control Optim 49(2):363–382
14. Polyakov A, Efimov D, Perruquetti W (2015) Finite-time and fixed-time stabilization: Implicit Lyapunov function approach. Automatica 51(1):332–340
15. Liberzon D (2003) Switchings in systems and control. Birkhauser, Boston
16. Zimenko K, Polyakov A, Efimov D, Perruquetti W (2020) Robust feedback stabilization of linear mimo systems using generalized homogenization. IEEE Trans Autom Control 65(12):5429–5436
17. Pazy A (1983) Semigroups of linear operators and applications to partial differential equations. Springer
18. Balas M (1988) Finite-dimensional controllers for linear distributed parameter systems: exponential stability using residual mode filters. J Math Anal Appl 133:283–296
19. Curtain RF, Zwart H (2012) An introduction to infinite-dimensional linear systems theory. Springer Science & Business
20. Prieur C, Trelat E (2019) Feedback stabilization of a 1-d linear reaction-diffusion equation with delay boundary control. IEEE Trans Autom Control 64(4):1415–1425
21. Sontag ED (1989) A 'universal' construction of artstein's theorem on nonlinear stabilization. Syst & Control Lett 13:117–123

Appendix

This chapter briefly surveys some very basic notions and results of set and operator theory, Banach and Hilbert spaces, functional analysis, measure theory and integration. We refer the reader, for example, to [1–8] for more deals about presented results.

A Elements of Set Theory and Spaces

A *set* Ω is a collection of distinct objects (elements or members of the set). Examples of mathematical sets are sets of numbers, sets of functions, sets of sets, etc. If an element a belongs to a set Ω we write $a \in \Omega$. A set of elements $a, b, c, \ldots$ is denoted $\{a, b, \ldots\}$. If all elements of a set A belong to a set Ω we use the notation $A \subset \Omega$. If a set has no elements, it is called *empty set* and denoted $\emptyset$. A set is countable *countable set* if all its members can be numbered as by natural numbers $1, 2, \ldots$. A countable set may have infinite number of elements. The power set of X is denoted 2^X is defined as a set of all possible subsets of X.

The *Cartesian product* of sets A and B is a set denoted $A \times B$ and defined as a set of pairs of elements from A and B, respectively,

$$A \times B = \{(a, b) : a \in A, b \in B\}.$$

The set $A \cap B$ is called the *intersection of sets* A and B and defined as a set of elements belonging simultaneously to A and B. If the sets A and B have no common elements then $A \cap B = \emptyset$.

The set $A \cup B$ is called the *union of sets* A and B and defined as a set of elements belonging, at least, to one of sets A, B.

The set $A \backslash B$ is called *complement* (or *set-theoretic difference*) of sets A and B and defined as a set of elements that are members of A but not members of B.

A. Polyakov, *Generalized Homogeneity in Systems and Control Volume II*, Communications and Control Engineering,
https://doi.org/10.1007/978-3-031-77496-6

Mathematical space is a set of mathematical objects (e.g., numbers, functions, sets, etc.) with some additional structures.

A *function* (*mapping*, *operator*) f between sets X and Y is denoted $f : X \mapsto Y$ and defined as a relation that associates to every element of the set X exactly one element of the set Y.

A *set-valued function (mapping, operator)* F between sets X and Y is denoted $F : X \mapsto 2^Y$ and defined as a relation that associates to every element of the set X many elements of Y.

A.1 Field

A *field* is a set $\mathcal{F}$ together with two operations on $\mathcal{F}$: addition and multiplication. Addition is a function $\mathcal{F} \times \mathcal{F} \to \mathcal{F}$ denoted $\alpha + \beta$ for $\alpha, \beta \in \mathcal{F}$, i.e., $(\alpha, \beta) \to \alpha + \beta$. The result of addition is called sum. Multiplication is a function $\mathcal{F} \times \mathcal{F} \to \mathcal{F}$ denoted $\alpha\beta$ (or $\alpha \cdot \beta$) for $\alpha, \beta \in \mathcal{F}$, i.e., $(\alpha, \beta) \to \alpha\beta$. The result of multiplication is called the product. These operations are required to satisfy the so-called field axioms:

Associativity: $\alpha + (\beta + \gamma) = (\alpha + \beta) + \gamma$, and $\alpha(\beta\gamma) = (\alpha\beta)\gamma$.

Commutativity: $\alpha + \beta = \beta + \gamma$, and $\alpha\beta = \beta\alpha$.

Existence of zero and identity: there exist two different elements denoted 0 and 1 in $\mathcal{F}$ such that $\alpha + 0 = \alpha$ and $\alpha \cdot 1 = \alpha$.

Additive and multiplicative inverses: for every $\alpha \in F$ there exists an element in F, denoted $-\alpha$ such that a $\alpha + (-\alpha) = 0$; for every $\alpha \neq 0$ in $\mathcal{F}$ there exists an element in $\mathcal{F}$ denoted by α^{-1} or $1/\alpha$ such that $\alpha \cdot \alpha^{-1} = 1$.

Distributivity of multiplication over addition: $\alpha(\beta + \gamma) = \alpha\beta + \alpha\gamma$.

The standard examples of fields are real and complex numbers.

A.2 Topological Space

A *topological space* is an ordered pair (X, τ), where X is a set and a mapping $\tau : X \mapsto 2^X$ satisfies the following axioms:

- if $N \in \tau(x)$ then $x \in N$.
- if $\exists N \in \tau(x) : N \in S$ then $S \in \tau(x)$, where $S \subset X$.
- $N_1, N_2 \in \tau(x) \Rightarrow N_1 \cap N_2 \in \tau(x)$.
- $\forall N \in \tau(x), \exists M \in \tau(x) : N \in \tau(y), \forall y \in M$.

For any $x \in X$ the set $\tau(x)$ is a collection of subsets of X. An element of $\tau(x)$ is called *neighborhood of* x.

The topological space is a mathematical space allowing definitions of continuity, connectedness, and convergence.

For example, a function $f : X \to Y$ between topological spaces is called continuous if for every $x \in X$ and every neighborhood M of $f(x)$ there is a neighborhood N of x such that $f(N) \subset M$.

A.3 Metric Space

A function $\rho : M \times M \to [0, +\infty)$ is a metric (a distance) on a set M if

- $\rho(a, b) = 0 \Leftrightarrow a = b$;
- $\rho(a, b) = \rho(b, a)$;
- $\rho(a, b) \le \rho(a, c) + \rho(c, b)$.

A pair (M, ρ) is a *metric space*.

The metric space (M, ρ) is a *topological space*. Indeed, for any point $x \in M$ we can define the set $B(x, r) := \{y \in M : \rho(x, y) < r\}$, which is called the open ball of radius $r > 0$ around $x \in M$. These open balls form the base for a topology on M. A *neighborhood of a point* x is any subset of M that contains an open ball around x.

A set $Q \in M$ is said to be *open* if $\forall x \in Q, \exists r > 0 : B(x, r) \subset Q$. A sequence $\{x_n\} \in \mathbb{M}, n = 1, 2, \dots$ in the metric space M is said to *converge to the limit* $x \in M$ if for any $\varepsilon > 0$ exists a natural number N such that $\rho(x_n, x) < \varepsilon$ for all $n > N$. A set $K \subset M$ is said to be *closed* if every sequence in K, that converge to a limit in M, has this limit in K.

A set $D \subset M$ is *dense* in a set $Q \subset M$ if for any $x \in Q$ and for any $\varepsilon > 0$ there exists $y \in D$ such that $\rho(x, y) < \varepsilon$.

A metric space (set) M is said to be *compact* if any sequence in M has a subsequence that converge to an element in M.

A sequence $\{x_n\} \in M$ is said to be a *Cauchy sequence* in M if for any $\varepsilon > 0$ there exists a natural N such that $\rho(x_n, x_m) < \varepsilon$ for all $n, m > N$.

A metric space M is said to be *complete* if any Cauchy sequence converges to an element $x \in M$. For any metric space M there exists a complete metric space $\overline{M}$, which contains M as a dense subspace. The set M is called *completion* of M. It can be constructed as a set of equivalence classes of Cauchy sequences in M. We say that two Cauchy sequences $\{x_n\}$ and $\{x_n'\}$ belong to the same *equivalence class* x of Cauchy sequences in M if

$$\lim_{n \to +\infty} \rho(x_n, x_n') = 0.$$

Since a constant sequence $x_n := z \in M$ is a Cauchy sequence as well, then we conclude that any element z of M belongs to some equivalence class of Cauchy sequences in M. The set $\overline{M}$ composed of all equivalence classes of Cauchy sequences in M is a complete metric space (the *completion* of M) with the metric $\rho : \overline{M} \times \overline{M} \mapsto [0, +\infty)$ given by

$$\rho(x, y) = \lim_{n \to +\infty} \rho(x_n, y_n),$$

where $\{x_n\} \in x$, $\{y_n\} \in y$ and $x, y \in \overline{M}$.

A metric space M is called *separable* if it contains a countable dense subset.

The Cartesian product $M = M_1 \times M_2$ of metric spaces (M_1, ρ_1) and (M_2, ρ_2) is also a metric space (M, ρ) with $\rho = (\rho_1^p + \rho_2^p)^{1/p}$, $p \geq 1$.

A.4 Vector Space

A *vector space* over a field $\mathcal{F}$ is a set $\mathbb{V}$ together with two operations. Elements of $\mathbb{V}$ are called vectors. Elements of $\mathcal{F}$ are commonly called scalars. The first operation is a vector addition $\mathbb{V} \times \mathbb{V} \to \mathbb{V}$ denoted $v + w$ for $v, w \in \mathbb{V}$, i.e., $(v, w) \to v + w$. The result of addition is called sum of two vectors. The second operation is a multiplication on a scalar $\mathcal{F} \times \mathbb{V} \to \mathbb{V}$ denoted as αv (or $\alpha \cdot v$) for $\alpha \in \mathcal{F}$ and $v \in \mathbb{V}$, i.e., $(\alpha, v) \to \alpha v$. The set V and the operations of addition and multiplication must satisfy following axioms.

Associativity of addition: $u + (v + w) = (u + v) + w$.

Commutativity of addition: $u + v = v + u$.

Identity element of addition: There exists the zero element $\mathbf{0} \in \mathbb{V}$ such that $v + \mathbf{0} = v$ for all $v \in \mathbb{V}$.

Inverse elements of addition: For any $v \in \mathbb{V}$ there exists $-v \in \mathbb{V}$ such that $v + (-v) = \mathbf{0}$.

Compatibility of scalar multiplication: $\alpha(\beta v) = (\alpha\beta)v$.

Identity element of scalar multiplication: $1v = v$, where 1 denotes the identity in $\mathcal{F}$.

Distributivity of scalar multiplication with respect to vector addition: $\alpha(u + v) = \alpha u + \alpha v$.

Distributivity of scalar multiplication with respect to field addition: $(\alpha + \beta)v = \alpha v + \beta v$.

For example, the set $\mathbb{R}^n = \mathbb{R} \times \mathbb{R} \times \cdots \times \mathbb{R}$ is a vectors space with operations of addition and multiplications defined in the componentwisely. The set $C([0, 1], \mathbb{R})$ of continuous functions $[0, 1] \mapsto \mathbb{R}$ is a vector space as well.

Let $\mathbb{V}$ and $\mathbb{X}$ be two vector spaces over the field $\mathcal{F}$. A function (mapping, operator) $f : \mathbb{V} \mapsto \mathbb{X}$ is said to be *linear* if

$$f(\alpha x + \beta y) = \alpha f(x) + \beta f(y)$$

for all $\alpha, \beta \in \mathcal{F}$ and all $x, y \in \mathbb{V}$. If $\mathbb{X} = \mathcal{F}$ then f is called a *linear functional*.

A *linear subspace* of $\mathbb{V}$ is a subset $\tilde{\mathbb{V}} \subset \mathbb{V}$ being again a vector space with the same addition and multiplication operations.

The set of all possible linear combinations of the vectors $e_1, e_2, \ldots, e_k \in \mathbb{V}$ is called *linear hull*, denoted as $\operatorname{span}\{e_1, e_2, \ldots, e_k\}$ and given by

$$\operatorname{span}\{e_1, e_2, \ldots, e_k\} := \{\alpha_1 e_1 + \cdots + \alpha_k e_k : \alpha_i \in \mathcal{F}, i = 1, 2, \ldots, k\}.$$

The vectors $\{e_1, e_2, \dots, e_k\}$ are said to be *linearly independent* if the linear equation

$$\sum_{i \in \Gamma} \alpha_i e_i = \mathbf{0}$$

with respect to $\alpha_i \in \mathcal{F}$ has the unique solution $\alpha_i = 0, \forall i \in \Gamma$, where Γ is a set of indexes (finite or infinite).

A set $\{e_i\}_{i \in \Gamma}, e_i \in \mathbb{V}$ is said to be a *basis in a vector space* $\mathbb{V}$ if $\{e_i\}$ are linearly independent and $\mathbb{V} = \text{span}\{e_i\}_{i \in \Gamma}$, where Γ is a set of indexes. A number of elements of Γ is called the *dimension of a linear vector space* $\mathbb{V}$ and denoted $\dim(\mathbb{V})$. For instance, $\dim(\mathbb{R}^n) = n$ and $\dim(C([0, 1]; \mathbb{R})) = \infty$.

Obviously, $\text{span}\{e_1, e_2, \dots, e_k\}$ is a linear subspace of $\mathbb{V}$. If the vectors $e_1, e_2, \dots, e_k$ are linearly independent then $\dim(\text{span}\{e_1, e_2, \dots, e_k\}) = k$.

A.5 Convex Sets

A set Ω of a vector space $\mathbb{V}$ over the field of real numbers is **convex** if a segment joining two given points from Ω belongs to Ω:

$$\forall x_1, x_2 \in \Omega \quad \Rightarrow \quad \lambda x_1 + (1 - \lambda) x_2 \in \Omega, \forall \lambda \in [0, 1].$$

If $A, B \subset \mathbb{R}^n$ are convex sets then the sets

$$A \cap B \quad \text{and} \quad A \dot{+} B$$

are convex, where $\dot{+}$ denotes the geometric sum of sets (see Notation).

Let $x_i \in \mathbb{V}, i = 1, 2, \dots, N$. The vector

$$\alpha_1 x_1 + \alpha_2 x_2 + \dots + \alpha_N x_N \in \mathbb{V}, \quad \alpha_i \in \mathbb{R}$$

is said to be *convex combination* of the vectors x_i if

$$\alpha_i \geq 0 \quad \alpha_1 + \alpha_2 + \dots + \alpha_N = 1, \tag{A.1}$$

where N is a natural number.

Let Ω be a subset of a vector space V. A smallest convex closed set containing Ω is called *convex hall* of Ω and denoted $\text{co}(\Omega)$.

A.6 Normed Vector Space

Let $\mathbb{V}$ be a vector space over a field of real $\mathcal{F} = \mathbb{R}$ or complex $\mathcal{F} = \mathbb{C}$ numbers. A function $\|\cdot\| : \mathbb{V} \mapsto [0, +\infty)$ is a *norm in a vector space* $\mathbb{V}$ if

- *positive definiteness*: $\|v\| \geq 0$ for all $v \in \mathbb{V}$ and $\|v\| = 0 \Leftrightarrow v = \mathbf{0}$;
- *homogeneity*: $\|\lambda x\| = |\lambda|\|x\|$ for all $\lambda \in \mathcal{F}$;
- *triangle inequality*: $\|v + u\| \leq \|v\| + \|u\|$ for all $v, u \in \mathbb{V}$.

A pair $(\mathbb{V}, \|\cdot\|)$ is a *normed vector space*.

For example, the set $\mathbb{S}^2$ of symmetric matrices 2×2 is a normed vector space with the norm $\|A\| = \sqrt{\text{trace}(A^2)}$, where trace$(A)$ denotes the trace (i.e., the sum of diagonal elements) of the matrix A. A possible basis in $\mathbb{S}^2$ is $e_1 = \left(\begin{smallmatrix} 1 & 0 \\ 0 & 0 \end{smallmatrix}\right), e_2 = \left(\begin{smallmatrix} 0 & 0 \\ 0 & 1 \end{smallmatrix}\right), e_3 = \left(\begin{smallmatrix} 0 & 1 \\ 1 & 0 \end{smallmatrix}\right)$ and $\dim(\mathbb{S}^2) = 3$.

Any normed vector space is a metric space. The norm $\|\cdot\|$ introduces the distance (metric) $\|u - v\|$ between elements $u, v \in \mathbb{V}$. The balls $B(v, r) = \{u \in \mathbb{V} : \|v - u\| < r\}$ define the so-called strong (or norm) topology on $\mathbb{V}$ consisting of all sets which can be written as a union of balls $B(v, r)$. The spheres in $\mathbb{V}$ are give by $S(v, r) = \{u \in \mathbb{V} : \|x - u\| = r\}$.

Let V_1 and V_2 be two normed vector spaces. A mapping $f : \mathbb{V}_1 \mapsto \mathbb{V}_2$ is said to be continuous at $x \in V_1$ if there exists $\omega \in \mathcal{K}$ such that $\|f(x) - f(x_0)\|_{V_2} \leq \omega(\|x - x_0\|_{V_1})$. If $V_2 = \mathbb{R}$ the the mapping f is called *functional*. A functional $f : V \mapsto \mathbb{R}$ is said to be continuous if it is continuous at any point of $\mathbb{V}$. A functional f is linear if $f(\alpha x + \beta y) = \alpha f(x) + \beta f(y), \forall \alpha, \beta \in \mathcal{F}$ and $\forall x, y \in \mathbb{V}$.

Let $\mathbb{V}$ be a normed vector space over a field $\mathcal{F}$ of real or complex numbers. A set of all continuous linear functional $\mathbb{V} \mapsto \mathcal{F}$ is called the *dual space* to $\mathbb{V}$ and denoted $\mathbb{V}^*$. The dual space defines the so-called *weak topology* on $\mathbb{V}$ as follows: a sequence $\{x_n\}$ in $\mathbb{V}$ converges to $x \in \mathbb{V}$ in the weak topology if $\phi(x_n) \to \phi(x)$ as $n \to +\infty$ for all $\phi \in \mathbb{V}^*$.

A.7 Banach Space

Real (complex) *Banach space* $\mathbb{B}$ is a complete normed vector space over the field of real (resp., complex) numbers.

A normed vector space is a Banach space if and only if $\sum_{n=1}^{+\infty} \|x_n\| < +\infty$ implies $\sum_{n=1}^{+\infty} x_n$ converges in $\mathbb{B}$, where $x_n \in \mathbb{B}$. Any normed vector space $\mathbb{V}$ has a completion $\mathbb{B}$ being a Banach space.

The examples of Banach spaces are

- ℓ^p—the space of sequences $\alpha = \{\alpha_i\}, \alpha_i \in \mathbb{R}$ such that $\sum_i |\alpha_i|^p < +\infty$ with the norm $\|\alpha\|_p = (\sum_i |\alpha_i|^p)^{1/p}$.
- $C(\Omega, \mathbb{V})$—a space of continuous functions $\Omega \mapsto \mathbb{V}$ with the norm $\|u\|_C = \sup_{x \in \Omega} \|u(x)\|_{\mathbb{V}}$, where $\mathbb{V}$ is a normed vector space.

Hahn–Banach Separation Theorem Let $\mathbb{B}$ be a real Banach space. Let $E_1 \subset \mathbb{B}$ and $E_2 \subset \mathbb{B}$ be linear subspaces of $\mathbb{B}$ such that

$$\operatorname{int} E_1 \neq \emptyset, \quad (\operatorname{int} E_1) \cap E_2 = \emptyset.$$

Then there exists a linear continuous functional $f : \mathbb{B} \mapsto \mathbb{R}$ such that

$$\sup_{u \in E_1} f(u) \leq \inf_{y \in E_2} f(y)$$

and

$$f(u) < \inf_{y \in E_2} f(y), \quad \forall u \in \operatorname{int} E_1.$$

A.8 Hilbert Space

A *Hilbert space* $\mathbb{H}$ is a Banach space with the norm

$$\|x\| = \sqrt{\langle x, x \rangle}$$

induced by an inner product $\langle \cdot, \cdot \rangle$. A function $\langle \cdot, \cdot \rangle : \mathbb{H} \times \mathbb{H} \mapsto \mathcal{F}$ defined on a vector space $\mathbb{H}$ over a field $\mathcal{F}$ of real ($\mathcal{F} = \mathbb{R}$) or complex ($\mathcal{F} = \mathbb{C}$) numbers is said to be an *inner product* on $\mathbb{H}$ if

- $\langle x, x \rangle > 0$ for all $x \in \mathbb{H}$ and $\langle x, x \rangle = 0 \Leftrightarrow x = \mathbf{0}$;
- $\langle x, y \rangle = \langle y, x \rangle$ (or $= \overline{\langle y, x \rangle}$ if $\mathcal{F} = \mathbb{C}$) for all $x, y \in \mathbb{H}$;
- $\langle \alpha x_1 + \beta x_2, y \rangle = \alpha \langle x_1, y \rangle + \beta \langle x_2, y \rangle$ for all $\alpha, \beta \in \mathcal{F}$ and $x_1, x_2, y \in \mathbb{H}$;

If $\mathbb{H}$ is a Hilbert space then $\forall u, v \in \mathbb{H}$ the following relations

$$\begin{aligned} &(\textit{Parallelogram identity}): \quad && \|u + v\|^2 + \|u - v\|^2 = 2\left(\|u\|^2 + \|v\|^2\right), \\ &(\textit{Cauchy–Schwarz inequality}): && |\langle u, v \rangle| \leq \|u\| \cdot \|v\| \end{aligned}$$

hold. Vectors $x, y \in \mathbb{H}$ are said to be orthogonal $x \perp y$ if $\langle x, y \rangle = 0$. For $x \perp y$ the *Pythagorean identity* $\|x + y\|^2 = \|x\|^2 + \|y\|^2$ holds.

Hilbert Projection theorem says that for every point $x \in \mathbb{H}$ and every nonempty closed convex set $K \subset \mathbb{H}$ there exists a unique $y \in K$ such that $\|x - y\| = \min_{z \in K} \| x - z\|$. The corresponding vector y is called the orthogonal projection of x on K.

If $\{e_i\}, i \in I$ is an orthonormal family ($\langle e_i, e_j \rangle = 0$ if $e_i \neq e_j$, $i, j \in I$ and $\langle e_i, e_i \rangle = 1$) then the *Bessel's inequality*

$$\|x\|^2 \geq \sum_{i \in I} |\langle x, e_i \rangle|^2$$

holds.

If B be an orthonormal basis of $\mathbb{H}$, i.e., a set of orthonormal vectors such that span B is dense in $\mathbb{H}$, then the *Parseval's identity*

$$\|x\|^2 = \sum_{e_i \in B} |\langle x, e_i \rangle|$$

holds. If the Hilbert space is separable then B is a countable set and for any $x \in \mathbb{H}$ we have

$$\left\| x - \sum_{i=1}^{+\infty} \langle x, e_i \rangle e_i \right\| = 0.$$

The series $\sum_{i=1}^{+\infty} \langle x, e_i \rangle e_i$ is called *Fourier series.*

Let H^* be a dual space to $\mathbb{H}$. *Riesz Representation theorem* says that for every element $u \in \mathbb{H}$ there exists a unique element $\phi_u \in \mathbb{H}^*$ given by

$$\phi_u(x) = \langle x, u \rangle.$$

This means that the convergence $x_n \to x$ in the weak topology of $\mathbb{H}$ is defined as

$$\langle x_n, y \rangle \to \langle x, y \rangle \text{ as } n \to +\infty \quad \forall y \in \mathbb{H}.$$

The *Euclidean space* is the finite-dimensional Hilbert space. The space $\mathbb{R}^n$ is the common example of the finite-dimensional Hilbert space (Euclidean space). The inner product in $\mathbb{R}^n$ is defined as $\langle x, y \rangle = x^\top P y$, where $x, y \in \mathbb{R}^n$ and $P \in \mathbb{R}^{n \times n}$ is a positive definite symmetric matrix.

B Elements of Operator Theory

B.1 Operators in Banach Spaces

Let $\mathbb{B}$ be a real or complex Banach space and $A : \mathcal{D}(A) \subset \mathbb{B} \mapsto \mathbb{B}$ with the domain $\mathcal{D}(A)$ be a linear operator, i.e.,

$$A(\alpha z + \beta y) = \alpha A(x) + \beta A(y), \quad \forall x, y, \in \mathcal{D}(A), \forall \alpha, \beta \in \mathcal{F}.$$

The linear operator $A : \mathbb{B} \mapsto \mathbb{X}$ is said to be bounded if

$$\|A\| = \sup_{u \neq \mathbf{0}} \frac{\|Au\|_{\mathbb{X}}}{\|u\|_{\mathbb{B}}} < +\infty,$$

where $\mathbb{X}$ is a normed vector space.

For example, any linear operator $\mathbb{R}^n \mapsto \mathbb{R}^n$ is bounded and defined by a matrix $A \in \mathbb{R}^{n\times n}$ as $x \to Ax$, where Ax is the matrix product (with $A \in \mathbb{R}^{n\times n}$ and $x \in \mathbb{R}^{n\times 1}$).

The differential operator $A := \frac{d}{dx} : C^1(\mathbb{R}; \mathbb{R}) \subset C(\mathbb{R}; \mathbb{R}) \mapsto C(\mathbb{R}; \mathbb{R})$ is the classical example of a linear unbounded operator in an infinite-dimensional space. The unboundedness follows from the following example: $\|A\phi_n\| \to +\infty$ as $n \to +\infty$ for $\phi_n = \sin(nx) \in C(\mathbb{R}; \mathbb{R}), \forall n \in \mathbb{N}$ with $\|\phi_n\| = 1$.

The linear operator $A : \mathbb{B} \mapsto \mathbb{X}$ is bounded if and only if it is continuous.

The *graph of a function* $f : \mathcal{D}(f) \subset \mathbb{B} \mapsto \mathbb{X}$ is the set

$$G(f) = \{(x, f(x)) : x \in \mathbb{B}\}.$$

Closed Graph Theorem If $\mathbb{B}$ and $\mathbb{X}$ are Banach spaces, and $A : \mathbb{B} \mapsto \mathbb{X}$ is a linear operator, then A is continuous if and only if its graph is closed in the space $\mathbb{B} \times \mathbb{X}$.

The set $\mathcal{L}(\mathbb{B}; \mathbb{X})$ of linear bounded operators $\mathbb{B} \mapsto \mathbb{X}$ is a Banach space with the operator norm $\|\cdot\|$ given above.

Let $f : \mathbb{B} \mapsto \mathbb{V}$ be a mapping between two Banach spaces $\mathbb{B}$ and $\mathbb{V}$. A linear bounded operator $Df(x) \in \mathcal{L}(\mathbb{B}; \mathbb{V})$ is said to a **Frechét derivative** of f at the point $x \in \mathbb{B}$ if

$$\frac{\|f(x+y) - f(x) - Df(x)y\|_{\mathbb{V}}}{\|y\|_{\mathbb{B}}} \to 0 \quad \text{as} \quad \|y\|_{\mathbb{B}} \to 0.$$

The mapping f is Frechét differentiable at x if there exists $Df(x) \in \mathcal{L}(\mathbb{B}; \mathbb{V})$ satisfying the above limit. For $\mathbb{B} = \mathbb{R}^n$ and $\mathbb{V} = \mathbb{R}^m$ the Frechét derivative becomes the conventional (strong) derivative $\frac{\partial f(x)}{\partial x} \in \mathbb{R}^{m\times n}$ of f at $x \in \mathbb{R}^n$.

If the mappings $f : \mathbb{B} \mapsto \mathbb{V}$ and $g : \mathbb{V} \mapsto \mathbb{X}$ are Frechét differentiable then the Frechét derivative $Dg(f(x)) \in \mathcal{L}(\mathbb{B}; \mathbb{X})$ of the composition $g(f)$ admits the *chain rule*:

$$Dg(f(x)) = Dg(y)|_{y=f(x)} \circ Df(x),$$

where $Dg(y) \in \mathcal{L}(\mathbb{V}; \mathbb{Y})$ is the Frechét derivative of g at the point $y \in \mathbb{V}$ and $Df(x) \in \mathcal{L}(\mathbb{B}; \mathbb{V})$ is the Frechét derivative of f at the point $x \in \mathbb{B}$.

Let $\mathbb{B}, \mathbb{X}, f : \mathbb{B} \mapsto \mathbb{X}$ be, as before, and $y \in \mathbb{B}$. The vector $Df(x; y) \in \mathbb{X}$ is said to be **Gateaux derivative** of f at the point $x \in \mathbb{B}$ in the direction $y \in \mathbb{B}$ if

$$Df(x; y) = \lim_{h\to 0} \frac{f(x+hy) - f(x)}{h}.$$

If f is Frechét differentiable then $Df(x; y) = Df(x)y$. The *chair rule* for Gateaux derivative is

$$D(g \circ f)(x; y) = Dg(f(x); Df(x, y)).$$

For $\mathbb{Y} = \mathbb{R}$ the left(right)-hand lower/upper Gateaux derivatives can be introduced analogously to Dini derivatives.

$$D^+ f(x; y) = \limsup_{h \to 0^+} \frac{f(x + hy) - f(x)}{h}.$$

$$D_+ f(x; y) = \liminf_{h \to 0^+} \frac{f(x + hy) - f(x)}{h}.$$

$$D^- f(x; y) = \limsup_{h \to 0^-} \frac{f(x + hy) - f(x)}{h}.$$

$$D_- f(x; y) = \liminf_{h \to 0^-} \frac{f(x + hy) - f(x)}{h}.$$

Obviously, if $\mathbb{Y} = \mathbb{R}$ and $Df(x; y)$ exists then $Df(x; y) = D^+ f(x; y) = D_+ f(x; y) D^- f(x; y) = D_- f(x; y)$.

Banach–Steinhaus Theorem (*Uniform boundedness principle*) Let $\mathbb{B}$ be a Banach space and $\mathbb{X}$ be a normed vector space. If $\Xi \subset \mathcal{L}(\mathbb{B}; \mathbb{X})$ and

$$\sup_{A \in \Xi} \|Ax\|_{\mathbb{X}} < +\infty, \quad \forall x \in \mathbb{B}$$

then

$$\sup_{A \in \Xi, \|x\|=1} \|Ax\|_{\mathbb{X}} = \sup_{A \in \Xi} \|A\| < +\infty.$$

By definition, the dual space $\mathbb{B}^*$ to $\mathbb{B}$ is the space of all linear bounded functionals $\mathbb{B} \mapsto \mathbb{R}$ on $\mathbb{B}$.

An operator $F : \mathcal{D}(F) \subset \mathbb{B} \mapsto \mathbb{X}$ is said to be *closed* if the limits $x_n \to x$ and $F(x_n) \to y$ as $n \to \infty$ imply $x \in \mathcal{D}(F)$ and $F(x) = y$. The operator $F : \mathcal{D}(F) \subset \mathbb{B} \mapsto \mathbb{X}$ is closed if and only if $G(F)$ is closed subset of $\mathbb{B} \times \mathbb{X}$. If $A : \mathcal{D}(A) \subset \mathbb{B} \mapsto \mathbb{X}$ is a linear closed operator then $\mathcal{D}(A)$ is a Banach space with the *graph norm*: $\|x\|_A = \|x\| + \|Ax\|$.

The operator $F : \mathcal{D}(F) \subset \mathbb{B} \mapsto \mathbb{X}$ is *densely defined* if $\mathcal{D}(F)$ is dense $\mathbb{B}$.

We say that $f : \mathbb{B} \mapsto \mathbb{X}$ is a Lipschitz mapping if it satisfy the *Lipschitz condition*:

$$\exists L > 0 : \|f(x) - f(y)\|_{\mathbb{X}} \leq L\|x - y\|_{\mathbb{B}}, \quad \forall x, y \in \mathbb{B}.$$

We say that a real Banach space $\mathbb{B}$ has the *Radon–Nikodym property* if every Lipschitz mapping $f : \mathbb{R} \mapsto \mathbb{B}$ is differentiable almost everywhere. All Hilbert spaces have the Radon–Nikodym property.

Banach Fixed-Point Theorem Let $F : \mathbb{B} \mapsto \mathbb{B}$ and there exists $q \in [0, 1)$ such that

$$\|F(x) - F(y)\| \leq q\|x - y\|, \quad \forall x, y \in \mathbb{B}.$$

Then there exists a unique point $x^* \in \mathbb{B}$ such that $T(x^*) = x^*$.

The above theorem also holds for a complete metric space $\mathbb{X}$ provided that the contraction condition is given by

$$\rho(F(x), F(y)) \leq q\rho(x, y) \quad \forall x, y \in \mathbb{X},$$

where $q \in (0, 1)$ and ρ is a metric in $\mathbb{X}$.

B.2 Operators in Hilbert Spaces

Let $\mathbb{H}$ be a real or complex Hilbert space and $A : \mathcal{D}(A) \subset \mathbb{H} \mapsto \mathbb{H}$ be a linear densely defined operator.

The operator $A^* : \mathcal{D}(A^*) \subset \mathbb{H} \mapsto \mathbb{H}$ with the domain $\mathcal{D}(A^*)$ is said to be adjoint to A if

$$\langle Ax, y\rangle = \langle x, A^* y\rangle, \quad \forall x \in \mathcal{D}(A), \forall y \in \mathcal{D}(A^*).$$

The adjoint operator to a linear densely defined operator A always exists and densely defined. The operator A is self-adjoint if $A = A^*$.

Implicit Function Theorem Let $\mathbb{X}$ and $\mathbb{Y}$ be normed vector spaces, let $\mathbb{R}$ be a Euclidean space, let $Q : \mathbb{X} \times \mathbb{Y} \mapsto \mathbb{R}$, let $(\hat{x}, \hat{y}) \in \mathbb{X} \times \mathbb{Y}$, let A be a linear bounded operator from $\mathbb{X}$ into $\mathbb{R}$ and B be a linear bounded operator from $\mathbb{Y}$ into $\mathbb{R}$ such that

- Q is continuous in a neighborhood of $(\hat{x}, \hat{y})$,
- (A, B) is the differential of ϕ at the point $(\hat{x}, \hat{y})$, i.e.,

$$\lim_{\delta \to 0^+} \frac{\sup_{|x-|<\delta, |y-\hat{y}|<\delta} |\phi(x, y) - A(x - \hat{x}) - B(y - \hat{y}) - \phi(\hat{x}, \hat{y})|}{\delta} = 0,$$

then there exists a neighborhood $U \subset \mathbb{X}$ of $\hat{x}$, a mapping $V : U \mapsto \mathbb{Y}$ and a continuous mapping C from $\mathbb{X}$ into $\mathbb{Y}$ such that

- $V(\hat{x}) = \hat{y}$;
- $Q(x, V(x)) = \phi(\hat{x}, \hat{y})$ for all $x \in U$;
- $A + BC = 0$,

where C is a differential of V at the point $\hat{x}$, i.e.,

$$\lim_{\delta \to 0^+} \frac{\sup_{|x-\hat{x}|<\delta,} \|V(x) - V(\hat{x}) - C(x - \hat{x})\|}{\delta} = 0.$$

C Elements of Measure Theory and Integration

C.1 Measure Space

Let $\mathbb{X}$ be set and $2^{\mathbb{X}}$ is a power set (a set of all subsets).

> A subset Σ of $2^{\mathbb{X}}$ is called a *σ-algebra* over $\mathbb{X}$ if
>
> - $X \in \Sigma$;
> - if $A \in \Sigma$ then $X \backslash A \in \Sigma$;
> - if $A_1, A_2, A_3, \ldots \in \Sigma$ then $A_1 \cup A_2 \cup A_3 \ldots \in \Sigma$.

If (X, τ) is a topological space then σ algebra generated by τ is called Borel σ-algebra.

Let Σ be a σ-algebra over $\mathbb{X}$. A function $\mu : \Sigma \mapsto [0, +\infty]$ is said to be a *measure* if

- $\mu(A) \geq 0$ for all $A \in \Sigma$;
- $\mu(\emptyset) = 0$;
- if $A_i \in \Sigma, i = 1, 2, \ldots$ and $A_i \cap A_j = \emptyset$ for $i \neq j$ then

$$\mu\left(\bigcup_{i=1}^{+\infty} A_i\right) = \sum_{i=1}^{+\infty} \mu(A_i).$$

The triple $(\mathbb{X}, \Sigma, \mu)$ is called the *measure space*.

Any measure defined on Borel σ-algebra is called Borel measure.

Lebeasgue measure $\mu_{\mathbb{R},L} : 2^{\mathbb{R}} \mapsto [0, +\infty]$ of a set $S \subset \mathbb{R}$ is defined as follows:

$$\mu_{\mathbb{R},L}(S) = \inf_{\{\mathcal{I}_k\}: S \subset \bigcup_k \mathcal{I}_k} \sum_k \ell(\mathcal{I}_k),$$

where $\mathcal{I}_n = (a_k, b_k)$, $a_k, b_k \in \mathbb{R}$, $a_k < b_k$ and $\ell(\mathcal{I}_n) = b_k - a_k$.

A set $S \in \mathbb{R}$ is Lebeasque measurable if and only if it satisfies *Carathéodory condition*

$$\mu_{\mathbb{R},L}(A) = \mu_{\mathbb{R},L}(A \cap S) + \mu_{\mathbb{R},L}(A \cap (\mathbb{R} \backslash S)), \quad \forall A \in 2^{\mathbb{R}}.$$

A collection $\Sigma_{\mathbb{R},L}$ of Lebesgue measurable set is σ-algebra over $\mathbb{R}$ and $(\mathbb{R}, \Sigma_{\mathbb{R},L}, \mu_{\mathbb{R},L})$ is a measure space.

We say that some property holds *almost everywhere* on S if a set $\Delta \subset S$ on which it does not hold has the measure zero. An example of the *set of Lebesgue measure zero* is a countable set of points from S.

The same constructions can be repeated for the Lebesgue measure space $(\mathbb{R}^n, \Sigma_{\mathbb{R}^n,L}, \mu_{\mathbb{R}^n,L})$ considering $I_k = (a_k^1, b_k^1) \times \cdots \times (a_k^n, b_k^n) \in \mathbb{R}^n$ and $\ell(I_k) = (b_k^1 - a_k^1)\ldots(b_k^n - a_k^n)$.

C.2 *Measurable Function*

Let (X, Σ_x, μ_x) and (Y, Σ_y, μ_y) be two measure spaces. A function $f : X \mapsto Y$ is said to be measurable as mapping (X, Σ_x, μ_x) on (Y, Σ_y, μ_y) if

$$\{x : f(x) \in A_y\} \in \Sigma_x, \quad \forall A_y \in \Sigma_y.$$

In particular, a function $f : X \mapsto \mathbb{R}$ is said to be Lebesgue measurable if for any $\gamma \in \mathbb{R}$ the set

$$\{x : f(x) > \gamma\}$$

is Lebesgue measurable.

Any continuous function is measurable. A sum of measurable functions defined on the same measure spaces is a measurable function.

Let (X, Σ_x, μ_x), (Y, Σ_y, μ_y) and (Z, Σ_z, μ_z) be three measure spaces. If functions $f : X \mapsto Y$ and $g : Y \mapsto Z$ are measurable then the composition $g \circ f : X \mapsto Z$ is a measurable function. The following theorem shows certain closeness of measurable and continuous functions.

Luzin Theorem Let $f : [a, b] \mapsto \mathbb{R}$ be a measurable function. Then, for every $\varepsilon > 0$ there exists a compact set $S \subset [a, b]$ such that f restricted to S is continuous almost everywhere and $\mu(S) > b - a - \varepsilon$.

C.3 *Riemann Integration*

Let $f : [a, b] \mapsto \mathbb{R}$. We say that f is Riemann integrable if there exists a number $s \in \mathbb{R}$ such that for any $\varepsilon > 0$ there exists $\delta > 0$ and a partition $a = x_0 < x_1 < \ldots < x_n = b$, $x_{i+1} - x_i < \delta$ such that

$$\left| \sum_{i=1}^{n} f(t_i)(x_i - x_{i-1}) - s \right| < \varepsilon$$

for some $t_i \in [x_i, x_{i+1}]$.

The number s is called the Riemann integral of f and denoted $\int_a^b f(x)dx$.

All continuous and piece-wise continuous functions are Riemann integrable. Any monotone function $[a, b] \mapsto \mathbb{R}$ is Riemann integrable as well.

Linearity: $\int_a^b \alpha f_1(x) + \beta f_2(x)dx = \alpha \int_a^b f_1(x)dx + \beta \int_a^b f_2(x)dx$ for any Riemann integrable $f_1 : [a, b] \mapsto \mathbb{R}$ and $f_2 : [a, b] \mapsto \mathbb{R}$ and any $\alpha, \beta \in \mathbb{R}$.

C.4 Lebesque Integration

Let (E, Σ, μ) be the Lebesque measure space, where $E \subset \mathbb{R}$. If $f : E \mapsto \mathbb{R}_+$ then, by definition, *Lebesque integral* of f is

$$\int_E f d\mu := \lim_{a\to+\infty} \int_0^a f^\mu(t)dt,$$

where $f^\mu(t) = \mu(\{x : f(x) > t\})$ and $\int_0^a$ is the Riemann integral. Notice that the function f_μ is monotone so it is Riemann integrable on any compact $[0, a]$.

If $f : E \mapsto \mathbb{R}$ then

$$\int_E f d\mu := \int_E f^+ d\mu - \int_E f^- d\mu$$

where $f^+(x) = f(x)$ if $f(x) \geq 0$ and $f^+(x) = 0$ if $f(x) < 0$, $f^-(x) = f(x)$ if $f(x) \leq 0$ and $f^-(x) = 0$ if $f(x) > 0$.

The *linearity* mentioned for the Riemann integral also takes the place for the Lebesgue integral as well as the *monotonicity*: if $f \leq g$ then

$$\int_E f d\mu \leq \int_E g d\mu.$$

Lebesgue's Theorem about the Differentiability of Monotone Functions Let f be a monotone function on (a, b). Then f is differentiable almost everywhere on (a, b).

Fatou's Lemma If $\{f_k\}$ is a sequence of non negative Lebesgue measurable functions then

$$\int \liminf_{k\to+\infty} f_k d\mu \leq \liminf_{k\to+\infty} \int f_k d\mu.$$

Dominated Convergence Theorem Let $\{f_k\}$ be a sequence of Lebesgue measurable functions having point-wise f and there exist a Lebesgue integrable function $g \geq |f_k|$ for all k then

$$\lim_{k \to +\infty} \int f_k d\mu = \int f d\mu.$$

The Lebesgue integral can also be defined for functions $\mathbb{R}^n \mapsto \mathbb{R}^m$ and generalized for some maps in Banach spaces (see the next section).

Change of Variable Theorem in Lebesgue Integral Let $(\mathbb{R}^n, \Sigma, \mu)$ be a Lebesgue measure space, $E \subset \mathbb{R}^n$ and $\nu : E \mapsto \mathbb{R}^n$ be a C^1 one-to-one mapping such that $\det(\nabla \nu)$ is separated from zero on E. If $f : \nu(E) \mapsto \mathbb{R}^m$ is a Lebesgue measurable function then

$$\int_{\nu(E)} f d\mu = \int_E f(\nu) \det(\nabla \nu) d\mu.$$

The set E in the latter theorem may be unbounded (e.g., $E = \mathbb{R}^n$). For a Lebesgue integrable function $f : \mathbb{R}^n \mapsto \mathbb{R}^m$ in this monograph we also use the following notation

$$\int_{\mathbb{R}^n} f(x) dx.$$

Notice that under this notation, the Change of Variable theorem in Lebesgue integral can be reformulated as follows:

$$\int_{\mathbb{R}^n} f(x) dx = \int_{\mathbb{R}^n} f(\nu(x)) \det(\nabla \nu(x)) dx,$$

where $\nu \in C^1(\mathbb{R}^n; \mathbb{R}^n)$ is an invertible mapping such that

$$\inf_{x \in \mathbb{R}^n} |\det(\nu(x))| > 0.$$

C.5 *Bochner Integration*

Let (E, Σ, μ) be a measure space and $\mathbb{B}$ be a Banach space. Let us denote

$$S_k(x) = \sum_{i=1}^{k} \chi_{E_i}(x) b_i,$$

where E_i are disjoint members of the σ-algebra Σ, b_i are elements of $\mathbb{B}$ and χ_{E_i} is the *indicator function* of E_i (i.e., $\chi_{E_i}(x) = 1$ if $x \in E_i$ and $\chi_{E_i}(x) = 0$ if $x \notin E_i$).

The function S_k given above is called the simple Bochner integrable function and its integral, by definition, is given by

$$\int_E S_k(x)d\mu = \sum_{i=1}^{k} \mu(E_i)b_i \in \mathbb{B}.$$

A function $f : E \mapsto \mathbb{B}$ is Bochner integrable if there exists a sequence of simple functions S_k such that

$$\lim_{k\to+\infty} \int_E \|f - S_k\| d\mu = 0.$$

The above identity contains the by definition, the *Bochner integral* of f is

$$\int_E f d\mu = \lim_{k\to+\infty} \int_E S_k d\mu.$$

Properties. A function $f : E \mapsto \mathbb{B}$ is Bochner integrable if and only if

$$\int_E \|f\| d\mu < +\infty.$$

If A is a linear bounded operator $\mathbb{B} \mapsto \mathbb{B}$ then

$$A \int_E f d\mu = \int_E Af d\mu.$$

The above identity also holds if $A : \mathcal{D}(A) \subset \mathbb{B} \mapsto \mathbb{B}$ is a closed linear operator, $f : E \mapsto \mathcal{D}(A)$ and Af is continuous. If f is Bochner integrable then

$$\left\| \int_E f d\mu \right\| \le \int_E \|f\| d\mu.$$

For the Bochner integral of a function $f : (a, b) \mapsto \mathbb{B}$ we also use the notation $\int_a^b f(t)dt$. If $E \subset \mathbb{R}^n$ and $B = \mathbb{R}^m$ then the Bochner integral becomes the Lebesgue integral of a function $f : E \subset \mathbb{R}^n \mapsto \mathbb{R}^m$.

Newton–Leibniz Formula If $f \in C^1([a, b]; \mathbb{B})$ then

$$\int_a^b \dot{f}(t)dt = f(b) - f(a).$$

Moreover, if $\mathbb{B} = \mathbb{H}$ is a Hilbert space then the Newton–Leibniz formula is fulfilled for locally Lipschitz continuous functions too.

The Dominated Convergence theorem (see above) remains valid for Bochner integrable functions.

D Elements of Functional Analysis

Function spaces are the common examples of infinite-dimensional Banach and Hilbert spaces.

D.1 Spaces of Strongly Differentiable Functions

- $C(X; Y)$ is a space of continuous functions $X \mapsto Y$, where X and Y are subsets of normed vector spaces. The function space C is a Banach space with the so-called uniform norm topology:

$$\|f\|_C = \sup_{x \in X} \|f(x)\|_Y, \quad f \in C(X; Y).$$

- $C^{0,\alpha}(X; Y)$ with $\alpha \in (0, 1]$, is a space of Hölder continuous functions:

$$C^{0,\alpha}(X; Y) = \left\{ f \in C(X, Y) : \exists k \geq 0,\ \|f(x) - f(y)\|_Y \leq k \|x - y\|_X^\alpha,\ \forall x, y \in X \right\}.$$

  For example, the function $x \mapsto x^\beta$ belongs to $C^{0,\alpha}([0, 1]; \mathbb{R})$ for any $\alpha \in (0, \beta]$. Moreover, for $\Omega \subset \mathbb{R}^n$ it holds $C^{0,\beta}(\Omega; \mathbb{R}^m) \subset C^{0,\alpha}(\Omega; \mathbb{R}^m)$ provided that $0 < \alpha < \beta \leq 1$.
- $C^q(X; Y)$ is a space of strongly continuously differentiable functions $X \mapsto Y$, at least, up to the order n, where $X \subset \mathbb{R}^n$ and $Y \subset \mathbb{R}^m$. This means

$$\frac{\partial^k f}{\partial x_1^{n_1} \partial x_2^{n_2} \ldots \partial x_n^{n_k}} \in C(X, Y), \quad \forall k = 0, 1, \ldots, q, \quad \forall n_i : \sum_{i=1}^n n_i = k,$$

  for any $f \in C^q(X; Y)$. The function space C^p is a Banach space with the norm

$$\|f\|_{C^q} = \sum_{k=0}^{q} \sum_{n_1+\dots+n_k=k} \left\| \frac{\partial^k f}{\partial x_1^{n_1} \partial x_2^{n_2} \dots \partial x_n^{n_k}} \right\|_C .$$

The following embedding $C^{q_1} \subset C^{q_2}$, obviously, holds for $q_1 \geq q_2$.

- $C_c(X; Y)$ is a vector space of continuous functions with compact supports in $X \subset \mathbb{R}^n$, i.e., $f(x) \neq \mathbf{0}$ only for x from a compact subset of X.
- $C^\infty(X; Y)$ is a vector space of infinitely smooth functions on X.
- $C_c^\infty(X; Y)$ is a vector space of infinitely smooth functions with compact supports in $X \subset \mathbb{R}^n$.
- $C_0(\mathbb{R}^n; \mathbb{R}^m)$ is a vector space of continuous functions, which vanish at infinity.
- *Schwartz space* $\mathbb{S}(\mathbb{R}^n, \mathbb{R})$ is a space of infinitely smooth functions whose derivatives are rapidly decreasing:

$$\mathbb{S}(\mathbb{R}^n; \mathbb{R}) = \left\{ f \in C^\infty(\mathbb{R}^n; \mathbb{R}) : \forall \alpha, \beta \in \mathbb{N}_*^n, \left\| x^\alpha D^\beta(x) \right\|_C < +\infty \right\},$$

where $N_* = \mathbb{N} \cup \{0\}$, $\mathbb{N}_*^n = \mathbb{N}_* \times \dots \times \mathbb{N}_*$, $x = (x_1, \dots, x_n)^\top \in \mathbb{R}^n$, $\alpha = (\alpha_1, \dots, \alpha_n)^\top \in \mathbb{N}_*^n$, $\beta = (\beta_1, \dots, \beta_n)^\top \in \mathbb{N}_*^n$, $x^\alpha = x_1^{\alpha_1} \cdot x_2^{\alpha_2} \cdot \dots \cdot x_n^{\alpha_n}$ and $D^\beta f(x) = \frac{\partial f^{\beta_1+\dots+\beta_n}(x)}{\partial x_1^{\beta_1} \partial x_2^{\beta_2} \dots \partial x_n^{\beta_n}}$.

Arzelà–Ascoli Theorem Let $\{f_k\}$ be a sequence of functions $f \in C([a, b]; \mathbb{R})$. If this sequence is *uniformly bounded*:

$$\exists M > 0 \ : \ \|f_k\|_C < M, \ \forall k,$$

and *equicontinuous*:

$$\forall \varepsilon > 0, \ \exists \delta > 0 \ : \ |x - y| < \delta \Rightarrow |f_n(x) - f_n(x)| < \varepsilon, \ \forall n$$

then there exists a subsequence $\{f_{n_k}\}$ and $f \in C([a, b]; \mathbb{R})$ such that

$$\|f_{n_k} - f\|_C \to 0 \quad \text{as} \quad k \to +\infty.$$

For example, a sequence of Lipschitz functions with the same Lipschitz constant $L > 0$ for all n satisfies the latter theorem.

Integration by parts. If $u, v \in C^1([a, b], \mathbb{R})$ then

$$\int_a^b u \frac{\partial v}{\partial x} dx = uv|_a^b - \int_a^b \frac{\partial u}{\partial x} v \, dx.$$

Let Ω be an open bounded set in $\mathbb{R}^n$ with a smooth boundary Γ. If $u \in C^1(\Omega, \mathbb{R})$ and $v \in C^1(\Omega, \mathbb{R})$ then

$$\int_\Omega u \operatorname{div}(v) \, dx = \int_\Gamma uv \cdot \nu_\Omega \, d\Gamma - \int_\Omega \nabla u \cdot v \, dx, \tag{A.2}$$

where ν_Ω is the outward unit normal vector to the boundary of Ω.

Heine–Cantor Theorem If a function $f : \mathbb{R}^n \mapsto \mathbb{R}^m$ is continuous on an open set $\Omega \subset \mathbb{R}^n$ then f is uniformly continuous[a] on any compact from Ω.

[a] A function $f : \mathbb{B} \mapsto \mathbb{V}$ is uniformly continuous on a set $C \subset \mathbb{B}$ if there exists a *modulus of continuity* $\omega \in \mathcal{K}$ such that $\|f(x) - f(y)\|_{\mathbb{V}} \leq \omega(\|x - y\|_{\mathbb{B}})$ for all $x, y \in C$.

Rademacher Theorem If a function $f : \mathbb{R}^n \mapsto \mathbb{R}^m$ is locally Lipschitz continuous[a] on an open set $\Omega \subset \mathbb{R}^n$ then f is differentiable almost everywhere on Ω.

[a] A function $f : \mathbb{B} \mapsto \mathbb{V}$ is locally Lipschitz continuous on an open set $\Omega \subset \mathbb{B}$ is if satisfy Lipschitz condition on any closed ball embedded in Ω.

D.2 Lebesgue Spaces L^p

The space $L^p(\Omega, \mathbb{R})$ is a space of Lebesgue measurable functions $\Omega \subset \mathbb{R}^n \mapsto \mathbb{R}$ for which the absolute value of p-th power is Lebesgue integrable. The space $L^p(\Omega, \mathbb{R})$ with $1 \leq p \leq +\infty$ is a Banach space with the norm

$$\|f\|_p = \left(\int_\Omega |f|^p d\mu\right)^{\frac{1}{p}}$$

due to *Minkowski inequality*

$$\|f + g\|_p \leq \|f\|_p + \|g\|_p.$$

The norm for $p = \infty$ is given by $\|f\|_\infty = \text{ess sup}_{x\in\Omega} |f(x)|$.

The *Hölder inequality*

$$\|fg\|_1 \leq \|f\|_p \|g\|_q, \quad 1/p + 1/q = 1, \quad q, p \geq 1$$

implies that the dual vector space to L^p is given by integrating against functions in L^q.

The Lebesgue space $L^2(\Omega, \mathbb{R})$ is a Hilbert space with the inner product

$$\langle f, g\rangle_{L^2} = \int_\Omega fg d\mu, \quad f, g \in L^2(\Omega; \mathbb{R}).$$

The space $L^p(\Omega; \mathbb{R}^m)$ can be defined similarly using the Lebesgue measurable functions $\Omega \subset \mathbb{R}^n \mapsto \mathbb{R}^m$.

The vector space $L^1_{loc}(\mathbb{R}^n; \mathbb{R}^n)$ contains locally Lebesque integrable (on any compact set $\Omega \subset \mathbb{R}^n$) functions.

D.3 Weakly Differentiable Functions. Sobolev Spaces $W^{k,p}$ and H^p

Let $f \in L^1((a,b);\mathbb{R})$. We say that $Df \in L^1((a,b);\mathbb{R})$ is a *weak derivative* of f if

$$\int_a^b f\frac{d\phi}{dx}d\mu = -\int_a^b Df\,\phi d\mu, \quad \forall\phi \in C_c^\infty((a,b);\mathbb{R}).$$

Notice that, $\phi \in C_c^\infty((a,b);\mathbb{R})$, in particular, means that $\lim_{x\to a}\phi(x) = \lim_{x\to b}\phi(x) = 0$. Hence, if $f \in C^1([a,b];\mathbb{R})$ then, using integration by part we conclude that the latter integral identity holds only for $Df := \frac{df}{dx}$. In other words, the weak derivative coincides with the strong derivative if the function is differentiable in the conventional sense.

Considering functions of n variables and arbitrary order of the derivative, we derive $D^\alpha f \in L^1(\Omega;\mathbb{R})$ is the αth weak derivative of $f \in L^1(\Omega;\mathbb{R})$ if

$$\int_\Omega f\cdot D^\alpha\phi d\mu = (-1)^\alpha\int_\Omega D^\alpha f\cdot\phi d\mu, \quad \forall\phi \in C_c^\infty(\Omega;\mathbb{R}),$$

where $\alpha = (\alpha_1,\ldots,\alpha_n)$, $\alpha_i \in 0, 1, 2, \ldots$. and

$$D^\alpha\phi := \frac{\partial^{|\alpha|}\phi(x)}{\partial x_1^{\alpha_1}\ldots\partial x_n^{\alpha_n}}, \quad |\alpha| = \alpha_1 + \cdots + \alpha_1, \quad x \in \mathbb{R}^n.$$

By definition, $D^0 f = f$. For $D^1 f$ we also use the notation ∇f.

We say that a function $f \in L^1(\Omega;\mathbb{R})$ is weakly differentiable up to the order k if it has weak derivatives $D^\alpha f \in L^1(\Omega;\mathbb{R})$ for all $\alpha : |\alpha| \le k$.

The set of functions $f \in L^1(\Omega;\mathbb{R})$, which are weakly differentiable, at least, up to the order n and which the weak derivatives belong to L^p space, is denoted $W^{k,p}$ and called *Sobolev space*. The Sobolev space $W^{k,p}$ is a Banach space with the norm

$$\|f\|_{W^{k,p}} = \sum_{|\alpha|\le k}\|D^\alpha f\|_{L^p}.$$

A function $(a,b) \overset{f}{\mapsto} \mathbb{R}^n$ belongs to the space $W^{k,p}_{loc}((a,b);\mathbb{R})$ if its restriction $(a',b') \overset{f}{\mapsto} \mathbb{R}$ belongs to $W^{k,p}((a',b');\mathbb{R})$ for any $(a',b') \subset \operatorname{int}(a,b)$.

For $p = 2$ we use the notation $H^k := W^{k,2}$. The Sobolev space H^k is a Hilbert space with the inner product

$$\langle f, g\rangle = \sum_{|\alpha|\le k}\langle D^\alpha f, D^\alpha g\rangle_{L^2}, \quad f, g \in H^k.$$

By definition, $H^0 = L^2$.

In the natural component-wise manner one can be proposed an extension of L^p, $W^{k,p}$ and H^p spaces to vector-valued functions $\Omega \subset \mathbb{R}^n \mapsto \mathbb{R}^m$.

The space $W_0^{1,p}$ is a completion of C_c^∞ with respect to the norm $\|\cdot\|_{W^{1,p}}$.

Poincaré Inequality If $u \in W_0^{1,p}(\Omega)$ then

$$\|u\|_{L^p} \leq C\|D^1 u\|_{L^p},$$

where C is independent of u and Ω is an open bounded connected set with a Lipschitz boundary.

The integration by parts formula (A.2) also holds for $u \in H^1(\Omega;\mathbb{R})$ and $v \in H^1(\Omega;\mathbb{R}^n)$.

The Sobolev space $W^{k,p}$ for a real $k \in [0,+\infty)$ and $1 < p < +\infty$ is defined as follows

$$W^{k,p}(\mathbb{R}^n;\mathbb{R}) = \left\{f \in L^p(\mathbb{R}^n;\mathbb{R}) : \mathcal{F}^{-1}[(1+\xi^2)^{k/2}\mathcal{F}(f)] \in L^p(\mathbb{R}^n;\mathbb{R})\right\},$$

where $\mathcal{F}$ is the *Fourier transform*

$$(\mathcal{F}[f])(\xi) = \int_{\mathbb{R}_n} f(x)e^{-\mathrm{i}2\pi\xi^\top x}dx, \quad \xi \in \mathbb{R}^n$$

and $\mathcal{F}^{-1}$ is the *inverse Fourier transform*

$$(\mathcal{F}^{-1}[\hat{f}])(x) = \int_{\mathbb{R}^n} \hat{f}(\xi)e^{\mathrm{i}2\pi\xi^\top x}d\xi.$$

The norm in such Sobolev space is defined as

$$\|f\|_{W^{k,p}} = \left\|\mathcal{F}^{-1}\left[(1+\xi^2)^{k/2}\mathcal{F}(f)\right]\right\|_{L^p}.$$

The inverse and fractional Laplacians given by $(-\Delta)^\alpha$ with $\alpha \in \mathbb{R}$ are considered in Example 2.7.

Plancherel Lemma The operator

$$x \mapsto \frac{\partial}{\partial x_i}\frac{\partial}{\partial x_j}\Delta^{-1}x$$

is a bounded operator from L^2 to L^2 for every i, j. Moreover, it is a bounded operator from H^k to H^k for every $k \geq 0$.

Hölder Inequality Let $1 \leq p, q \leq +\infty$, $\frac{1}{p} + \frac{1}{q} = 1$ and $\Omega \subset \mathbb{R}^n$. Then for $u \in L^p(\Omega)$, $v \in L^q(\Omega)$ it holds

$$\int_{\Omega} |uv|dx \leq \|u\|_{L^p} \|v\|_{L^q}.$$

Gagliardo–Nirenberg–Sobolev Inequality Let $1 \leq p < n$. Then there exists a constant $\rho > 0$ (depending only on p and n) such that

$$\|u\|_{L^{p^*}} \leq \rho \|Du\|_{L^p}, \quad \forall u \in C^1(\mathbb{R}^n),$$

where $p^* = \frac{np}{n-p}$ is the *Sobolev conjugate* of p.

Hardy–Littlewood–Sobolev Fractional Integration Theorem Let $\alpha \in (0, n)$ and $1 < p < q < +\infty$. Let $I_\alpha = (-\Delta^{-\alpha/2})$, where Δ is a Laplacian on $\mathbb{R}^n$. Then for q satisfying

$$\frac{1}{q} = \frac{1}{p} - \frac{\alpha}{n}$$

there exists $C > 0$ dependent only on p such that

$$\|I_\alpha f\|_{L^q} \leq C \|f\|_{L^p}.$$

This implies *Sobolev embedding theorem*:

$$\frac{1}{p} - \frac{k}{n} = \frac{1}{q} - \frac{\ell}{n} \quad \Rightarrow \quad W^{k,p}(\mathbb{R}^n) \subset W^{\ell,q}(\mathbb{R}^n),$$

where $k \geq 0$, $\ell \geq 0$ and $p \geq 1, q \geq 1$

E Supporting Formulas

E.1 Some Integral Inequalities

Grönwall–Bellman Inequality Let $a \in \mathbb{R}$ and $\alpha, \beta : \mathbb{R} \mapsto \mathbb{R}$ be real-valued continuous functions. If β is non negative and a function $u : \mathbb{R} \mapsto \mathbb{R}$ satisfies the inequality

$$u(t) \leq \alpha(t) + \int_a^t \beta(\tau)u(\tau)d\tau, \quad \forall t \geq a$$

then

$$u(t) \leq \alpha(t) + \int_a^t \alpha(s)\beta(s)e^{\int_s^t \beta(\tau)d\tau} ds, \quad \forall t \geq a.$$

If, additionally, the function α is non-decreasing, then

$$u(t) \leq \alpha(t) e^{\int_a^t \beta(\tau) d\tau}, \quad \forall t \geq a.$$

Bihari–LaSalle Inequality Let $w : (0, +\infty) \mapsto (0, +\infty)$ be continuous non-decreasing function and β be, as before, non negative continuous function. If for some $\alpha \in \mathbb{R}$ and some function u the following inequality

$$u(t) \leq \alpha + \int_0^t \beta(\tau) w(u(\tau)) d\tau, \forall t \geq 0$$

holds, then

$$u(t) \leq g^{-1} \left(g(\alpha) + \int_0^t \beta(\tau) d\tau \right), \quad \forall t \in [0, T],$$

where

$$g(z) = \int_{z_0}^{z} \frac{dx}{w(x)}, \quad z \geq 0, \quad z_0 > 0,$$

g^{-1} is the inverse of g and T is chosen such that $G(\alpha) + \int_0^t \beta(\tau) d\tau$ belongs to the domain of g^{-1}.

E.2 Hermite Functions and Polynomials

The *Hermite polynomials* are defined as follows:

$$h_i(x) = (-1)^i e^{x^2} \frac{d^i e^{-x^2}}{dx^i}, \quad x \in \mathbb{R}, \quad i = 0, 1, \ldots,$$

e.g., $h_0 = 1$, $h_1 = x$, $h_2 = 4x^2 - 2$, $h_3 = 8x^3 - 12x$, etc.
Some properties of h_i:

- $h_i(-x) = (-1)^i h(x)$;
- $h_{i+1}(x) = 2x h_i(x) - 2i h_{i-1}(x)$;
- $\int_{-\infty}^{+\infty} e^{-x^2} h_i(x) h_j(x) dx = \begin{cases} 2^i i! \sqrt{\pi} & \text{if } i = j; \\ 0 & \text{if } i \neq j; \end{cases}$
- $\sum_{i=0}^{\infty} \frac{t^i h_i(x)}{i!} = e^{2xt - t^2}, \quad \forall x, t \in \mathbb{R}.$

The *Hermite functions* are defined as follows:

$$\phi_i(x) = (2^i i! \sqrt{\pi})^{-\frac{1}{2}} e^{-\frac{x^2}{2}} h_i(x), \quad x \in \mathbb{R}, \quad i = 0, 1, \ldots.$$

Some properties of ϕ_i:

- ϕ_i is an orthonormal basis in $L^2(\mathbb{R})$:

$$\int_{-\infty}^{+\infty} \phi_i(x)\phi_j(x)dx = \begin{cases} 1 \text{ if } i = j; \\ 0 \text{ if } i \neq j; \end{cases}$$

- $\frac{d}{dx}\phi_i(x) = \sqrt{\frac{i}{2}}\phi_{i-1}(x) - \sqrt{\frac{i+1}{2}}\phi_{i+1}(x).$

E.3 Γ-Function

The Γ-function $\Gamma : \mathbb{C} \mapsto \mathbb{C}$ is defined as follows:

$$\Gamma(z) = \int_0^\infty t^{z-1}e^{-t}dt, \quad \Re(z) > 0.$$

Some properties of Γ:

- $\Gamma(n) = n!$ for all $n \in \mathbb{N}$;
- $\Gamma(z+1) = z\Gamma(z)$ for all $z \in \mathbb{C}$;
- $\Gamma(\alpha) = \int_0^1 \left(\log \frac{1}{s}\right)^{\alpha-1} ds$ for any $\alpha > 0$;
- $\Gamma(\alpha)\Gamma(\beta) = \Gamma(\alpha+\beta) \int_0^1 s^{\alpha-1}(1-s)^{\beta-1}ds$ for all $\alpha > 0$ and all $\beta > 0$.

References

1. Yosida K (1980) Functional analysis. Springer, Berlin, Heidelberg, New York
2. Stein E (1970) Singular integrals and differentiability properties of functions. Princeton University Press
3. Maz'ja V (1985) Sobolev spaces. Springer, Berlin, Heidelberg, New York
4. Filippov AF (1988) Differential equations with discontinuous right-hand sides. Kluwer Academic Publishers
5. Rudin W (1991) Functional analysis. McGraw-Hill, Inc.
6. Conway JB (1994) A course in functional analysis. Springer
7. Driver BK (2003) Analysis tools with applications. Springer
8. Bogachev VI (2007) Measure theory. Springer

Index

A. Polyakov, *Generalized Homogeneity in Systems and Control Volume II*, Communications and Control Engineering,
https://doi.org/10.1007/978-3-031-77496-6

The manufacturer's authorised representative in the EU is Springer Nature Customer Service Centre GmbH, Europaplatz 3, 69115 Heidelberg, Germany. If you have any concerns regarding our products, please contact ProductSafety@springernature.com

Printed and bound by CPI Group (UK) Ltd, Croydon, CR0 4YY
30/09/2026
02225588-0015